The Practice of
NMR Spectroscopy

with Spectra–Structure Correlations for Hydrogen-1

The Practice of NMR Spectroscopy

with Spectra–Structure Correlations for Hydrogen-1

Nugent F. Chamberlain

Analytical Research Laboratory
EXXON Research and Engineering Company
Baytown, Texas

PLENUM PRESS · NEW YORK AND LONDON

Library of Congress Cataloging in Publication Data

Chamberlain, Nugent F   1916-
    The practice of NMR spectroscopy, with spectra-structure correlations for
hydrogen-1.

    Bibliography: p.
    1. Nuclear magnetic resonance spectroscopy. 2. Hydrocarbons — Spectra.
3. Nuclear magnetic resonance spectroscopy — Charts, diagrams, etc.   I. Title.
QC762.C43   538'.3   74-11479
ISBN 0-306-30766-9

To my wife, Barbara, whose patience and understanding made this work easier.

Preface

The purpose of this book is to present systematic methods, backed by comprehensive data correlations, for evaluating NMR spectra and measuring the spectral parameters, and for using these data to determine molecular structure and to analyze or characterize multicomponent systems. This volume is intended as a handbook to be used in the laboratory or office rather than as a reference book for the library. It is intended to supplement rather than replace text books. The data correlations are sufficiently detailed and comprehensive to be used effectively by the experienced chemist and spectroscopist, but the explanations and procedures are presented at a level which, it is hoped, will reach the undergraduate student and the laboratory technician.

The correlations and procedures presented herein are the result of 15 years of analytical NMR experience in the author's laboratory. The chemical shift and coupling constant correlations have been effective in providing rapid access to masses of data, and the typical spectra have been very helpful in pattern recognition.

The correlations are based on very extensive blocks of data, as indicated in the following table:

Number of chemical shifts included 24,000
Number of coupling constants included . 10,000
Number of typical spectra presented 400
Number of compounds studied 10,000

These data have been cross-checked against each other and against theory to reduce errors.

It is hoped that the extended usage and broad data base have resulted in correlations which are comprehensive and accurate, and which will be useful for some time to come.

Acknowledgments

The author sincerely appreciates the backing of the management of Esso Research and Engineering Company in supporting this work and in granting permission to publish it. He also wishes to acknowledge gratefully the assistance of a number of people without whose help, over a period of years, this book would not have been possible. Thanks are due to K. W. Bartz, Clayton Heathcock, J. J. R. Reed, F. C. Stehling, and R. B. Williams for permission to use some of their published and unpublished work. Thanks are due also to Theo Hines and T. J. Denson, whose expert operation of the NMR spectrometers produced many of the high-quality spectra used in these studies. Thanks are also due to D. R. Butler for reproducing the reduced spectra used in the figures and illustrations, and to Willadene Hines, A. J. Woods, and Lila Lill for the superb drafting work which has made the figures and charts usable. Finally, thanks are due to Bernice Alvis, Jerry McKinney, Carolyn Julian, and Gene Henderson who have typed a number of versions of this work over the past few years.

The author is further indebted to S. H. Hastings, J. J. R. Reed, and R. B. Williams for reading the manuscript and offering constructive criticism.

Most of the spectra used in this volume are of compounds purchased from commercial suppliers. Several, however, were synthesized by local chemists. Their contributions are gratefully acknowledged as follows: The compounds used for Figures 22, 23, and 24 were synthesized by W. H. Starnes, E.R.E. Co., those for Figures 26 and 27 by N. P. Neureiter, E.R.E. Co., and the one for Figure 52(g) by E. Lazzari, U. of Texas Dental Branch, Houston.

Contents

Lists of Correlations xvii

    Index Charts xvii

    Chemical Shift Summary Charts xvii

    Chemical Shift Detailed Charts xviii

    Coupling Constant Summary Charts xx

    List of Plates (Typical Spectra) xxi

    List of Figures xxii

    List of Tables... xxiii

Chapter 1

Introduction 3

    I. General 3

    II. Spectral Characteristics 4

        A. Multiplicity 4

        B. Chemical Shift 7

            1. Definition and Measurement 7

            2. Causes 8

            3. Uses 9

        C. Band Intensity 10

        D. Coupling Constant 12

        E. Band Width and Shape 13

Chapter 2

Producing NMR Data 15

    I. Sample Preparation 15

        A. Sample Requirements 15

            1. State 15

            2. Phases 15

            3. Temperature 17

            4. Concentration 17

            5. Paramagnetic Materials 17

B. Solvents 17
1. Requirements 17
2. Selection 20
C. Removing Extra Phases 21
D. Suggestions for Sampling Handling 21
II. Data Production 22
A. Recording the Spectrum 22
1. Calibration of the X-Axis 22
2. Signal-to-Noise 24
3. Resolution 25
4. Phasing 26
5. Saturation 28
6. Aspect Ratio 29
7. Integral 30
B. Identifying Special Features or Groups 31
1. Sidebands 31
2. Satellites 32
3. Readily Exchangeable Hydrogens 33
4. Use of Chemical Derivatives 35
5. Complexes 36
C. Measuring the Spectral Characteristics 40

Chapter 3
Analytical Procedures 45

I. Determination of Molecular Structure 45
A. Advantages and Limitations of H^1 NMR 45
B. Determining an Unknown Structure 45
C. Confirming a Proposed Structure 51
II. Quantitative Multicomponent Analysis 53
A. Advantages and Limitations of H^1 NMR 53
B. Recommended Procedure 54
C. Solvent Separations 56
III. Characterization 56
A. Definition 56
B. Characterization by Inspection 56
C. Characterization of Saturated Hydrocarbons 59
D. Characterization of Olefinic Hydrocarbons 62
E. Characterization of Aromatic Hydrocarbons 63
F. Characterization of Polymers 67
1. General 67
2. Regular Homopolymers 68
3. Irregular Homopolymers 74
4. Copolymers 74

Chapter 4
Chemical Shift Correlations 75

I. Introduction 75
II. Descriptions of the Shift Charts 76

III. Useful Generalizations 80

    A. Chemical Shift versus Substituent Electronegativity
       and Size 81

       1. For Saturated Chains 81

       2. For Olefin Groups 82

       3. For Aromatic Rings 83

    B. Chemical Shift versus Hybridization and Substitution
       of the Principal Substituent Atom 84

       1. For Chain Methyls 84

       2. For Olefin Groups 86

       3. For Aromatic Rings 86

    C. Shift Ranges for Functional Groups 87

    D. Effect of Steric Factors on Chemical Shift 87

       1. *Cis-Trans* Isomerism 87

       2. Nonequivalence 88

       3. "Crowding" and "Shielding" in Paraffinic Chains 89

       4. Steric Effects in Saturated Ring Systems 90

    E. Chemical Shift versus Ring Size 90

    F. Strong Intramolecular Hydrogen Bonds 91

Chapter 5
Coupling Constant Correlations 93

    I. Introduction 93

    II. Description of the Charts 94

    III. Useful Generalizations 96

       A. Coupling Constant vs. Isotope 96

       B. Coupling Constant vs. Hybridization, Bond Order,
          or Bond Length 97

       C. Coupling Constant vs. Substituent Electronegativity 98

         1. One-Bond Couplings 98

         2. Two-Bond Couplings 98

         3. Three-Bond Couplings 98

       D. Effect of Steric Factors on the Coupling Constant 99

         1. *Cis-Trans* Isomerism 99

         2. Basal–Axial Couplings in SF$_5$ Compounds 99

         3. Nonequivalence 99

         4. Steric Effects in Saturated Ring Systems 99

       E. Effect of Number of Bonds Between Coupled Nuclei 100

         1. For Saturated Chains 100

         2. For Aromatic Rings 101

         3. For Olefinic Chains 101

       F. Couplings Transmitted Through Heteroatoms 102

       G. Coupling Constant vs. Ring Size 102

       H. Virtual Coupling 103

Chapter 6
Typical NMR Spectra 105

    I. Introduction 105

    II. Hydrocarbon Spectra 105

    A. Paraffins 105
    B. Cycloparaffins 107
    C. Olefins 107
        1. Olefinic Hydrogens 107
        2. Nonolefinic Hydrogens 107
    D. Aromatics 109
  III. Substituted Hydrocarbons 110
    A. Monosubstituted Saturated Aliphatic Chains 110
    B. The X—CH₂—CH₂—Y Group 110
    C. The Substituted Vinyl Group 111
    D. Substituted Benzene Rings 111
        1. Monosubstituted Benzenes 112
        2. Disubstituted Benzenes 112
        3. Trisubstituted Benzenes 113
        4. Tetrasubstituted Benzenes 114
        5. Penta- and Hexasubstituted Benzenes 115
        6. Condensed Ring Aromatics 115
    E. Phosphorus Compounds 115

Chemical Shift Correlation Charts 117
  The Index Chart I1 118
  The Summary Charts S1 to S35 120
  The Detailed Charts 1 to 89 190

Coupling Constant Correlation Charts 293
  The Index Chart I2 294
  The Summary Charts J1 to J38 296

Typical Spectra Plates 1 to 59 335

References

    I. Introduction 403
    A. Organization 403
    B. Selection and Processing of Data 403
   II. Text References 404
  III. Acknowledgments 404
  IV. Data References—Major Sources 405
    A. Textbooks 405
    B. Catalogs of NMR Spectra 405
    C. Tabulations of NMR Data 405
    D. Chemical Shift Charts 405
    E. Theoretical Spectra 405
    F. Spectral Pattern Guide 406
   V. Data References—Scientific Papers 406
    A. Hydrocarbons 406
    B. Oxygen Compounds 406
    C. Halogen Compounds 407
    D. Nitrogen Compounds 407

E. Sulfur Compounds 408

F. Phosphorus Compounds 410

G. Coupling Constants 411

H. Strong Intramolecular Hydrogen Bonds 412

Appendix—Filing System for NMR Spectra 413

Author Index 417

Subject Index .... .421

Lists of Correlations

Index Charts

Chart I1. Index to and Condensed Summary of Chemical Shifts 118
Chart I2. Index to and Condensed Summary of Coupling Constants 294

Chemical Shift Summary Charts

Chart S1. Summary of Chemical Shifts of $CH_3-C-C-X$, CH_3-C-X and CH_3-CX_n 121
Chart S2. Summary of Chemical Shifts of $CH_3-(C=X, C\equiv X, \phi,$ ⬡$_x$, P, Hal.) 123
Chart S3. Summary of Chemical Shifts of $CH_3-(S, N, O)$ 125
Chart S4. Summary of Chemical Shifts of $R-C-CH_2-C-C-X$ and $R-C-CH_2-C-X$ 127
Chart S5. Summary of Chemical Shifts of $R-C-CH_2-CX_n$ and $Y_nC-CH_2-CX_n$ 129
Chart S6. Summary of Chemical Shifts of $R-C-CH_2-(C=X, C\equiv X, \phi, P, O, Hal.)$ 131
Chart S7. Summary of Chemical Shifts of $R-C-CH_2-(S, N)$ 133
Chart S8. Summary of Chemical Shifts of $X-C-CH_2-(C=Y, C\equiv Y, \phi, P)$ 135
Chart S9. Summary of Chemical Shifts of $X-C-CH_2-(S, N)$ 137
Chart S10. Summary of Chemical Shifts of $X-C-CH_2-(O, F, Cl, Br, I)$ 139
Chart S11. Summary of Chemical Shifts of $X-CH_2-(C=Y, C\equiv Y)$ 141
Chart S12. Summary of Chemical Shifts of $X-CH_2-(\phi, P, F, Cl, Br, I)$ 143
Chart S13. Summary of Chemical Shifts of $X-CH_2-(S, O)$ 145
Chart S14. Summary of Chemical Shifts of $X-CH_2-N$ and some CH 147
Chart S15. Summary of Chemical Shifts of $X-C-CHY_2$ and $X-CH(Z)-Y$ 149
Chart S16. Summary of Chemical Shifts of $X-C-CH=CH-Y$, $X-C=C=CH-Y$, $X-C=CH_2$ 151
Chart S17. Summary of Chemical Shifts of $X-CH=CR_2$ and $X-C=CH-Y(A)$ 153
Chart S18. Summary of Chemical Shifts of $X-C=CH-Y(B)$, $X-C-C\equiv CH$, and $X-C\equiv CH$ 155
Chart S19. Summary of Chemical Shifts of $X-CH=C(Y)Z$ and $X-CH=(O, NY, SY_2)$ 157
Chart S20. Summary of Chemical Shifts of CH_2 and CH in Cyclics 159
Chart S21. Summary of Chemical Shifts of $=CH_2$ and =CH in Cyclics 161
Chart S22. Summary of Chemical Shifts of Aromatic H in $(\phi)_n$, $\phi(C-X)_n$, $\phi-X$ 163
Chart S23. Summary of Chemical Shifts of ortho H in ortho-Disubstituted Benzenes 165
Chart S24. Summary of Chemical Shifts of meta H in ortho-Disubstituted Benzenes 167
Chart S25. Summary of Chemical Shifts of o^2 H in meta-Disubstituted Benzenes 169

xvii

Chart S26. Summary of Chemical Shifts of ortho H in meta-Disubstituted Benzenes 171
Chart S27. Summary of Chemical Shifts of m^2 H in meta-Disubstituted Benzenes 173
Chart S28. Summary of Chemical Shifts of H in para-Disubstituted Benzenes (A) 175
Chart S29. Summary of Chemical Shifts of H in para-Disubstituted Benzenes (B) 177
Chart S30. Summary of Chemical Shifts of H in para-Disubstituted Benzenes (C) 179
Chart S31. Summary of Chemical Shifts of OH, PH and SH 181
Chart S32. Summary of Chemical Shifts of NH in Amines and N(+) Salts 183
Chart S33. Summary of Chemical Shifts of NH in Amides and Imides 185
Chart S34. Summary of Chemical Shifts of OH in Intramolecular H-Bonds 187
Chart S35. Summary of Chemical Shifts of NH in Intramolecular H-Bonds 189

Chemical Shift Detailed Charts

Chart 1.   General Shifts for Acyclic Hydrocarbons 190
Chart 2.   Precise Shifts for Paraffinic Methyl Triplets 191
Chart 3.   Precise Shifts for Paraffinic Methyl Doublets 192
Chart 4.   Precise Shifts for Nonequivalent Methyls in Isopropyl Groups 193
Chart 5.   Precise Shifts for Paraffinic Methyl Singlets 194
Chart 6.   Precise Shifts for Paraffinic Methylenes 195
Chart 7.   Precise Shifts for Paraffinic Methines 196
Chart 8.   Precise Shifts for Nonolefinic H in Aliphatic Mono-Olefins 197
Chart 9.   Precise Shifts for Olefinic H in Aliphatic Mono-Olefins 198
Chart 10. General Shifts for Cycloparaffinics and Alicyclic Olefins 199
Chart 11. Tentative Shifts for Alkyl Cyclopentanes 200
Chart 12. General Shifts for Cyclic Olefins 201
Chart 13. General Shifts for Uncondensed Aromatics 202
Chart 14. General Shifts for Aromatic Naphthenes (uncondensed aromatic rings) 204
Chart 15. General Shifts for Condensed Ring Aromatics 206
Chart 16. General Shifts for Aromatic Olefins, (Nonconjugated and Cyclic) 207
Chart 17. General Shifts for Aromatic Olefins, Conjugated 208
Chart 18. General Shifts for Oxygenated Alkanes 210
Chart 19. General Shifts for Carboxyalkenes, Nonconjugated 212
Chart 20. General Shifts for Carboxyalkenes, Conjugated 214
Chart 21. General Shifts for Unsaturated Aliphatic Aldehydes and Ketones 216
Chart 22. General Shifts for Unsaturated Aliphatic Alcohols and Ethers 218
Chart 23. General Shifts for Unsaturated Aliphatic Ethers and Esters 219
Chart 24. General Shifts for Acetylenic Acids, Esters and Ethers 221
Chart 25. General Shifts for Acetylenic Alcohols 222
Chart 26. General Shifts for Oxygenated Cycloalkanes 223
Chart 27. General Shifts for Cyclic Esters and Ketones 224
Chart 28. General Shifts for Cyclic Ethers 225
Chart 29. General Shifts for Furans 226
Chart 30. General Shifts for Alkyl-Aryl Ketones, Ethers, Esters 227
Chart 31. General Shifts for Aralkyl Acids and Ketones 228
Chart 32. General Shifts for Aralkyl Alcohols and Ethers 229
Chart 33. General Shifts for Aromatic Acids and Esters 230
Chart 34. General Shifts for Aromatic Aldehydes and Ketones 231
Chart 35. General Shifts for Phenyl Esters and Ethers 232

Chart 36. Precise Shifts for Alkoxy Benzenes 233

Chart 37. General Shifts for Phenols 234

Chart 38. General Shifts for Haloalkanes and Halocycloalkanes 235

Chart 39. General Shifts for Fluoro- and Iodoalkenes and Vinyl Halides 236

Chart 40. General Shifts for Chloroalkenes 237

Chart 41. General Shifts for Bromoalkenes 238

Chart 42. General Shifts for Chlorinated Aromatics 239

Chart 43. General Shifts for Brominated Aromatics 240

Chart 44. Precise Shifts for Halobenzenes and Haloalkyl Benzenes 241

Chart 45. General Shifts for Aliphatic Amines and Their Salts 242

Chart 46. General Shifts for Aromatic Amines 243

Chart 46.1 Precise Shifts for Protonated Aromatic Amines 244

Chart 47. General Shifts for Cyclic Imines 245

Chart 48. General Shifts for Aliphatic Amides 246

Chart 49. General Shifts for Methyl Groups Alpha to Amide Nitrogen 247

Chart 50. General Shifts for Acrylamides and Aralkyl Amides 248

Chart 51. General Shifts for Alkyl-Aryl Amides 250

Chart 52. General Shifts for Aromatic Amides 251

Chart 53. General Shifts for Cyclic Amides and Imides 252

Chart 54. General Shifts for Other Amide Types 253

Chart 55. General Shifts for Acyclic Imines and Semicarbazones 254

Chart 56. General Shifts for Hydrazones 255

Chart 57. General Shifts for Oximes 256

Chart 58. General Shifts for Nitriles 257

Chart 59. General Shifts for Nitro Compounds 258

Chart 60. General Shifts for Pyrroles 259

Chart 61. General Shifts for Hydrocarbon Substituted Pyridines 260

Chart 62. General Shifts for Pyridine Derivatives 261

Chart 63. General Shifts for Other Nitrogen Compounds 262

Chart 64. General Shifts for Alkyl Thioacids, Thioesters and Thiones 263

Chart 65. General Shifts for Alkyl Thiols, Sulfides, and Sulfonium Compounds 264

Chart 66. General Shifts for Alkyl Sulfoxide Derivatives and Sulfones 265

Chart 67. General Shifts for Alkyl Sulfonic Acids and Their Derivatives 266

Chart 68. General Shifts for Other Alkyl Sulfur Compounds and Thiolic H 267

Chart 69. General Shifts for Alkenyl Thiols and Sulfides (A) 268

Chart 70. General Shifts for Alkenyl Thiols and Sulfides (B) 269

Chart 71. General Shifts for Alkenyl Aryl Sulfides 270

Chart 72. General Shifts for Alkenyl Thioacetates, Sulfones, and Other 271

Chart 73. General Shifts for Aralkyl Thiols and Sulfides 272

Chart 74. General Shifts for Aralkyl Sulfur Compounds Except Thiols and Sulfides 273

Chart 75. General Shifts for Styryl Sulfides, Thioacetates and Sulfonium Salts 274

Chart 76. General Shifts for Aralkenyl Sulfones 275

Chart 77. General Shifts for Phenyl Sulfur Compounds and Alkyl Aryl Sulfonium Ylides 276

Chart 78. General Shifts for Phenyl Sulfonyls and Saturated Sulfur Heterocyclics 277

Chart 79. General Shifts for Thiophenes 278

Chart 80. General Shifts for Phosphines 279

Chart 81. General Shifts for Phosphine Oxides (and Sulfides) 281

Chart 82. General Shifts for Alkyl Groups in Phosphorus Esters and Amides 282

Chart 83. General Shifts for Alkenyl Groups in Phosphorus Esters 284
Chart 84. General Shifts for Acetylenic Phosphorus Compounds 286
Chart 85. General Shifts for Aromatic Phosphorus Acid Groups 287
Chart 86. General Shifts for Acyl and Acylmethyl Phosphonates 288
Chart 87. General Shifts for Cyclic Esters and Alkyl Amides of Phosphorus 289
Chart 88. General Shifts for Alkyl Phosphorus Halides 290
Chart 89. General Shifts for Phosphonium Salts 291

Coupling Constant Summary Charts

Chart J1.   Hydrogen-Hydrogen Coupling Constants, H–C–H Series (A) 296
Chart J2.   Hydrogen-Hydrogen Coupling Constants, H–C–H(B) and H–C(=)–H 297
Chart J3.   H–C–H Coupling Constants vs. Ring Size 298
Chart J4.   Hydrogen-Hydrogen Coupling Constants, H–C–C–H Series (A) 299
Chart J5.   Hydrogen-Hydrogen Coupling Constants, H–C–C–H Series (B) 300
Chart J6.   Hydrogen-Hydrogen Coupling Constants, H–C(=)–C–H Series 301
Chart J7.   Hydrogen-Hydrogen Coupling Constants, H–C=C–H Series (A) 302
Chart J8.   Hydrogen-Hydrogen Coupling Constants, H–C=C–H Series (B) 303
Chart J9.   Hydrogen-Hydrogen Coupling Constants, H–C(=)–C(=)–H, H–C≡C–H and 4-Bond 304
Chart J10. Hydrogen-Hydrogen Coupling Constants, Nonaromatic 5- to 9-Bond 305
Chart J11. Hydrogen-Hydrogen Coupling Constants for Benzenes 306
Chart J12. Hydrogen-Hydrogen Coupling Constants for Condensed Aromatics 307
Chart J13. Hydrogen-Hydrogen Coupling Constants for Heterocyclic Aromatics 308
Chart J14. Hydrogen-Hydrogen Coupling Constants, Aromatic 4- to 6- Bond 309
Chart J15. Hydrogen-Hydrogen Couplings Through Oxygen 310
Chart J16. Hydrogen-Hydrogen Couplings Through Nitrogen (A) 311
Chart J17. Hydrogen-Hydrogen Couplings Through Nitrogen (B) 312
Chart J18. Hydrogen-Hydrogen Couplings Through Sulfur 313
Chart J19. Hydrogen-Carbon-13 Coupling Constants, H–C Series (A) 314
Chart J20. Hydrogen-Carbon-13 Coupling Constants, H–C Series (B) 315
Chart J21. Hydrogen-Carbon-13 Coupling Constants, H–C= and H–C≡ 316
Chart J22. Hydrogen-Carbon-13 Coupling Constants, 2– and 3– Bond 317
Chart J23. Hydrogen-Carbon-13 Coupling Constants for Aromatics 318
Chart J24. Hydrogen-Carbon-13 Coupling Constants for Heterocompounds 319
Chart J25. Hydrogen-Fluorine Coupling Constants, H–C–F Series 320
Chart J26. Hydrogen-Fluorine Coupling Constants, H–C–C–F Saturated 321
Chart J27. Hydrogen-Fluorine Coupling Constants, H–C–C–F Unsaturated and Long Range 322
Chart J28. Hydrogen-Fluorine Coupling Constants for Aromatics 323
Chart J29. Hydrogen-Fluorine Couplings Through Heteroatoms 324
Chart J30. Hydrogen-Nitrogen Coupling Constants, H–N Series 325
Chart J31. Hydrogen-Nitrogen Coupling Constants, H–C–N Series 326
Chart J32. Hydrogen-Nitrogen Coupling Constants, H–C–C–N and Other 327
Chart J33. Hydrogen-Phosphorus Coupling Constants, Summary and H–P Series 328
Chart J34. Hydrogen-Phosphorus Coupling Constants, H–C–P Series 329
Chart J35. Hydrogen-Phosphorus Coupling Constants, H–C–C–P Series 330
Chart J36. Hydrogen-Phosphorus Coupling Constants, H–C–O–P Series 331
Chart J37. Hydrogen-Phosphorus Coupling Constants, H–C–S–P and H–C–N–P 332
Chart J38. Hydrogen-Phosphorus Coupling Constants, Aromatic and Long Range 333

List of Plates (Typical Spectra)

Plate 1.   Three Basic Paraffinic CH_3 NMR Patterns (60 MHz) 337

Plate 2.   Alkyl CH_3 NMR Triplet Patterns vs. Chain Length (60 MHz) 338

Plate 3. CH_3 NMR Patterns for Monomethyl Paraffins (60 MHz) 339

Plate 4.   NMR Patterns Resulting from Nonequivalent Methyls in Isopropyl Groups (60 MHz) 340

Plate 5.   Alkyl CH_2 NMR Patterns vs. Chain Length (60 MHz) 341

Plate 6.   Alkyl CH and CH_2 NMR Patterns vs. Chain Length (60 MHz) 342

Plate 7.   The Isolated Paraffinic Ethyl Group (60 and 100 MHz) 343

Plate 8.   NMR Spectra of n-Paraffins (100 MHz) 344

Plate 9.   NMR Spectra of 2-Methyl Paraffins (100 MHz) 345

Plate 10.   NMR Spectra of 3-Methyl Paraffins (100 MHz) 346

Plate 11.   NMR Spectra of 4- and 5-Methyl Paraffins (100 MHz) 347

Plate 12.   NMR Spectra of Some Important Dimethyl Paraffins (100 MHz) 348

Plate 13.   NMR Spectra of Paraffins with Nonequivalent Methyls in Isopropyl Groups (100 MHz) 349

Plate 14.   Typical NMR Spectra of Paraffins with Single Branches Longer Than Methyl (100 MHz) 350

Plate 15.   Representative NMR Spectra of Cycloparaffins (100 MHz) 351

Plate 16.   Characteristic Spectral Patterns of Olefinic Hydrogens (60 MHz) 352

Plate 17.   Characteristic Spectral Patterns for α-CH_3 Groups (60 MHz) 353

Plate 18.   Characteristic Spectral Patterns of β, γ, and δ, δ^+ Methyl Groups (60 MHz) 354

Plate 19.   Characteristic Spectral Patterns of α-Methylene Groups (60 MHz) 355

Plate 20.   NMR Spectra of the Saturated Part of Linear α-Olefins (60 MHz) 356

Plate 21.   NMR Spectra of the Vinyl Group in Aliphatic Olefins (100 MHz) 357

Plate 22.   Spectra of Olefinic H in 1,1-Disubstituted Ethylenes (100 MHz) 358

Plate 23.   Spectra of Olefinic H in 1,2-Disubstituted Ethylenes (A) (100 MHz) 359

Plate 24.   Spectra of Olefinic H in 1,2-Disubstituted Ethylenes (B) (100 MHz) 360

Plate 25.   Spectra of Olefinic H in Trisubstituted Ethylenes (100 MHz) 361

Plate 26.   Spectra of $CH_3(R)C=C$ (100 MHz) 362

Plate 27.   Spectra of $CH_3-C=C-R$ in Linear Olefins (100 MHz) 363

Plate 28.   Spectra of $CH_3-C=C-R$ and $(CH_3)_2$ $C=C-R$ in Branched Olefins (100 MHz) 364

Plate 29.   NMR Spectra of Saturated Part of C_3-C_5 Linear α-Olefins (100 MHz) 365

Plate 30.   NMR Spectra of the Saturated Part of C_6-C_{14} Linear α-Olefins (100 MHz) 366

Plate 31.   NMR Spectra of the Alkyl Part of n-Alkylbenzenes (100 MHz) 367

Plate 32.   NMR Spectra of the C_1-C_3 Alcohols (60 MHz) 368

Plate 33.   NMR Spectra of the Butanols (60 MHz) 369

Plate 34A. NMR Spectra of the Pentanols (60 MHz) 370

Plate 34B. NMR Spectra of the Pentanols (60 MHz) 371

Plate 35.   NMR Spectra of C_6-C_{12} Linear Alcohols (60 MHz) 372

Plate 36.   NMR Spectra of the C_1-C_3 Alcohols (100 MHz) 373

Plate 37.   NMR Spectra of the Butanols (100 MHz) 374

Plate 38A. NMR Spectra of the Pentanols (100 MHz) 375

Plate 38B. NMR Spectra of the Pentanols (100 MHz) 376

Plate 39.   NMR Spectra of C_6-C_{12} Linear Alcohols (100 MHz) 377

Plate 40.   NMR Spectra of the n-Butyl Group with Various Chemical Shifts (60 and 100 MHz) 378

Plate 41.   Typical NMR Spectra of the $X-CH_2-CH_2-Y$ Group (60 and 100 MHz) 379

Plate 42.   Spectra of the Vinyl Group with Various Substituents (A) (60 and 100 MHz) 380

Plate 43.   Spectra of the Vinyl Group with Various Substituents (B) (60 and 100 MHz) 381

Plate 44.   Typical NMR Spectra of Monosubstituted Benzene Rings (60 and 100 MHz) 382

Plate 45. Typical NMR Spectra of ortho-Disubstituted Benzene Rings (60 and 100 MHz) 383

Plate 46. Complete Interpretation of an ABCD System 384

Plate 47. Typical NMR Spectra of meta-Disubstituted Benzene Rings (60 and 100 MHz) 385

Plate 48. Typical NMR Spectra of para-Disubstituted Benzene Rings (60 and 100 MHz) 386

Plate 49. Typical NMR Spectra of 1,2,3-Trisubstituted Benzene Rings (100 MHz) 387

Plate 50. Typical NMR Spectra of 1,2,4-Trisubstituted Benzene Rings (60 and 100 MHz) 388

Plate 51. NMR Spectra of Some Tri- and Tetrasubstituted Benzene Rings (100 MHz) 389

Plate 52. NMR Spectra of Some Tetra- and Pentasubstituted Benzene Rings (100 MHz) 390

Plate 53. NMR Spectra of Condensed Ring Aromatics, Unsubstituted (100 MHz) 391

Plate 54. Similar Spectra for an Alkylphosphine and an Alkane (100 MHz) 392

Plate 55. NMR Spectra of Tri-n-Octylphosphine (60 MHz) 393

Plate 56. Tri-n-Octylphosphine in $CDCl_3$ + Trifluoroacetic Acid (60 MHz) 394

Plate 57. Effect on Their NMR Spectra of Converting Phosphines to Phosphine Oxides (60 MHz) 395

Plate 58. Phosphorus-Hydrogen Spin Coupling Helps Identify Phosphorus Esters (60 and 100 MHz) 396

Plate 59. Spectra of Alpha Alkyl Groups in Phosphorus Esters 397

Plate 60. Widening of Ortho Hydrogen Multiplet by Spin Coupling to Phosphorus
in Monosubstituted Benzene (60 MHz) 398

Plate 61. The Very Wide H-P Doublet 399

List of Figures

Figure 1. High Resolution H^1 NMR Spectrum, Showing Characteristics Used in Analysis 4

Figure 2. Secondary Magnetic Fields Generated by Interatomic Electron Currents
in Closed Molecular Paths 9

Figure 3. Chemical Shift-Molecular Structure Correlations for Two Compound Types 10

Figure 4. High Resolution H^1 NMR Spectrum, Showing Electronic Integral Measurement
of Band Intensity 10

Figure 5. Measuring Intensity of Small Band on Skirt of Large Band 11

Figure 6. Band Widths and Shapes 13

Figure 7. Peak Doubling by Two-Phase Sample 16

Figure 8. Spectrum of the CD_2H Group 20

Figure 9. Calibration of A-60 Spectrometer with p-Anisaldehyde 23

Figure 10. Effect of Noise on NMR Spectrum 24

Figure 11. NMR Spectrum of Ethylcyclohexane at 100 MHz 25

Figure 12. Effect of Phasing on the NMR Signal 27

Figure 13. Effect of Phasing on the Integral 28

Figure 14. Effect of Aspect Ratio on Visibility of Band Features 29

Figure 15. Good Aspect Ratio for Different Parts of a Complex Spectrum 30

Figure 16. Spinning Sidebands and C^{13} Satellites 32

Figure 17. C^{13} and Si^{29} Satellites of TMS 33

Figure 18. Identifying OH Band by Adding Acid 34

Figure 19. Eliminating Interference by Shifting OH Band 35

Figure 20. Inducing Additional Shifts with $Eu(DPM)_3$ 38

Figure 21. Separating Spectra of Alcohols with $Eu(DPM)_3$ 39

Figure 22. High Resolution NMR Spectrum to be Interpreted 41

Figure 23. Identifying Bands and Multiplets 42

Figure 24. Measuring Shifts, Couplings and Intensities 42

Figure 25. Use of the Chemical Shift Index Chart 49

Figure 26.   Confirming a Proposed Structure 51
Figure 27.   Identifying Impurities 53
Figure 28.   Spectrum of Mixture for Quantitative Analysis 55
Figure 29.   60 MHz NMR Spectra of Some Petroleum Fractions 57
Figure 30.   60 MHz NMR Spectra of Typical Detergent Alkylates 58
Figure 31.   100 MHz Spectrum of a Typical Lubricating Oil 59
Figure 32.   Characterization of Paraffin Mixtures 60
Figure 33.   Characterization of an Olefin Mixture 62
Figure 34.   Characterization of a Detergent Alkylate 64
Figure 35.   Characterization of Aromatic Hydrocarbons 66
Figure 36.   NMR Spectra of Polymers of 4-Methyl-1-Pentene 67
Figure 37.   NMR Spectra of Polyisobutylene and of Poly-3-Methylbutene-1 68
Figure 38.   NMR Spectrum of Polybutene-1 69
Figure 39.   NMR Spectrum of a Polyisoprene 69
Figure 40.   NMR Spectrum of a Polyisoprene 70
Figure 41.   NMR Spectrum of Polypentadiene-1,3 71
Figure 42.   NMR Spectrum of a Poly(2-Vinylpyridine) 72
Figure 43.   NMR Analysis of an Alkylene Oxide Copolymer 72
Figure 44.   NMR Spectrum of Copolymer of Isobutylene and Isoprene 73
Figure 45.   Methyl Chemical Shift vs Chain Length for $X-(CH_2)_n-CH_3$ 79
Figure 46.   Chemical Shift of CH_3-X vs Electronegativity
             of the Attached Atom of X 80
Figure 47.   Chemical Shift of $Alkyl-CH_2-X$ vs Electronegativity
             of the Attached Atom of X 81
Figure 48.   Chemical Shift of $X-CH=CR_2$ vs Electronegativity and Covalent Radius
             of the Attached Atom of X 82
Figure 49.   Chemical Shift of $X-CR=CH_2$ vs Electronegativity and Covalent Radius
             of the Attached Atom of X 83
Figure 50.   Chemical Shift of $\phi-X$ vs Electronegativity and Covalent Radius
             of the Attached Atom of X 84
Figure 51.   Some Important Chemical Shift Trends 85
Figure 52.   Magnetic Asymmetry 88
Figure 53.   Chemical Shift vs Ring Size 90
Figure 54.   Magnitude Trends of Hydrogen Coupling Constants, 1- and 2- Bond 95
Figure 55.   Magnitude Trends of Hydrogen Coupling Constants, 3- Bond 96
Figure 56.   Coupling Constant vs Number of Bonds Between Coupled Atoms, for Aliphatics 100
Figure 57.   Coupling Constant vs Number of Bonds Between Coupled Atoms, for Benzene Rings 101
Figure 58.   H—C—H Coupling Constant vs Ring Size 102

List of Tables

Table 1.   Spin–Spin Multiplet Rules 5
Table 2.   Number of Lines Produced by Various Hydrogen Spin Systems 6
Table 3.   Reference Standards for Hydrogen-1 NMR 7
Table 4.   Characteristics of the Most Useful Solvents for H' NMR 18
Table 5.   Solvents for High- and Low-Temperature Work 21
Table 6.   Solvents for Special Applications 22
Table 7.   Complexes of Eu and Pr Used as Lanthanide Shift Reagents 37

Table 8.   Percent Methyl Hydrogens in Paraffins (Acyclic) 61
Table 9.   Characterization of a Detergent Alkylate 64
Table 10. Pertinent Isotope Characteristics 97
Table 11. Coupling Constant vs Bond Order 97
Table 12. Coupling Constant vs Hybridization 97

The Practice of NMR Spectroscopy

CHAPTER 1

Introduction

I. GENERAL

Although it is assumed that the reader has been exposed to the elementary theory of NMR and to the operation of an NMR spectrometer, a brief review of some of the basic concepts and definitions will indicate the point of view used in this book and clarify some of the definitions. The discussion is confined to the hydrogen-1 isotope because this is by far the most generally used and, consequently, far more data are available for it than for any other isotope. This wealth of data, in turn, leads to the most accurate and comprehensive set of spectra—structure correlations.

Nuclear magnetic resonance is the spectrometric technique that utilizes the energy transitions of atomic *nuclei* precessing in a static *magnetic* field in *resonance* with an alternating (rotating) magnetic field. The instrument required to observe the phenomenon consists of a dc or permanent magnet to supply the static field, a radio-frequency generator to supply the alternating field, a probe which can hold the sample in the proper position relative to these two fields, a radio receiver to detect the signals generated by the hydrogen nuclei in the sample, and a recorder to make a permanent record of the spectrum produced. The homogeneity of the static field must be high enough to permit observation of the closely spaced bands and fine structure which indicate the chemical environment in which each hydrogen nucleus is located. A spectrum which shows this fine structure is called a high-resolution spectrum (as opposed to a wide-line spectrum in which the chemical environmental features are obscured by the broad line width).

When a sample containing hydrogen is placed in the static magnetic field, each hydrogen nucleus will precess at a frequency determined by the magnetic field it actually experiences. This field, in turn, is determined by the electronic, and therefore the chemical, environment of the nucleus. Thus the variety of chemical environments that exist in a molecule will produce a spectrum of precession frequencies that will indicate the chemical nature of the various parts of the molecule. The remaining problem is to observe this spectrum of frequencies.

There are two general methods of observing the spectrum. The first consists of varying the static magnetic field or the radio frequency slowly so as to bring groups of like nuclei into resonance, one group at a time, and recording the signal produced by each. This *continuous wave* (CW) method may be likened to using a narrow-pass filter to analyze the frequency components of a complex wave. It is a slow but effective process and is the method most used at present.

The second method consists of bringing all the nuclei into resonance at once and then decomposing the resulting complex wave into its component frequencies by Fourier analysis. This is popularly called *Fourier transform* MNR. It is a couple of orders of magnitude faster and/or more sensitive than the CW method, but the instrumentation required is between two and ten times as expensive. This method is particularly attractive when great sensitivity is required, as in carbon-13 NMR.

Both methods produce the same plot of signal amplitude *vs.* magnetic field strength (or frequency) as shown in Figure 1. This spectrum is presented according to the conventions standardized by ASTM and generally agreed

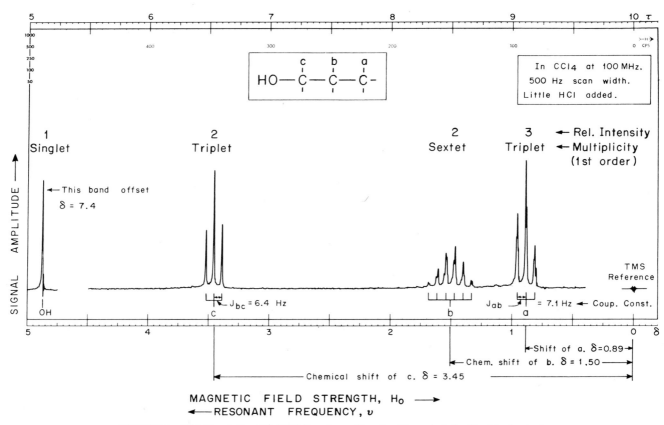

FIGURE 1. High-Resolution H¹ NMR Spectrum, Showing Characteristics Used in Analysis

to by NMR spectroscopists throughout the world. In these conventions, signal amplitude increases upward, magnetic field strength increases to the right, and position along the magnetic field axis is measured in parts per million from the resonance of tetramethylsilane as zero. The position scale, with numbers increasing positively to lower field, is generally called the δ scale (lower scale in Figure 1). Another scale which has been used widely is the τ scale, shown at the top of Figure 1. This scale is also divided into parts per million, but the TMS resonance is 10 instead of 0 and the numbers decrease toward lower field.

II. SPECTRAL CHARACTERISTICS

There are five characteristics of an NMR spectrum that are used in analytical work. These are the multiplicity, the chemical shift, the band intensity, the coupling constant, and the band width and shape. All are defined or illustrated in Figures 1 and 4, and each will be further discussed in the following paragraphs. Another charac-

teristic, the relaxation time indicated roughly by the line width, is used only qualitatively in most analytical work at present. It may become more important in the future as the Fourier transform spectrometers make it possible to measure true relaxation times more routinely.

A. MULTIPLICITY

The multiplicity of a band is the number of lines in that band. It is the result of indirect spin–spin interaction and is determined by the number and the spin quantum number of other nuclei spin-coupled to the nuclei which produced the band. Simplified rules for recognizing spin–spin multiplets are outlined in Table 1, and the numbers of lines produced by various hydrogen spin systems are detailed in Table 2.

The true multiplicity of some bands cannot be determined by inspection because many of the lines will not be resolved. Furthermore, the total number of lines in complex bands can be so great as to be meaningless without detailed calculation of the theoretical spectrum

TABLE 1. Spin–Spin Multiplet Rules

I. Definitions

Multiplet A...The resonance band produced by a single group of equivalent nuclei. Equivalent nuclei all have the same chemical shift and have the same sets of coupling constants to all other nuclei.

Group B........All the nuclei of a single species which are spin-coupled, with identical coupling constants, to the nuclei producing multiplet A.

Multiplet B...The resonance band produced by any equivalent subgroup of Group B. There is a separate multiplet for each chemical shift of the B nuclei.

J....................The coupling constant, in hertz between nuclei A and B.

Δ.................The separation, in Hz, between the centers of multiplets A and B. This is the chemical shift difference between the bands, converted to hertz.

I_B.................The spin quantum number of the B nuclei.

First-order
    multiplet....The simplest multiplet that can be produced by a spin system. Mathematically, first-order multiplets are those produced when there are only diagonal terms in the matrix representation of the secular equation describing the spin system. They are described in practical terms in Section II of these rules.

II. First-Order Multiplets. $J/\Delta < 0.03$

A. Number
    1. Number of peaks in multiplet A produced by n equivalent nuclei in group B is equal to $2n_B I_B + 1$.
    2. When $I_B = \frac{1}{2}$ (as with H^1, C^{13}, F^{19}, P^{31}), the number of peaks in A is equal to $n_B + 1$.
    3. When $I_B = 1$ (as with deuterium and N^{14}), the number of peaks in A is equal to $2n_B + 1$.

B. Spacing
    Separations between peaks in multiplet A will be equal, and will be equal to those in multiplet B. The magnitude of the spacing, in hertz, is equal to the coupling constant J.

C. Intensities (areas under peaks)
    1. When $n = 1$, all peaks are of equal intensity.
    2. When $n > 1$, the peaks will be distributed symmetrically about the midpoint of the group.
    3. When $I_B = \frac{1}{2}$, peak intensities will be:

Multiplet A	Peak intensities	No. B nuclei	Multiplet A	Peak intensities	No. B nuclei
Doublet	1:1	1	Sextet	1:5:10:10:5:1	5
Triplet	1:2:1	2	Septet	1:6:15:20:15:6:1	6
Quartet	1:3:3:1	3	Octet	1:7:21:35:35:21:7:1	7
Quintet	1:4:6:4:1	4	Nonet	1:8:28:56:70:56:28:8:1	8

    4. When $I_B = 1$, multiplet intensities will be:

Multiplet A	Peak intensities	No. B nuclei
Triplet	1:1:1	1
Quintet	1:2:3:2:1	2
Septet	1:3:6:7:6:3:1	3
Nonet	1:4:10:16:19:16:10:4:1	4

Multiplets produced by spin coupling of nuclei of different species, such as hydrogen and deuterium, should be first order if all the nuclei in each group are equivalent. J/Δ is much less than 0.03.

III. Non-First-Order Multiplets. $J/\Delta > 0.03$

A. *When $0.03 < J/\Delta < 0.15$*
    The multiplets may still be interpreted broadly as first order in number and spacing of peaks. Intensities of the inner peaks (on the sides between the two multiplets) are enhanced at the expense of the outer peaks, destroying the symmetry of the multiplets.

B. *When $J/\Delta > 0.15$*
    Multiplets may no longer be first order in number, spacing, or intensity of peaks. Additional peaks may appear. Mathematical analysis is frequently required for complete interpretation. When $J/\Delta > 2$ virtual coupling may appear (Chapter 5, Section IIIH).

C. *When $J/\Delta = \infty$ or 0*
    1. When $\Delta = 0$, only one peak is produced (except when J is very large; i.e., > 100 Hz).
    2. When $J = 0$, one peak is produced for each chemical shift.

TABLE 2. Number of Lines Produced by Various Hydrogen Spin Systems

Number of H's n	Spin system[a]	First-order spectra Multiplets[b]	First-order spectra Max. number of lines[c]	Non-first-order spectra, max. number lines[d]
1	A	1	1	1
2	A_2	1	1	1
2	AB	2,2	4	4
3	A_3	1	1	1
3	AB_2	3,2	5	9
3	ABC	4,4,4	12	15
4	A_4	1	1	1
4	AB_3	4,2	6	16
4	A_2B_2	3,3	6	18
4	AA'BB'	Cannot be 1st order	—	26
4	AB_2C	6,4,6	16	—
4	ABC_2	6,6,4	16	—
4	ABCD	8,8,8,8	32	56
5	A_5	1	1	1
5	AB_4	5,2	7	25
5	A_2B_3	4,3	7	33
5	A_2B_2C	6,6,9	21	—
5	A_2BC_2	6,9,6	21	—
5	A_2BCD	8,12,12,12	44	—
5	AB_2CD	12,8,12,12	44	—
5	ABCDE	5 times 16	80	210
6	ABCDEF	6 times 32	192	792
7	ABCDEFG	7 times 64	448	3003

[a] Each letter represents a different chemical shift. Each subscript indicates the number of hydrogens having the shift indicated by the letter. Thus, the AB_2 system has one hydrogen with shift A and two with shift B. The spectrum of an A_2B system is the mirror image of that of an AB_2 system. In the AA'BB' system, A and A' have the same chemical shift but different couplings to B ($J_{AB} \neq J_{A'B}$). Similarly for B and B'.

[b] The first-order multiplets are shown in the same sequence as the letters of the spin system. 1 = singlet, 2 = doublet, etc.

[c] Assuming all possible couplings are nonzero. The maximum number of first-order lines = $n(2^{n-1})$.

[d] Maximum number non-first-order lines = $\sum_{n+1} P_i P_j$ where P_i and P_j are the relative intensities (transition probabilities) of adjacent lines in a multiplet, and $n+1$ is the multiplet to be used. Thus, for $n = 3$ the quartet line intensities, 1:3:3:1, are used:

$$\sum_4 P_i P_j = (1)\,(3) + (3)\,(3) + (3)\,(1) = 15$$

The maximum number of lines calculated from the above formula applies when all shifts and all couplings in the system are different and all possible couplings are nonzero (AB, ABC, ABCD, etc.). When some shifts or some couplings are identical (A_2B, AA'BB', A_2B_2, etc.) the maximum number of lines is reduced. For such systems the maximum number has been determined from calculations of theoretical spectra[T43]. The number of observed lines may be less than maximum in any case because of accidental overlap or insufficient resolution.

(see Table 2). The multiplicity that is really of value in interpreting a spectrum by inspection is the first-order multiplicity. Consequently, first-order multiplicity is emphasized in all interpretations in this book.

The characteristics of first-order multiplets are outlined in Table 1 and are illustrated by many typical spectra presented in this book. It is not necessary that a multiplet be completely first order to be interpreted as

such. First-order interpretations can be imposed on non-first-order multiplets rather easily in many cases, as shown in Figure 1. Band c is nearly first order, but bands a and b show considerable higher-order splitting and some intensity deviations. Yet the first-order interpretations of bands a and b are easily made by following the envelopes of the major line groups rather than the individual lines. These interpretations are shown as spin multiplet diagrams (the "pitchforks") underneath the bands in Figure 1, and the multiplicity so obtained is listed above each band.

B. CHEMICAL SHIFT

1. Definition and Measurement

The chemical shift is the displacement of a resonance signal along the magnetic field coordinate resulting primarily from the magnetic effects of the chemical environment of the nuclei producing that signal. It is directly proportional to the basic strength of the static magnetic field or to the basic spectrometer frequency. Because of this dependence, the chemical shift *values* determined at 60 MHz would differ from those determined at 100 MHz if they were expressed in field- or frequency-dependent units. To avoid this variation in values among the different frequencies in use, chemical shifts should always be expressed in the field-independent dimensionless units of parts per million. To reduce the size of the numbers and increase the accuracy, the chemical shift is measured from a reference point in the spectrum rather than from the absolute zero of magnetic field or frequency.

As shown in Figure 1, the chemical shift of a particular band is measured as the displacement (in parts per million along the magnetic field axis) of the center of that band from the center of the reference resonance. The values are expressed in delta units, as read from the lower scale. They can just as easily be expressed in tau units by reading from the top scale. The delta scale has become the more popular one in the U.S. and may be adopted worldwide in the future. There is a sizable body of literature using the tau scale already in existence, however, so that it is necessary to be able to convert from one to the other rapidly. These conversions are simply

$$\tau = 10 - \delta$$
$$\delta = 10 - \tau$$

The reference resonance is a convenient means of calibrating the magnetic field (or frequency) for each sample. For molecular structure determinations this calibration should be at the sample molecule to eliminate variations due to factors other than molecular structure. In such cases it is necessary to dissolve the reference molecule in the sample solution. When it is desired to include the effects of such factors as the bulk magnetic susceptibility of the sample, the total influence of solvents, etc., the reference is placed in a separate tube concentric with the sample solution.

Reference standards useful for hydrogen resonance work are listed in Table 3. Tetramethylsilane has been selected by the ASTM as the primary reference for

TABLE 3. Reference Standards for Hydrogen-1 NMR

Compound (Abbreviation)	Formula	m.p. °C	b.p. °C	d 20/4	Isotopic purity[a]	Hydrogen bands Group[b]	δ	τ
Tetramethylsilane (TMS)[c]	$(CH_3)_4 Si$	—	26.5	0.65	—	CH_3 (s)	0.000	10.000
Hexamethyl disiloxane (HMDS)[d]	$[(CH_3)_3 Si]_2 O$	−59	100.4	—	—	CH_3 (s)	0.04	9.96
Sodium 3-trimethylsilyl-1-propane sulfonate (DSS)[e]	$(CH_3)_3 Si—CH_2—CH_2 —CH_2—SO_3—Na$	Solid Salt	—	—	—	CH_3 (s) 1—CH_2 (m) 2—CH_2 (m) 3—CH_2 (m)	0.00 0.6 1.8 2.9	10.00 9.4 8.2 7.1
Sodium 3-trimethylsilyl-propionate-d₄ (TSP)[f]	$(CH_3)_3 Si—CD_2—CD_2 —COO—Na$	Solid Salt	—	—	99.5	CH_3 (s)	0.00	10.00

[a] The percent replacement of hydrogen by deuterium in the deuterated groups.
[b] The functional group which produces the observed band. The multiplicity of the band is indicated in parenthesis; s = singlet; m = complex multiplet.
[c] Primary reference standard for room temperature and below.
[d] Can be used as reference up to 180°C.
[e] Reference for water solutions. The CH_2 bands can interfere with weak sample bands.
[f] Reference for water solutions.

hydrogen work. All chemical shifts should be referred to TMS regardless of the standard actually used in the experiment.

The low boiling point and low water solubility of TMS make it less than universally acceptable, however. A good standard for high-temperature work is hexamethyldisiloxane (HMDS). Although its boiling point is 100.4°C, it has been used successfully to 180°C at less than 10 vol. % concentration in samples sealed with plastic caps.

The resonance of HMDS is only slightly lower than that of TMS, at $\delta = 0.04$. This difference is small enough to ignore for most practical purposes. This compound is not as readily available as it might be, however. The only supplier the author has found is K and K Laboratories, Inc.

The silicone oils could be used at still higher temperatures, if needed. Their resonances are at about the same place as those of HMDS.

Two similar compounds have been found satisfactory as standards for water solutions. These are sodium 3-trimethylsilyl-1-propanesulfonate (DSS) and the deuterated form of sodium 3-trimethylsilylpropionate (TSP). TSP is recommended because it does not have any significant bands other than the reference peak. DSS has several low-intensity bands which sometimes interfere with weak sample bands. The resonance positions of both reference peaks are so close to those of TMS that the difference can be neglected.

2. Causes

Chemical shifts are caused by secondary magnetic fields set up by several types of electron currents induced into the molecule by the applied static magnetic field. These secondary fields strengthen or diminish the applied field so that the field at each nucleus varies according to the electron configuration in its vicinity. This spreads the resonances out along the magnetic field axis. The electron configuration, in turn, determines the chemical characteristics of the atom in which the nucleus is located, so that the secondary fields, and therefore the chemical shifts, are directly related to the chemical nature of the atom.

To simplify discussion, the electron currents may be arbitrarily divided into the following types:[T38]

1. Local diamagnetic currents. These currents are in spherical electron clouds which permit them to flow freely in any plane. Then, following the rules of magnetic induction, these induced currents flow so that the resulting magnetic fields always directly oppose the applied field. This causes shifts to higher applied field, decreasing δ. These currents, and therefore the shifts they cause, are proportional to the electron density. For this reason, when the local diamagnetic currents are dominant, substituents which reduce the electron density will cause shifts to lower field and those which increase electron density will cause shifts to higher field.

2. Paramagnetic currents, approximately localized. These results from a mixing of the ground and excited electronic states of the atom, and account for the fact that electron clouds are seldom actually spherical. These currents cause shifts to lower applied field, increasing δ. They are usually considered less important in producing H^1 shifts than the other types of currents.

3. Local currents around neighboring atoms. The currents described in 1 and 2 above affect neighboring atoms as well as the atoms they circulate around, although to a lesser extent. When the neighboring atom is a large one such as Br or I, the shifts their currents induce into neighboring hydrogen may be significant.

4. Interatomic currents in closed molecular paths. The π electrons in aromatic rings and in aliphatic double bonds and triple bonds will circulate around allowed closed paths as shown in Figure 2. The magnetic field generated by each current is restricted to certain directions with respect to the molecule, and therefore produces different degrees and directions of shift, depending on the spatial orientation of the hydrogen atom of interest with respect to the field. In the aromatic ring the induced field always opposes the applied field over the plane of the ring and assists it at the edge of the ring. Therefore, hydrogens held over the plane of the ring will resonate at higher field (H_b) and those near the edge of the ring will resonate at lower field (H_a and the methyl group). Hydrogens held at points where the induced field is at right angles to the applied field (H_c) will have unchanged shifts.

These three possibilities can be represented by regions of upfield shift (cones) and regions of downfield shifts (doughnuts) with a surface of unchanged shift between them. This is indicated for the double-bond and triple-bond groups in Figure 2 (b) and (c).

Chemical shift correlation data indicate that the resonances of hydrogen groups near C=X groups are usually shifted downfield. Only in rigid condensed ring

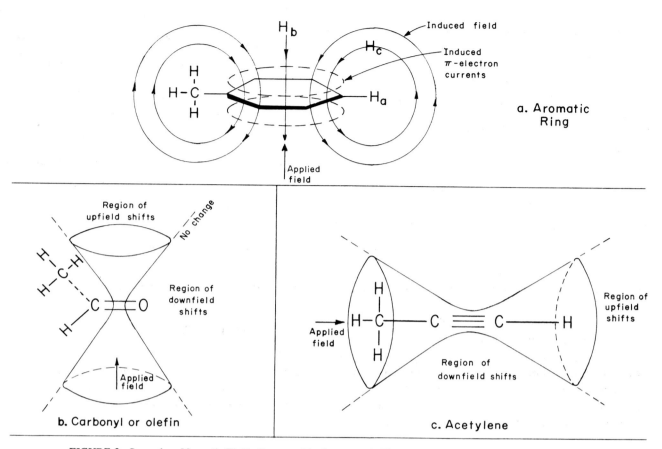

FIGURE 2. Secondary Magnetic Fields Generated by Interatomic Electron Currents in Closed Molecular Paths

compounds (such as norbornene) has an upfield shift of some resonances been observed by the author. It seems that it is difficult to hold a hydrogen within the upfield shift region of C=X in nonrigid systems.

On the other hand, hydrogen groups attached to C≡X groups are shifted upfield by the induced field. Groups which are farther removed from C≡X appear to be little affected. It may be difficult to hold a group in the downfield shift region of C≡X.

There may be interatomic currents flowing in saturated systems, also. They could be the result of the magnetic anisotrophy of single bonds, which is frequently postulated to account for the shifts in saturated hydrocarbons.

3. Uses

Chemical shifts are used to identify the functional groups which can produce the various bands in a spectrum and to determine their spatial positions to a first level. This is done by means of correlations of chemical shift with molecular structure, such as shown in Figure 3. In this correlation, the chemical shift range of each functional group is plotted against its spatial position with respect to the group causing the primary shift (the reference group). A clear distinction between the compound types shown can be made from the shifts of the groups alpha to the reference or from the presence or absence of an aldehydic hydrogen shift. The alpha group shifts also indicate reasonably well whether the group is a methyl, methylene, or methine.

Within each line of a chart section the shifts of groups beta and gamma to the reference are clearly differentiated, but the distinction becomes much less clear if all three lines are considered. Thus, if one can determine from other information whether a band is produced by a methyl, methylene, or methine, then from the chemical shift he can determine the spatial position of that group with respect to the reference. The additional information required is supplied by the band intensity or multiplicity, so that NMR tends to be self-

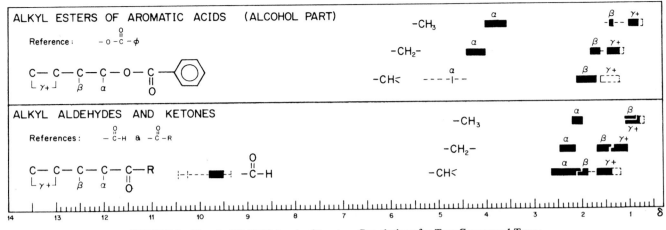

FIGURE 3. Chemical Shift-Molecular Structure Correlations for Two Compound Types

sufficient in determining molecular structure in many cases.

Hydrogen chemical shifts are stable and predictable, so that their correlations with structure can be indirect. That is, it is not necessary to establish a correlation for each compound, but only for a compound type, including a reasonable number of steric configurations. This is a major advantage of NMR in the determination of structure. The indirect chemical shift correlations cou-

pled with the absolute intensity measurements (next section) is a powerful combination.

C. BAND INTENSITY

The intensity of an NMR band is defined as the area under that band between the signal trace and the baseline. It is usually measured as the distance between horizontal lines or inflection points of the electronic integral of the band, as shown in Figure 4. It can also be

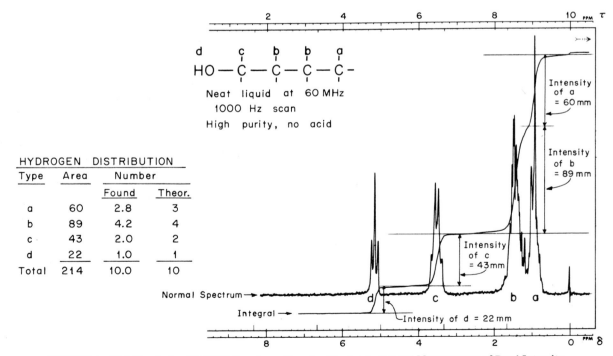

HYDROGEN	DISTRIBUTION		
Type	Area	Number	
		Found	Theor.
a	60	2.8	3
b	89	4.2	4
c	43	2.0	2
d	22	1.0	1
Total	214	10.0	10

FIGURE 4. High-Resolution H¹ NMR Spectrum, Showing Electronic Integral Measurement of Band Intensity

measured by planimeter, counting squares, or cutting out the band and weighing the paper.

Figure 4 shows the practical accuracy of routine integral measurements. It is good enough for determinations of molecular structure and ordinary quantitative analyses, but not good enough for precise quantitative work. The greatest error is encountered between overlapping bands where the signal does not return to the baseline and the integral does not flatten out. One then measures to the inflection point between the bands,

aided by the valley in the normal spectrum. This assumes that the area on the far side of the valley is the same for the two bands, an assumption which is usually not quite true. Nevertheless, the inflection method provides an objective and reproducible way to measure the areas of overlapping bands.

In the special case of a small band lying on the skirt of a much larger band (Figure 5), it is clear that a substantial portion of the area between inflection points really belongs to the larger band. One objective way to

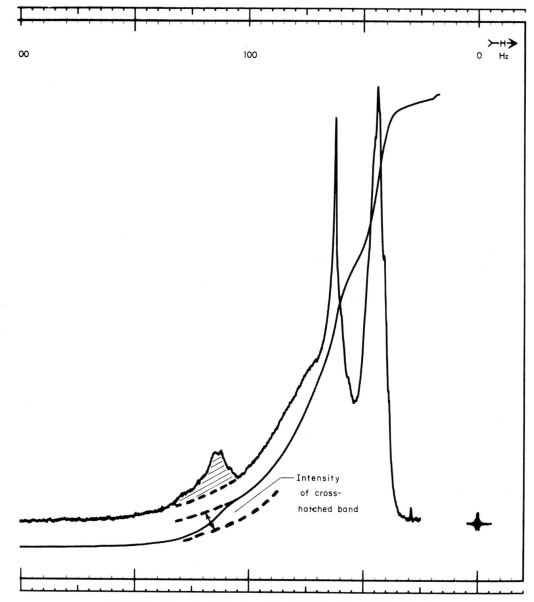

FIGURE 5. Measuring Intensity of Small Band on Skirt of Large Band

12

improve the area measurement is to continue the integral curve of the larger band skirt, at the same curvature, past the integral inflections on each side of the small band, then measure the distance between these curves. This method is shown in Figure 5.

If greater accuracy is desired, band overlap must be reduced by using a higher spectrometer frequency or, if resolution is limiting, by achieving higher resolution. For greater accuracy for nonoverlapping bands, the integrals must be run with great care on an expanded vertical scale.

The intensity of an NMR band is directly proportional to the number of nuclei contributing to it, provided saturation is negligible or is the same for all bands. For hydrogen spectra, it is usually easy to set instrumental conditions so that relative saturation between bands is negligible because the relaxation times are reasonably short and tend to equalize somewhat when the sample is in solution. It is important that the instrument operator set the instrumental conditions properly for each sample, as described in Chapter 2, Section IIA.

This absolute relationship of intensity to the number of nuclei is a tremendous asset for NMR. It means that the relative number of hydrogens contributing to several bands can be determined *without calibration*. It further means that if the absolute number of hydrogens represented by any band of a spectrum is known, the absolute number represented by every band is known. For this reason, it is possible to determine with complete confidence the hydrogen distribution of a new compound for which no calibration data can be available, the only limitation being band overlap. This capability is of great importance to the determination of molecular structure and to quantitative multicomponent analysis.

D. COUPLING CONSTANT

The spin–spin coupling constant is a measure of the specific interaction of one nucleus with another that gives rise to a splitting of the resonance bands produced by each nucleus. It is independent of the applied magnetic field (and the spectrometer frequency) and could be expressed in field or frequency units for all spectrometers. It is generally agreed, and standardized by ASTM, that it will be expressed in hertz. This coupling constant can be measured precisely in only a very few spectra. In most cases it must be determined by calculations that are too slow and too laborious to be useful in day-to-day analytical work. A practical definition of the coupling constant that is more useful in structural studies and quantitative analysis is simply the measured separation between adjacent "first-order" peaks in a spin–spin multiplet.

The practical definition is illustrated in Figure 1. Band c is nearly a first-order triplet, so that the coupling constant between b and c, J_{bc}, is the separation, in hertz, between adjacent peaks. Bands a and b, on the other hand, show many higher-order peaks, so that the first-order multiplet diagrams must first be drawn for them. The peaks of band a form a group of three composite peaks which resemble a triplet as defined by the first-order spin-multiplet rules (Table 1). The first-order multiplet diagram for a triplet is therefore drawn under this band, and the separation between adjacent bars in this diagram is the practical value of the coupling constant, J_{ab}. The peaks of band b form a group of six composite peaks which resembles a first-order sextet. The first-order multiplet diagram for a sextet is therefore drawn under this band. The separation between adjacent bars in this diagram is an average of the values measured from the other two bands. This value is as good as the other two for analytical purposes and could be used for the other two if they could not be measured.

Spin coupling is caused by the indirect intramolecular interaction of nuclear magnetic moments through the bonding electrons. Unlike the direct nuclear moment interactions mentioned in Chapter 2, Section IA-1, the indirect ones are not averaged out by molecular tumbling or sample spinning. The coupling constant varies with a number of factors which are directly related to the detailed structure of the molecule. These factors are discussed qualitatively in the following paragraphs. A detailed qualitative and quantitative discussion is given by Jackman and Sternhell,[T28] pp. 269–356.

1. Isotope. The largest couplings are those between hydrogen and the nucleus of the atom to which it is directly bonded. The effect of isotope on the coupling constant can be indicated by the maximum observed values of these one-bond couplings, as shown in the table below:

H^1-N^{14}	= 91 Hz max
H^1-N^{15}	= 96 Hz max
H^1-C^{13}	= 270 Hz max
H^1-H^1	= 280 Hz
H^1-F^{19}	= 615 Hz
H^1-P^{31}	= 745 Hz max
H^1-Sn^{119}	= about 1900 Hz max

2. Number of bonds between coupled nuclei. In general, coupling constants decrease with increasing numbers of bonds between the coupled nuclei. For freely rotating saturated chains the decrease is rapid, and coupling becomes insignificant beyond three or four intervening bonds. A similar rapid decrease is observed for aromatic ring couplings. In unsaturated linear systems, however, the coupling tends to attenuate less rapidly. Significant coupling has been observed over as many as nine bonds in conjugated systems.

3. Spatial positions of coupled nuclei. In rotationally restricted systems, such as saturated rings, the relative spatial positions of the coupled nuclei may exert more influence on the coupling than does the number of bonds between them. This effect of spatial position can be even more pronounced in unsaturated systems. This effect has been attributed to the degree of alignment of the atomic orbitals. The special case of the effect of vicinal bond angle between the coupled nuclei has been worked out by Karplus.[T30] The coupling constant is maximum at $0°$ and $180°$ and minimum at $90°$. There may also be a dependence on geminal angle between the coupled nuclei, but this dependence has not been demonstrated satisfactorily.

4. Bond order or length. Correlations indicate that higher-order and therefore shorter bonds give greater coupling constants.

5. Electronegativity of Adjacent Groups. Increasing electronegativity of an adjacent group tends to increase coupling constants in the positive direction. The actual peak separation may either decrease or increase, however, depending on the sign of the constant.

In summary, if the coupling constant is dependent on the degree of alignment and overlap of atomic orbitals, then anything which affects either of these factors will affect the coupling constant. Because of the foregoing relationship, coupling constants help determine the relative spatial positions of hydrogens and give some clues about the nature of the groups they are in or the kinds of atoms they are spin-coupled to.

E. BAND WIDTH AND SHAPE

The width of a resonably smooth and symmetrical band has been defined by agreement as the distance between the sides of the band at half the band height. This definition is illustrated in Figure 6, band *b*. This band width may include the effects of relaxation time, field inhomogeneity, unresolved chemical shifts, and unresolved multiplicity. If the band is known to consist of a single line, such as the TMS band (ignoring the Si^{29} satellites), then the width reflects only the effects of relaxation time and field inhomogeneity. If the natural line width is known to be very narrow (long relaxation

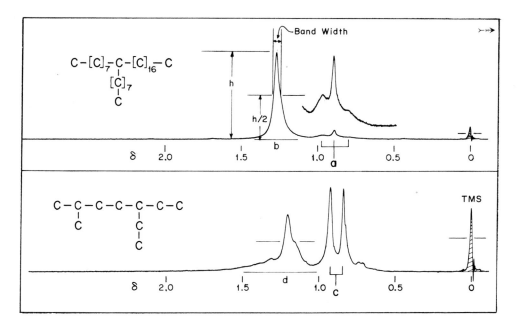

FIGURE 6. Band Widths and Shapes

time), as is the case with TMS, then the line width is a measure of field inhomogeneity. If the relaxation time is fairly short, however, then the line width at half height is inversely proportional to the relaxation time.

If the band is not smooth and symmetrical, the width at half height has no real meaning, as illustrated by band *d* in Figure 6. This band obviously includes some unresolved chemical shifts and multiplicities. In such cases the shape of the band is of more use than the width in analytical work.

If the band is fairly symmetrical but not smooth, it may be interpretable as a first-order multiplet. This is the case with bands *a* and *c* in Figure 6. Band *a* can be interpreted as a triplet, and band *c* can be interpreted as a doublet. Then the peak separation is used as the coupling constant, and the band width is ignored.

If the spectrometer is operated in the CW (scanned) mode, the *ringing* (or *wiggle beats*) which results from fairly fast scanning indicates the relative widths of narrow lines—the more wiggle beats, the narrower the line. However, this characteristic is lost when the spectrometer is operated in the pulse-Fourier transform mode or scanned at very slow speeds.

When all known information about a band is considered, the band width and shape can contribute significantly to the interpretation of the spectrum in terms of molecular structure. For instance, in Figure 6, the few wiggle beats on the upfield sides of the TMS lines indicate that the field homogeneity is good. The difference in shapes of bands *b* and *d* show that they are produced by different structures, and that band *d* probably represents shorter chains.

The shape of a well-defined multiplet is also useful in determining the groups to which it is coupled. In Figure 1, band *c*, the upfield peak is slightly taller than the downfield peak. In a first-order triplet these two peaks would be the same height. Therefore, this multiplet is not quite first order. It is said to be enhanced in the upfield direction. This enhancement shows that the hydrogens producing this band are spin-coupled to some which produce a band 20 or more coupling constants away upfield. Similarly, the downfield enhancement of band *a* shows that the hydrogens producing it are spin-coupled to hydrogens which produce a band not very far downfield. The other fine structure on band *a* indicates that it is probably produced by part of a paraffinic chain. In fact, this particular band shape indicates the normal propyl and similar groups (Plates 1, 3A, 5, 7A) and would permit the identification of such a group even if the other two bands could not be observed. This kind of information is helpful in determining the components of mixtures.

Producing NMR Data

I. SAMPLE PREPARATION

A. SAMPLE REQUIREMENTS

1. State

If hydrogen atoms are held close to one another in rigid or viscous systems, severe line broadening will result from their direct magnetic interactions (the spin–spin relaxation time, T_2, is short). The band separation and fine structure necessary to chemical structure determination cannot be observed under such conditions. These "direct dipole" interactions are eliminated when the hydrogen atoms are in constant rapid motion with respect to one another, leaving only the "indirect dipole" interactions which produce spin–spin multiplets. Consequently, the observation of the hydrogen resonance at high resolution requires that the sample molecule be in relatively free motion in all its segments. This usually means that the sample must be a gas or a liquid, although in a few cases useful spectra can be obtained from gels. The most convenient state is the liquid, so that both gases and solids are usually dissolved in suitable solvents for observation. Solids may also be melted.

Techniques are now being developed for the observation of solid samples at high resolution, but they are not yet generally available for hydrogen resonances. Carbon-13 and perhaps other low-abundance isotopes can be observed at high resolution in some solids with modern methods. For the present, however, it must be assumed that samples to be observed by hydrogen resonance must be liquids of low viscosity.

2. Phases

The sample should be a single phase. The presence of a second phase, whether liquid or solid, can lead to line broadening, false peaks, and sometimes to a doubling of all the peaks in the spectrum. This last phenomenon is shown in Figure 7. Two spectra of the same sample of sulfonated polyethoxylated phenol are shown, one (a) made from a clear single-phase solution and the other (b) from a hazy two-phase system. All peaks except the water peak are doubled in the spectrum of the hazy sample (b). The addition of acetic acid-d_4 to the hazy sample caused the water to coalesce into a separate phase which was removed. This left a clear solution which gave the clean spectrum (a). The residual water peak was moved downfield by the acid so that it does not show in spectrum (a).

If the interface between two separated phases is within the detection range of the receiver coil of the spectrometer, the spectrum will be distorted. A vortex of air frequently causes doubling of the peaks as in Figure 7. It is essential, then, that the sample extend beyond the detection range both above and below the receiver coil. A sample depth of 2 or 3 cm in a 5-mm tube is sufficient if the center of the sample is at the center of the receiver coil.

If less sample must be used, vortex formation can be avoided by slow spinning, or can be eliminated by placing a plastic plug in contact with the top of the liquid. If the plastic has nearly the same magnetic susceptibility as the sample, the interface will not be troublesome. This is

16

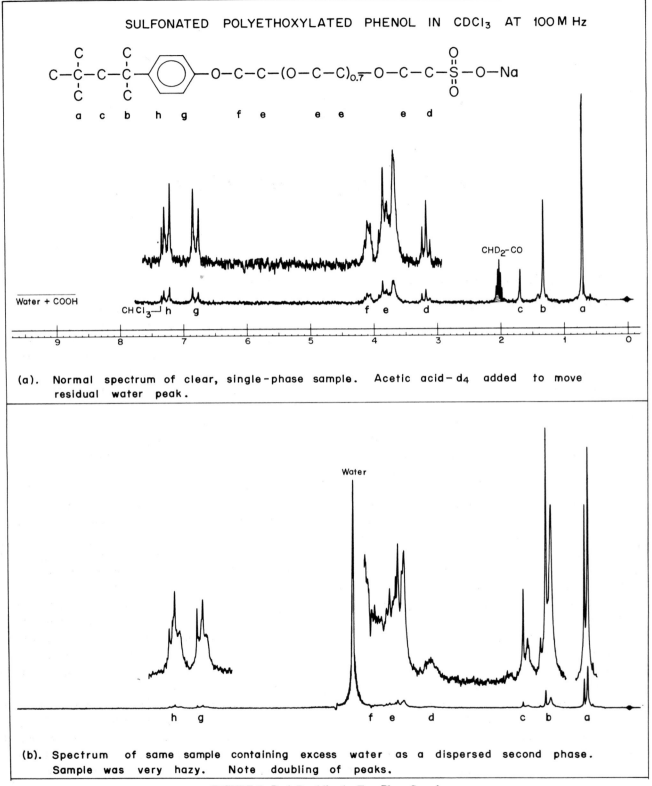

(a). Normal spectrum of clear, single-phase sample. Acetic acid–d₄ added to move residual water peak.

(b). Spectrum of same sample containing excess water as a dispersed second phase. Sample was very hazy. Note doubling of peaks.

FIGURE 7. Peak Doubling by Two-Phase Sample

the reason for using nylon plugs with chloroform solutions. Whenever small samples are used, however, one must be wary of distortion of the spectrum by the inteface.

On the other hand, if the interface is outside the detection range of the receiver coil, two separated liquid phases can be observed independently in the same tube. If a variable temperature probe is used, it may be necessary to remove the heater-sensor to permit the sample tube to be pushed down far enough to place the upper phase in the receiver coil.

3. Temperature

The normal temperature in the probe of a modern spectrometer is about 30–35°C. This will provide adequate motion for small molecules, but not for high polymers. The highest resolution for polymers in the 100,000 MW range requires sample temperatures of 150–200°C. Also, elevated temperatures are sometimes needed to increase sample solubility or decrease viscosity. On the other hand, significant slowing of the rotation of small molecules requires temperatures of −60 to −150°C. To fulfill all these needs, then, the sample temperature must be controllable at will from −150 to +200°C. This range is provided in many commercial spectrometers.

4. Concentration

The useful sample concentration is limited on the low side by the sensitivity of the spectrometer and on the high side by viscosity and by the phenomenon of radiation damping. Modern CW spectrometers operating at 30 to 60 MHz will provide adequate signal-to-noise ratio (S/N) for sample concentrations of about 5% to 100% (neat liquid), provided the sample viscosity will permit. Spectrometers operating at 100 MHz and above can provide adequate S/N for concentrations of 0.05 to 1%, but may be limited to a top concentration of 10 to 20% by radiation damping, which broadens peaks and cuts their height when the signals are too intense.

The S/N for CW operation can be increased tenfold or more by time-averaging accumulated signals in a computer. This requires considerable time, but permits observation of samples at 1/10 the concentration required for a single scan. Pulse-Fourier transform systems can increase the S/N about tenfold in the time usually required for a single CW scan, and may increase it several hundredfold if time is available. The pulse-Fourier sys-

tem thus extends the sensitivity of NMR into the upper microanalytical range.

5. Paramagnetic Materials

Samples to be observed at high resolution must be free of all materials which generate strong magnetic fields within the sample bulk. In other words, the sample must be diamagnetic. Paramagnetic and ferromagnetic materials such as free radicals, paramagnetic ions, iron, nickel, or cobalt particles, etc., must be removed or rendered diamagnetic. Filtration, HCl extraction, and ion exchange can reduce the problem, as discussed later.

B. SOLVENTS

1. Requirements

Because nearly all present high-resolution NMR spectrometers require liquid samples of low viscosity, and some require that the sample concentration not exceed about 20%, it is necessary that all gases and solids and at least some liquids be dissolved in suitable solvents before observation. The solvent, in turn, must be a liquid of low viscosity which will (1) dissolve an adequate amount of the sample, (2) not produce resonances which will interfere with the sample bands, (3) not attack the sample chemically, and (4) retain these properties in the temperature range over which the sample is to be studied. For work requiring accurate chemical shift measurements the solvent should also be magnetically isotropic and should not complex with or otherwise associate with any particular part of the sample molecule. The ideal solvent for general work would be a chemically inert, hydrogen-free, magnetically isotropic liquid of low melting point and high boiling point, which would dissolve a wide variety of compound types. When desired, magnetic anisotropy would be introduced by anisotropic complexing agents (shift reagents) as discussed later.

As the sensitivity of NMR spectrometers has increased it has become necessary to check all solvents for impurities which show up at maximum sensitivity. It is also necessary to check the more volatile solvents for impurities which are left behind when the solvent is evaporated. In general, solvent grades especially purified for spectrometry or chromatography should be used. Regular ACS pure-grade compounds are frequently unsatisfactory.

TABLE 4. Characteristics of the Most Useful Solvents for H^1 NMR

Compound (Abbreviation)	Formula	m.p., °C	b.p., °C	d 20/4	Isotopic purity^a	Hydrogen bands Group^b	δ	τ	Notes
Solvents without Hydrogen									
Carbon tetrachloride	CCl_4	−23	76.8	1.585	−	None	−	−	General solvent for hydrocarbons and many nonhydrocarbons. **Preferred when precise chemical shift measurements are needed.**
Carbon disulfide	CS_2	−110.8	46.3	1.260	−	None	−	−	Good solvent for hydrocarbon polymers and asphaltic materials. Gives best resolution for given concentration. *Hazardous*—toxic and highly flammable. Ignition temperature 120°C, flash point −30°C.
Tetrachloroethylene (TCE)	$Cl_2C = CCl_2$	−22	121	1.623	−	None	−	−	Not significantly better than CCl_4. May be unstable at high temperatures.
Hexachlorobutadiene (HCBD)	$Cl_2C = CCl—CCl = CCl_2$	−21	215	1.682	−	None	−	−	Solvent for hydrocarbon polymers at elevated temperatures. Limited by reactivity.
Tetrachlorothiophene (TCT)		29	233	1.704	−	None	−	−	Stable solvent for hydrocarbon polymers up to 200°C.
Deuterated Solvents									
Acetic acid-d$_4$ (HAc)	CD_3COOD	17	119	1.05	99.5	CD$_2$H(p) / COOH(s)	1.8–2.1 / 4.5–12.5	7.9–8.2 / (−2.5)–5.5	Useful for shifting OH bands and promoting hydrogen exchange.
Acetone-d$_6$ (acet.)	$CD_3—CO—CD_3$	−95	56	0.79	99.5	CD$_2$H(p)	2.0–2.2	7.8–8.0	General solvent for polar compounds. Can cause appreciable change in some shifts.
Acetonitrile-d$_3$	CD_3CN	−46	80	0.79	99	CD$_2$H(p)	2.0	8.0	Solvent for polar compounds. Lower purity and higher price than acetone.
Benzene-d$_6$	C_6D_6	6	80	0.88	99.5	CH(s)	7.2	2.8	Solvent for aromatic compounds, including aromatic polymers. May change chemical shifts and markedly alter spectrum.

Solvent	Formula	m.p. (°C)	b.p. (°C)	Density	Atom % D[a]	Band[b]			Remarks
Chloroform-d	CDCl₃	−64.1	61	1.49	99.8 / −100	CHCl₃ (s)	7.3 / −7.5	2.5 / −2.7	Good general solvent for most types of compounds. Slow hydrogen exchange frequently increases CHCl₃ peak.
Deuterium oxide	D₂O	3.8	101.4	1.105	99.8 / −100	H₂O (s, sharp to broad)	4.7 / −5.2	4.8 / −5.3	For water soluble materials. Exchangeable hydrogens in solute normally merge with H₂O to produce single peak.
Dimethyl-d₆ sulfoxide (DMSO)	CD₃–SO–CD₃	19	190	1.10	99.5 / −100	CD₂H(p)	2.5	7.5	For difficulty soluble materials.
Hexafluoroacetone deuterate (HFAD)	CF₃–CO–CF₃ · 1.6D₂O	< 27	57 at 93 mm	—	99.5	H₂O (s, sharp to broad)	5.0	5.0	Solvent for polyformaldehyde, polyamides, polyacrylonitrile, polyvinyl alcohol, polyesters, etc.
Methanol-d₄	CD₃OD	−100	65.4	0.82	99	CD₂H(p) / OH (s)	3.1 / −3.5 / 4–7	6.5 / −6.9 / 3–6	Useful at low temperatures and for special problems. Too impure and expensive for general use.
Methylene-d₂ chloride	CD₂Cl₂	−97	40	1.34	99	CDH(t)	5.3	4.7	Useful at low temperatures.
Pyridine-d₅	C₅D₅N	−42	116	0.98	99	α-CH(~s) / β-CH(~s) / γ-CH(~s)	8.5 / 7.0 / 7.3	1.5 / 3.0 / 2.7	For difficulty soluble acidic or aromatic materials. Produces marked changes in chemical shift in many cases.

The normal hydrogenated versions of these solvents are useful in selected cases, at considerably less expense.

Solvents with Hydrogen

Solvent	Formula	m.p. (°C)	b.p. (°C)	Density	Atom % D[a]	Band[b]			Remarks
Ortho-Dichlorobenzene	o-C₆H₄Cl₂	−17	179	1.305	—	om-CH(m) / mp-CH(m)	7.4 / 7.15	2.6 / 2.85	Popular solvent for nonaromatic hydrocarbon polymers.
Trifluoroacetic acid	CF₃COOH	−15.3	72.4	1.535	—	COOH	4.5 / −12.5	(−2.5) / −5.5	Solvent for nitrogen compounds and other materials which are readily protonated. Useful for shifting OH bands. Tends to produce marked changes in chem. shift. May esterify alcohols and decompose some compounds.
Hexamethylphosphoramide (HMPA)	[(CH₃)₂N]₃P=O	6 to 8	> 100	1.03	—	CH₃ (d)	2.45	7.55	Solvent for amides, hydrazides and others which may be difficult to dissolve.

[a] Atom % deuterium, as manufactured. HAc, D₂O, HFAD, and CH₃OD will all pick up hydrogen from water absorbed from the air, and thus show increasing amounts of hydrogen with time of use. They will also exchange deuterium for exchangeable hydrogens in the sample, although this is not necessarily a problem. CDCl₃ will exchange part of its deuterium for hydrogen in some samples, producing a much larger CHCl₃ peak than the solvent alone would produce.

[b] The functional group which produces the observed band. The multiplicity of the band is indicated in parenthesis; s = single; d = doublet; t = triplet; p = pentet (quintet); m = complex multiplet.

2. Selection

No solvent that is ideal for all applications is available, but it is possible to find one which approaches the ideal for a particular application. The solvents that have been found most useful for hydrogen NMR are listed in Table 4. They are grouped according to their hydrogen content and are listed alphabetically by name within each group. The name, abbreviation (in parentheses) when different from the empirical formula, empirical formula, important physical characteristics, and hydrogen resonance band positions and shapes are listed, along with a brief description of the applications where each solvent is normally used. This information should enable the reader to select the most suitable solvent for his particular application. It may be helpful, however, to regroup the solvents according to other characteristics which are important in specific cases.

a. General-Use Solvents

Carbon tetrachloride is the most nearly ideal solvent for room temperature work. It is chemically inert, hydrogen-free, and magnetically isotropic and will dissolve a wide variety of hydrocarbons and many non-hydrocarbons. It will also accept the anisotropic shift reagents. Because of its isotropy it has essentially no effect on the chemical shifts of solutes and is the preferred solvent when shifts are to be measured precisely. It is limited to a rather narrow temperature range around room temperature, however, and it does not dissolve many highly polar or ionic compounds or high-molecular-weight polymers.

Chloroform (as $CDCl_3$) is the most frequently used solvent for nonhydrocarbons because it dissolves such a wide variety of compounds. It is neither as inert nor as isotropic as CCl_4, but it will accept the shift reagents. It can also be used at lower temperatures than CCl_4. One of its worrisome drawbacks is that it will sometimes exchange its deuterium for exchangeable hydrogens in a sample, resulting in a larger than normal chloroform peak and a somewhat smaller exchangeable hydrogen peak from the sample. This can be troublesome when the sample contains aromatics because the chloroform peak is in the aromatic region of the spectrum.

Acetone (as $CD_3-CO-CD_3$) is a good solvent for polar compounds which are not very soluble in chloroform. It is, however, an anisotropic molecule which can complex with some sample molecules to produce appreciable changes in chemical shifts (compared to those obtained in CCl_4 or $CDCl_3$). For this reason it is not recommended when chemical shifts are to be measured precisely, but may be helpful in reducing band overlap or exposing some multiplets more clearly. It is also useful at low temperatures. See Figure 8 for residual spectrum.

Deuterium oxide (D_2O) is the preferred solvent for many water-soluble materials. It is also a good extractive solvent for separating hydrophilic from hydrophobic components. The deuteriums will readily exchange with any exchangeable hydrogens in the sample, usually producing a single peak in the water region, $\delta = 4.7-5.2$.

Dimethyl sulfoxide (as $CD_3-SO-CD_3$) is useful for samples which are difficult to dissolve in the other general solvents. Like acetone, it is an anisotropic molecule which will complex with some sample molecules and produce significant chemical shift changes. It will also complex with alcoholic and possibly phenolic OH hydrogens in such a way as to isolate them from the remainder of the sample and solvent molecules. This isolation stops rapid hydrogen exchange and restores spin coupling between the OH and neighboring hydrogens. DMSO can be used at high temperatures, increasing sample solubility still further. See Figure 8 for residual spectrum.

b. Special-Use Solvents

Tables 5 and 6 group the solvents according to various special applications.

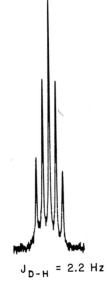

FIGURE 8. Spectrum of the CD_2H Group: Typical of residual hydrogen in CD_3-COOH, $CD_3-CO-CD_3$, CD_3-CN, $CD_3-SO-CD_3$, and CD_3-OD

$J_{D-H} = 2.2$ Hz

TABLE 5. Solvents for High- and Low-Temperature Work

Solvents for High-Temperature Work

Solvent	b.p. °C	Remarks
Tetrachlorothiophene	233	Stable, inert, no hydrogen. Good for hydrocarbon polymers.
Hexachlorobutadiene	215	Limited by stability and reactivity, but useful for hydrocarbon polymers. No hydrogen.
Dimethyl sulfoxide	190	General solvent for use at high temperatures. Appears to be resonably inert, but could promote reactions in some mixtures. Deuterated form preferred.
o-Dichlorobenzene	179	Stable, inert, contains hydrogen. Popular solvent for nonaromatic hydrocarbon polymers.

Solvents for Low-Temperature Work

Solvent	b.p. °C	Remarks
Carbon disulfide	(−110.8)	Has lowest melting point, but is hazardous because it is toxic and highly flammable. Reactive with some species, especially amines. Useful with hydrocarbons.
Methanol	(−100)	The deuterated form is expensive but is useful at very low temperatures.
Methylene chloride	(−97)	Popular low-temperature solvent. Deuterated form is expensive.
Acetone	(−95)	Should be useful at low temperatures if it can be kept dry.
Chloroform	(−64)	Useful to moderately low temperatures. Good general solvent.

C. REMOVING EXTRA PHASES

As stated earlier, extra phases in a sample can broaden the lines and even produce false lines. All emulsions and suspensions should be broken, and the interface between the consolidated phases should be moved out of the detection range of the receiver coil. Solids should be removed.

It is good practice to filter all sample solutions through a glass wool plug in a medicine dropper to remove sizable solid particles. Glass wool is preferred to filter paper because it retains much less of the solution. Suspensions and less stable emulsions can be broken down conveniently by centrifuging. This gives good separations without the risk of losing either sample or solvent through evaporation or through absorption in a filter. The more stable emulsions can frequently be broken down by heating or by acidifying, followed by centrifuging. Single magnetic particles, which can cause capricious changes in the spectrum, can sometimes be removed by pulling them up above the liquid with a magnet.

In special cases one may wish to study gels or other forms of multiphase systems at as high a resolution as possible. One should then study a number of different portions of the sample in different tubes, and perhaps under different conditions of concentration, temperature and liquid phase (solvent), to make sure the observed spectrum is a true one. Sometimes remarkably useful results can be obtained from such samples.

D. SUGGESTIONS FOR SAMPLE HANDLING

It is convenient to make up the sample solution in a small screw cap vial where it can be agitated, heated,

TABLE 6. Solvents for Special Applications

Solvent	Application
Acetic acid	Useful acidifying agent, for shifting OH bands and promoting exchange.
Acetonitrile	For polar compounds. May be effective in some cases where acetone is not suitable.
Benzene-d$_6$	An anisotropic solvent useful for changing chemical shifts. Useful for dissolving aromatic compounds.
Carbon disulfide	Good solvent for hydrocarbon polymers and asphaltic materials. Gives the best resolution for a given concentration, possibly because its solutions have lower viscosity and fewer agglomerates of sample molecules. Seems to discourage micelle formation. *Toxic and highly flammable.*
Hexafluoroacetone deuterate	A better solvent for some polymeric materials such as polyformaldehyde, polyamides, polyesters, polyacrylonitrile, polyvinyl alcohol.
Pyridine-d$_5$	For acidic or aromatic materials which are difficult to dissolve. It is also an anisotropic complexing agent which can produce marked changes in chemical shifts.
Trifluoroacetic acid	Solvent for nitrogen compounds and other materials which are readily protonated. Strong acidifying agent for shifting OH bands, breaking emulsions, protonating amines, phosphines, etc. May esterify alcohols and may decompose some samples. Tends to produce marked changes in chemical shift as compared to CCl$_4$ or CDCl$_3$.

extracted, centrifuged, etc. It can then be transferred from this vial to a sample tube through a glass-wool plug in a medicine dropper or through some other suitable filter. A 1-dram vial will hold enough solution for a 5-mm NMR tube and allow room for the above operations. Larger vials are useful for solvent extraction.

If sample is scarce, it is wise to try the lower boiling solvents first on a portion of the sample. These solvents can be removed easily by evaporation in a stream of dry nitrogen or under vacuum, and the next solvent can be applied to the same portion of sample. In this way the trial and error part of the procedure can be carried out on only one portion of the sample, saving the rest for the final work with the best solvent and conditions.

If there is plenty of sample, time and expensive deuterated solvents can be saved in many cases by trying different portions of sample in the fully hydrogenated forms of the solvents to be used. Then only those solvents which prove effective can be used in their deuterated forms in the final runs.

II. DATA PRODUCTION

A. RECORDING THE SPECTRUM

Good NMR interpretation requires, first of all, a good spectrum. The several characteristics which make a good spectrum should be provided at the time the spectrum is run, although it is possible to improve some defects later by the use of computer methods. The desirable characteristics are an accurate X-axis scale, a good signal-to-noise ratio, adequate resolution, proper phasing, minimum saturation, good aspect ratio, and an accurate integral. Although the latest routine-type instruments have simplified the procedure a great deal, some judgment must still be exercised to produce a good spectrum. Even more judgment must be used on the more complex research instruments. In either case, the operator must look for each of the required characteristics to make sure it is incorporated into the final product.

1. Calibration of the X Axis

Modern commercial NMR spectrometers have the scanning mechanism coupled directly to the recorder X axis so that horizontal pen position is directly related to magnetic field or frequency. This permits the use of charts precalibrated on the X axis for direct reading of chemical shifts and easy measurement of peak separations. These calibrations are subject to change with time. Furthermore, although the charts have linear markings, the field–pen relationship is not linear for all spectrometers. The most accurate chemical shift measurements thus require precise calibration of the nonlinear scales and periodic checking of the calibrations of all scales.

The most accurate calibration method utilizes modulation sidebands from a strong sharp peak, such as that of benzene, to provide equally spaced check points all along the scale. This method is somewhat time-consuming, however. For the periodic calibration checks it is more convenient to use a standard sample solution.

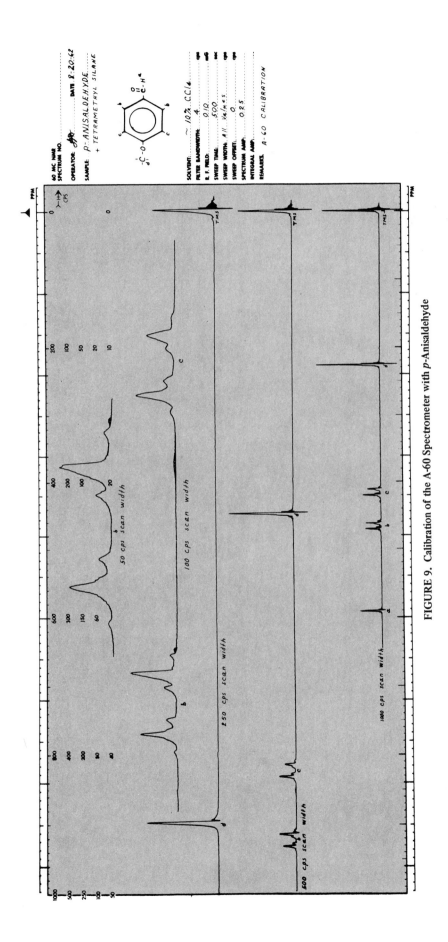

FIGURE 9. Calibration of the A-60 Spectrometer with *p*-Anisaldehyde

One such solution, proposed by Jungnickel,[T29] consists of a number of compounds which provide peaks over the greater part of the hydrogen spectral region. The author found a single compound, *p*-anisaldehyde, that also provided peaks over the hydrogen region, spaced so that all scales of the A-60 and similar spectrometers can be checked with it. In each case a master spectrum of the calibration sample may be obtained on a spectrometer accurately calibrated by the modulation sideband method. Thereafter, scans of this sample under the same conditions on the same type of spectrometer can be compared by direct overlay with the master spectrum to check and correct the calibration. A master spectrum of *p*-anisaldehyde for the A-60 spectrometer is shown in Figure 9.

2. Signal-to-Noise Ratio

A low signal-to-noise ratio, S/N, causes some signal peaks to be lost in the noise, or can cause some noise peaks to be mistaken for signal peaks. In either case,

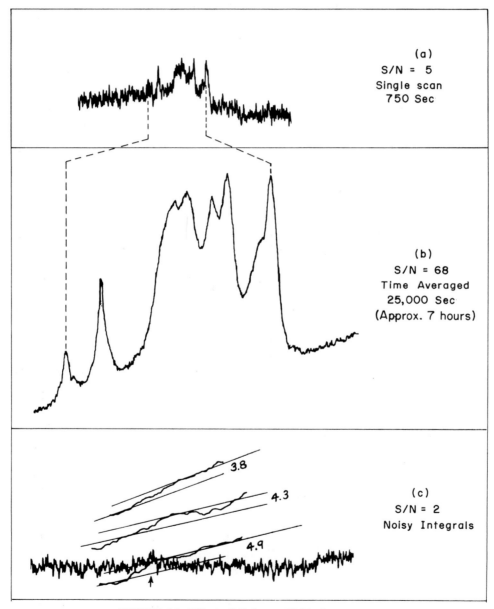

FIGURE 10. Effect of Noise on NMR Spectrum

errors in interpretation can result. The problem is shown in Figure 10. Spectrum (a) shows a band at a S/N of 5. It is clear that a band is present, but its shape and intensity are uncertain. In spectrum (b) this same band is presented at a S/N of 68 as the result of time-averaging. The shape is now clearly observable and the intensity is measurable.

In spectrum (c) another band, centered at the arrow, is presented at the very low S/N of 2. Even the presence of this band is uncertain, and the integrals are so noisy that intensity measurement is very difficult. Repeated integration of this region indicates that a band is present and gives at least a rough idea of its intensity. It would be much better, however, to increase the S/N of this band than to base conclusions on the evidence presented here.

Increasing S/N requires getting more molecules into the observing region of the spectrometer or accumulating the signals over a sufficiently long period so that the random noise will average to a lower value. The extent of the observing region of the spectrometer can be determined by reducing the height of liquid in the sample until the S/N begins to decrease. CW spectrometers in common use today can effectively observe about 3 cm of sample in a 5-mm tube. If scarcity of sample limits the height to about 1 cm, it is advisable to use a plastic plug on top of the liquid to prevent vortex formation. If smaller samples must be run, commercial microcells can provide some S/N improvement.

If the observing volume is filled, increasing S/N requires a higher concentration of sample molecules in the solution. If solubility is limiting, a different solvent and/or a higher temperature are the obvious remedies.

If the number of sample molecules cannot be increased, and a more sensitive spectrometer is not available, time-averaging should be considered. This requires hours of repeated scanning for CW spectrometers, but the time can be reduced by a factor of about ten with Fourier transform systems.

3. Resolution

NMR spectral resolution is defined by the ASTM as the magnitude of the minimum band width that can be observed in a given spectrometer. By analogy, the resolution of a sample is the width of the narrowest line that can be observed in that sample. This is usually the width of the TMS line. It can be observed on the spectrum unless TMS is used as the internal lock compound. Then it must be observed in the unlocked mode on an oscilloscope or in a separate scan. Good resolution is needed to permit observation of the spin multiplets, and, in some cases, to provide the band separation needed for good

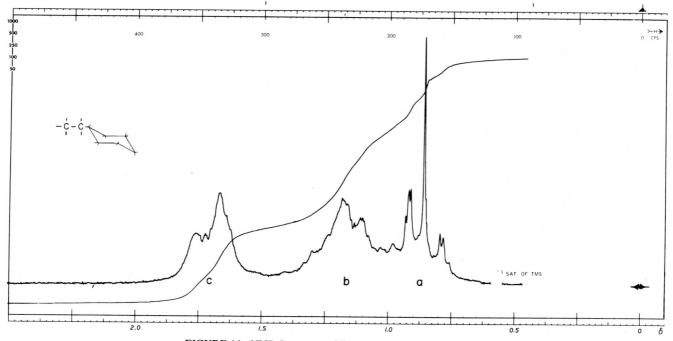

FIGURE 11. NMR Spectrum of Ethylcyclohexane at 100 MHz

intensity measurements. Different portions of a spectrum may have different resolution, as shown in Figure 11. Band *a* has a sharp central line and reasonably good resolution in the wings. Bands *b* and *c,* however, are broad and show little fine structure.

Poor resolution can be caused by the magnetic field, by the sample solution, or by the sample molecule. If the magnetic field homogeneity is poor, *all* parts of the spectrum will show poor resolution. Field homogeneity can be checked readily by observing a standard sample, such as a mixture of TMS and chloroform, which produces sharp singlets. Until the spectrometer will produce this spectrum at high resolution, it will not produce any spectrum at high resolution.

If the sample solution itself is responsible for poor resolution, again *all* parts of the spectrum will have poor resolution. This can be due to:

1. Excessive viscosity. This can be corrected by adding more solvent, raising the temperature, changing solvents, or a combination of these.

2. Paramagnetic materials such as dissolved oxygen, metal particles, ions, and metal chelates. The most universal of these materials is dissolved oxygen. This will have little effect on hydrocarbon samples, but can substantially limit the resolution of nonhydrocarbons, particularly oxygenated compounds. It is reduced fairly well by bubbling nitrogen through the sample, but if resolution on the order of 0.2 Hz is desired, oxygen must be removed by freeze-thaw degassing.

Metal particles should have been removed during sample preparation, but some may have been missed. They should be removed by centrifuging or filtering.

Paramagnetic ions can be removed by ion exchange, preferably by substituting hydrogen. Nonparamagnetic ions will cause some loss of resolution, but are much less troublesome than the paramagnetic ones.

Metals held as chelates or as fine suspensions can be removed in many cases by washing the solution with concentrated HCl. This, of course, requires that the solvent be immiscible with water.

3. Extra phases. These should have been removed during sample preparation, but sometimes they form afterwards. They can be removed as described under sample preparation.

If the sample molecule is responsible for poor resolution, the reference peak will be sharp, and some of the sample peaks may be sharp, as in Figure 11. Poor resolution in the remaining parts of the spectrum is then probably due to restricted rotation of parts of the sample molecule. Restriction of rotation reduces averaging of the chemical shifts among the various rotamers, and may reduce some relaxation times. The reduced averaging of chemical shifts results in the production of many more lines which may not be resolvable. The rate of motion may also be such that the lines are broadened, reducing the resolution further. Reduction of relaxation times also produces line broadening. All these effects together produce the broadened bands shown in Figure 11.

Segmental rotation can be increased by raising the temperature of the sample, changing solvents if necessary. In the case of high-molecular-weight polymers, it may be necessary to go to 150 or 200°C to achieve usable resolution.

In addition to the foregoing three causes, poor resolution can be produced by excessive radiofrequency amplitude and by radiation damping. Excessive rf amplitude broadens the peaks by saturation. Because saturation increases with increasing relaxation time, this will broaden the sharpest peaks first. Highest resolution therefore requires low rf amplitudes.

Radiation damping is broadening due to saturation of the receiver coil by excessive signal. It therefore affects the tallest peaks first. This is corrected by reducing the sample concentration.

Resolution will also be reduced if the scan rate in the CW mode is too fast. Because the rf amplitude which can be tolerated is a function of the scan rate, the rf must be reduced when the scan rate is reduced. Some practice in juggling these two factors is required to achieve the highest resolution.

4. Phasing

The NMR experiment produces two types of signals simultaneously, as indicated in Figure 12. One is in phase with the alternating magnetic field (the rf field) and the other is 90° out of phase with this field, so that it is possible to select one over the other by means of a phase-sensitive detection system. The absorption-mode signal (a) has the single-maximum symmetrical shape that is easy to identify as a peak, provides an easily located band center, and integrates easily. The dispersion-mode signal (b), on the other hand, is a dual-maximum band that can lead to confusion in interpretation, and it does not integrate precisely. It is desirable,

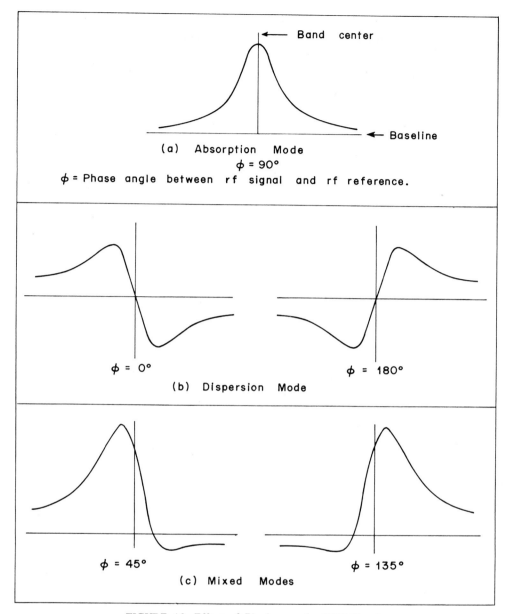

FIGURE 12. Effect of Phasing on the NMR Signal

therefore, to select the absorption-mode signal for analytical work.

Selection of the absorption-mode signal is made with the phasing control on a CW spectrometer. In Fourier transform spectrometers the phasing is adjusted for the transformed spectrum by computer manipulation. Figure 12 shows the effect on the signal from a sharp band as the phasing control (or computer parameter) is rotated through a phase angle of 180°. The signal varies from a pure dispersion mode at 0° (b) through a pure

absorption mode at 90° (a) to the oppositely phased dispersion mode at 180°. At all intermediate angles the signal will contain a mixture of the two modes, as indicated for 45° and 135° (c). The angle values given are illustrative only. Actual values for a given control dial depend on the magnetic properties of the sample and the wiring of the spectrometer.

Accurate analytical work, particularly accurate integration, requires that the pure absorption mode be selected. It is apparent from Figure 12(c) that mixed-mode signals have peaks which are displaced from the true band center, leading to errors in measuring chemical shift. The part of the mixed-mode signal which lies below the baseline will integrate negatively, leading to serious intensity errors.

Furthermore, because the actual phase setting required to select the absorption mode depends on the magnetic properties of the sample, the phase must be adjusted for each sample run. For most spectrometers operated in the field-sweep mode, the phasing will be constant throughout a single run, but this is not necessarily true for frequency-swept scans. The phasing does not always stay the same for all gain settings, either. In short, the phase should be set or checked before each scan, preferably by setting the baseline to equal levels immediately on each side of a sharp peak.

The integral is more sensitive to phase settings than the absorption signal, as shown in Figure 13. In (a) the phasing is adjusted properly. The baselines on each side of the overlapped bands are the same height, and the slope of the integral is the same at both top and bottom. In (b) the integral is sloped at the bottom but is flat at the top, indicating incorrect phasing. It is somewhat harder to see the difference in baseline levels, however. In (c) the integral slopes in opposite directions at top

and bottom, and the difference in baseline levels is readily observed.

Figure 13 also touches on the problem of adjusting the phase for overlapping bands. When one of the bands is tall and sharp and the other is low and fairly narrow, as in the figure, the tall band will dominate the baselines and will serve quite well for phase adjustment. If one or both bands are low and broad, however, phasing is difficult and imprecise. The integral of a low band overlapping a tall band may be in error by 50% or more because of inaccurate phasing. In such cases the introduction of a compound giving a sharp peak should be considered.

5. Saturation

Application of the alternating (rf) field, H_1, to a sample gives rise to two competing phenomena. It causes net upward transitions between the magnetic energy levels, thus producing the absorption signal. These upward transitions tend to reduce the difference in populations between the energy levels, however, thus reducing the signal. This reduction in population difference is called saturation. At the same time, spin lattice relaxation is causing net downward transitions which tend to restore the energy level populations to thermal equi-

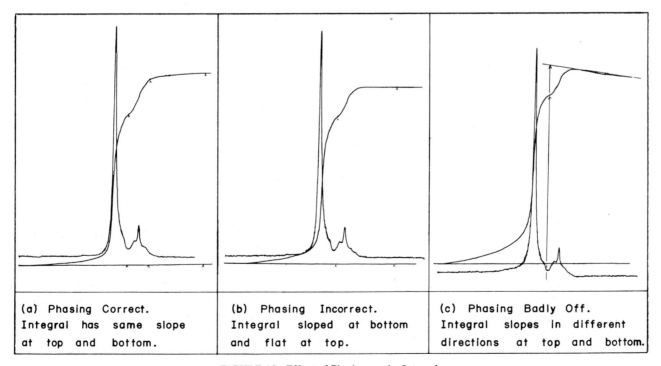

| (a) Phasing Correct. Integral has same slope at top and bottom. | (b) Phasing Incorrect. Integral sloped at bottom and flat at top. | (c) Phasing Badly Off. Integral slopes in different directions at top and bottom. |

FIGURE 13. Effect of Phasing on the Integral

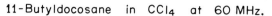

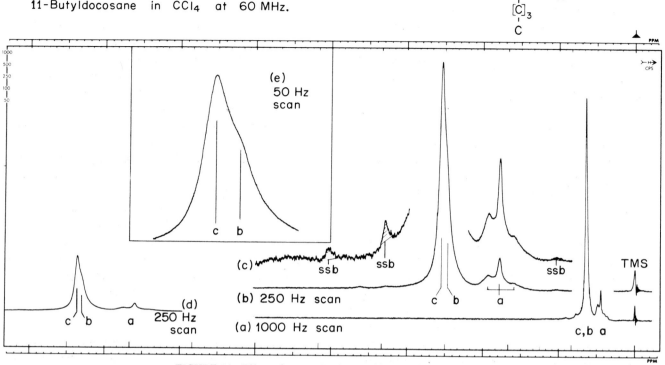

FIGURE 14. Effect of Aspect Ratio on Visibility of Band Features

librium. The result of these processes is that as H_1 is increased the absorption signal increases from zero to a maximum and then decreases to zero.

The degree of saturation of a given signal depends on the amplitude of H_1, the scan rate, and the relaxation times of the nuclei contributing to the signal. For fixed H_1 and scan rate, nuclei with longer relaxation times experience more saturation and more loss of signal. Thus, if H_1 is too high, bands with longer relaxation times will lose significantly more signal than those with shorter relaxation times, producing errors in observed intensities. The sharper peaks will also be significantly broadened.

Because of this potential intensity error and peak broadening, it is good practice to use the lowest value of H_1 consistent with good S/N. In no case should H_1 be higher than the value required to give maximum S/N. This value is determined by measuring S/N as H_1 is increased in increments at a fixed scan rate. For most work where S/N is not critical, it is considered wise to set H_1 at about half the value required for maximum S/N.

Even if the degree of saturation is significant, as it is

when maximum S/N is achieved, intensity errors will be significant only if there are sizable differences in relaxation time. Fortunately, the relaxation times of hydrogens in molecules in solution tend to equalize, so that if H_1 is kept at the value for maximum S/N or below there is little likelihood that intensity errors will be important. The author has never seen a case where saturation has led to significant errors in hydrogen intensities when H_1 is kept at reasonable values as outlined above. Such errors have been reported for pulsed experiments, however, when proper H_1 adjustment was difficult to determine.

It is the responsibility of the spectrometer operator to see that H_1 is kept at a proper value. The accuracy of calibration of the H_1 setting device is subject to considerable deviation with time on some spectrometers and should be checked at least roughly when the standard sample is run.

6. Aspect Ratio

For easy interpretation of an NMR spectrum it is desirable that all the important features of the spectrum be readily visible to the eye. All significant bands should be clearly raised above the noise and the baseline. All resolved peaks should be clearly separated from each other and spaced so that the eye can pick them out without difficulty. Poorly resolved or overlapping bands should be presented so that the "shoulders" and other features are readily discernible. In other words, the bands should be clearly visible with a good aspect (height-to-length) ratio. At the same time, an overall view is needed to give a feel for the extent and complexity of the entire spectrum and to show the relation of all the bands to one another. These requirements of good spectral presentation are frequently overlooked, causing the interpreter to accept incomplete information or rerun the spectrum.

It is not usually possible to present the whole spectrum in all its detail at the proper aspect ratio in a single scan of practical dimensions. Consequently, it is frequently necessary to present an overall spectrum and then present different sections of it separately, expanding the horizontal scan or the vertical gain or both to achieve the proper aspect. Examples of this type of presentation are shown in Figures 14 and 15. In Figure 14, scan (a) at 1000 Hz shows that the compound has no bands outside the paraffinic region. It is not possible to get a good idea of the structure of the bands from this scan, however. Scan (b) at 250 Hz gives a better view of the bands but does not show the shoulder on band c very well. Scan (c), also at 250 Hz but at greater vertical gain, shows the spinning sidebands more clearly and gives band a an aspect ratio more nearly like that used in the typical spectra of methyl groups in this book. Scan (d) presents bands b and c at an aspect ratio which makes band b much easier to see. Band a, however, is almost lost in the baseline at this aspect ratio. In scan (e) the horizontal scale has been expanded to show band b even more clearly.

In Figure 15 scan (a) shows the overall spectrum and scans (b) through (d) are the expansions. Scan (b) shows bands a, b, and c well enough to compare with the typical spectra of linear olefins presented later in this book. Scan (c) shows band e adequately, but band d needs further expansion as in scan (d).

7. Integral

Because the integral is the primary measure of band intensity, it should be run with the care necessary to

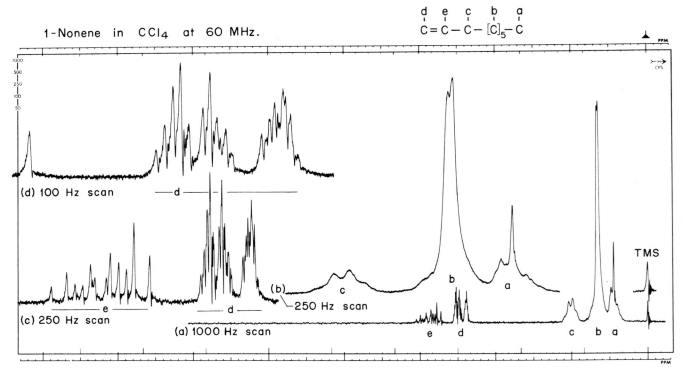

FIGURE 15. Good Aspect Ratio for Different Parts of a Complex Spectrum

achieve the desired accuracy. Several trial partial or total integrals may be needed to check the following factors:

1. *Baseline drift.* This should be corrected to as close to zero as is reasonable. The operating manual for the spectrometer should explain the method for doing this.
2. *Phasing.* The necessity and the method for correcting this have been discussed earlier in this chapter in Section IIA-4.
3. *Total integral height.* This should span the entire page for maximum accuracy in reading.
4. *S/N.* If the integral is noisy it will be necessary to run several, or to time-average the spectrum.

After these factors are set satisfactorily, the integral can be recorded. If the integral of any band is too small to measure accurately, rerun that part at increased integral gain or at slower scan rate. This will expand the integral vertically by the direct ratio of the gains or the scan times.

B. IDENTIFYING SPECIAL FEATURES OR GROUPS

1. Sidebands

Any periodic variation in the strength of the static magnetic field produces a corresponding periodic variation in the precession frequencies of the nuclei under observation. This latter variation is a frequency modulation of the signal, and is accompanied by the usual frequency modulation sidebands. Such sidebands are used intentionally in the stabilization systems of many NMR spectrometers, and are used for absolute calibration of the horizontal scale. Similar sidebands which arise from spinning of the sample, from vibrations of the probe, or from an ac component in the magnet current, however, interfere with observation of the spectrum and must be identified or eliminated.

Spinning sidebands (ssb), which are the most common form of these spurious signals, arise from rotation of the sample through field gradients in the XZ plane. If the homogeneity in the XZ plane is reasonably good and the spinner does not wobble, ssb have the following characteristics which permit them to be identified in the spectrum:

1. *Position.* Symmetrically spaced on each side of the parent centerband. Every band in the spectrum will have ssb, although they may not all be visible at the gain used.

2. *Spacing.* Spacings between the centerband and the first sidebands, between the first and second sidebands, etc. are equal. The value is a whole multiple of the spin rate, and is usually equal in hertz to the spin rate in revolutions per second. This fact provides the most positive method of identification. If the spin rate is increased the sidebands will move outward, and if it is decreased they will move inward. Also, once the spacing is established for one band, it is known for all the bands in that spectrum.

3. *Intensity.* For most commercial spectrometers, the field homogeneity can be adjusted so that the first sidebands are less than 2% of the intensity of the parent centerband. Under these conditions second and higher sidebands are not usually visible in a normal scan, but may be seen in vertical expansion if the S/N is over 100. Under normal conditions the intensities of corresponding ssb will be equal; that is, first ssb will have equal intensities, second ssb will also have equal intensities, etc. The intensities of the second ssb will usually be lower than those of the first ssb. If the field has a bad gradient or if the spinner wobbles too much, however, these intensity relationships can be upset. Spinning-sideband intensities are derived from the parent band and should be included in measurements of parent-band intensity.

4. *Shape.* Spinning-sidebands have the same shape as their parent band. This is helpful in differentiating ssb from adjacent or overlapping bands.

Figure 16 shows a simple spectrum in which spinning sidebands have been allowed to rise to about 3% of the intensity of the parent bands. The ssb are identified according to their parent bands. The spin rate is 41 rps and S/N is over 200.

Figure 14 shows both the first and second ssb of bands *b* and *c* in the vertically expanded scan (c). Here the spin rate is 22.5 rps and S/N is 95.

The size of the spinning sidebands is determined by the degree of variation in magnetic field experienced by the sample as it makes a revolution. This, in turn, is determined by the magnetic field gradients in the X and Z directions, by the amount of camber or eccentricity in the sample tube, by the amount of wobble in the spinner turbine, and by vibration of the probe body or insert. Unusual field gradients can be caused by paramagnetic dirt on the sample tube, in the probe insert or body, or on the magnet pole faces. Camber and eccentricity in the

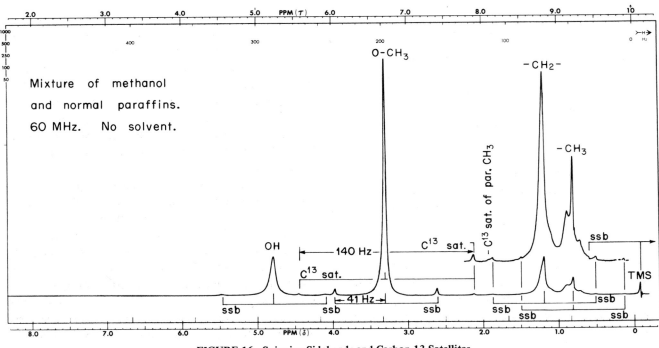

FIGURE 16. Spinning Sidebands and Carbon-13 Satellites

sample tube can be checked by rolling the tube down a plate glass mirror. Both defects will show up as variations in light transmitted by the tube as it rolls. The wobble of a loose turbine or the binding of a tight or dirty turbine will transmit vibrations to the probe which may be large enough to cause ssb. This is especially true when the probe insert is fastened at each end by elastic cement. Also, if the probe body is loose on its mount, it can pick up vibrations from the spinner. A clean magnet, probe, and spinner turbine, clean straight concentric sample tube, and a tight probe mount are as important as proper magnet shimming in reducing ssb.

The size of ssb is also a function of spin rate. As the spin rate increases, the ssb move outward and lose intensity. At the same time the vibration due to spinning may increase, increasing the ssb. The operator must strike the best balance between these two opposing factors when setting the spin rate.

Sidebands which are independent of spin rate may be caused by floor vibrations picked up by a loose probe, or by an ac ripple on the magnet current. The latter, of course, will be spaced at a multiple of the power line frequency and indicates a need for power supply repair.

2. Satellites

A few elements whose principal isotopes are non-magnetic (spin = 0) have one or more isotopes in low abundance with spin of ½. These low-abundance isotopes will spin-couple with attached hydrogens to produce extra bands called satellites, which are prominent enough to require interpretation. These satellites are approximately symmetrically spaced on each side of the band made by the hydrogens attached to the non-magnetic isotope of the element. The spacing between the satellite band centers is the coupling constant between hydrogen and the magnetic isotope. The intensities of satellite bands are equal, and each is equal to ½ the natural abundance of the magnetic isotope times the parent-band intensity. Satellite bands do not necessarily have the same shape as the parent band because they introduce the possibility of additional spin multiplicity which was degenerate in the parent band.

The satellites most frequently encountered are those due to C^{13}. The spacing between satellites will be equal to the proper coupling constant given in Charts J19 to J24. Because the natural abundance of C^{13} is 1.1%, the intensity of each satellite will be 0.55% of the intensity of the parent band about which they are spaced.

Figure 16 shows the C^{13} satellites for the methyl peak of methanol. These satellites are singlets like the parent peak because this methyl is isolated from other hydrogen-containing groups and there is no new H—H multiplicity introduced by the C^{13}—H splitting. The spacing of the satellites is 140 Hz, in the range predicted in Chart J19, line 14. The intensity of each is about 0.5% of the intensity of the methyl peak, as predicted from the natural abundance of C^{13}.

One C^{13} satellite of the paraffin methyl band is observed in Figure 16. It overlaps one ssb of the methylene band. The other satellite is off scale. Satellites of the methylene band are not shown at the gain used in Figure 16. They will be broader than the parent band because of additional H—H multiplicity.

Figure 17 shows the C^{13} and Si^{29} satellites of tetramethylsilane. Again, the C^{13} satellites have the predicted spacing (Chart J19, line 2) and intensity. The Si^{29} satellites are much more closely spaced because they are coupled across two bonds, through the carbon. Their combined intensities total about 5% of that of the parent peak, as predicted from the natural abundance of Si^{29}.

Although spinning sidebands are a nuisance which should be eliminated if possible, C^{13} satellites can be helpful in analytical work. The multiplicity and the C^{13}—H coupling constant measured from them indicate something about the structure of the parent band (see Chapter 4).

3. Readily Exchangeable Hydrogens

Hydrogens which can exchange with each other rapidly resonate over a wide region of the spectrum. Their band shapes can also vary from broad to sharp because of variation in exchange rate. Because of these characteristics such hydrogens frequently cannot be identified by either chemical shift or band shape. They can be identified, however, by the changes in shift or band shape which can be induced in their resonance bands by additional experiments. Hydrogens which fall into this class include those in the OH of alcohols, phenols, water, and acids, in the NH of amines, and sometimes in the SH of thiols.

Hydrogens exchanging rapidly between two or more chemical environments will have a chemical shift which is the weighted average of the shifts due to each environment separately. Therefore, if a new average environment can be created by adding something to the solution, the band due to the exchanging hydrogens can be shifted to a new location, thus revealing its presence and

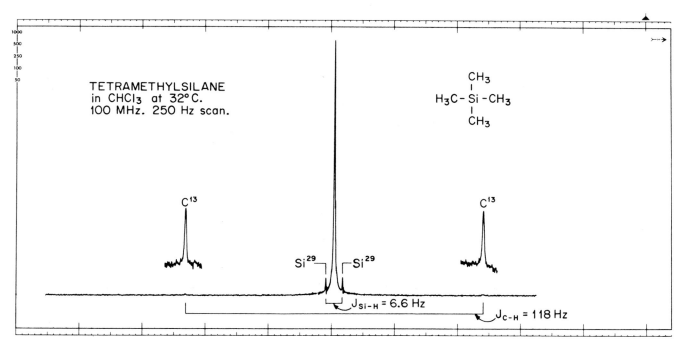

FIGURE 17. Carbon-13 and Silicon-29 Satellites of TMS

former chemical shift. A new average environment can be conveniently created by the addition of a small amount of an acid (such as acetic acid-d$_4$ or trifluoro-acetic acid), causing a shift to lower field (higher δ value) of the bands produced by all but COOH and SH hydrogens. COOH bands can be moved upfield by the addition of D$_2$O. SH bands move only in basic solutions.

If the hydrogens are exchanging very slowly they can spin-couple to their nearest neighbors and produce fine structure on both bands. The addition of acid speeds up the exchange of OH and NH and reduces or destroys the coupling and the associated fine structure. This speedup also sharpens bands broadened by intermediate exchange rates.

The identification of OH hydrogen by the addition of acid is illustrated in Figure 18. The OH hydrogen is exchanging slowly in the neat liquid so that spin coupling to the α-CH$_2$ occurs. The addition of acetic acid-d$_4$ shifts the OH band downfield and reduces it to a singlet. At the same time the α-CH$_2$ band is reduced from a quartet to a triplet by elimination of the spin coupling to the OH.

If the sample is in a water-immiscible solvent, the exchangeable hydrogen can be removed by exchange with D$_2$O. The extraction can be carried out right in the NMR tube. This method is useful if the sample reacts adversely with acid. The water layer should be scanned, also, to account for other water-soluble components that may have been removed.

Interference between an OH or NH band and another band in the spectrum can be eliminated by the band shifting or removal methods just described. This is shown in Figure 19.

If hydrogen exchange is severely hindered by bulky flanking groups (as in 2,6-di-t-butylphenol) the addition of acid may cause an OH peak to broaden but not move. This is a good indication of a severely hindered OH. The acid should be added in very small increments to permit observation of the broadening of the OH band in steps. Too much acid will broaden the band beyond detection and make it appear that it has moved.

Thiolic (SH) hydrogen exchanges so slowly in neutral or acidic solutions that it is usually spin-coupled to its nearest neighbors. In basic solutions, however, the exchange rate is increased so that the spin coupling is eliminated and the band can be shifted. The addition of amines will bring about the desired changes.[T39]

Heating the sample about 30°C will cause a clearly observable upfield shift in the resonances of simple alchoholic or phenolic OH or water. This is attributed to weakening of the weak hydrogen bonds normally present in these compounds.

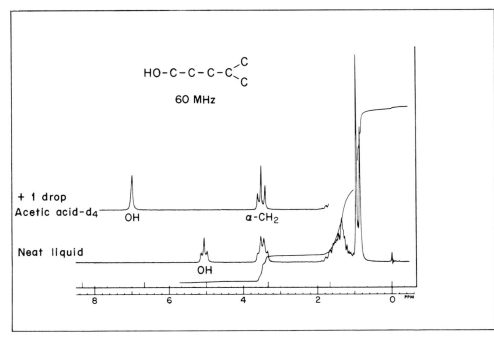

FIGURE 18. Identifying OH Band by Adding Acid

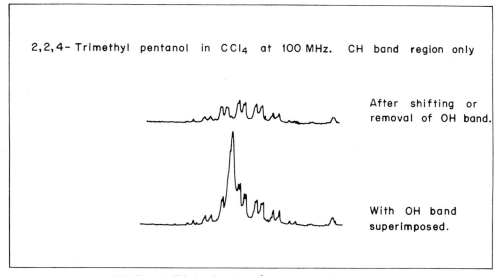

FIGURE 19. Eliminating Interference by Shifting OH Band

4. Use of Chemical Derivatives

a. Esters

Identification of the alcoholic OH function and determination of its position can be appreciably simplified in some cases by converting it to the ester. This conversion also provides data for determining the average molecular weight of polyalkylene glycols. The most convenient reagent for this conversion appears to be trichloroacetyl isocyanate.[T23] This reaction is complete in a few seconds for primary and secondary alcohols, and in two or three minutes for tertiary alcohols. It can be carried out in the NMR sample tube.

The reaction (shown for the primary alcohol) is

$$Cl_3C-CO-N=C=O + R-CH_2-OH \rightarrow$$
$$Cl_3C-CO-NH-CO-O-CH_2-R$$

It produces three important changes in the spectrum: (1) The resonance of the alpha group is shifted downfield about 0.5 to 1.0 ppm for CH_2 and about 1.0 to 1.5 ppm for CH. This frequently moves this resonance out from under other interfering bands so that it can be observed and measured. (2) The OH hydrogen, which has an uncertain chemical shift, is converted to an imide NH which has a narrow range of shifts near $\delta = 10$ (9.8 to 10.6 actually shown in the paper). The NH is exactly equivalent in intensity to the OH it replaces, so that the OH becomes identifiable and measurable. (3) Water OH is converted to trichloroacetamide ($Cl_3C-CO-NH_2$) with the NH_2 band occurring near $\delta = 7.8$, well removed from

the imide NH band. This effectively separates the water OH from the alcoholic OH, eliminating the interference which normally occurs between these two. The NH_2 band is not a reliable measure of the water present in the sample, however, because the reagent absorbs water very fast, producing its own trichloroacetamide.

This reaction also occurs with phenols, requiring two or three minutes. Because phenols have no alpha hydrogen groups they could be confused with tertiary alcohols, unless the aromatic ring resonances can be used to resolve the uncertainty. Changes brought about in the aromatic bands by the reaction may not be great enough to show positively that phenols are present. Phenols must therefore be considered an interfering material in an alcohol analysis of this type.

A second reagent which has been used for the esterification of alcohols is dichloroacetic anhydride.[T3] This reaction also produces a shift in the resonances of the alpha groups and substitutes a stable resonance for the OH resonance. The dichloroacetyl peak, a singlet, occurs in the range of δ equal to about 5.8 to 6.0 in $CDCl_3$ and δ equal to about 6.6 to 6.9 in DMSO. This peak derived from different alcohols is reported to move toward higher field (lower δ) in the order primary to secondary to tertiary, with those from tertiary alcohols clearly separated from those from the other two types.

Trifluoroacetic acid added to a solution of alcohols to shift the OH band may also esterify part of the alcohol. The resulting small new band representing alpha groups in the esters is further indication of the presence of alcohols but cannot be used as a quantitative measure. The reaction does not usually proceed very far at room temperature.

Thiols can be esterified with trichloroacetyl isocyanate,[T10] with advantages similar to those listed for alcohols. In thiols the SH hydrogen usually has a stable chemical shift and it is spin-coupled to hydrogens of the alpha group. The SH band is frequently overlapped by other bands in the spectrum, however. Conversion to the thiocarbamate eliminates both the spin coupling and the band overlap, simplifying the spectrum significantly. The reaction in this case is

$$Cl_3C-CO-N=C=O + R-CH_2-SH \rightarrow$$
$$Cl_3C-CO-NH-CO-S-CH_2-R$$

The NH band occurs near $\delta = 10$, as in the case of alcohols.

b. Salts

The addition of a positive charge to an atom produces a marked downfield shift of hydrogens near it. Consequently, those atoms which can be readily protonated can be identified by the resulting shift of the resonances of attached hydrogens or of alpha groups. Thus amines and phosphines can be identified by converting them to salts with acetic or trifluoroacetic acid.

The protonation of amines results in a downfield shift of 0.5 to 1.0 ppm of the resonances of α-CH$_3$, α-CH$_2$, or α-CH groups. In aliphatic amines the magnitude of the shift tends to increase in the order primary to secondary to tertiary. For this reason, and because nitrogens in different environments will protonate at different pH's, titration of a mixture of amines or of a polyamine will provide additional structural or kinetic information.[T16,T41,T42]

Sometimes the hydrogens of the NH$_4$+ and NH$_3$+ groups are spin-coupled to the nitrogen, producing a triplet with a coupling constant of about 52 Hz, and with the three peaks about equal in height. This has not been observed for the NH$_2$+ or NH+ groups, and *may* be a further indication of NH$_4$+ or NH$_3$+. NH$_x$+ hydrogens may also spin-couple to their nearest neighboring hydrogens. This coupling can be observed clearly when the nitrogen is not also coupled to these same neighbors. The nitrogen–hydrogen coupling can be controlled by varying the acid strength, and perhaps the sample concentration and temperature. Presumably the electric fields set up by the mobile ions in the solution control

the relaxation of the nitrogen spin through its electric quadrupole.

The protonation of phosphines produces about the same shifts of resonances of alpha groups as in amines. This can be important, because the spectra of alkyl phosphines are difficult to distinguish from the spectra of similar hydrocarbons. After protonation the phosphonium salts become readily distinguishable.

Although amines and phosphines are soluble in organic solvents, their salts may not be. If the addition of acid to a sample produces a solid precipitate, it may be necessary to dissolve this precipitate in water or to repeat the entire experiment in water to observe the characteristic shifts.

c. O-Methyloximes

Because NMR does not observe the carbonyl group directly, it would be helpful if the carbonyl could be tagged with an observable group. Such tagging can be done with *O*-methyl hydroxylamine[T21] which converts the carbonyl of ketones to *O*-methyloximes. The *O*-methyl group resonance at $\delta = 3.8$–4.0 is a measure of the carbonyl content of the sample. The preparation and separation of the methyloximes requires a number of hours, however, so that this procedure is attractive in only a few special cases.

5. Complexes

Complexing the sample molecule with a suitable agent can enhance the chemical shifts in a manner similar to that obtained by chemical derivatization. It can also change the rotamer population and therefore the measured average coupling constants, and it can slow the rate of exchange in some cases so as to restore spin coupling where exchange had destroyed it.

a. Lanthanide Shift Reagents

The most important complexing agents are themselves complexes of lanthanide series metals, popularly called the lanthanide shift reagents. Although these reagents were only discovered in 1969,[T26] developments have been very rapid because of their effectiveness in producing large additional chemical shifts in a variety of compound types.[T11,T34]

The most popular of the lanthanide shift reagents have been the complexes of europium and praseodymium (M = Eu or Pr) shown in Table 7. These reagents are used in organic solvents, preferably in the

TABLE 7. Complexes of Eu and Pr Used as Lanthanide Shift Reagents

Abbreviation	Name	Structure
M(DPM)$_3$ or M(thd)$_3$	Tridipivalatomethanato-M(III) or Tris(2,2,6,6-tetramethylheptane-3,5-dione) M (III)	$$M(C{-}\underset{\underset{C}{\|}}{\overset{\overset{C}{\|}}{C}}{-}C{-}\underset{O}{\overset{\|}{C}}{-}C{-}\underset{O}{\overset{\|}{C}}{-}\underset{\underset{C}{\|}}{\overset{\overset{C}{\|}}{C}}{-}C)_3$$
M(fod)$_3$	Tris(1,1,1,2,2,3,3-heptafluoro-7,7-dimethyloctane-4,6-dione) M(III)	$$M(F{-}\underset{F}{\overset{F}{\|}}{C}{-}\underset{F}{\overset{F}{\|}}{C}{-}\underset{F}{\overset{F}{\|}}{C}{-}\underset{O}{\overset{\|}{C}}{-}C{-}\underset{O}{\overset{\|}{C}}{-}\underset{\underset{C}{\|}}{\overset{\overset{C}{\|}}{C}}{-}C)_3$$
M(fod)$_3$-d$_{27}$	Deuterated M(fod)$_3$	$$M(F{-}\underset{F}{\overset{F}{\|}}{C}{-}\underset{F}{\overset{F}{\|}}{C}{-}\underset{F}{\overset{F}{\|}}{C}{-}\underset{O}{\overset{\|}{C}}{-}C{-}\underset{O}{\overset{\|}{C}}{-}\underset{\underset{CD_3}{\|}}{\overset{\overset{CD_3}{\|}}{C}}{-}CD_3)_3$$

absence of water or acids which decompose them. Although any of these reagents may cause shifts of specific hydrogens in either direction, in general Eu complexes cause shifts to lower field and Pr complexes cause shifts to higher field. Also, the Eu complexes seem to produce the more advantageous shifts in most cases, so that they have been the most popular.

The effect of a Eu reagent on an alcohol is illustrated in Figure 20. The upper spectrum is that of 1-octadecanol in CDCl$_3$ at 100 MHz, showing the five bands which can normally be observed at this frequency. The resonances of groups *b* to *f* are observed only as a single sharp band at δ = 1.25. The lower spectrum shows the dramatic increases in chemical shifts of this same sample induced by Eu(DPM)$_3$. Bands produced by groups *e* through *g* are now completely separated so that intensity, multiplicity, and coupling constants can all be measured. Bands produced by groups *c* and *d* are observable, although not measurable. The OH band has been moved off scale downfield. This spread of bands permits the direct and positive identification of six of the methylene groups and the methyl, whereas the normal spectrum permitted direct observation of only two of the methylenes and the methyl.

Figure 21 shows the value of Eu(DPM)$_3$ in separating the spectra of two alcohols in a mixture so that major features of the structure of each compound could be identified. The upper spectrum shows the normal spectrum of the mixture in CCl$_4$ at 100 MHz. Only the distortion of the alpha-methylene band (*h,q*) gives any indication that more than one compound is present. It is apparent that one of the compounds is a linear alcohol

(the OH band is under one of the upfield bands), but it is not at all clear what the other compound is. The addition of Eu(DPM)$_3$ (lower spectrum) makes possible the resolution of this uncertainty. The spectra of the two compounds are separated from each other and spread out so that resonances of many of the groups can be observed directly. The features of the spectrum of 1-octadecanol can be identified with the aid of Figure 20. This permits crosshatching of the remaining bands to provide the spectrum of the unknown compound. The two OH bands (not shown in Figure 21) indicate that a second alcohol is present, and *q* and *p* indicate that it is branched at the beta-carbon. The shapes and apparent intensities of the other bands indicate that one and perhaps both branches are longer than five carbons. Although this is certainly not a complete structural analysis, it is adequate to explain the anomalous behavior of the sample which prompted the analysis.

Eu(DPM)$_3$ was the first successful shift reagent reported and was the first one commercially available. It has found wide application for shifting the resonances of alcohols, amines, oximes, ethers, esters, ketones, sulfoxides, phosphine oxides, cyanides, and related compounds. It does not shift the resonances of halides, indoles, or alkenes, however. Its own resonance bands are outside the normal range of hydrogen resonances, the principal band being at δ = −1 to −2. Its effectiveness is limited by low solubility in some cases, and is eventually limited by the line broadening it causes at the higher concentrations.

Eu(fod)$_3$ is more soluble than Eu(DPM)$_3$ and more strongly affects the less basic substances like ethers and

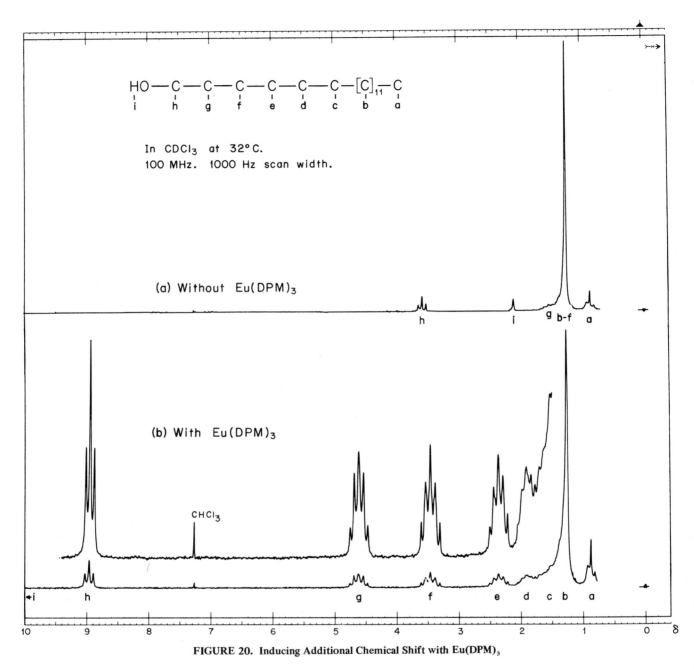

FIGURE 20. Inducing Additional Chemical Shift with Eu(DPM)₃

esters. Consequently, it can produce greater shifts with less line broadening in some cases, and will be the preferred reagent for many applications. The principal disadvantage seems to be that it produces a peak at $\delta = 1$ to 2 which interferes with the resonances of paraffinic chains. Deuterated Eu(fod)₃ was made to reduce this interference.

The lanthanide shifts can be produced in aqueous solutions by the perchlorates of Eu and Pr.[T25] In this case Pr(ClO₄)₃ shifts the resonances downfield, and is the preferred reagent.

The mechanism of action of the lanthanide shift reagents seems to be well established as a complexation of the reagent with the sample molecule to establish a more or less definite spatial relationship between the two, followed by direct dipole interaction between electrons from the paramagnetic metal atom and the hydrogens of the sample molecule. The degree of interaction

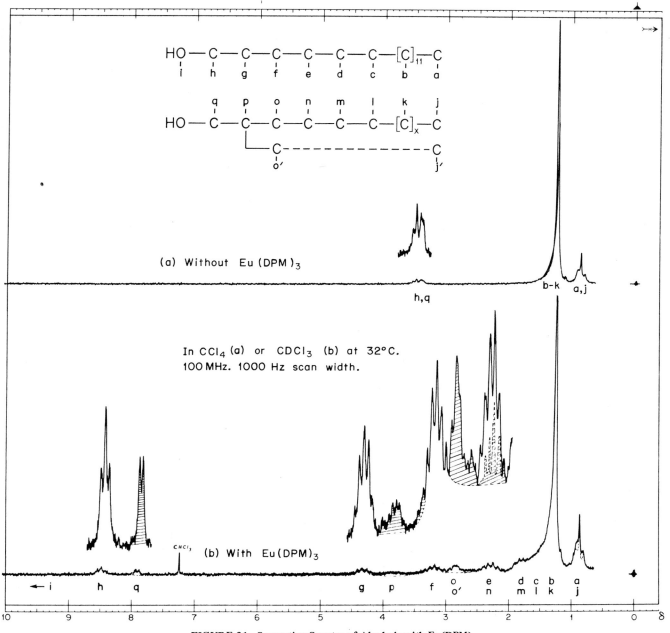

FIGURE 21. Separating Spectra of Alcohols with Eu(DPM)$_3$

with any given hydrogen atom depends on the hydrogen–metal distance and the angle the line between them makes with the principal axis of the reagent complex. Because of this angle–distance relation, the shifts induced by the shift reagents have been used to deduce the three-dimensional structures of some complex molecules.[T2,T15] If this application proves valid, it will supplement X ray in providing the complete structures of the smaller biologically active molecules.

Compounds of different types, and to some extent perhaps compounds of the same type but of quite different structures, will have different affinities for the shift reagents. For this reason the spectra of different compounds in a mixture will be affected at different rates and times. Under these circumstances, it seems advisable to add the shift reagent in several small amounts, observing the spectrum after each addition. In this way the optimum band separation for each band can be observed, and more information about the various compound types may be obtained.

The complexed and uncomplexed portions of a sample exchange rapidly so that each hydrogen type produces a single band even though complexing is incomplete. Each band moves in a single direction as more shift reagent is added, until complexing is complete. After that the addition of more shift reagent does not produce any further shifting.

b. Solvent Complexes

Some of the solvents listed in Table 4 will complex with some compounds so as to modify the spectra in helpful ways. Dimethyl sulfoxide complexes with alcoholic OH (presumably by hydrogen bonding) in such a way as to shift the OH band downfield and markedly slow hydrogen exchange.[T13] Several beneficial things then happen:

1. OH and water resonances become separated.
2. The resonances of OH's in different environments in polyols become separated.
3. The OH hydrogens spin-couple with their nearest neighbors, provided the solution is near neutral pH. Either acids or bases can promote exchange enough to decouple the OH's again.

Under these circumstances the different kinds of OH hydrogens can be measured separately, and the OH band multiplicity will show whether it is in a primary (triplet), secondary (doublet), or tertiary (singlet) position (the methanol OH band will be a quartet). Even steric differences may be detectable. This can be a big help in the analysis of compounds containing alcoholic OH.

Dry hexamethylphosphoramide causes phenolic OH bands to fall into characteristic chemical shift regions according to the structure of the phenol.[T17] This is helpful in resolving the composition of mixtures of phenols.

Pyridine complexes mildly with alcohols so as to cause the resonance of the alpha group to move downfield a little. This can be helpful in separating bands in ether-alcohols.[T37]

Benzene and pyridine can complex with compounds containing carbonyl or olefin groups, (presumably through the π electrons) so as to produce significant shifts in the spectral bands of the sample. The observation of such compounds in CCl_4 or $CDCl_3$ and then benzene or pyridine can give significantly more information than either observation alone. Benzene can also shift the resonances of saturated compounds at times so that better band separation is achieved. It can cause some bands to overlap more, however.

Because of all these solvent effects, it is wise to observe difficult spectra in several different types of solvents to take advantage of all possible help.

C. MEASURING THE SPECTRAL CHARACTERISTICS

This section presents a systematic procedure for obtaining the numerical values for the spectral characteristics used in analysis. This is by no means the only way to proceed, but it is a logical method that should ensure that all characteristics will be considered and that the maximum information will be obtained from the spectrum.

A spectrum to be interpreted is presented in Figure 22. It is the spectrum of a solid sample dissolved in CCl_4 and run at 60 MHz at 32°C. Before proceeding with the measurement, we should check the quality of this spectrum according to the criteria of Section IIA. The calibration cannot be checked, but it will be assumed to be satisfactory unless the measured chemical shifts appear to be in error when the spectrum is interpreted in terms of molecular structure. The S/N ratio is good enough to show the sample bands clearly and measurably, yet the baseline noise shows that the recorder is "live" enough to pick up small bands. The TMS and sample bands show that the resolution is good, and the "ringing" on the TMS band shows that saturation is not high enough to produce any significant line broadening. This probably means that saturation is not a significant factor in intensity measurements, either. The flat baseline on each side of the tall peaks shows that phasing is good and will not introduce any errors into intensity measurements. The aspect ratios of the upfield bands are good in the 1000-Hz scan, but it is necessary to run the downfield bands at 250-Hz scan to make their fine structure clearly visible. Finally, the integral looks good, although a little drift or "wander" in it may mar the accuracy of the intensity measurements. This spectrum is of good quality.

Now the special features discussed in Section IIB should be considered. At the vertical gain used, this spectrum does not show any sidebands, and C^{13} satellites

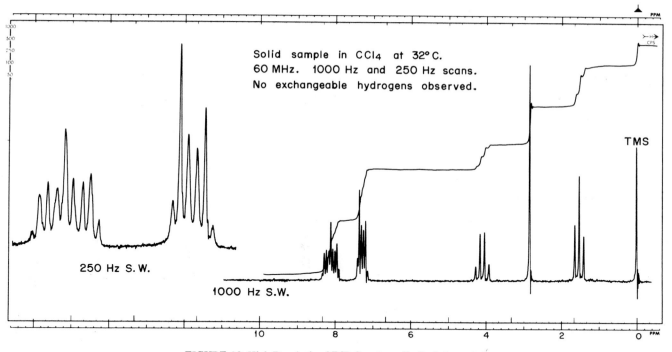

Solid sample in CCl₄ at 32°C.
60 MHz. 1000 Hz and 250 Hz scans.
No exchangeable hydrogens observed.

TMS

250 Hz S. W.

1000 Hz S.W.

FIGURE 22. High-Resolution NMR Spectrum To Be Interpreted

are lost in the baseline. No readily exchangeable hydrogens are observed, and there is no indication at this time that special chemical derivatives or complexes are needed. The spectrum appears ready for measurement.

As discussed in Chapter 1, Section II, the characteristics to be measured are the multiplicity, chemical shift, band intensity, and coupling constant, with qualitative interpretations derived from the band widths and shapes. The procedure recommended is (1) to identify the multiplets and locate their centers of area, (2) to locate the centers of remaining bands which cannot be identified as pseudo-first-order multiplets, and (3) to measure the chemical shifts, coupling constants, and intensities.

Figure 23 shows the identification of bands and multiplets of the spectrum used as an example. In addition to the TMS peak, the spectrum has five well-separated groups of resonances which will be tentatively considered separate bands. Using the spin-multiplet rules of Table 1, band *a* can be identified as a triplet slightly enhanced toward downfield, band *b* is a narrow singlet, and band *c* is a quartet slightly enhanced toward upfield. Bands *d* and *e* are not first-order multiplets but they do have well-defined fine structure. They may perhaps be identified by comparison with typical spectra presented

in Chapter 4. Band *d* is very much like the upfield half of the AA'BB' type systems shown in Plate 53, spectra (a) to (c). Band *e* does not look like the downfield half of such a system, but it does look like a separate band which could be spin-coupled to band *d*. The two could be a four-spin system which has been modified from an AA'BB' system by substitution. The foregoing reasoning permits the drawing of the spin multiplet and band center diagram of Figure 23.

There remains to measure the shifts, couplings, and intensities, as shown in Figure 24. The chemical shifts can be read with sufficient accuracy for most purposes directly from the precalibrated scale at the bottom of the chart. Each shift is read at the center of area of its band.

Each coupling constant is measured as the separation in hertz between the adjacent peaks of the corresponding multiplet. A convenient way to do this is to set dividers at the extreme ends of a multiplet, transfer this distance to the frequency scale to obtain the total width in hertz, and then divide by the number of spaces included in the multiplet. This gives the *average* peak separation in a single measurement, and provides better accuracy. The only couplings which can be measured directly in Figure 24 are those for bands *a* and *c*. They prove to be the

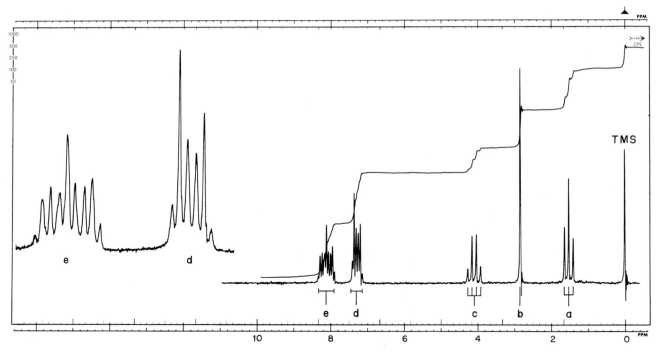

FIGURE 23. Identifying Bands and Multiplets

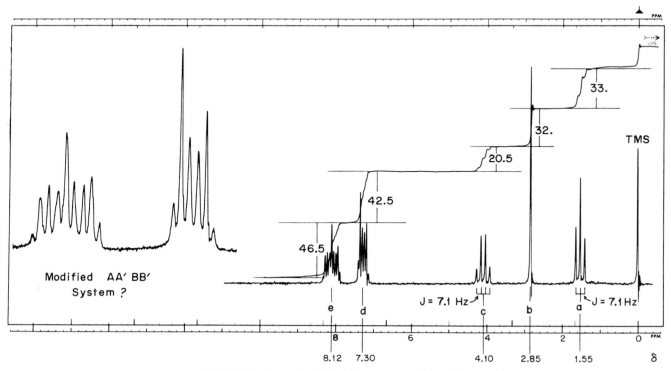

FIGURE 24. Measuring Shifts, Couplings, and Intensities

same, 7.1 Hz. Bands *d* and *e* are too far from first order for their peak separations to be translated easily into coupling constants.

Intensities are measured as the separations in millimeters between the horizontal parts of the integral. The drift in the integral becomes more apparent when the straight lines are drawn through the horizontal parts. Potential errors are indicated for bands *e* and *a* because of this drift.

All the characteristic data have now been derived from the spectrum. It is now possible to proceed with its interpretation in terms of molecular structure.

CHAPTER 3

Analytical Procedures

I. DETERMINATION OF MOLECULAR STRUCTURE

A. ADVANTAGES AND LIMITATIONS OF H^1 NMR

Hydrogen NMR is particularly well suited to the determination of molecular structure because:

1. It shows the distribution of hydrogen according to chemical environment, without the necessity for direct calibration of either chemical shift or intensity for each sample.
2. The correlation of chemical shift with molecular structure is good, and the correlation of coupling constant with structure is useful.
3. Hydrogen occurs in most organic compounds.
4. Hydrogen is the most easily detected stable isotope.

On the debit side, hydrogen has the smallest range of chemical shifts, making it difficult to separate the resonances of closely similar hydrogens. Also, high-resolution hydrogen resonances of adequate sensitivity can be obtained at present only from liquid samples of low viscosity.

B. DETERMINING AN UNKNOWN STRUCTURE

This section presents a systematic procedure for interpreting a spectrum of an unknown compound in terms of the most probable molecular structure. The procedure is designed to obtain the maximum information from the NMR spectrum alone. This information can then be checked independently by other means. This procedure is made possible by the extensive correlation charts and typical spectra presented following Chapter 6.

These correlations and spectra make possible a search of large blocks of NMR data in a reasonable time, so that a number of alternative structures can be considered. The final result can then include all the structures covered by these correlations which can reasonably be expected to produce the given spectrum.

The recommended procedure is as follows:

1. Measure the spectral characteristics as described in Chapter 2, Section IIC.

2. Look for recognizable patterns, using the typical spectra of Plates 1 to 59 or spectra from your own collection. These may reveal partial or complete carbon–hydrogen groups. Note the groups so identified.

3. From the band intensities and the hydrogen counts of identified structural groups, determine the number of hydrogrens contributing to each band or group.

4. From the chemical shifts, determine the larger structural groups that can be present. Use the index, summary, or detailed chemical shift charts, depending on what is already known about the sample.

5. Determine which structural groups are consistent with the measured coupling constants. Use the index or summary coupling constant charts, as needed. This may be conveniently done along with steps 2, 3, and 4 instead of as a separate step.

6. From the set of structural groups so determined, construct the most likely molecule (or molecules) that is consistent with all the known facts. If more or less than one complete molecule is possible from the NMR data, additional data from infrared or mass spectra or elemental analyses will usually resolve the ambiguity. Be

e to take possible impurities into account. If desirable, use the techniques of Chapter 2, Section IIB, to
identify special groups or to shift bands to reduce interference.

EXAMPLE: The data presented in Figure 24 will
serve as an example of the use of this procedure. These
data represent the end of step 1. The next step is to look
for recognizable patterns. Comparison with the spectrum
of ethanol in Plate 36 indicates that bands a and c are
the resonances of an ethyl group with appreciable chemical shift. The triplet-quartet pair is almost identical to
that of ethanol, and the coupling constant of 7.1 Hz is
close to the value of 6.8 Hz given for ethanol. This is an
H-C-C-H coupling, and Chart J4 (line 4) shows that both
the above values are within the range of 5.8 to 8.0
observed for ethyl groups. It is not possible to tell from
the pattern alone whether bands a and c represent a
single ethyl group or two or more identical ethyl groups.
In any case band a represents multiples of three hydrogens and band c represents multiples of two hydrogens.

It was determined from Plate 53 that bands d and e
could represent a modified $AA'BB'$ system in a condensed ring aromatic group. If this is true, then bands d
and e both represent multiples of two hydrogens.

The intensity measurements can now be used with
the numbers of hydrogens in the identified structural
groups to determine the relative numbers of hydrogens
contributing to all the bands. The numbers of hydrogens
are in direct ratio to the area. It is recommended that at
this time the pertinent data be set down in a table for
ease in handling.

Band	Area	Number of H's		System
		Found	Theoretical	
		(a)	(b)	
a	33	3.0	3.0	Multiple of 3	Ethyl group
b	32	2.9	3.0		Methyl group?
c	20.5	1.9	2.0	Multiple of 2	Ethyl group
d	42.5	3.9	4.0	Multiple of 2	Mod. AA'BB'
e	46.5	4.2	4.0	Multiple of 2	Mod. AA'BB'
Total	174.5	15.9	16.0	

[a] Preliminary calculation using band a as the basis. Found =
(3/33) (Area of band to be calculated).
[b] Rounding off to the nearest whole number.

If this is a single compound, all hydrogen counts
must be whole numbers. If band a represents a single
methyl group, then the groups producing bands d and e

must have four hydrogens each. Also, band b is found to
represent three hydrogens. A sharp singlet representing
three hydrogens is most likely a methyl group, so this
speculation is included in the table.

Next, determine from the chemical shifts the larger
structural groups that can be present. This amounts to
determining the atoms or groups probably causing the
chemical shifts and their relation to the groups already
identified. It also includes identifying groups producing
bands not yet interpreted. In the example being used we
have identified, at least tentatively, a carbon-hydrogen
group for each band. It now remains to identify any
groups which do not contain hydrogen and to tie the
whole molecule together.

The most information about groups that cause shifts
is usually gleaned from those bands that are shifted the
most. For this reason it is best to start the shift analysis
with the low-field bands. In the example, these are bands
e and d. To make sure that all the possibilities covered
by this book are considered for band e, lay a straight
edge on Index Chart I1 at $\delta = 8.12$, as indicated in
Figure 25. Every bar crossed by this straight edge indicates a structural group which could have given rise to
this chemical shift. The indicated group is identified at
the left side of the chart. These possibilities will be
accepted or rejected on the basis of the other information available about this band.

It has already been found from the shapes and multiplicities of bands e and d that they could be part of a
condensed ring aromatic system. This possibility is confirmed by the chemical shift, which crosses the bars for
several kinds of aromatic systems, including condensed
ring ones (Figure 25, lines 58–65). Another possibility
appears to be the X—CHY_2 group (line 37), but when the
referenced summary chart, S15 (lines 21–48), is examined it is found that a shift of 8.12 does not cross or
even approach any bars. The X—CHY_2 group is therefore
ruled out.

The next possibility is 1,2-disubstituted ethylenes
(line 49). Plates 21 to 25 show that the only olefin
group which can produce the observed multiplicity and
approach the observed band shape is R—CH_2—CH=CH—
CH_2—R. The referenced summary charts, however, show
that this group will not produce the observed shift. This
rules out not only the 1,2-disubstituted ethylenes but
also the groups indicated by lines 50–53 in Figure 25.

Furans and thiophenes (line 56) are ruled out because the chemical shift of band e requires that they be

substituted by nonalkyl substituents (Chart S21 and Chart 29). Like the olefins, these would not produce the observed multiplicity.

Pyridines, etc. (line 57) are ruled unlikely because single aromatic rings, no matter how substituted, were not found to produce the pattern of band e. (Plates 44 to 52). A condensed ring pyridine system would be required.

The last possibility indicated by the chemical shift is various exchangeable hydrogens (lines 66–77). These are ruled out because no exchangeable hydrogen was found in the sample.

After considering all the possibilities covered by this book, it is concluded that band e arises from an aromatic group, probably in a condensed ring system. Because the shift of band d includes essentially the same group possibilities as band e (Figure 25) and because bands e and d are both required to produce the observed pattern from an aromatic system, it is concluded that both bands are part of the same condensed ring system.

Continuing with the hypothesis that lowest-field bands are most informative, the next study should be of band c. Because band c has been identified as the methylene of an ethyl group, only the shifts of ethyl groups will be considered. This confines the study to lines 12–21 of Chart I1 (Figure 25).

The first possibility encountered is R—C—CH$_2$—halogen (line 18). If the sample under study is a single compound, the atom attached to the ethyl group must have another bond which can tie the ethyl to the rest of the molecule. This rules out halogens, which have only one bond.

The next possibility is R—C—CH$_2$—S (line 19). Examination of Chart S7 shows that the only group with this shift is R—C—CH$_2$—S(X)=CY$_2$. This group has nonequivalent CH$_2$ hydrogens, however, whereas our CH$_2$ hydrogens are equivalent. Thus, this possibility is ruled out.

Line 20 indicates that R—C—CH$_2$—N should be considered. Chart S7 shows the only possibilities are some alkyl nitrosamines and N-alkyl amides. It seems likely that the ethyl group must eventually be attached to the aromatic ring system. If so, the nitrosamines are not acceptable because they have no aromatic attachment. The N-alkyl amides are attached to an aromatic ring, but they cause the CH$_2$ hydrogens to be nonequivalent. The nonequivalence rules out the amides.

A "near miss" on Chart S7 is R—C—CH$_2$—N+(pyridini-

um). As mentioned earlier, a condensed ring pyridine system would be required to produce bands d and e.

The last possibility is R—C—CH$_2$—O—X (line 21). Chart S6 (line 19), shows that X can be phenyl (or other aromatic), CO—(R,phenyl,NR$_2$), N=CR$_2$, SO$_2$—Y, CS—Y, S+Y$_2$, P(Z)Y$_2$, or NO$_2$. The aromatic group is acceptable. The P(Z)Y$_2$ group can be ruled out because it would split band c into two clear quartets (Chart J36, line 5) whereas only one quartet is observed. NO$_2$ is ruled out because it does not have another bond for attachment to an aromatic ring. The other groups must be checked by search of the detailed charts. The index to these charts is on the back of Chart S6. The result of this search is shown in the following table:

Group		Status	Chart (line)
R—C—CH$_2$—O	—aromatic	Accepted	30(5)
	—CO—ϕ	Accepted	18(9)
	—CO—R	Rejected	No aromatic connection
	—N=C(R)ϕ	Accepted	57(12)
	—CS—Y	Rejected	64(8)
	—SO$_2$—ϕ	Accepted	67(5)
	—S+(Y)ϕ	Rejected	68(5) this is uncertain
	—CO—N(R)ϕ	Accepted	Should be similar to 18(6)

As the result of the search of the CH$_2$ data, five possible groups have been found to fit X in the formula R—CH$_2$—O—X-aromatic. A further search of the charts listed in the preceding table shows that all these groups also fit the CH$_3$ shift as well as can be determined. Thus, we conclude that any of the five—O—X—groups might lie between the ethyl group and the aromatic ring.

Only band b remains to be identified. It has already been postulated that it arises from a methyl group, but we do not yet know what that methyl is attached to. Chart I1 (Figure 25) indicates it can be attached to an aromatic ring directly or through phosphorus, sulfur, nitrogen, or oxygen groups (lines 6–11). The halogens are ignored again because they cannot be attached to both the methyl and the ring. Phosphorus groups can also be ruled out because they would split the methyl band into a doublet (Chart J36, line 2). The four remaining types of groups will be examined on the summary charts.

Chart S2 shows that the methyl can indeed be attached to a condensed ring aromatic (line 10). It could

CHART I 1 (top)

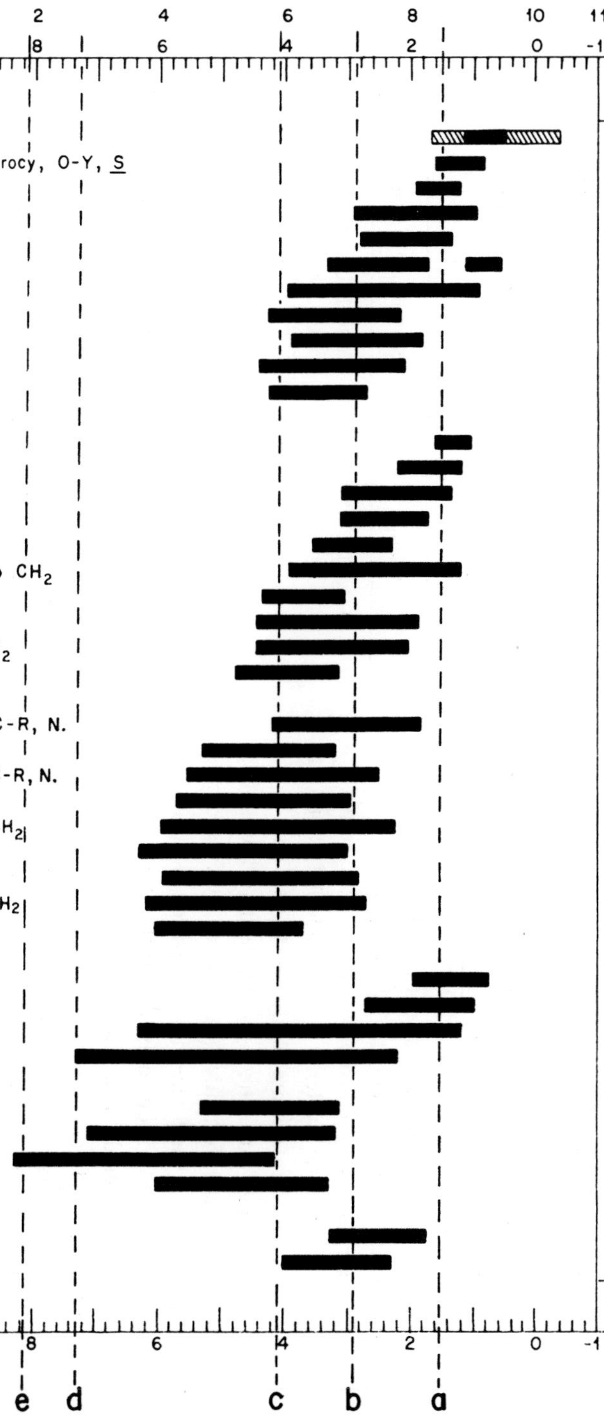

48

LINE	Group	Notes	SUMMARY CHART (LINE)	NO. CPDS.
1	$CH_3-C-C-X$	X : Any group	S1 (2-4)	1292
2	CH_3-C-X	X: C=Y, C≡Y, φ, NR_2, $\overset{+}{NR_3}$, N=Y, P(Z)Y_2, Heterocy, O-Y, $\underline{S}$	S1 (5,6,11,14,16,19,20,22,24,25)	1177
3	CH_3-C-X	X: F, Cl, Br, I, NO_2, $-\overset{+}{N}$(Pyrioinium), PR_3^+	S1 (21,23,26-29)	46
4	$CH_3-C(X)_{2,3}$	X: Combinations of the above groups.	S1 (7-10,12,13,15,-18,24,27-29)	101
5	$CH_3-(C=X, C=C-Y)$	X: C-Y, N-Y, O, S	S2 (1-6,11-17)	596
6	CH_3-Aromatic	Includes 5 & 6-membered heterocyclic aromatics	S2 (7-10,18)	786
7	$CH_3-\underline{P}$	$\underline{P}$: Any phosphorus group with P bonded to CH_3	S2 (19-24)	55
8	CH_3-CF,Cl,Br,I		S2 (25)	4
9	$CH_3-\underline{S}$	$\underline{S}$: Any sulfur group with S bonded to CH_3	S3 (1-13)	208
10	$CH_3-\underline{N}$	$\underline{N}$: Any nitrogen group with N bonded to CH_3	S3 (14-32)	306
11	CH_3-O-X		S3 (33,34)	1139
12	$R-C-CH_2-C-R$		S4 (1)	790
13	$R-C-CH_2-C-X$		S4 (2-16)	920
14	$R-C-CH_2-C(X)-Y$ or $Y-C-CH_2-C-X$		S5	163
15	$R-C-CH_2-(C=X, C=Y)$	X: C-Z, N-Z, O, S. Y: C-Z, N.	S6 (1-11)	422
16	$R-C-CH_2-$Aromatic	Includes CH_2 on pyridinium ring	S6 (12-14)	114
17	$R-C-CH_2-\underline{P}$	$\underline{P}$: Any phosphorus group with P bonded to CH_2	S6 (15-17)	46
18	$R-C-CH_2-(F,Cl,Br,I)$		S6 (21 23)	35
19	$R-C-CH_2-\underline{S}$	$\underline{S}$: Any sulfur group with S bonded to CH_2	S7 (1-11)	189
20	$R-C-CH_2-\underline{N}$	$\underline{N}$: Any nitrogen group with N bonded to CH_2	S7 (12-27)	229
21	$R-C-CH_2-O-X$		S6 (18-20)	525
22	$X-C-CH_2-(C=Y, C≡Y, Arom., \underline{P}, \underline{S}, \underline{N})$	Y: C-Z, O. Y': C-R, N.	S8, S9	620
23	$X-C-CH_2-(O-Y,F,Cl,Br,I)$	Y: C-Z, N-Z, O. Y': C-R, N.	S10	636
24	$X-CH_2-(C=Y, C≡Y')$		S11	489
25	$X-CH_2-$Aromatic		S12 (1-19)	356
26	$X-CH_2-\underline{P}$	$\underline{P}$: Any phosphorous group with P bonded to CH_2	S12 (20-39)	78
27	$X-CH_2-(F,Cl,Br,I)$		S12 (40-54)	141
28	$X-CH_2-\underline{S}$	$\underline{S}$: Any sulfur group with S bonded to CH_2	S13 (1-33)	232
29	$X-CH_2-\underline{N}$	$\underline{N}$: Any nitrogen group with N bonded to CH_2	S14 (1-34)	246
30	$X-CH_2-O-Y$		S13 (34-50)	228
31	$(R-C)_2 CH-C-R$		S14 (35-37)	156
32	$(R-C)_2 CH-C-X$		S14 (38)	—
33	$(R-C)_2 CH-X$		S14 (39)	—
34	$R-C-CH(X)-Y$		S14 (40)	—
35	$(X-C)_2 CH-Y$		S15 (1-3)	17
36	$X_n C-CHY_2$		S15 (4-20)	60
37	$X-CHY_2$		S15 (21-48)	100
38	$X-CH(Y)-Z$		S15 (49-59)	52
39	$X-C=C≡CH$		S18 (36-40)	35
40	$X-C≡CH$		S18 (41-45)	23

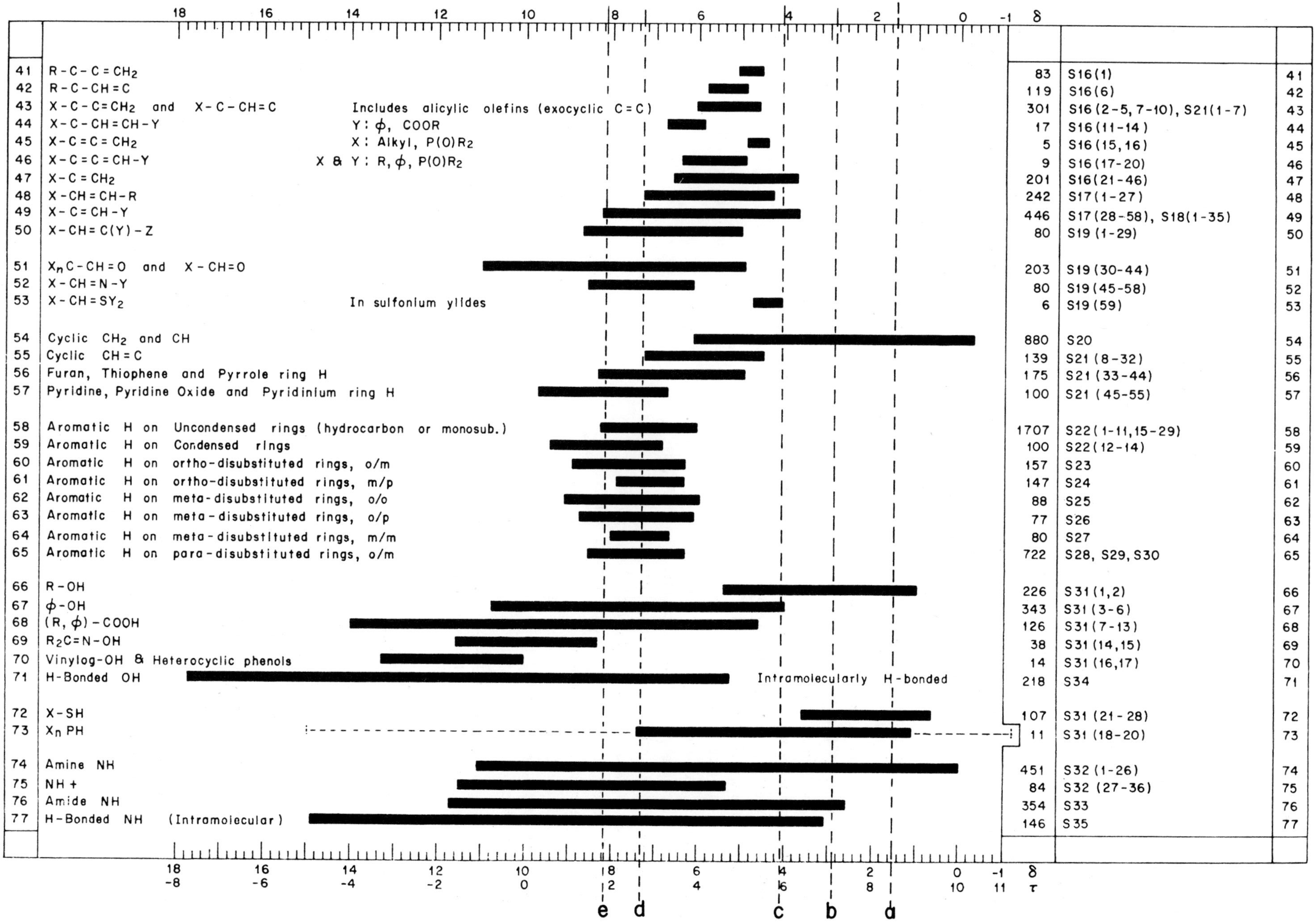

41	41	83	S16 (1)
42	42	119	S16 (6)
43	43	301	S16 (2-5, 7-10), S21(1-7)
44	44	17	S16 (11-14)
45	45	5	S16 (15, 16)
46	46	9	S16 (17-20)
47	47	201	S16 (21-46)
48	48	242	S17 (1-27)
49	49	446	S17 (28-58), S18(1-35)
50	50	80	S19 (1-29)
51	51	203	S19 (30-44)
52	52	80	S19 (45-58)
53	53	6	S19 (59)
54	54	880	S20
55	55	139	S21 (8-32)
56	56	175	S21 (33-44)
57	57	100	S21 (45-55)
58	58	1707	S22 (1-11, 15-29)
59	59	100	S22 (12-14)
60	60	157	S23
61	61	147	S24
62	62	88	S25
63	63	77	S26
64	64	80	S27
65	65	722	S28, S29, S30
66	66	226	S31 (1,2)
67	67	343	S31 (3-6)
68	68	126	S31 (7-13)
69	69	38	S31 (14,15)
70	70	14	S31 (16,17)
71	71	218	S34
72	72	107	S31 (21-28)
73	73	11	S31 (18-20)
74	74	451	S32 (1-26)
75	75	84	S32 (27-36)
76	76	354	S33
77	77	146	S35

41	R-C=C=CH₂
42	R-C-CH=C
43	X-C=C=CH₂ and X-C-CH=C
44	X-C=C=CH-Y
45	X-C=C=CH₂
46	X-C=C=CH-Y
47	X-C=CH₂
48	X-CH=CH-R
49	X-C=CH-Y
50	X-CH=C(Y)-Z
51	Xₙ C-CH=O and X-CH=O
52	X-CH=N-Y
53	X-CH=SY₂
54	Cyclic CH₂ and CH
55	Cyclic CH=C
56	Furan, Thiophene and Pyrrole ring H
57	Pyridine, Pyridine Oxide and Pyridinium ring H
58	Aromatic H on Uncondensed rings (hydrocarbon or monosub.)
59	Aromatic H on Condensed rings
60	Aromatic H on ortho-disubstituted rings, o/m
61	Aromatic H on ortho-disubstituted rings, m/p
62	Aromatic H on meta-disubstituted rings, o/o
63	Aromatic H on meta-disubstituted rings, o/p
64	Aromatic H on meta-disubstituted rings, m/m
65	Aromatic H on para-disubstituted rings, o/m
66	R-OH
67	φ-OH
68	(R,φ)-COOH
69	R₂C=N-OH
70	Vinylog-OH & Heterocyclic phenols
71	H-Bonded OH
72	X-SH
73	Xₙ PH
74	Amine NH
75	NH +
76	Amide NH
77	H-Bonded NH (Intramolecular)

Includes alicyclic olefins (exocyclic C=C)

Y: φ, COOR
X: Alkyl, P(O)R₂
X & Y: R, φ, P(O)R₂

In sulfonium ylides

Intramolecularly H-bonded

FIGURE 25. Use of the Chemical Shift Index Chart

49

also be attached to certain single ring aromatics, but these have already been ruled out.

Chart S3 indicates that the sulfur group which best fits the data is $CH_3-SO_2-\phi$ (line 10). This group is acceptable. The groups shown on lines 4 and 12 are set aside because they do not include aromatics.

Chart S3 further indicates that a number of nitrogen groups associated with aromatics could produce the observed chemical shift of the methyl (lines 15, 16, 20), but they all require additional substituents. The only ones which can be accepted are those which could carry the ethoxy group as the other substituent. This limits the choice to $CH_3-N(\phi)-CO-O-C-C$, and possibly $CH_3-N(\phi)-SO_2-O-C-C$. Acceptance of these groups also means that the aromatic ring system can be substituted in only one position.

One other group accepted in connection with the ethyl groups was $C-CH_2-O-N=C(R)\phi$. For this to be acceptable now, R must be methyl. Chart S2 (lines 11 to 13) indicates this group will not provide the necessary shift for the methyl, and the whole group is rejected.

Other chemical shift bars on Chart S3 which include the shift of the methyl (lines 21, 22, 24, 29, and possibly 30) are produced by groups not associated with aromatics. These groups are therefore rejected.

This study has indicated four groups which could be the source of band *b*, and has eliminated one group previously accepted as a possible source of band *c*. This also completes the determination of the functional groups which could be present in the sample molecule. All the groups accepted at this point are listed in the following table:

Band	Number of H's	Structural groups
a	3	$CH_3-C-O-X$ $X = \phi, CO-\phi, SO_2-\phi, CO-N(CH_3)\phi$, where ϕ is a condensed aromatic system
b	3	CH_3-X $X = \phi, SO_2-\phi, N(\phi)-SO_2-O-C-C$, $N(\phi)-CO-O-C-C$
c	2	$C-CH_2-O-X$ X is the same as for band *a*
d	4	Upfield part of modified AA'BB' condensed ring system
e	4	Downfield part of modified AA'BB' condensed ring system

The problem now is to put all these groups together into a single believable molecule.

The key to the molecular structure is the ring structure. Band *d* closely resembles the upfield part of the three ring systems in Plate 53, spectra (a) to (c). Bands *d* and *e* represent four hydrogens each, indicating that there must be two identical rings, as in the systems of Plate 53. Also, the aromatic system must have one or two substitutable positions in addition to the pseudo AA'BB' system in order to accommodate the other groups found. The only system which fulfills these requirements is anthracene, spectrum (b), which has two extra substitutable positions. Substitution of both *c* positions is necessary to eliminate band *c* from the spectrum, and the substituents should be unlike to introduce asymmetry into the two remaining bands. Like substituents would leave the remaining bands nearly symmetric, as in spectrum (c).

The requirement of two substituents on the aromatic system eliminates the three groups which include both the methyl and ethyl units and produce only one ring substituent. The final set of structures derived from the NMR data is

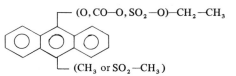

where $-SO_2-X$ is not on both positions at the same time. These possibilities can be narrowed further by infrared or mass spectra or elemental analysis. The structures actually selected in the light of such additional data was

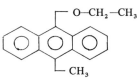

This procedure for determining a completely unknown structure seems somewhat lengthy and tedious when described in step-by-step fashion. Actually, with experience, one learns to combine steps and to consult other analytical techniques at intermediate stages so as to shorten the process. There are times, however, when other available techniques shed little light on a structure problem and NMR must solve essentially the whole problem. It is at these times that the complete procedure as illustrated here is needed.

C. CONFIRMING A PROPOSED STRUCTURE

In the more common case, there is enough known from the method of synthesis of a sample, or its source, or its commercial description to narrow the choices of structure down to a few reasonable possibilities. In such cases the job for NMR is to confirm or reject these proposed structures. This is a relatively simple task which can be accomplished in the following steps:

1. Measure the spectral characteristics as described in Chapter 2, Section IIC.

2. Using the typical spectra as a guide, look for key patterns that should appear in the spectrum of the proposed compound.

3. Using the appropriate detailed chemical shift charts, determine whether the chemical shifts are consistent with the proposed structure.

4. Using the coupling constant summary charts, determine whether the observable coupling constants are consistent with the proposed structure.

5. See if the band intensities are consistent with the hydrogen counts for the various groups in the proposed structure.

6. If *all* the spectral characteristics of the proposed structure are found, it is acceptable to NMR. If not,

adjust the proposed structure to correct any minor difference encountered. Positions of substitution may need rearranging, ethers may turn out to be esters, sulfides may turn out to be sulfoxides or sulfones, etc.

7. Any bands left over after the spectrum of the proposed structure has been accounted for must be checked to see if they are impurities or a part of an extended structure which includes the original proposal. The relative intensities of the extra peaks are usually a prime clue as to whether they can be part of the same molecule as the proposed structure. Impurities to be considered are isomers of the proposed structure, unreacted starting materials, products of side reactions, and solvents used in the synthesis or separation steps.

EXAMPLE: This example involves confirmation of the structure of a sample thought to be isobutyl chloromethyl sulfone. An impure sample was chosen to illustrate the practical problems encountered. The spectrum with its characteristics measured is shown in Figure 26. No exchangeable hydrogen was observed in the region covered by this spectrum.

The proposed compound should produce the spectrum of an isobutyl group with appreciable chemical shift and a singlet at still lower field representing the

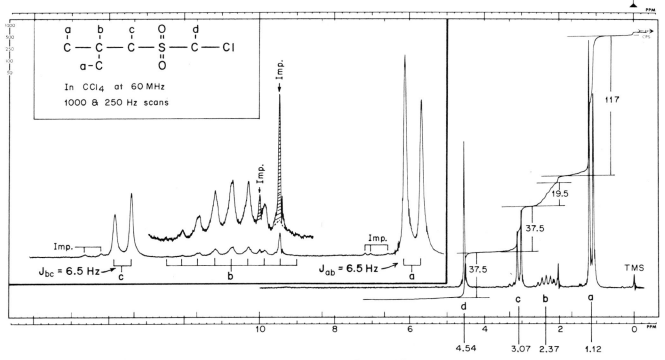

FIGURE 26. Confirming a Proposed Structure

chloromethyl group. Comparison of the sample spectrum with that of isobutyl alcohol in Plate 37 shows that sample bands *a, b,* and *c* could have been produced by the sample isobutyl group. Both the patterns and the coupling constants are correct. There are some extra peaks, however, which probably come from impurities, because their intensities are obviously out of line with the intensities of the isobutyl group. Band *d* could easily be the singlet expected from the chloromethyl group. Thus, all the bands expected from the proposed compound have been located, and assignments of bands to specific hydrogen groups can be made as in Figure 26.

The chemical shifts of the isobutyl group are checked from Chart 66 (lines 11 to 13). The shift of the chloromethyl group is checked from Chart S12 (line 46) and Chart S13 (line 26). The summary charts must be used for this shift check because there is no detailed chart covering the SO_2-CH_2-Cl system. The coupling constants have already been found acceptable, but they can be confirmed by Chart J4 (lines 1 and 2). Band intensities should be converted to numbers of hydrogens, using the total number of hydrogens in the proposed structure as the basis:

Band	Intensity	Number of H's	
a	117.0	6.1	6
b	19.5	1.0	1
c	37.5	1.9	2
d	37.5	2.0	2
Total	211.5	11.0	11

$$\text{Number H} = \frac{11.0}{211.5} \text{ (Intensity)}$$

From this tabulation it is obvious that the spectrum is consistent with the proposed structure in every characteristic. The proposed structure is accepted.

The extra peaks mentioned earlier do not have the characteristics of either spinning sidebands or C^{13} satellites. Their intensities are way out of line with anything identified with the proposed structure. It appears, therefore, that they are impurities. It would be helpful to know at least the major impurities to assist in deciding about separation or further use. This identification is considered in Figure 27.

The sample was known to have contacted acetic acid, and the peak labeled *e* has just the right shape and position to be produced by the acetic acid methyl. It is therefore assigned to this methyl. This assignment could be proved by adding more acetic acid to the sample, resulting in the growth of peak *e* but no change in shape.

Band *f* appears to be a doublet with the same coupling as band *c*. It is possible that it is produced by another isobutyl group. Band *h* does not appear from either its coupling or its intensity to be a part of the same isobutyl, however. The remainder of the bands that go with band *f* must be hidden under bands *a* and *b*. If so, then the impurity isobutyl is similar to that in the principal sample compound. A search of the applicable charts indicates that this impurity is likely to be diisobutyl sulfone or isobutyl sulfonamide.

Peak *g* can be due to a number of methyl compounds, but further identification would be difficult without additional clues. Bands *h* and *i* are also difficult to identify because there are so many possibilities. The 7.0-Hz coupling is larger than that observed in typical methyl doublets, so that band *h* may not be a doublet. Further knowledge of the procedure used to synthesize this compound would be helpful in identifying these remaining impurities, but this information is not avail-

All the observed and theoretical data can now be tabulated for ease of comparison.

Band	Multiplicity		Chemical shift		Coupling		Number of H's	
	Observed	Theoretical	Observed	Theoretical	Observed	Theoretical	Observed	Theoretical
a	Doublet	Doublet	1.12	1.00–1.28	6.5	~6.4	6	6
b	Nonet?	Nonet	2.37	2.0–2.5	6.5	~6.4	1	1
c	Doublet	Doublet	3.07	2.75–3.50	6.5	~6.4	2	2
d	Singlet	Singlet	4.54	4.48–5.18	–	–	2	2

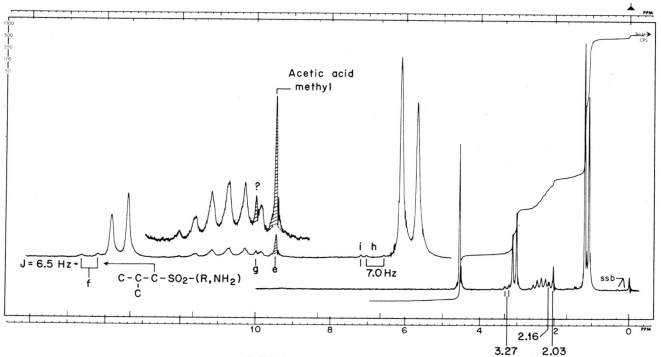

FIGURE 27. Identifying Impurities

able. Further analysis of this sample really becomes a problem in multicomponent analysis, which is discussed in the next section.

II. QUANTITATIVE MULTICOMPONENT ANALYSIS

A. ADVANTAGES AND LIMITATIONS OF H¹ NMR

In general, gas chromatography, mass spectrometry, or infrared spectroscopy are better tools for multicomponent anlysis than is NMR. At times, however, NMR is the preferred technique for the following reasons:

1. Intensity measurements are absolute, which is a tremendous advantage when calibration of the other methods is not reasonable.

2. NMR can observe samples which are sealed in glass tubes under atmospheres of the experimenter's choice. This is an advantage when the sample is sensitive to water vapor, oxidation, contamination, etc.

3. NMR can sometimes identify the components of unfamiliar mixtures as well as give their quantitative distribution.

4. NMR is sometimes faster than the other methods.

5. Like infrared spectroscopy, but unlike gas chromatography, mass spectrometry, and chemical methods, NMR can examine a sample without separating or other-

wise disturbing the components. This is an advantage when the sample amount is limited and to be used for other purposes after analysis.

These advantages make NMR a very useful quantitative technique for simple mixtures that produce at least as many well-separated and measurable bands as there are components. There are some limitations to the technique which make it unsuitable for the analysis of other mixtures, however. These are:

1. The NMR spectrum sometimes has a smaller number of bands than the sample has components. In such cases, of course, it is not possible to set up enough independent simultaneous equations to resolve the mixture.

2. Band overlap sometimes prevents the separate measurement of some components. This is especially serious in complex mixtures. In such cases the separation techniques like gas chromatography and mass spectrometry are better.

3. The sensitivity of many NMR spectrometers is too low to detect components in low concentrations. This limitation is being reduced by the pulse-Fourier transform instruments, and it may not be serious in the future except for trace analysis.

In some cases limitations might be overcome by

using direct calibration, as is now used for other spectroscopic methods. In general, however, NMR does not seem to be as suitable for this type of work as is infrared spectroscopy.

Limitations 1 and 2 can both be overcome in some cases by the use of chemical shift reagents or by operating at higher frequencies. They can also be overcome by using a separation method to simplify the mixture before NMR analysis. One should always use the best method or combination of methods available for each analytical job.

B. RECOMMENDED PROCEDURE

The following procedure for carrying out multicomponent analyses by NMR is recommended for simple mixtures that produce spectra with well-separated bands. There should be as many measurable bands as there are components to be measured. The calculations always give the distribution in mole percent because NMR counts hydrogens by number, not by weight. This means that each component must be identified as a distinct molecule of known structure, or at least of known average structure and molecular weight. If only one or two components of a more complex mixture are to be measured, it is necessary to know the average number of hydrogens per molecule in the rest of the components, and to know the distribution of these hydrogens among the NMR bands. With this information, the analysis depends only on the relative areas of the bands of the sample spectrum, and is independent of the gain and resolution of the spectrometer system, the concentration of the sample, and, to some extent, the degree of saturation of the resonances. This independence, in turn, speeds and simplifies the analysis.

The recommended procedure is as follows:

1. Measure the spectral characteristics as described in Chapter 2, Section IIC.

2. Identify the components as in this chapter, Section IB or C. Assign all the hydrogens in each component to the proper bands in the spectrum.

3. Set up an array showing all the equations, derivable from the spectrum, which can assist in the analysis. A sample array is described below.

4. Solve those equations necessary to a complete analysis. Use equations unique to a component where possible.

5. Solve the remaining equations as checks. Do not

expect perfect checks. The NMR intensities are not generally good enough to provide perfect solutions to the equations. Some judgment may be needed to select the best solutions in unusual cases.

The array of equations is recommended to prevent mistakes from oversight and to provide maximum use of the extra equations for checking the analysis. These calculations are simple and straightforward, and one is tempted to use short cuts to save time. Such short cuts frequently lead to errors because of band overlap that was overlooked or hydrogens that were not considered. Error also results from the use of percentage hydrogen distribution along with percentage area distribution. The array recommended helps ensure that *all* the hydrogens in each component are accounted for, and that they are counted by number and not by percentage.

The recommended array is:

Number of H's	Area	Component 1	Component 2	...	Component n
H_{as}	$= kA_a$	$= H'_{a1}M_1$	$+ H'_{a2}M_2$	$+ \cdots +$	$H'_{an}M_n$
H_{bs}	$= kA_b$	$= H'_{b1}M_1$	$+ H'_{b2}M_2$	$+ \cdots +$	$H'_{bn}M_n$
.	.	.	.	.	.
.	.	.	.	.	.
.	.	.	.		.
.		.	.		
H_{ms}	$= kA_m$	$= H'_{m1}M_1$	$+ H'_{m2}M_2$	$+ \cdots +$	$H'_{mn}M_n$
H_s	$= kA_t$	$+ H'_1M_1$	$+ H'_2M_2$	$+ \cdots +$	H'_nM_n

Where: Subscripts $a, b, \ldots, m$ refer to NMR bands. Subscripts $1, 2, \ldots, n$ refer to components. Subscript s refers to the whole sample. H_{as} is the number of hydrogens in the effective sample volume contributing to band a. It is directly proportional to A_a, the area of band a. H_s is the number of hydrogens in the effective sample volume contributing to the entire spectrum. H'_{a1} is the number of hydrogens in one molecule of component 1 which contribute to band a. M_1 is the number of molecules of component 1 in the effective sample volume. H'_1 is the total number of hydrogens in one molecule of component 1. A_a is the area of band a. A_t is the total area of the spectrum produced by the sample components. It excludes the area of the reference, solvent, and any other noncomponent bands. k is the proportionality constant between hydrogen number and area. Because it cancels out in the calculations, it is taken as 1.

Each line of this array simply equates the total number of hydrogens contributing to a given band to the sum of the number contributed by each of the com-

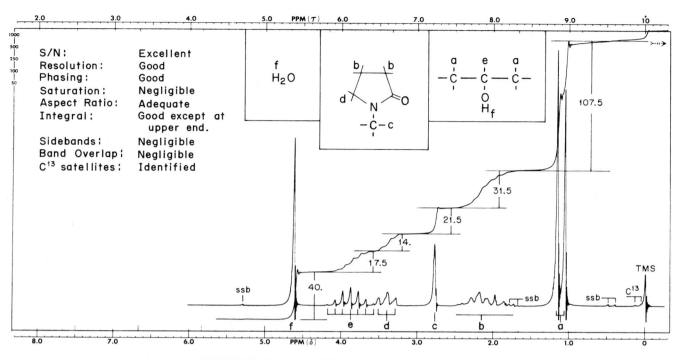

FIGURE 28. Spectrum of Mixture for Quantitative Analysis

ponents. The last line also shows the total number of hydrogens in each component to be compared with the true number so that none will be overlooked. These equations are solved as necessary to obtain $M_1, M_2, \ldots M_n$, and the sum of these, M_t. The mole percent of component 1 is then $(M_1/M_t)(100)$.

The column headed Number H's is shown only to make the equations clear. It is not used in the actual data array.

EXAMPLE: Use of the foregoing procedure will be illustrated by solution of the simple problem represented by Figure 28. The sample consists of a mixture of three known compounds. The spectrum consists of six well-separated bands, with band overlap only in the case of the exchangeable hydrogens (band f). This spectrum is an ideal type for quantitative analysis by NMR.

The first two steps in the procedure are shown completed in Figure 28. All spectral characteristics are measured and listed, and all components are identified and their hydrogens assigned to the bands. In addition, the quality of the spectrum is assessed in terms of the factors discussed in Chapter 2, Section IIA.

The recommended data array for this problem is now set up:

Band	Area		Isopropanol		Amide		Water
a	107.5	=	$6M_1$	+	$0M_2$	+	$0M_3$
b	31.5	=	$0M_1$	+	$4M_2$	+	$0M_3$
c	21.5	=	$0M_1$	+	$3M_2$	+	$0M_3$
d	14.0	=	$0M_1$	+	$2M_2$	+	$0M_3$
e	17.5	=	$1M_1$	+	$0M_2$	+	$0M_3$
f	40.0	=	$1M_1$	+	$0M_2$	+	$2M_3$
All	232.0	=	$8M_1$	+	$9M_2$	+	$2M_3$

$M_1 = 107.5/6 = 17.9 \quad M_1 = 17.5/1 = 17.5 \quad \text{Avg.} = 17.7$

$M_2 = 31.5/4 = 7.9 \quad M_2 = 21.5/3 = 7.2 \quad \text{Avg.} = 7.6$

$M_3 = (40-17.7)/2 = 22.3/2 = 11.2$

$M_t = 17.7 + 7.6 + 11.2 = 36.5$

$X_1 = 17.7/36.5 = 0.485 = 48.5 \text{ mole \%}$

$X_2 = 7.6/36.5 = 0.208 = 20.8 \text{ mole \%}$

$X_3 = 11.2/36.5 = 0.307 = 30.7 \text{ mole \%}$

A final tabulation of the results may be as follows:

Component	Mole %	Mol. Wt.	Weight	Weight %
2-Propanol	48.5	60.1	2914.85	52.7
N- Methyl-2-pyrrolidone	20.8	99.1	2061.28	37.3
Water	30.7	18.0	552.60	10.0
	100.0		5528.73	100.0

C. SOLVENT SEPARATIONS

If the mixture is too complex for direct analysis by NMR, a rough separation of the components by solvent extraction or dual solvent separation may be helpful. This sometimes cleanly separates an active material from a hydrocarbon solvent or an inert or insoluble base. At other times it changes the ratio of components so that it becomes possible to determine which bands go together. Weathering in a stream of nitrogen or partial distillation on a vacuum line can remove volatile components so that nonvolatile ones can be analyzed. In all these cases, the spectrum of the total sample should be run, followed by the extraction or weathering, then the spectra of the fractions or residue should be run. After the components are identified and the separated portions analyzed, the overall analysis can be determined from the total sample spectrum.

Any of the solvents listed in Table 4 may be useful for extraction of solid samples. For liquid samples, the dual immiscible solvent systems D_2O-CCl_4 or $D_2O-CDCl_3$ have been successful. The procedure developed by Mr. T. J. Denson of the author's laboratory is:

1. Dissolve the sample in a mixture of 50 vol. % D_2O and 50 vol. % CCl_4 or $CDCl_3$. Shake vigorously.

2. If an emulsion forms, try to separate it in a centrifuge. If the emulsion is too stable to break this way, add acid slowly until it appears to be breaking, then finish with the centrifuge. After the addition of acid there will sometimes be three phases.

3. Run the NMR spectra of all phases, if possible.

III. CHARACTERIZATION

A. DEFINITION

Some samples with which the analyst must deal are so complex that the determination of the complete multicomponent distribution is either not reasonable or not helpful. For instance, some rather simple petroleum fractions have been found to contain at least 150 individual compounds, and the more complex ones have been shown to have over 2000 components. Although these mixtures could be separated by chromatography and their components identified by combinations of techniques, the resulting data would not be usable in most practical problems. Such an analysis is simply not worth the time and expense except in rare cases.

On the other hand, the quantitative distribution of the *types* of compounds or the chemical functional groups present is of real help to the researcher who is concerned with the utilization of such fractions. This determination of the chemical *character* of a sample is called characterization. It indicates the type of behavior one can expect from a sample, or indicates its general contents, at minimum effort and expense. It is a useful way to examine reaction mixtures, petroleum fractions, polymers, and any other complex sample or complex molecule for which a complete analysis is not practical or not necessary.

Characterization is more meaningful if more than one technique is used on a sample. For samples which might contain a wide variety of compound types, infrared spectroscopy and NMR make a good combination. Mass spectrometry can provide additional useful data. For samples containing only hydrocarbons, the most useful combination seems to be NMR and mass spectrometry, with infrared spectroscopy providing additional information about olefin types. In any case, preliminary separations by chromatography will make the characterizations more meaningful.

B. CHARACTERIZATION BY INSPECTION

The simplest NMR characterizations can be made by inspection of the spectrum, without any of the usual detailed measurements. This requires familiarity with the chemical shifts of major groups and with the band shapes normally associated with various compound types. This information is included in the charts and plates in this book.

In Figure 29 the pertinent chemical shift data bars are drawn under the spectra to simplify the inspection characterization of these petroleum fractions. The paraffinic crude (spectrum A) consists, on the average, of moderately branched paraffinic chains with a small amount of aromatic rings which are mostly uncondensed. There are some methyl groups on the rings, and

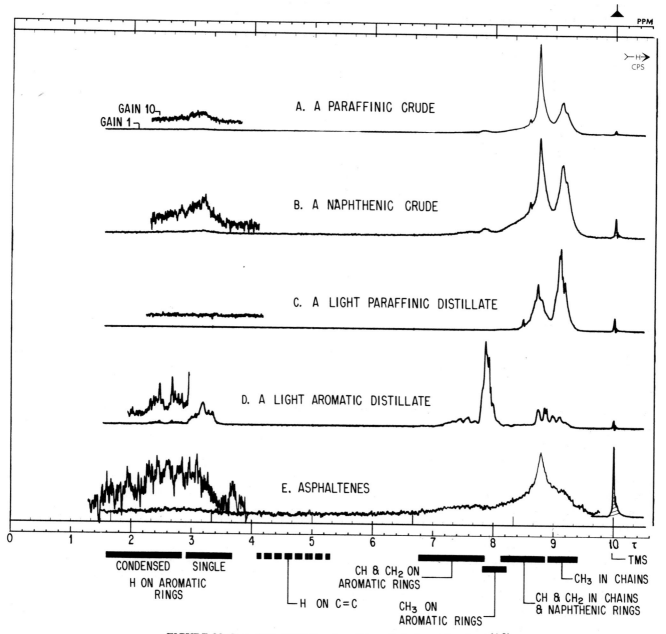

FIGURE 29. Sixty-MHz NMR Spectra of Some Petroleum Fractions (A2)

undoubtedly some of the chains are attached to the rings. There is probably a significant amount of naphthenes (cycloparaffins) in this crude also. There is no measurable amount of olefins.

The naphthenic crude (spectrum B) has much branchier chains, as indicated by the larger methyl band. It also has more naphthenes, as indicated by the wider skirt on the methylene band. It also has more aromatics, and

more hydrogens on groups alpha to aromatics. No olefins are detectable.

The light paraffinic distillate (spectrum C) is both low molecular weight and highly branched. It contains no olefins or aromatics, and probably has little or no naphthenes.

The light aromatic distillate (spectrum D) consists of alkyl benzenes and unsubstituted naphthalene. The ben-

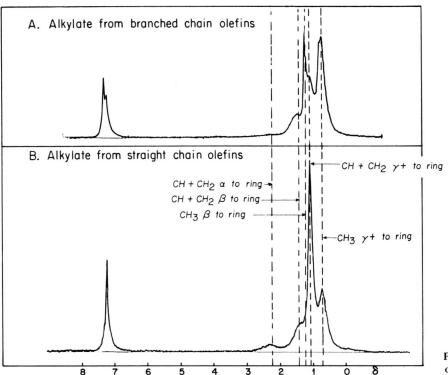

A. Alkylate from branched chain olefins

B. Alkylate from straight chain olefins

CH + CH₂ γ+ to ring

CH + CH₂ α to ring
CH + CH₂ β to ring
CH₃ β to ring

CH₃ γ+ to ring

FIGURE 30. Sixty-MHz NMR Spectra of Typical Detergent Alkylates (A2)

zenes are substituted with methyls, ethyls, and some normal propyls.

The asphaltenes (spectrum E) contains alkyl aromatics. The aromatics include both condensed and uncondensed rings. The alkyl groups may be mostly cycloalkyl, but the line broadening caused by paramagnetic components makes this judgment debatable.

Figure 30 shows spectra of typical detergent alkylates that can be characterized by inspection. Spectrum A shows bands for aromatic hydrogens and for groups beta and gamma to the ring, but very little hydrogen on groups alpha to the ring. The sharp peak at the position of methyls beta to the ring indicates geminal methyls on the alpha carbon. This is consistent with the lack of hydrogens on this carbon. Therefore, this is a highly branched alkyl chain with the benzene ring substituted, most of the time, at a tertiary position. The broadening of the aromatic band also is consistent with this interpretation.

In spectrum B the aromatic band is sharper, there is considerably more hydrogen on alpha groups, the sharp peak corresponding to beta methyls has been replaced with a sharp peak corresponding to a long chain of

methylenes gamma plus to the ring. The methyls gamma plus to the ring are reduced in intensity, also. This indicates that this sample consists of straight or very lightly branched chains with the benzene ring substituted at a primary or at least a secondary position.

A more detailed characterization of these detergent alkylates could be attempted by considering the band intensities. If the intensities could be measured accurately, one could even draw a meaningful structure for the "average molecule." The band overlap is so great, however, that such a structure would be subject to considerable inaccuracy. The band overlap is significantly reduced at 100 MHz, making the more detailed characterization reasonable. It is presented in Section E of this chapter.

The last spectrum to be characterized by inspection is in Figure 31. This is the spectrum of a typical high-quality "paraffinic" lubricating oil. It consists of long lightly to moderately branched paraffinic chains, with very little aromatics. Width of the skirt of the methylene band indicates there are probably some naphthenes present. The other bands are from the additives rather than the base hydrocarbon stock.

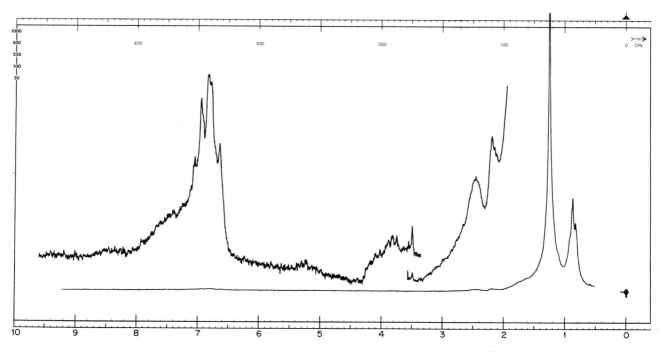

FIGURE 31. One-Hundred-MHz Spectrum of a Typical Lubricating Oil

C. CHARACTERIZATION OF SATURATED HYDRO-CARBONS

Mixtures of saturated hydrocarbons which are essentially free of olefins and aromatics are best characterized by a combination of mass spectrometry and NMR. Mass spectrometry shows the distribution of acyclic paraffins, single-ring cycloparaffins, and two- to six-ring condensed cycloparaffins. NMR shows the average branchiness of the aliphatic chains, and provides some information about the lengths of the chains between branches. The initial separation of the paraffins from nonparaffins is usually made by chromatography, although molecular sieves are sometimes used to separate only the linear paraffin fraction.

The NMR spectra of paraffin mixtures usually have only two separately measurable bands. One is from the methyl hydrogens and the other is from the remaining (nonmethyl) hydrogens (Figure 32). The quantitative distribution of the two hydrogen types plus the band shape correlations of Plates 1 to 14 are used to provide the NMR characterization information. The information about branchiness can be expressed in two ways: as the percent methyl hydrogen or as the ratio of methyl hydrogen to nonmethyl hydrogen. The percent methyl hydrogen, in turn can be combined with average molecular weight to estimate the average number of branches or the average number of methyl groups per molecule. If the mixture is known to consist of normal paraffins only, the percent methyl hydrogen translates directly into average carbon number of the fraction. These estimates are speeded by use of Table 8, which shows the percent methyl hydrogen in acyclic paraffins according to carbon number and number of branches or number of methyls per molecule. Molecular weight and boiling range are also given for each carbon number to assist in the estimates involving these quantities.

The NMR parts of the characterizations of three paraffin mixtures are shown in Figure 32. Spectrum (a) is that of a mixture of normal paraffins run at 100 MHz. There is very little overlap between the methyl and methylene bands, and the break point for measuring the integrals is taken as the lowest point of the valley between the two bands. The percent methyl hydrogen is 12.0 which, from the "Normal" column of Table 8, indicates an average carbon number of 24 and an average molecular weight of about 340. The sharp methylene band indicates chain lengths of 12 or greater, in agreement with the average molecular weight (Plate 8).

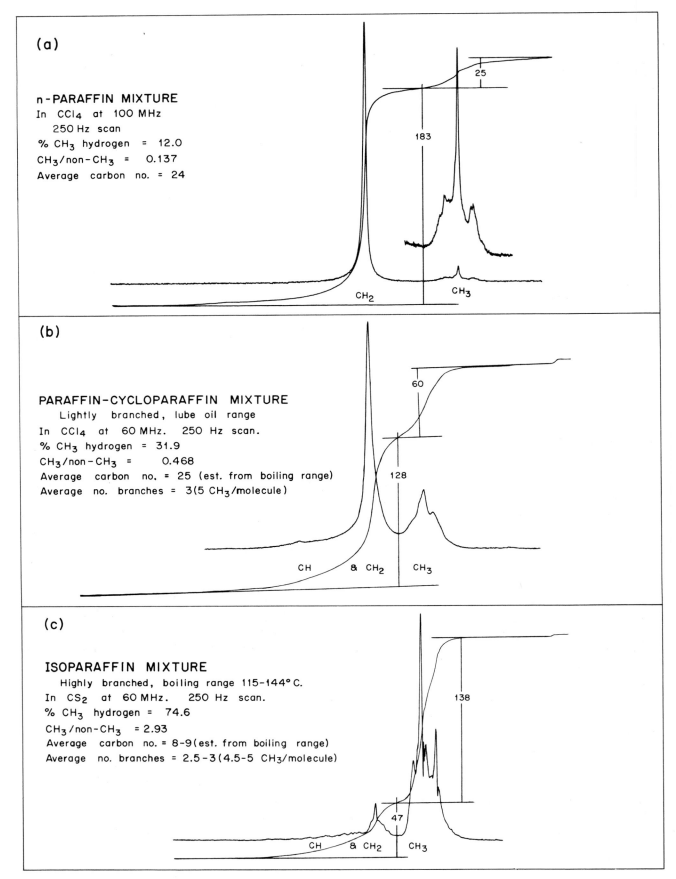

(a)

n-PARAFFIN MIXTURE
In CCl₄ at 100 MHz
   250 Hz scan
% CH₃ hydrogen = 12.0
CH₃/non-CH₃ = 0.137
Average carbon no. = 24

(b)

PARAFFIN-CYCLOPARAFFIN MIXTURE
   Lightly branched, lube oil range
In CCl₄ at 60 MHz. 250 Hz scan.
% CH₃ hydrogen = 31.9
CH₃/non-CH₃ = 0.468
Average carbon no. = 25 (est. from boiling range)
Average no. branches = 3(5 CH₃/molecule)

(c)

ISOPARAFFIN MIXTURE
   Highly branched, boiling range 115-144° C.
In CS₂ at 60 MHz. 250 Hz scan.
% CH₃ hydrogen = 74.6
CH₃/non-CH₃ = 2.93
Average carbon no. = 8-9(est. from boiling range)
Average no. branches = 2.5-3(4.5-5 CH₃/molecule)

FIGURE 32. Characterization of Paraffin Mixtures

TABLE 8. Percent Methyl Hydrogens in Paraffins (Acyclic)

Number C	Number of H's	Boiling range,[a] °C	Mol. Wt.	% CH$_3$ Hydrogens						
				Normal 2 Me	1 Br. 3 Me	2 Br. 4 Me	3 Br. 5 Me	4 Br. 6 Me	5 Br. 7 Me	6 Br. 8 Me
2	6	−88.6	30.07	100						
3	8	−42.1	44.09	75.0						
4	10	−11.8 to −0.5	58.12	60.0	90.0					
5	12	9–36	72.14	50.0	75.0	100				
6	14	50–68	86.18	42.9	64.3	85.7				
7	16	79–98	100.21	37.5	56.3	75.0	93.8			
8	18	106–126	114.23	33.3	50.0	66.7	83.3	100		
9	20	122–151	128.26	30.0	45.0	60.0	75.0	90.0		
10	22	147–174	142.29	27.3	40.9	54.5	68.2	81.8	95.5	
11	24	196	156.31	25.0	37.5	50.0	62.5	75.0	87.5	100
12	26	216	170.34	23.1	34.6	46.1	57.7	69.2	80.7	92.3
13	28	235	184.37	21.4	32.1	42.9	53.6	64.3	75.0	85.7
14	30	254	198.39	20.0	30.0	40.0	50.0	60.0	70.0	80.0
15	32	271	212.42	18.8	28.1	37.5	46.9	56.2	65.6	75.0
16	34	287	226.45	17.6	26.5	35.3	44.1	52.9	61.8	70.6
17	36	302	240.48	16.7	25.0	33.3	41.7	50.0	58.3	66.7
18	38	316	254.50	15.8	23.7	31.6	39.5	47.3	55.2	63.2
19	40	330	268.53	15.0	22.5	30.0	37.5	45.0	52.5	60.0
20	42	344	282.56	14.3	21.4	28.6	35.7	42.9	50.0	57.1
21	44	357	296.58	13.6	20.5	27.3	34.1	40.9	47.7	54.5
22	46	369	310.61	13.0	19.6	26.1	32.6	39.1	45.7	52.2
23	48	380	324.64	12.5	18.8	25.0	31.3	37.5	43.8	50.0
24	50	391	338.66	12.0	18.0	24.0	30.0	36.0	42.0	48.0
25	52	402	352.69	11.5	17.3	23.1	28.8	34.6	40.4	46.1
26	54	412	366.72	11.1	16.7	22.2	27.8	33.3	38.9	44.4
27	56	422	380.75	10.7	16.1	21.4	26.8	32.1	37.5	42.9
28	58	432	394.77	10.3	15.5	20.7	25.9	31.0	36.2	41.4
29	60	441	408.80	10.0	15.0	20.0	25.0	30.0	35.0	40.0
30	62	450	422.83	9.7	14.5	19.4	24.2	29.0	33.9	38.7
32	66	468	450.88	9.1	13.6	18.2	22.7	27.3	31.8	36.4
34	70	483	478.93	8.6	12.9	17.1	21.4	25.7	30.0	34.3
36	74	498	506.99	8.1	12.2	16.2	20.3	24.3	28.4	32.4
38	78	512	535.04	7.7	11.5	15.4	19.2	23.1	26.9	30.8
40	82	525	563.10	7.3	11.0	14.6	18.3	22.0	25.6	29.3

[a] Range for all known paraffins given for C¹⁰ and below. The highest value is always for the normal paraffin. Above C¹⁰ only the boiling point of the normal paraffin is given. From R. C. Wilhoit and B. J. Zwolinski, *Handbook on Vapor Pressure and Heats of Vaporization of Hydrocarbons and Related Compounds*, Texas A & M Research Foundation (1971).

Spectrum (b) is that of a mixture of paraffins and cycloparaffins, as indicated by the wide skirt of the nonmethyl band. The presence of cycloparaffins coupled with the smaller chemical shift obtained at 60 MHz means that there will be appreciable overlap between the methyl and nonmethyl bands, making the methyl band appear larger than it should. Rather than introduce the uncertainty of estimating this overlap, it seems better to accept it temporarily and make the break point for integral measurement at the lowest point in the valley between the bands, as shown. The overlap error can then be taken into account in the final estimate of branchiness. This sample has 31.9% methyl hydrogen, and from the fact that it boils in the lubricating oil range, its average carbon number is estimated to be about 25. In Table 8 at a carbon number of 25, 32% methyl hydrogen falls between 3 and 4 branches. In view of the overlap error, the number of branches is estimated as 3. This estimate, however, assumes that there are methyls at each end of the main chain, an assumption which may not be true of the cycloparaffins. It seems more meaningful in this case to report the number of methyls per molecule, which is 5. The sharp top of the nonmethyl band indicates that an appreciable proportion of long-chain segments is present.

Spectrum (c) is that of an isoparaffin mixture which contains essentially no normal paraffins or cycloparaffins. Even at 60 MHz the separation between methyl and nonmethyl bands is good. The methyl hydrogen constitutes 74.5% of the total hydrogen, and the boiling range indicates an average carbon number of 8 to 9 (Table 8). At a carbon number of 8, the average number of branches would be 2.5. At a carbon number of 9, there would be an average of 3 branches. These correspond to 4.5 and 5 methyls per molecule, respectively. The broad nonmethyl band indicates short chains, probably no more than 2 or 3 carbons between branches (Plate 6). This is consistent with 8- to 9-carbon chains with 2 or 3 branches.

It is possible to use total hydrogen content (which can also be measured by NMR) and the average molecular weight to calculate the average number of cycloparaffin rings per molecule. This value can then be combined with the branchiness information obtained from the NMR hydrogen distribution to produce a structure for the "average molecule" in the mixture.[T12] This procedure is of some value in gross structure or processing studies, but it is not nearly so useful as the more detailed information provided by the combination of NMR and mass spectrometry.

D. CHARACTERIZATION OF OLEFINIC HYDROCARBONS

Mixtures of aliphatic olefins can be characterized by NMR provided they are free of saturates, aromatics, and cyclic olefins. It is difficult to obtain such a fraction, however, so that characterization frequently becomes a matter of determining the distribution of hydrogen in different olefin groups and in groups alpha to olefinic carbons. The groups are identified from the chemical shift correlations in Charts 1, 8, and 9 and the band shape correlations in Plates 16 to 30. The hydrogen distribution is then calculated from the integral as percent of the total hydrogen observed.

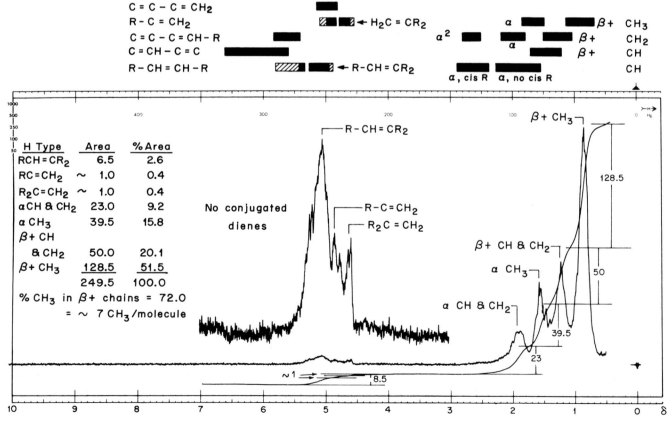

FIGURE 33. Characterization of an Olefin Mixture (100 MHz, CCl₄)

This procedure is illustrated in Figure 33, which shows the spectrum of a mixture of aliphatic olefins of high but unknown purity, run in CCl_4 at 100 MHz. The most common chemical shift correlations are drawn above the spectrum to make the band assignments clearer. The observed chemical shifts indicate there are no conjugated diolefins and little or no R—CH=CH—R groups present. There is also no observable number of C=C—CH_2—C=C groups which contain an α^2—CH_2. Considering the shapes and positions of the observed bands, and bearing in mind the groups ruled out above, the assignments are made as shown in Figure 33.

Break points in the integral are taken at the valleys between bands. Although appreciable error in the intensity measurements will result from the considerable band overlap, this method will nevertheless provide a picture of the nature of this sample and will permit valid comparisons with other samples characterized the same way. The final hydrogen distribution is shown at the left side of Figure 33. The percent methyl hydrogen in the saturate chains is also shown to indicate the branchiness of these chains. At the reported carbon number of 12 to 15, this translates to about 7 methyls per molecule (Table 8).

One possible flaw in this analysis is that tetrasubstituted ethylenes, $R_2C=CR_2$, are not observed in the olefinic region because they have no olefinic hydrogen. The hydrogen in groups alpha to C=C will be observed in the saturate region, however. If tetrasubstituted ethylenes are present, the average molecular weight calculated from the olefinic hydrogen as a basis will be too high, and the calculated bromine number will be too low. In the example shown in Figure 33, the average molecular weight is calculated as follows:

Area units per olefinic hydrogen per molecule
$= 6.5/1 + 1/2 + 1/2 = 7.5$
Average number of hydrogens per molecule
$= 249.5/7.5 = 33.3$
Assuming all mono-olefins, C_nH_{2n}, the average carbon number is 16.6
Reported carbon number for this sample was 12 to 15

These data indicate that the sample may contain tetrasubstituted ethylenes or paraffins. Mass spectrometry could resolve this question, as could bromine number. Infrared spectroscopy could check the assignments of other olefin types.

E. CHARACTERIZATION OF AROMATIC HYDRO-CARBONS

Mixtures of aromatic hydrocarbons which are free of saturates and olefins can best be characterized by a combination of mass spectrometry and NMR. Mass spectrometry provides the distribution of compound types (by fragmentation techniques) or the distribution of compound types with the distribution of molecular weights or carbon numbers within each compound type (by low-voltage techniques). NMR provides information about the nature of the alkyl chains attached to the aromatic rings. NMR can also provide an average characterization consisting of the distribution of hydrogen among the various functional groups present, and in special cases can provide a picture of the "average molecule or molecules." The functional groups are identified from the chemical shift correlations in Charts 13 to 15 and the band shape correlations in Plates 44 to 53. The hydrogen distribution is then calculated from the integral as percent of the total hydrogen observed. If the sample consists of a single compound type, it may then be worthwhile to deduce structures for one or two molecules which have the calculated hydrogen distribution and which also fit other information known about the sample.

The complete procedure is illustrated in Figure 34 and its accompanying data shown in Table 9. The sample is detergent alkylate made from benzene and propylene tetramer. In addition to the alkyl benzenes, it contains some unreacted olefin. The 60-MHz spectrum of this sample (Figure 30) did not permit the separate measurement of the various alkyl group hydrogens, but the 100-MHz spectrum of Figure 34 does permit reasonably accurate measurements. A 300-MHz spectrum would be even better.

The tabulated data show the functional groups assigned to each band. Only the groups associated with the alkyl benzenes are used, because the characterization is confined to this class of compounds. This makes it necessary to subtract from each band the amount of area contributed to it by the unreacted olefin. The proper area increments are calculated from the hydrogen distribution of the unreacted olefin after the olefin group resonances are assigned to the alkylate bands. The olefin spectrum and hydrogen distribution are presented in Figure 33. The complete correction calculation is shown under the "% area" heading in Table 9. The result is the

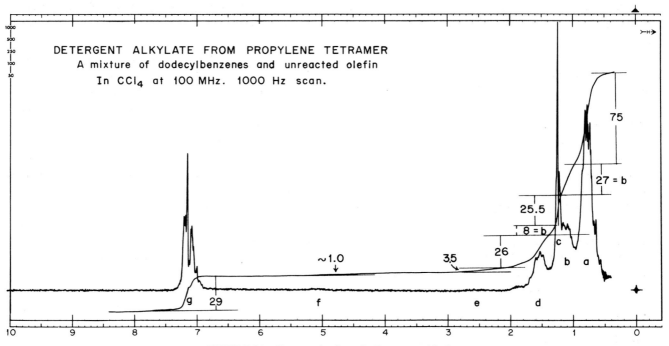

FIGURE 34. Characterization of a Detergent Alkylate

TABLE 9. Characterization of a Deterent Alkylate
(Use with Fig. 34)

Band	H Type	Area	% Area Total sample	Olefin[a]	Olefin free	Number of H's Olefin free	Calc.[b]	Number carbons Whole numbers	Structure I	II
a	γ+—CH₃	75.0	38.4	8.6	29.8	10.7	3.6	3–4	4	4
b	γ+—CH and CH₂	35.0	18.0	2.0	16.0	5.8	2.9–5.8	4–5	4	5
c	β—CH₃	25.5	13.1	1.4	11.7	4.2	1.4	1	1	1
d	β—CH and CH₂	26.0	13.3	4.2	9.1	3.3	1.6–3.3	1–2	2	1
e	α—CH and CH₂	3.5	1.8	–	1.8	0.6	0.3–0.6	½	1[c]	1
f	R—CH=CR₂	1.0	0.5	0.5	0	0	–	–	–	–
g	Aromatic H	29.0	14.9	–	14.9	5.4	6	6	6	6
	TOTAL	195.0	100.0	16.7	83.3	30.0		18	18	18

[a] Using data for the propylene tetramer-pentamer in Figure 33. Proportion of olefin areas to be used = 100(% olefin area in sample)/% olefin area in Figure 33 = 100(0.5)/3.0 = 16.7%.
[b] Number of carbons = number H/number H per carbon in groups producing band.
[c] A quaternary carbon:

Structure I: ... about 50%

Structure II: ... about 50%

Olefin structure:

hydrogen distribution, in % area, for the alkyl aromatics alone.

Because this sample consists of compounds of only one type and the range of carbon numbers is small, it seems worthwhile to determine the average structure. This is accomplished in the following steps:

1. From the hydrogen distribution calculate the number of hydrogens contributing to each band. Use the total number of hydrogens per molecule or other known value as the basis.

2. Calculate the number of carbons in the groups contributing to each band. This is simply the number of hydrogens divided by the number of hydrogens per carbon for the identified groups. When more than one group contributes to a band, a minimum and maximum number of carbons can be calculated.

3. From the calculated number of carbons per band, estimate the most likely whole number of carbons or range of whole numbers of carbons which contribute to the band. Use as a guide the average number of carbons per molecule and any other known values.

4. Devise the minimum number of structures that will fit the total collection of data.

The application of this procedure is also shown in Table 9. The number of hydrogens per band is calculated on the basis of a total of 30 hydrogens in the average molecule. The number of carbons per band is based on the average of 18 carbons per molecule and 6 carbons per aromatic ring. Knowing that the sample was produced from propylene tetramer, one can then devise the reasonable structures shown as I and II in Table 9. These could have been produced from the average olefin structure shown. Although these structures are probably much too regular and simple to represent very many of the actual molecules, they are a reasonable indication of the type of material in this sample.

A more limited and more generally applicable aromatic characterization is presented in Figure 35. The spectra shown are of three petroleum fractions which contain a wide variety of aromatic compounds, including both uncondensed and condensed ring types. The pertinent chemical shift correlations from Charts 13 to 15 are shown below the spectra to make the band assignments clearer. The assignments in the alkyl region simply include all the possibilities, but there is an arbitrary division between uncondensed and condensed aromatic ring hydrogens because of the overlap of these correlations. The lowest field shift in the single-ring category is that of benzene, which can be ignored in such samples

unless it shows as a single peak at about $\delta = 7.3$. The highest field shifts in the condensed-ring category are naphthalene beta hydrogens ortho to methyls. These should be present in minor amounts in heavy petroleum fractions, although they may be important in lighter fractions. Consideration of the foregoing facts, combined with experience in characterizing aromatic fractions, has led the author to select $\delta = 7.2$ as the break point between uncondensed and condensed ring hydrogens. As with other overlapping data, this will give consistent results, but the accuracy may vary somewhat from one sample to another.

Spectrum (a) of Figure 35 is for a mixture of medium-boiling aromatics. Only four bands can be measured separately, counting the arbitrary separation between uncondensed and condensed rings, so only four hydrogen types are calculated. This is the simplest of the aromatics characterizations. The quantitative data are supplemented by the qualitative observations that the alkyl chains are short, that the alpha groups are mostly methyls, and that the sample consists of alkyl benzenes and alkyl naphthalenes.

Spectrum (b) is for a mixture of heavier aromatics. It has seven bands which can be measured with reasonable accuracy. The assignments and hydrogen distribution among these bands are shown on the spectrum. This sample contains a wide variety of aromatic types, including phenanthrenes and kindred types, possibly some fluorenes or dihydroanthracenes, and alkyl benzenes. It probably contains alkyl naphthalenes, but these cannot be distinguished separately in this mixture. The alkyl chains are long and lightly branched, although there seem to be some methyls on the rings. There seem to be few naphthene (cycloparaffin) groups.

Spectrum (c) shows a fraction containing an appreciable amount of naphthenes, as indicated by the broad skirt on band b. The overlap of band c with this broad skirt of band b makes it desirable to use the parallel slope method of measuring the intensity of band c. This sample appears to contain moderately branched chains of moderate length in addition to the naphthene groups. There appear to be few methyls on the aromatic rings.

There may be times when average structural features of a sample are desirable, particularly when the sample is too heavy to volatilize in the mass spectrometer. Several schemes for determining these average properties have been worked out,[T12, T14, T24, T27, T31, T44, T45] each requiring elemental analysis, density, molecular weight, and sometimes infrared analysis in addition to NMR.

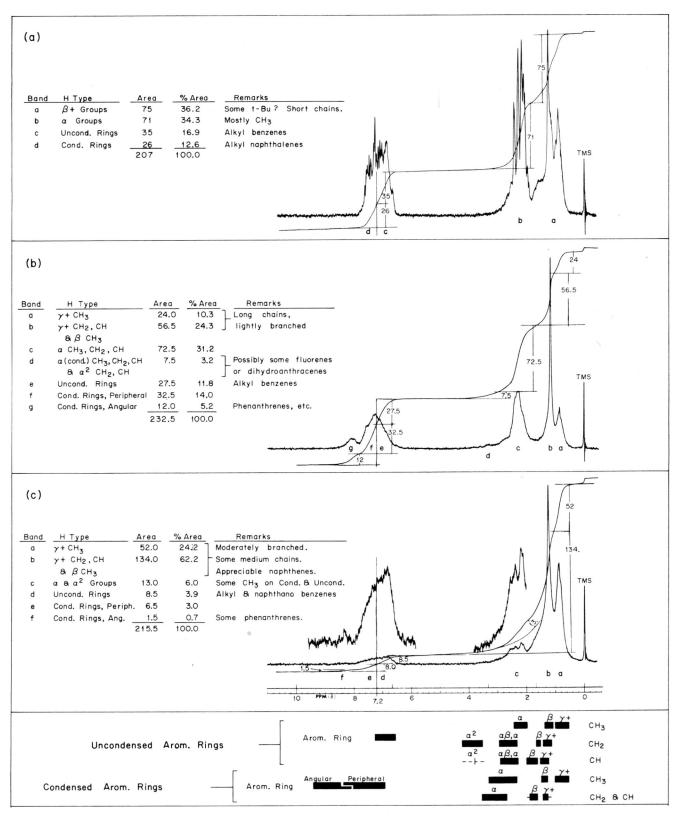

FIGURE 35. Characterization of Aromatic Hydrocarbons (All Spectra at 60 MHz in CC1₄)

(a)

Band	H Type	Area	% Area	Remarks
a	β+ Groups	75	36.2	Some t-Bu? Short chains.
b	α Groups	71	34.3	Mostly CH$_3$
c	Uncond. Rings	35	16.9	Alkyl benzenes
d	Cond. Rings	26	12.6	Alkyl naphthalenes
		207	100.0	

(b)

Band	H Type	Area	% Area	Remarks
a	γ+ CH$_3$	24.0	10.3	Long chains,
b	γ+ CH$_2$, CH & β CH$_3$	56.5	24.3	lightly branched
c	α CH$_3$, CH$_2$, CH	72.5	31.2	
d	α(cond.) CH$_3$, CH$_2$, CH & α^2 CH$_2$, CH	7.5	3.2	Possibly some fluorenes or dihydroanthracenes
e	Uncond. Rings	27.5	11.8	Alkyl benzenes
f	Cond. Rings, Peripheral	32.5	14.0	
g	Cond. Rings, Angular	12.0	5.2	Phenanthrenes, etc.
		232.5	100.0	

(c)

Band	H Type	Area	% Area	Remarks
a	γ+ CH$_3$	52.0	24.2	Moderately branched.
b	γ+ CH$_2$, CH & β CH$_3$	134.0	62.2	Some medium chains. Appreciable naphthenes.
c	α & α^2 Groups	13.0	6.0	Some CH$_3$ on Cond. & Uncond.
d	Uncond. Rings	8.5	3.9	Alkyl & naphthano benzenes
e	Cond. Rings, Periph.	6.5	3.0	
f	Cond. Rings, Ang.	1.5	0.7	Some phenanthrenes.
		215.5	100.0	

These methods are long and detailed and are covered thoroughly in the references. They will not be repeated here.

F. CHARACTERIZATION OF POLYMERS

1. General

NMR is especially well suited to determining the microstructure of polymers because of its ability to show the functional groups present and the length and branchiness of hydrocarbon chains. For polymers which can be dissolved in acceptable solvents, hydrogen NMR is superior to infrared and mass spectrometry and to pyrolysis-gas chromatography for showing the steric structure. Carbon-13 NMR has additional advantages in that its greater chemical shift can show even finer points of structure, and it can observe rubbery polymers in the solid state. For crystalline polymers that are insoluble, neither hydrogen nor C^{13} NMR is satisfactory at present because the resolution is reduced below the usable limit by the direct dipole interactions in the solid. This limitation may be overcome in the future, however, by the use of multiple pulse, adiabatic demagnetization, or other techniques that are now in the experimental stage.

The characterization of polymers by NMR is an important and extensively studied subject. A good re-

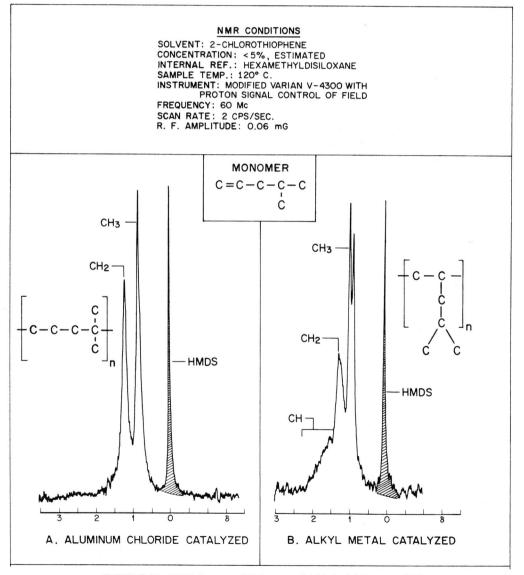

FIGURE 36. NMR Spectra of Polymers of 4-Methyl-1-Pentene (A5)

view of the field is presented in Bovey's latest book.[T7] This section will be confined to the use of hydrogen NMR to show the more obvious features of structure and to indicate some of the important inferences that can be derived therefrom. It will barely introduce the subject of polymer characterization, however, so that the reader should consult the literature further if a more comprehensive discussion is needed.

Limitations on the study of polymers by hydrogen NMR are the low solubility of many polymers, the high viscosity of the solutions, the low mobility of segments of the polymer molecule, and band overlap. All these limitations are minimized by running the sample at elevated temperatures, which increases the solubility, reduces viscosity, increases molecular mobility, and narrows the bands. Some band overlap may remain at high temperature, however. This can be reduced at times by changing solvents, changing temperature somewhat, or adding chemical shift reagents. Changing solvents can improve resolution or change unfavorable shifts or both. Changing temperature sometimes changes unfavorable shifts, also. Therefore, in general, it is advisable to study a polymer in several different solvents at several temper-

atures to make sure all the usable data have been derived from the spectra. Shift reagents function best at low temperatures, so that their use should be confined to those samples which can be run at room temperature. Even then, the reagent may not be as effective as it would be with a smaller molecule of similar structure, possibly because of steric hindrance in the larger molecule.

2. Regular Homopolymers

When a single monomer is polymerized to a chain of regularly repeating units, all the hydrogens of a given type in each unit have the same chemical shift and band shape. The NMR spectrum then consists of the simple spectrum of the repeating unit. This, in turn, indicates the monomer from which the polymer could have been made and shows the type of enchainment that took place.

Examples of this type of polymer are shown in Figure 36, the 60-MHz spectra of two different polymers of 4-methyl-1-pentene. Spectrum A shows only two sharp peaks, one in the methyl region and one in the methylene region, with equal areas. This means the re-

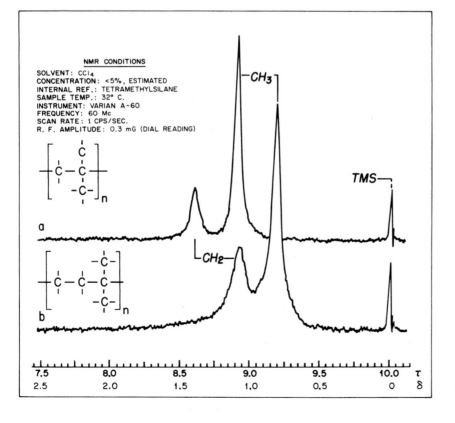

FIGURE 37. NMR Spectra of (a) Polyiso-butylene (b) Poly-3-Methylbutene-1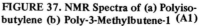

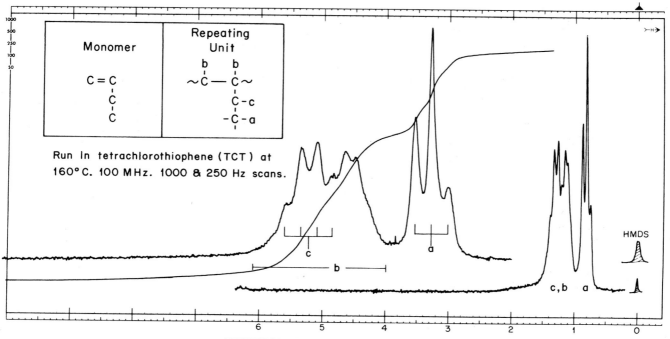

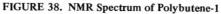

FIGURE 38. NMR Spectrum of Polybutene-1

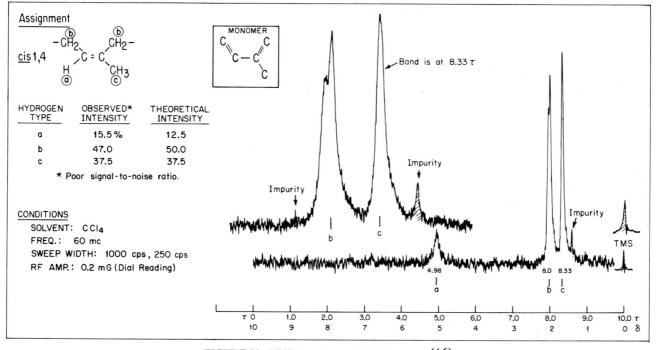

FIGURE 39. NMR Spectrum of a Polyisoprene (A6)

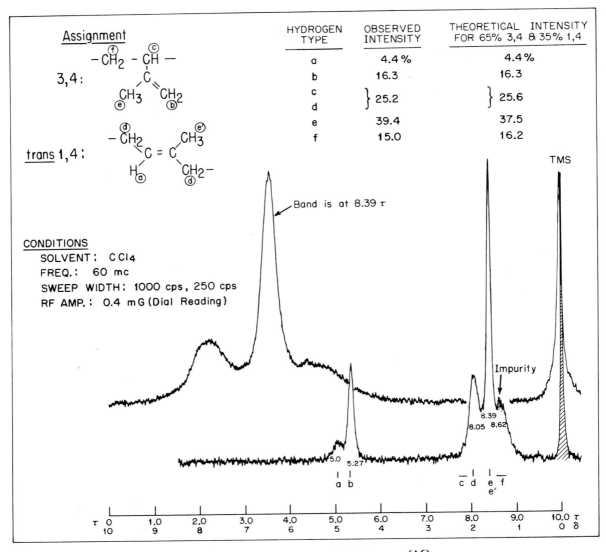

	HYDROGEN TYPE	OBSERVED INTENSITY	THEORETICAL INTENSITY FOR 65% 3,4 & 35% 1,4
	a	4.4%	4.4%
	b	16.3	16.3
	c	} 25.2	} 25.6
	d		
	e	39.4	37.5
	f	15.0	16.2

FIGURE 40. NMR Spectrum of a Polyisoprene (A6)

peating unit is composed of isolated methyls and isolated methylenes in the ratio of 2 methyls to 3 methylenes. Knowing that a 6-carbon monomer was used, one can quickly draw the structure shown in A. Even if one does not know the monomer, however, this structure is still the smallest one which can produce the spectrum. It is therefore the structure of the repeating unit and indicates that the polymer most likely comes from a single 6-carbon monomer. Another possibility, that it is a copolymer of ethylene and isobutylene with perfect alternation of monomer units, is most unlikely.

Because this polymer was made from 4-methyl-1-pentene, it is clear that enchainment took place at the 1 and 4 carbons (1,4 enchainment). It is conceivable that it could have been made from 4-methyl-1,3-pentadiene,

with 1,4 enchainment, and with the resulting unsaturated polymer being hydrogenated. Other possible monomers seem unlikely because of the limitations of known polymerization mechanisms.

Spectrum B has three bands, representing methyl, methylene, and methine groups. The doublet methyl band and the shift of the methine band, coupled with their relative intensities, indicates that these two are associated in an isopropyl group. This leads to the repeating structure shown in B. (The other methine resonance is under the methyl or methylene bands.) This structure shows that this polymer is made by 1,2 enchainment of the monomer.

Some care must be exercised in using the chemical shift correlations with polymer spectra. Steric factors in these large molecules can sometimes produce sizable additional shifts, throwing the polymer shifts outside the limits usually encountered with small molecules. For this reason, one should always check the charts to see if any special correlations are indicated for polymers (see Charts 5, 6, 13, 67). The steric effect in polyisobutylene

is shown in Figure 37. Both methyl and methylene resonances have been shifted about 0.6 ppm to low field of their usual positions. The corresponding resonances of poly-3-methylbutene-1 are in their normal positions.

The spectrum of a homopolymer of butene-1, which seems to have regular 1,2 enchainment is shown in Figure 38. The band assignments are shown on the structure diagram.

Figure 39 shows the spectrum of a regular *cis*-1,4 enchainment of polyisoprene to illustrate the characterization of an unsaturated polymer. The intensity measurements are hampered by poor signal-to-noise, but seem reasonable under the circumstances.

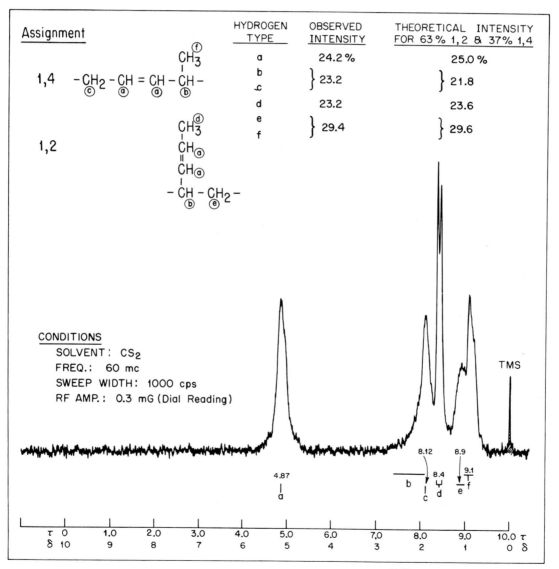

FIGURE 41. NMR Spectrum of Polypentadiene-1, 3 (A6)

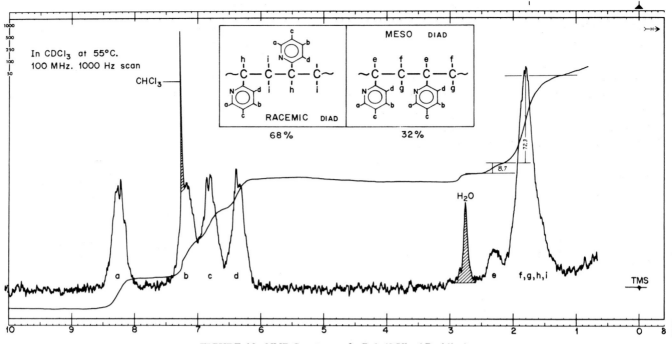

FIGURE 42. NMR Spectrum of a Poly(2-Vinyl Pyridine)

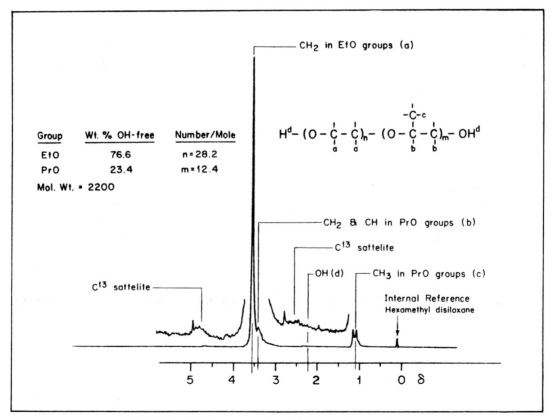

FIGURE 43. NMR Analysis of an Alkylene Oxide Copolymer

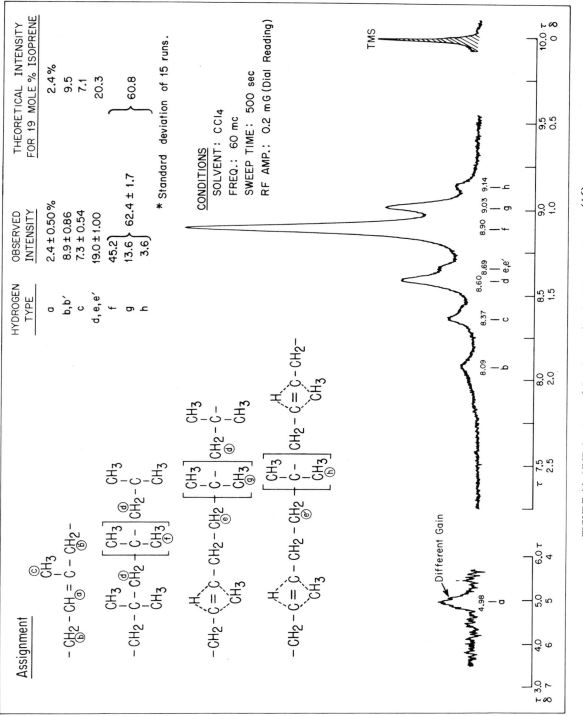

FIGURE 44. NMR Spectrum of Copolymer of Isobutylene and Isoprene (A6)

3. Irregular Homopolymers

When the units in a homopolymer do not repeat regularly, either because of varying steric arrangements of the monomer in the chain or because of varying enchainments, there are more magnetic environments in the chain than there are hydrogen types in the monomer. Although these magnetic environments may not all lead to separately resolved bands, the spectrum of an irregular polymer should have more bands or poorer resolution, or both, than the spectrum of a regular polymer of the same monomer. No single repeating unit will suffice to describe such a polymer. When the irregularity is due parimarily to varying enchainment, the structure corresponding to each type of enchainment can be used as a repeating unit. This is illustrated in Figure 40, which shows the spectrum of polyisoprene containing a mixture of 3,4 and *trans* 1,4 enchainments. This polymer probably also shows irregularity due to different steric arrangements of the monomers, but the shifts due to these differences are not resolved. The proportions of the two enchainments are calculated by the method described in Section IIB, with the key information supplied by the olefinic hydrogen bands. Only the final check results are shown in Figure 40.

Another example of a polymer for which enchainments dominate the characterization is shown in Figure 41. In this case the key information leading to analysis comes from the saturate region.

When the irregularity is due to varying steric arrangement of the monomer, the structures corresponding to each of the detectable arrangements can be used as a repeating unit. The description of each arrangement requires two monomer units (diads) or three (triads) or more (tetrads, pentads, etc.). In Figure 42, only the differences at the diad level are detectable. Band *e* is unique to the meso diad, permitting a calculation of the proportions of the two diads in the polymer.

4. Copolymers

The NMR spectra of copolymers or terpolymers generally reveal the types of monomers used and their proportions, provided the band overlap is not too great. If a polymer consists of blocks of homopolymers of the different monomers, its spectrum will simply be the sum of the spectra of the individual homopolymers, with some minor features due to the regions where the blocks tie together. If the monomers are distributed at random, or if they alternate in regular sequence, the spectrum will have major features which are unique to the copolymer. These unique features may consist of new bands which are not observed in the homopolymer spectra or different shifts or multiplicites for the homopolymer bands. In either case these features can be used to distinguish between block and nonblock copolymers, and to indicate some of the microstructure of the polymer.

The spectrum of a common type of copolymer is shown in Figure 43. It is the 60-MHz spectrum of a copolymer of ethylene oxide and propylene oxide. Band *c* is unique to propylene oxide, permitting calculation of the proportions of the monomers without regard to molecular weight. In this case, however, the OH end groups also permit the calculation of the average molecular weight. Because the OH band itself is subject to interference from water and any other exchangeable hydrogens that may be present, it is advisable to use a derivative, such as the carbamate described in Chapter 2, Section IIB-4a, to measure the OH intensity.

The analysis of a hydrocarbon copolymer is shown in Figure 44. In this case the proportions of the monomers are determined, and four segments of microstructure are indicated. The minor component, isoprene, seems to be fairly randomly distributed among the isobutylene units. The excess isobutylene is, as expected, in blocks of at least several units each.

CHAPTER 4

Chemical Shift Correlations

I. INTRODUCTION

The chemical shift has proved to be the most useful characteristic of the hydrogen resonance spectrum for indicating features of molecular structure. This is because, first, the chemical shift is sensitively and reliably related to major and minor aspects of structure. If solvent, concentration, and temperature are held constant, the chemical shifts measured on a stable spectrometer are reproducible to about one part per billion for samples of high purity. For ordinary work, where concentration and temperature are controlled only approximately, the shifts are usually reproducible to 0.02 to 0.05 ppm. This is sufficient precision to show the small variations in shift caused by differences in steric arrangement or differences in the lengths of short chains. These small variations show up as perturbations on the larger variations caused by differences in electronegativity of substituents, degree of saturation, "resonance interaction" of neighboring groups, or the size of small rings.

Secondly, the chemical shift can be measured with acceptable accuracy a greater percentage of the time than any other spectral characteristic. It can be measured for isolated bands even when the lack of fine structure prevents the measurement of multiplicity and coupling constant. It can also be estimated adequately for many overlapping bands where the overlap prevents the determination of intensity.

A priori calculations of chemical shifts of hydrogen in given structural environments, when they are possible, are not sufficiently accurate for analytical purposes.

Instead, it is necessary to use empirical correlations of shift with structure. Such correlations have now become sufficiently extensive, accurate, and detailed to be of great value in both theoretical and analytical work. (In this book the term *analytical work* includes the determination of molecular structure as well as quantitative analysis and characterization.)

To show all the structural differences that can be indicated by chemical shift it would be necessary to make all determinations at constant solvent, concentration, and temperature. Even then, the data would apply only to pure compounds or to mixtures of closely similar composition. This degree of control is justified in special cases, such as for the shifts of hydrocarbon groups, but is impractical for general work. Practical correlations must compromise among accuracy, detail, and speed of access to the data.

The most detailed and accurate data have been presented in extensive tabulations, where each single value can be explicitly set down along with a description of the conditions under which it was obtained. The first tabulation of this type was published by Tiers [14] More extensive and more recent ones have been published by Brügel,[18] Bovey,[19] Chamberlain and Reed,[25] Bhacca, *et al.*,[26] and Sadtler.[26] These tabulations are useful for very precise work on a limited number of compounds, but they are not suited to general analytical use. The access time is too long and the relation between shifts of various functional groups is difficult to visualize.

A more convenient form of presentation for analytical use is the bar graph. By giving up some detail and accuracy one can achieve very rapid access to massive amounts of data, and can thus ensure that all available correlations have been considered in making an interpretation. The relationship between the shifts of different groups is clearer, so that alternate interpretations can be checked rapidly. The flexibility of use is also greater. When the bar graphs are properly prepared, it is as easy to find the chemical shifts which will arise from a given structural group as it is to find the structural groups which could have given rise to an observed shift. Two separate tables are required to permit these two searches. Furthermore, bar graphs indicate the reliability of the correlations by showing the range of shifts for each group and the degree of overlap between adjacent ranges. They also show the trends of the shifts so that unavailable data can be predicted by interpolation, extrapolation, or analogy. The first such bar graph was published by Meyer *et al.*,[27] and extended versions have been published by a number of authors.[28–35] The bar graph can also be used to present precise data,[35] if one is willing to confine the correlation to a narrow group of compounds. Nukada *et al.*,[29] and Yamamoto *et al.*,[34] present a series of bar graphs of increasing detail to permit the user to make the compromise between detail and access speed which best suits his purpose.

The author has attempted to incorporate the best features of the various types of bar graphs into this book. A series of charts of increasing detail and complexity is used to speed the search at the desired level. Each set of charts serves as the index to the next set, so that little time is lost in transferring from the less complex to the more complex charts. The data are organized by functional group and by compound type, with the emphasis being changed from one set to another, depending on the purpose of the set. Experience with these charts indicates they are convenient and usable and will materially improve the speed, accuracy, and completeness of analytical work.

Although these correlations include most of the data now available, it has been necessary to omit some classes of compounds. Correlations for organometallic compounds have been omitted because data are somewhat scarce and spectral assignments are frequently uncertain. Correlations for compounds containing condensed saturate rings have been omitted because it does not seem possible at present to sort out the large steric effects on the chemical shifts. Correlations for complex hetero-

cyclic compounds have been omitted because of the scarcity of data.

II. DESCRIPTIONS OF THE SHIFT CHARTS

A. THE INDEX CHART, I1

Chart I1 presents a complete view of the hydrogen chemical shift correlations which are included in this work. It serves to call attention rapidly to the major classes of functional groups which are important to a given problem, and to eliminate those which can have no bearing on it. It provides the fastest access to all the data but shows the least detail.

This chart is arranged primarily by functional group in the order methyls, monosubstituted acyclic methylenes, disubstituted acyclic methylenes, mono- and disubstituted acyclic methines, trisubstituted acyclic methines, acetylenic H, acyclic olefinic H, CH=(O,N,S), nonaromatic and heterocyclic rings, aromatic rings, OH, SH, PH, and NH. Spaces separate the group categories from each other for ease of recognition. Within each category, major substituents are grouped according to their effect on the chemical shift or according to their primary element, whichever is more helpful in indicating possibilities. This arrangement is the same as that used for the major divisions of the summary charts, for which Chart I1 serves as the index.

The data are arranged in columns insofar as is reasonable. At the extreme left and right are line numbers which can be referred to in the text. They are duplicated at both sides of the chart to assist the reader in following a line across. Next is the identification of the functional group and the associated structure, by line structure in most cases but as word descriptions when line structures are not suitable. This is followed by definitions of substituents when appropriate, X and Y representing any group unless specifically identified. Next is a bar showing the range of shifts of the hydrogens which are explicitly identified in the functional group. The cross-hatched parts of the bar in line 1 refer to special situations which are encountered much less frequently than those represented by the solid part. The dotted portion of the bar in line 73 indicates uncertain values.

To indicate the amount of confidence the reader should place in a given shift range, the number of compounds from which that range was derived is shown in the third column from the right. These numbers show that some ranges are based on truly massive amounts of data (lines 1, 2, 11, 58, etc.) and therefore are considered very reliable. Other ranges, however, are based on sparse data (lines 8, 45, 46, 53, etc.) and should be used with some caution. These latter ranges may widen when more data are added.

The last column lists, for each line in the index chart, the summary chart and line where the corresponding data may be found in more detail. This gives the user access to the much more detailed correlations of the summary charts with a speed comparable to that with which data can be found on the index chart. Use of the index and summary charts in solving a structure problem is illustrated in Chapter 3, Section IB.

Both τ and δ scales are used on this chart and all other chemical shift charts in this book, so that the user can employ the system he prefers or can compare the data directly with data in either system in the literature. Identically subdivided scales are placed at top and bottom of each half to permit the alignment of a straight edge for accurate reading of the shift values (See Figure 25, Chapter 3).

B. THE SUMMARY CHARTS (S1 TO S35)

The purpose of the summary charts is to provide enough detail to permit the solution of many problems, while retaining fairly rapid access to all the data. They are organized so as to permit about equally rapid determination of the structural groups which could have given rise to an observed chemical shift and to the determination of the chemical shift of a given structural group. It is anticipated that the summary charts will be the most useful ones for the general determination of unknown structures.

The summary charts are arranged primarily by functional group in the same order as used in Chart I1. For each functional group there is a further subdivision by structural group, arranged to facilitate both the structure-to-shift and shift-to-structure searches. In this context a structural group is a functional group, such as methyl, together with the substituent or substituents which affect its shift and such other hydrocarbon groups as are needed to define the steric situation.

The data are presented in lines which are numbered on the extreme right. From the left, each line includes the structural group under consideration (this column is labeled "functional group," but the structure includes the substituent and auxiliary groups), identification of the substituents, and the bar showing the range of shifts

of the identified group. This is followed by the range expressed numerically in delta values, to speed the reading of values once a bar is located.

The next division shows the number of compounds from which the data were obtained, to indicate the degree of confidence which should be placed in the shift range. This is followed by the number of examples studied, which is the total number of data points actually included in the range. When this number exceeds the number of compounds, some of the compounds were run in more than one solvent. If the numbers are the same, however, it does not mean that all compounds were run in the same solvent. The solvents used are not specified on these charts, but are specified on the detailed charts which follow.

The data are presented at two levels of accuracy. Solid bars represent measured values, whereas dashed lines indicate ranges estimated by interpolation, extrapolation, or analogy. In these charts no distinction is made in the precision of the data. This distinction is reserved for the detailed charts.

On the page facing each summary chart is a cross index to the detailed charts from which the data were taken. This provides rapid access to the proper detailed chart if additional information is desired. In many structure determinations it is better to obtain a clue to the structure from the summary charts, then finish the identification and check the whole from the detailed charts. This is because the detailed charts are arranged by compound types primarily, showing the chemical shift information for all the function groups in a given compound type in the same box.

The summary charts contain much information not included in the detailed charts, however. In some cases the data are not presented efficiently by the format of the detailed charts. This is true of most of the data of Charts S23 to S30. In other cases the data became available after the detailed charts were finished, and revising the charts was not practical. The presence of data not included in the detailed charts is indicated by an X in the cross index.

When the functional group is directly connected to two substituents which significantly affect its chemical shift, the shift data are presented twice, once with each substituent as the "key." Thus the shift for $\phi—CH_2—C=C$ is included in Chart S11(2) with C=C as the key and in Chart S12(1) with ϕ as the key. This speeds the search for this shift by permitting the use of either key. It may also permit one to get a narrower range for this individual shift by comparing the two recorded ranges. In most cases the two ranges in which the shift of a given double-substituted group is plotted will not include the same additional pairs of substituents. The desired shift must then be in the region of overlap of the two ranges. In the example just given, S11(2), with a range of 3.39–2.82, includes four other pairs of substituents. On the other hand, S12(1), with a range of 3.35–2.96, includes the shift of one other pair. The shift of ϕ–CH$_2$– C=C must lie in the range 3.35–2.96, the overlap region. Sometimes this narrower range will simplify a decision to accept or reject a group as a possibility in a structure determination.

C. THE DETAILED CHARTS (1 TO 89)

Charts 1 to 89 present the chemical shift correlations in the greatest detail consistent with a reasonable access time. They are also organized in a way different from that of the index and summary charts to provide a new route of access to the data. It is expected that the detailed charts will be most useful for confirming proposed structures and for determining the effects of steric factors on chemical shift. They are the only charts which can effectively assist in the analysis of hydrocarbons.

The detailed charts are organized by compound type, with a further breakdown into individual functional groups within each compound type. All of the chemical shifts which can normally be associated with a single compound type are presented within a single section of a chart, and adjacent chart sections present all of the data for each of a series of related compound types. For this purpose the definition of compound type is made as narrow as necessary to permit showing the fine distinctions in structure which are reliably indicated by chemical shift differences. Each chart section is limited to data from a specified type of hydrocarbon skeleton (e.g., paraffins, aliphatic olefins, single-ring aromatics) or to a specified skeleton type substituted with a single type of nonhydrocarbon group (e.g., saturated aliphatic acids and esters, benzoic acids and esters, phenyl ethers, alicyclic ketones). With few exceptions, data for unlisted hetero-substituents are not shown; for example, Chart 37 (phenols) does not contain data for nitrophenols, chlorophenols, carboxy phenols, etc. In no case is a correlation influenced significantly by an unlisted substituent.

The format used varies somewhat with the nature and extent of the available data, but most of the general charts have comparable formats. Each general chart section is identified by the name of the compound type in the upper left corner. Underneath this is the reference identification, which is discussed later. The remainder of the left portion of the section is devoted to structure diagrams which define typical functional group and spatial designations used to identify the chemical shift ranges.

The center portion of the chart contains the bars used to indicate the ranges of chemical shifts of each functional group in each of its structural environments. To facilitate reading and referencing, these bars are organized into zones, subzones if needed, and numbered lines. Each zone includes all the shift ranges for a single functional group, and each bar in the zone is identified with a structural environment by a spatial designation placed above or below it (or, in a few cases, beside it).

The next column (second from right) shows the number of compounds and/or compound–solvent combinations from which the data were obtained, to indicate the confidence which should be placed in the shift range. In each line, these numbers are in the same order as are the bars (left number corresponds to left bar, etc.). In most charts, dashes are used to indicate the lack of data for dashed bars or lines, but in a few (column headed "No. S") some dashes have been omitted.

The molecules of each compound type contain one or more key functional groups which are responsible for the major chemical shifts. To show the influence of these key groups clearly, all other groups in each molecule are referred to them by spatial position (α, β, γ, etc.). These key groups are the reference groups identified in the upper left corner of each chart section. When two reference groups are required to describe a structural situation, an illustrative bar is used in the reference corner to show which reference is indicated by each of the spatial designators (above and below the bar). The single exception to the above format is the paraffins, for which no reference is used. For these compounds the shifts are correlated with chain length and branching (Charts 1 to 7).

A spatial designation indicates the position of the entire carbon–hydrogen group rather than the position of the hydrogen itself. Thus, the designation "α to the aromatic ring" refers to a CH, CH$_2$, or CH$_3$ group attached to the ring, but not to a hydrogen attached to the ring. Hydrogen attached to the ring is designated "H on ring" or "ring H." This nomenclature permits the consistent and separate designation of hydrogens in CH_3, CH_2, and CH groups regardless of the nature of the reference group. Proximity to more than one reference group of the same kind is indicated by superscripts (α^3, α^2, o^2) or by multiple symbols not separated by commas ($\alpha\beta$, op, oo'). Multiple symbols separated by commas indicate that the data bar applies to more than one position with respect to the reference. Thus (o, m, p) indicates that the data bar applies to groups in the ortho, meta, or para positions with respect to the reference group, whereas the group (o, om) indicates that the data block applies either to groups which are ortho to a single reference group or to those which are ortho to one reference and meta to another reference of the same kind.

In general, the influence of a reference group does not extend past the beta position (Figure 45); therefore, groups which are gamma and farther from the reference group are designated $\gamma+$ and designations such as $\alpha\gamma$ and $\beta\gamma$ are reported simply as α and β, respectively. The designation γ^2 is used in those few cases where it can be differentiated from $\gamma+$. In Charts 20, 21, and 23, however, the designation γ and $\delta+$ are used to show observable effects.

The effects of steric arrangements on chemical shifts are indicated by special notes, as in Chart 10 (line 1) or by special chart zones such as Chart 13, lines 12 and 13 vs. lines 14 and 15, and the "exterior" and "interior" designations of lines 16 and 17. The bars corresponding to *cis* and *trans* isomers are indicated by the addition of c or t to their spatial designations, as in Charts 20 to 22.

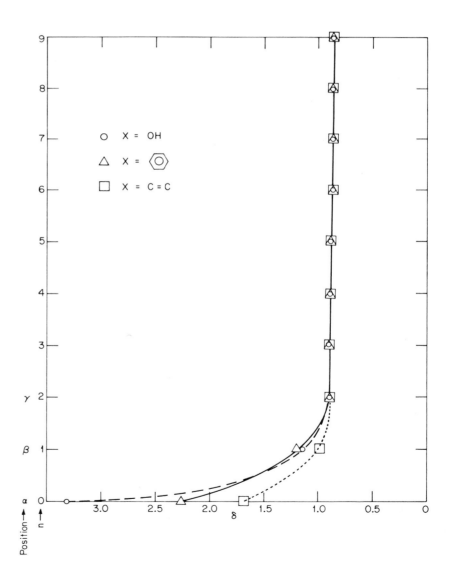

FIGURE 45. Methyl Chemical Shift vs. Chain Length for X–$(CH_2)_n$–CH_3

The most precise data reveal such fine distinctions in the effect of molecular structure on chemical shift that specific structure diagrams are usually needed to identify the bars on the charts (Charts 2 to 9).

Data in the charts are supplied at several different levels of precision and accuracy. The general charts present data which include scatter due to different solvents, concentrations, and temperatures. These data are useful for rapid comparison with samples under a wide variety of conditions, including multicomponent mixtures. The solvents used are listed on each chart. The precision charts present data obtained under conditions which minimize solvent effects. They provide the ultimate accuracy in shift–structure correlation, but the sample data must be obtained under the same conditions as the chart data. The solvent used and the concentration range for which the data are applicable are listed on each precision chart.

The solid black or crosshatched bars represent data actually measured from reliable spectra. Bars or portions of bars outlined in broken lines show ranges determined by analogy to reliable data for comparable functional groups. These ranges are considered reasonably accurate, although they have not been confirmed by actual measurements. Dashed and dotted lines (not bars) indicate ranges estimated from correlations but not supported directly by reliable data. The utility of the charts has been improved considerably by estimating the major portion of the missing data.

Reading of the positions of the bars is facilitated by placing a straight edge between corresponding scale divisions at the top and bottom of a chart. Vertical lines drawn between corresponding scale divisions speed the reading even more. When such lines are printed in black, however, they interfere severely with some of the small lettering. If such lines are desired, it is recommended that they be drawn in transparent color.

III. USEFUL GENERALIZATIONS

Although the organization of the chemical shift charts should permit the reader to discover all the correlations available for the solution of a given problem, there are certain generalizations which can simplify the task still further. These generalizations are concerned with the

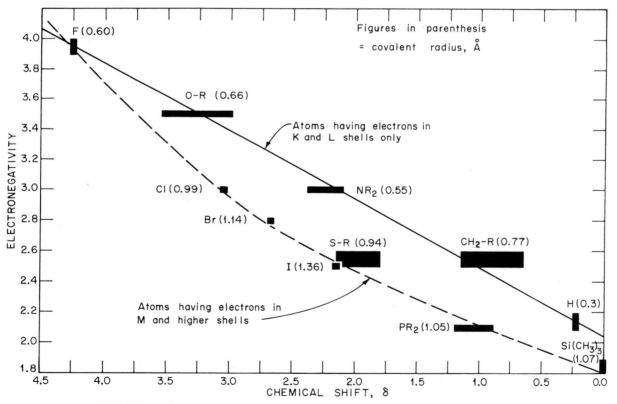

FIGURE 46. Chemical Shift of CH_3-X vs. Electronegativity of the Attached Atom of X

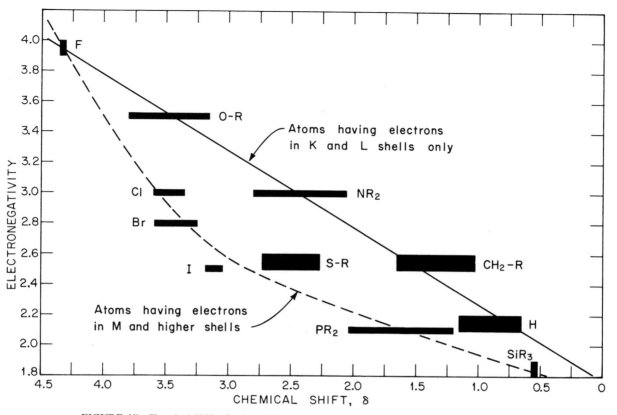

FIGURE 47. Chemical Shift of Alkyl–CH$_2$–X vs. Electronegativity of the Attached Atom of X

trends in shift with nature of the substituent, steric factors, ring size, etc. Some of the more useful of these trends will be discussed and illustrated in this section.

A. CHEMICAL SHIFT VS. SUBSTITUENT ELECTRO-NEGATIVITY AND SIZE

Although the intramolecular causes for the chemical shift are believed to be known (Chapter 1, Section IIB-2), it is not generally possible to determine to what extent each contributes to the shift of a given group in a given structural situation. It is helpful, however, to have at least a qualitative idea of the relative importance of certain measurable factors on the shift. Such knowledge enables one to estimate some relative shifts and shift changes without having to consult the charts. It can also be helpful in predicting which charts one should consult.

1. For Saturated Chains

Two factors which correlate with chemical shifts of many groups are the electronegativity* and size (co-

* Pauling values with some later modifications.

valent radius)† of that atom of a substituent which is directly attached to the group of interest. Figures 46 and 47 show the effect of these two factors on the shifts of methyls and methylenes in saturated chains. Because the ranges shown for each substituent include shifts for all causes, including solvent shifts and steric factors, it seems clear that electronegativity is the dominant factor for substituent atoms the size of carbon and smaller. For atoms larger than carbon, however, covalent radius appears to be a significant factor, although electronegativity still dominates. This can be interpreted to mean that for the smaller atoms the local diamagnetic current is the dominant cause of shifts. The withdrawal of charge by the electronegative atom reduces this current and reduces the shielding it has afforded the hydrogens. For the larger atoms the local currents around the substituent atoms themselves may make a significant contribution to the shift. In both cases, however, other factors may contribute to the shift, and the reader

† From atomic model data, Bronwill Scientific Division, Will Scientific, Inc., in Angstrom units.

should not take the foregoing discussion to mean that the shifts have been "explained."

2. For Olefin Groups

Figure 48 shows the shifts of the CH= group alpha to the substituent as functions of the electronegativity and covalent radius. It appears that both have some influence on the shift, but neither can be said to dominate. Both increasing electronegativity and increasing size of the substituent produce increasing deshielding (increasing delta values) for the alpha CH= group. This is similar to the effects on the saturated groups.

The effects differ drastically for the $=CH_2$ group beta to the substituent, however, as shown in Figure 49. In this case increasing electronegativity appears to produce increasing *shielding* of the group, the opposite of the effects on the saturated and alpha CH= groups. Furthermore, the correlation line for the halogens lies at higher delta values than the line for the rest of the substituents.

On the other hand, increasing covalent radius of the substituent produces increasing deshielding of the beta

$=CH_2$ group, as it did with the alpha CH= group. Again separate lines can be drawn for the halogens and for the rest of the substituents (the dashed lines), but a single line (solid) seems about as good in this case. It should also be noted that hydrogen fits the electronegativity correlation, but does not fit the atom size correlation.

The trends noted in Figure 49 can be interpreted as showing the effects of resonance interaction superimposed on the effects observed in Figures 46 to 48. (Here the term "resonance interaction" means the electronic charge delocalization concept used in organic chemistry. It is quite different from the magnetic resonance of the hydrogen nucleus or the electron). The C=C group can have a circulating interatomic current (see Figure 2) as well as the ability to enter into resonance interaction. If it is assumed that resonance produces a redistribution of charge in the system, then resonance would be expected to have a marked influence on chemical shift. Furthermore, if atoms with lone pairs of electrons, like O, S, and P, are assumed to donate part of their negative charge to the beta $=CH_2$ group during resonance interaction, then the increased shielding of

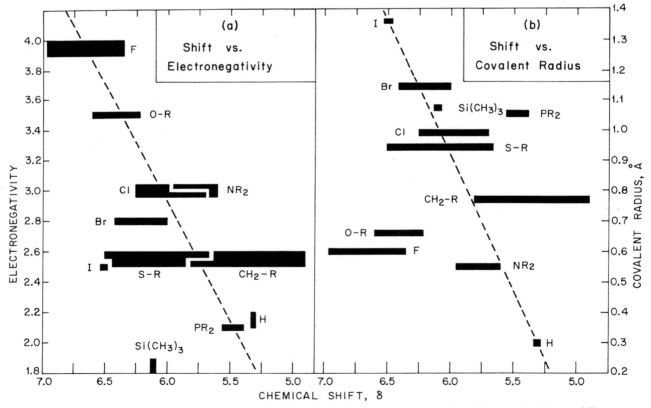

FIGURE 48. Chemical Shift of X–CH=CR$_2$ vs. Electronegativity and Covalent Radius of the Attached Atom of X

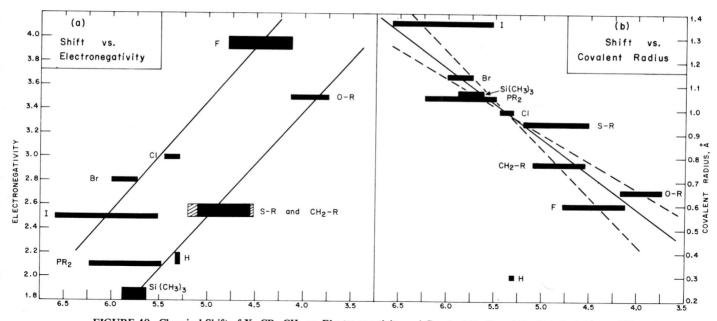

FIGURE 49. Chemical Shift of X–CR=CH₂ vs. Electronegativity and Covalent Radius of the Attached Atom of X

this group is expected. The halogens, with a common outer electronic configuration which is different from that of the other atoms, might very well have a different resonance interaction and therefore, as a class, have a different effect on the shift of the beta $=CH_2$ group. Both resonance and induction could affect the inter-atomic current as well as the local diamagnetic and paramagnetic currents, making the total effect quite complex. Hydrogen would not be expected to enter into resonance interaction at all, and would therefore not be expected to fit the pertinent correlation.

3. For Aromatic Rings

Figure 50 shows the shifts of aromatic ring hydrogens as functions of electronegativity and covalent radius of the substituent. The trends for the hydrogens ortho to the substituent are like those for $=CH_2$ groups. The trends for the meta hydrogens are also similar but much less pronounced. Trends for the para hydrogens, on the other hand, follow the meta hydrogen trends for the larger and less electronegative atoms, but curve toward the ortho hydrogen trends for the smaller and more negative atoms.

There is a separation of the shifts of the three types of hydrogens in some cases but not in others. When separation occurs because the ortho hydrogen band has moved downfield (to higher delta values), the meta and para hydrogen bands stay upfield and close together.

When separation is due to movement of the ortho hydrogen band upfield (to lower delta values), the para hydrogen band moves with the ortho band, leaving the meta band alone downfield. It is this difference in motion of the para hydrogen band that makes the curve in the trend of the para shift.

These trends can be interpreted as the combined effects of induction and resonance, as for the olefinic hydrogens. If the resonance interaction is assumed to produce a charge redistribution around the aromatic ring, affecting the ortho and para positions much more than the meta position, then the trends in both band movement and band separation are expected. Induction withdraws charge more strongly from the ortho than from the para position, accounting for the greater downfield shift of the ortho band. When charge is donated by resonance, the donation is opposed unequally by induction so that the net donation to the ortho and para positions is about the same. These charge changes may also affect the ring current (see Figure 2) and thus affect the shifts of all the hydrogens, including the meta. It is difficult to determine the contribution of each of the possible causes to the final shifts.

The shifts caused by fluorine fit the meta trends but not the ortho and para trends. The shifts caused by hydrogen fit the electronegativity correlation for meta and para shifts but do not fit the atom size correlations for any of the shifts. These anomalies can be attributed to a low degree of resonance interaction between these atoms and the ring.

Many studies have shown that these aromatic hydrogen shifts correlate quite well with the Hammett or Taft substituent constants, showing that the shifts are an indication of the chemical reactivities of the various hydrogens.

B. CHEMICAL SHIFT VS. HYBRIDIZATION AND SUBSTITUTION OF THE PRINCIPAL SUBSTITUENT ATOM

The discussion in Section A was concerned with substituent atoms which, when substituted themselves, were all connected by single bonds to alkyl groups. It is also helpful to see what happens when the attached atom of the substituent is connected by a variety of bond types to a variety of nonalkyl substituents. The more important trends of this type are shown in Figure 51. This illustration is divided into horizontal sections which show the shift trends due to changes in the substituents on carbon, phosphorus, sulfur, nitrogen, and oxygen. It is further divided into columns which

show the effects these substituent changes have on the shifts of aromatic, nonterminal olefinic, terminal olefinic, and chain methyl hydrogens.

1. For Chain Methyls

Line 1 of the far right (fourth) column (Figure 51) shows that the shifts of methyls alpha (crosshatched bar) and beta (solid bar) to CH_2-R in a saturated chain are identical. As the CH_2-R is changed to C=C the alpha methyl band moves downfield, but the beta methyl band is essentially unaffected (line 2). As X in C=X changes from C to N to O the alpha methyl band moves farther downfield (lines 2–4). This increasing shift is in the order of increasing electronegativity. The beta methyl band remains unaffected. Changing from carbonyl to carboxyl produces no further shift (line 5). Changing to C=S, however, produces further downfield shift of both the alpha and beta bands. This is similar to the data of Figure 46, where, although carbon and sulfur have the same electronegativity the shift produced by sulfur is greater than that produced by carbon.

Lines 7 to 9 of this same column show that both the

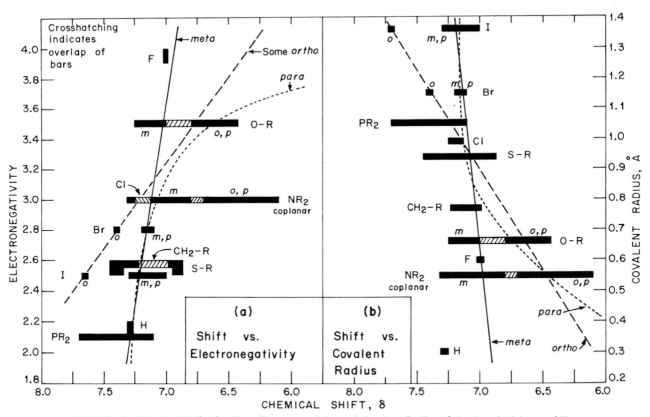

FIGURE 50. Chemical Shift of Φ-X vs. Electronegativity and Covalent Radius of the Attached Atom of X

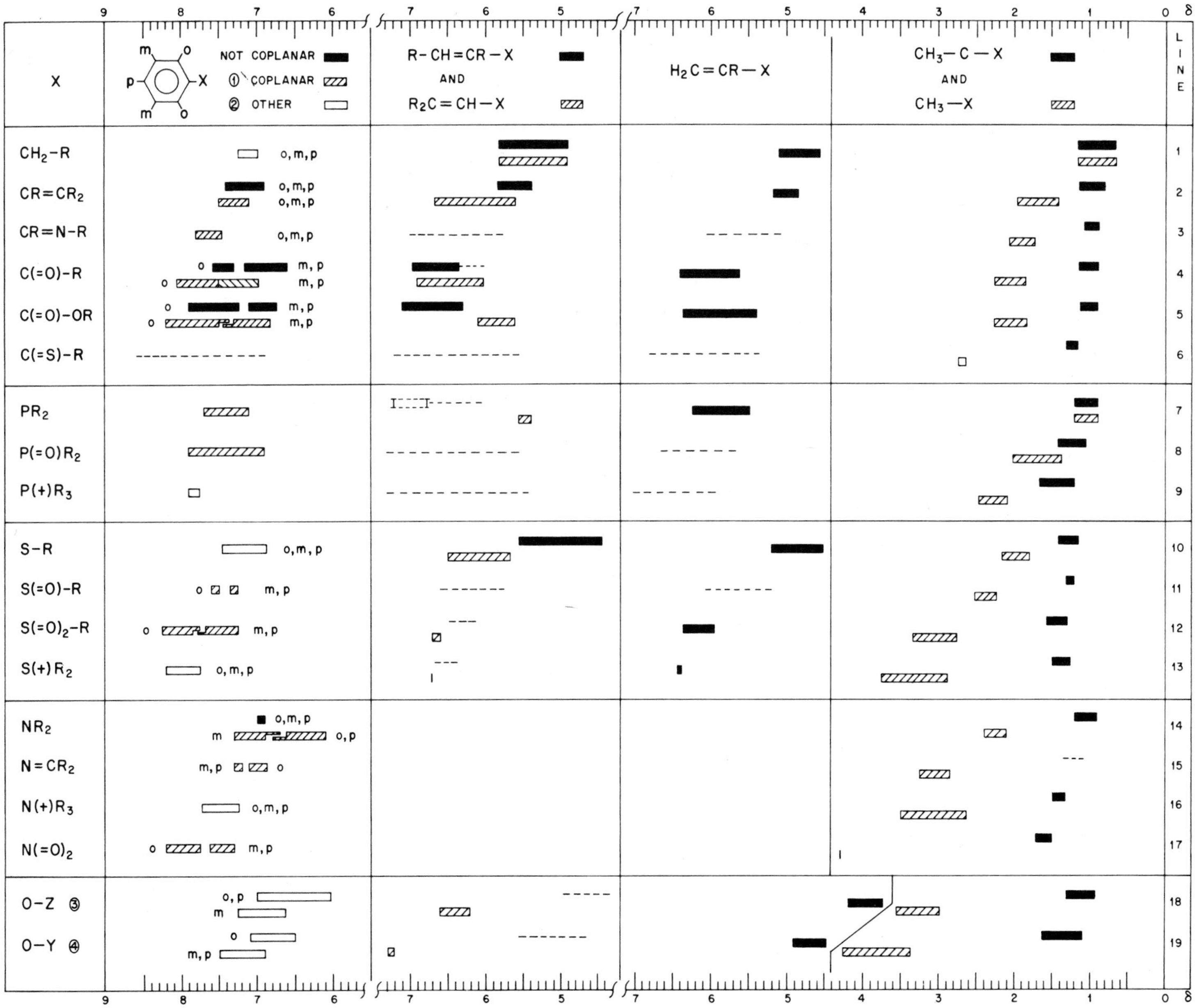

FIGURE 51. Some Important Chemical Shift Trends

1. Coplanar means the π electron systems or the p and π systems of the ring and substituent are oriented so as to approach maximum mutual resonance interaction.

2. The concept of planarity seems inappropriate in these cases. The resonance interaction may be weak or it may be little affected by orientation, or orientation effects may not be observable in the available examples.

3. Alcohols, aliphatic ethers and phenols (Z = H, alkyl). Includes, as exceptions, $CH_3-O-SO-R$ and $R_2C=CH-O-P(O)(OR)_2$.

4. Esters of carboxylic, phosphorus and sulfur acids $(Y = CO-R, CS-R, P(OR)_2, P(O,S)(OR)_2, SO_2-R, SO_3-R)$. Includes alkyl part of alkyl aryl ethers $(CH_3-O-\phi)$.

alpha and beta methyl shifts increase as the phosphorus substituent changes in the order P < P=O < P+. The positive charge is more effective than the (=O) in producing shift.

Lines 10 to 13 show similar trends for the sulfur groups. Shifts of alpha methyls are increased by the addition of (=O) or a positive charge to the sulfur. Furthermore, the addition of two doubly-bonded oxygens produces more shift than the addition of one. All the sulfur substituents cause a significant shift of the beta methyls with the SO_2 group producing the greatest shift.

The same methyl shift trends are noted for nitrogen substituents (lines 14–17). The addition of (=X) or a positive charge to the nitrogen increases the shift of both the alpha and beta methyls. For the alpha methyls, NO_2 produces greater shift than N+, whereas SO_2 produced less shift than did S+. The two sets of substituents produce shifts in the same order for the beta methyls, however. No simple rationalization is offered for this phenomenon at this time.

Lines 18 and 19 show that the change from alcohols or ethers to esters is accompanied by a downfield shift of about 0.5 ppm for alpha methyls and about 0.25 ppm for beta methyls. Similar shifts are observed for methylenes, also [Chart S6(18 and 19)]. This is helpful in differentiating these classes of compounds, and in detecting the esterification of alcohols during acid treatment.

2. For Olefin Groups

The data of column 3, through sparse, indicate that the shift trends for terminal olefinic hydrogens may be similar to those for methyls. A major exception is the effect of directly linked oxygen (lines 18 and 19). Here the bands are at higher field than for any other substituent. The band for vinyl esters is still at lower field than that for vinyl ethers, however.

In column 2 (lines 1 to 5), it is shown that the effects of the various carbon groups on the shifts of the beta nonterminal CH= (solid bars) are similar to their effects on the alpha methyl (column 4). Their effects on the shifts of the alpha =CH are substantially different, however. The result is that as the substituent changes through the C=X series the alpha and beta bands approach each other and then change relative places. This can be attributed to resonance interaction.

Line 7 indicates that PR_2 shifts the beta =CH more

than the alpha, but the data are so sparse this conclusion may not be warranted.

Lines 10 to 13 indicate that shifts due to the sulfur substituents increase steadily down the list, with shifts of alpha =CH always being greater than those of beta =CH. Here again, however, the data are sparse and the conclusion may be premature.

Lines 17 and 18 show that the change from alcohol or ether to ester causes a downfield shift of the alpha =CH. It is thought to cause a downfield shift in the beta =CH, also, by analogy to the terminal beta shifts (column 3). Note that for each of these oxygen substituents the extreme downfield position of the alpha band and the extreme upfield position of the terminal beta band makes the widest separation of vinyl group bands observed, about 2.5 ppm. This separation is so wide it is sometimes difficult to recognize these bands as arising from the same vinyl group.

3. For Aromatic Rings

In order to rationalize the shifts of aromatic ring hydrogens it is convenient to utilize the concept of coplanarity. This concept assumes that the degree of resonance interaction between the ring and a substituent having electrons in p or π orbitals depends on the degree of coplanarity of these orbitals and the π orbitals of the ring electrons. Then the difference in shift between coplanar and noncoplanar conformations should indicate some of the effect of resonance interaction on the shift. The coplanarity concept further assumes that if the substituent is free to rotate it will always assume a mean position such that the resonant orbitals are coplanar. Some kind of steric hindrance is necessary to force noncoplanarity. This should provide an objective way to determine whether substituents are coplanar with the ring or not.

The data of column 1 of Figure 51 are divided according to whether the substituent is free (assumed coplanar) or hindered severely (assumed noncoplanar) by additional substituents in ortho positions. These divisions were made with the aid of molecular models. A third category (other) is used when the data are not clear as to coplanarity or the substituent should not have any p or π orbitals. The data do correlate according to steric hindrance and therefore are consistent with the concept of coplanarity.

Lines 1 to 5 show that when the C=X substituent is coplanar (crosshatched bars) the shift of all the ring hydrogen bands is greater than when the substituent is not coplanar (solid bars). Line 14 shows that when NR_2 is coplanar the bands are separated and the shifts cover a broad range, but when it is not coplanar the bands are not separated and the shifts cover a narrow range.

The shifts increase in the order of listing of the substituents, as they do for most of the groups shown in Figure 51. Note that CO, SO, SO_2, and NO_2 all move the ortho band to low field, whereas $N=CR_2$ and O—Y (esters) move it to high field. NR_2 and O—Z (phenols and ethers) move both the ortho and para bands to high field. There is very little difference between the shifts of phenols and phenyl esters, and these compounds cannot be reliably distinguished by ring hydrogen shifts alone at this level of precision.

C. SHIFT RANGES FOR FUNCTIONAL GROUPS

It is helpful in forming a preliminary opinion about a spectrum to have in mind the broad ranges occupied by the shifts of the major classes of functional groups. These ranges may be divided as follows:

Shift Ranges of Major Groups, δ

Saturated groups (CH_3, CH_2, CH)

In paraffins	1.94–0.65
In substituted paraffins	8.15– (–0.41)
In cycloparaffins	2.16– (–0.28)
In substituted cycloparaffins including nonaromatic heterocyclics	6.09–0.22

Olefinic groups ($=CH_2$, =CH)

In acyclic olefins	5.88–4.42
In cyclic olefins	7.20–4.95
In substituted olefins, including CH=X	11.00–4.40
Acetylenic hydrogen (≡CH)	4.00–1.74
Aromatic ring hydrogen	
In alkyl benzenes	7.37–6.60
In polyphenyls, cyclophanes, arom. olefins and other substituted aromatics	8.06–4.05
In condensed ring aromatics	9.43–6.85
Thiolic hydrogen (SH)	3.62–0.64

Exchangeable hydrogens (OH, COOH, NH)

Not intramolecularly H-bonded	14.0–0.0
Intramolecularly H-bonded	17.7–2.55

These data are all presented in greater detail in Chart I1.

From these data it can be seen that the shifts of paraffins, cycloparaffins, acetylenic H, and thiolic H are in the high-field end of the spectrum, $\delta = 0$–4. The shifts of olefinic hydrogen in olefinic hydrocarbons are in the range $\delta =4$–7, and those for ring hydrogen in common aromatic hydrocarbons are in the range $\delta = 6.6$–9.5. The shifts of exchangeable hydrogens cover most of the hydrogen spectral range, with the greatest shifts of all being due to intramolecular hydrogen bonding. Considerable overlap of the ranges is introduced by substitution of the hydrocarbons.

When substitution, position, and steric factors are equal, the shift of a methylene will be greater (higher δ) than that of a corresponding methyl. This is indicated by the simpler sections of charts such as 18, 38, and 45. Isopropyl methines will have still greater shifts, but nonisopropyl methine shifts may substantially or completely overlap those of the methylenes. This overlap is shown in Chart 18 (3, 13, 17).

The shift ranges of acetylenic and of thiolic hydrogens are narrow enough to be helpful in identifying these hydrogens.

Exchangeable hydrogens obviously cannot be identified by chemical shift alone. Means for identifying them are discussed in Chapter 2, Section IIB-3.

The most extreme *upfield* hydrogen shifts are produced by hydrogens held inside the large rings of annulenes [Chart 13(16, 17)] and porphyrins. The shifts of annulene hydrogens at −60°C have been observed at the extreme values of $\delta = -3$ or $\tau =13$. Porphyrin shifts have been observed to $\delta = -4$ or $\tau = 14$. Scarcity of data prevents an estimate of the range of porphyrin hydrogen shifts.

D. EFFECT OF STERIC FACTORS ON CHEMICAL SHIFT

1. Cis-Trans Isomerism

In many cases the chemical shift indicates with a high degree of confidence whether the compound under study is the *cis* or *trans* isomer. In other cases the shift is different for the two isomers, but the difference is too small to be diagnostic or the data are inadequate to establish the ranges with the necessary precision. Illustrations of these two cases are found in the charts listed below.

Shift Diagnostic for Isomer

Olefins:	S16(21–24, 29, 30–35, 38–43)
	S17(8–11, 28–29, 38–39)
	S18(6–7, 16–25, 29, 30)
	S21(3–4, 6–7)
Nitrosamines:	S3 (22, 23)
	S7 (16, 17)
Acyclic Imines:	S19(14–15, 17 and 19, 46–47, 49–50, 52–53, 56–57)

Shift Not Diagnostic at Present

Olefins:	S16(11–14, 25–28, 37, 44–46)
	S17(1–7, 30–38, 40–58)
	S18(1–5, 8–15, 26–28, 31–35)
	S19(45, 48, 51, 54, 55, 58)

2. Nonequivalence

If a methylene group is free to spin, and if both hydrogens pass through exactly the same magnetic environments during the spin, the shifts of the two hydrogens will be identical. If, however, the two hydrogens experience different magnetic environments, either because of restricted spin or because the spinning environments are not identical, they will not have the same chemical shifts (except, occasionally, accidentally). In the latter case the hydrogens are said to be nonequivalent. The same situations applied to geminal methyl groups results in equivalent or nonequivalent methyls.

The configurations that produce magnetic asymmetry are illustrated in Figure 52. The prime requirement is that the atom to which the methylene is attached have a magnetically asymmetric environment. In the case of a carbon atom, this requires that the carbon have three different groups attached to it besides the methylene group. This is shown in Figure 52(a) and (b). In (a) is an extended view of this configuration, and in (b) is a view of the same configuration looking along the carbon-carbon bond. In (b) the methylene is represented

(a)   (b)

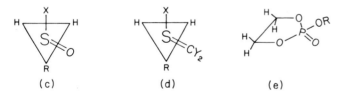

(c)   (d)   (e)

(a) to (e) Magnetically asymmetric groups

 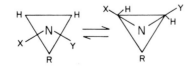

(f) Magnetically   (g) Magnetically symmetric when   **FIGURE 52. Magnetic Asymmetry**
    symmetric   nitrogen inversion is fast

by a triangle to differentiate it from the asymmetric carbon.

It must be emphasized at this point that a magnetically asymmetric carbon is not necessarily optically asymmetric. Optical asymmetry requires four different substituents on the same carbon, but magnetic asymmetry requires only three. One of the three can be identical with the methylene and its attached R group. It is, however, necessary that the R group not be hydrogen in the case of a methylene or not methyl in the case of geminal methyls. This R group might be considered to serve the purpose of the fourth group on an optically asymmetric atom, although this may be too simplistic a view.

Figure 52(c) shows the asymmetric sulfoxide configuration. Here the asymmetry is supplied by the angle of the oxygen. The empty space at the third position is just as effective in changing the magnetic field in that region as another substituent would be. This same reasoning can apply to the olefinic representation of sulfonium ylides (d).

Figure 52(e) shows nonequivalence caused by different environments in a rigid system. Here one methylene hydrogen sees the (=O) whereas the other sees the OR group. These two produce quite different magnetic environments.

The carbonyl group, unlike the sulfoxide group, has its oxygen aligned with the other substituent when viewed along the bond to the methylene. This results in a magnetically symmetric environment and the methylene hydrogens are equivalent.

The nitrogen atom is inherently asymmetric, but its rate of inversion is usually so fast that it appears symmetric to attached methylenes. This inversion is illustrated in Figure 52(g). If the inversion can be slowed by substitution or by cooling, nitrogen compounds should show nonequivalent hydrogens in appropriate cases.

Even though a configuration should produce magnetic asymmetry, the resulting nonequivalence in the chemical shifts may or may not be large enough to detect in the actual spectrum. It is helpful, therefore, to know when nonequivalence is usually observed. The configurations which almost always produce observable nonequivalence are shown in the charts indicated below.

Examples of Chemical Shift Nonequivalence

Paraffins	4(a to e)
Amides	S7(18), S12(7,8), S14(20,21)

Sulfites	S10(16), 74(7,8)
Cyclic sulfites	S20(29,30) and cyclic phosphates
Sulfonium ylides	S7(4), S12(19), S13(14,15)
Sulfoxides	S11(13,14), S13(19–21)

3. "Crowding" and "Shielding" in Paraffinic Chains

When two bulky groups such as *tert*-butyl or geminal methyl are separated by a single methylene the methyls on the bulky groups are, according to Stuart and Briegleb molecular models, forced to mesh together somewhat like the teeth on meshing gears. This "crowding" appears to hinder rotation of both bulky groups and may cause some distortion of the bond angles. It causes the bands produced by the methyl groups to shift downfield from their normal positions. This phenomenon is shown in Chart 3(a) vs. (b) and (d) vs. (e) and in Chart 5(a) and (b) vs. (c) and (f) and (g) vs. (h).

"Crowding" also causes a downfield shift of the band produced by a methylene between internal geminal methyl groups, as shown in Chart 6(c). Extreme crowding, as in polyisobutylene, causes extreme downfield shifting of both methyl and methylene bands, as shown in Chart 5(f) and Chart 6(c). If the methylene is between an isopropyl and a single methyl branch, however, its band will be shifted upfield, as in Chart 6(f).

If one of the "crowded" groups is an isopropyl or single methyl branch, the band produced by its methine is shifted downfield, as shown in Chart 7(a) and (b) vs. (d), and (h) vs. (g). On the other hand, a methine between the "crowded" groups will produce a band which is shifted upfield, as in Chart 7(f) vs. (g).

The effect of crowding on chemical shift may be due to the combined effects of the various bond anisotropies, modified by changes in rotamer populations and electron cloud distortions. This is purely speculative, however, because the actual causes of chemical shift in paraffinic groups have not been worked out.

If the two bulky groups are attached directly together instead of through an additional carbon, there is little hindrance to rotation and no significant bond angle distortion (again according to Stuart and Briegleb models). Under these conditions the methyl and isopropyl methine bands are usually shifted upfield, as though the groups had acquired additional magnetic shielding. This "shielding" effect is shown in Chart 3(c) vs. (b) and (f) vs. (e); in Chart 5(d) and (e) vs. (c), and (i) to (k) vs. (h);

and in Chart 7(e) vs. (d). Nonisopropyl methine bands, on the other hand, may be shifted either upfield or downfield, as shown in Chart 7(f), (i), and (j) vs. (g). Bands produced by single methyl branches are not shifted significantly unless there are shielding groups on both sides, as indicated by Chart 3(c) and note 3 to (b).

The effect of "shielding" can be speculatively attributed to the combined effect of the various bond anisotropies, without the modifications produced by crowding. The anomalous behavior of nonisopropyl methines and single methyl branches is not accounted for.

4. Steric Effects in Saturated Ring Systems

The effect of steric factors on the chemical shifts of the hydrogens of saturated ring systems is so great that it prevents useful correlation of the data from many of these systems. In the case of nonrigid systems such as cyclohexane, the rate of interconversion between allowed conformations can be markedly altered by substitution. A slowing of the interconversion rate reduces chemical shift averaging among the various hydrogens and greatly broadens the spectrum. Similar broadening of the spectrum accompanies the change from nonrigid to rigid systems, as in the change from *cis* to *trans*

decalin. Also, in the rigid systems long-range interactions which do not normally take place through bonds can take place through space, adding considerable complication to the spectra. The net result is that shifts due to steric factors are of the same magnitude as many of the shifts due to substitution, and it becomes very difficult to sort them all out. It is for this reason that no correlations are presented in this book for rigid multiring saturated systems, and few are presented for single-ring saturated systems.

The only attempt at detailed correlation for a cycloparaffin is presented in Chart 11. The data for methyl groups show that adjacent groups on the same side of the ring cause the methyl bands to shift to significantly higher field. It appears that two adjacent *cis* groups have twice the effect of one group. The data for ring CH_2 hydrogens show that when the ring is substituted with a single linear chain the ring hydrogens produce two well-separated bands. It has not been possible to assign specific hydrogens to each band, however.

E. CHEMICAL SHIFT VS. RING SIZE

The chemical shifts of ring hydrogens in nonaromatic systems vary with ring size. This variation is shown

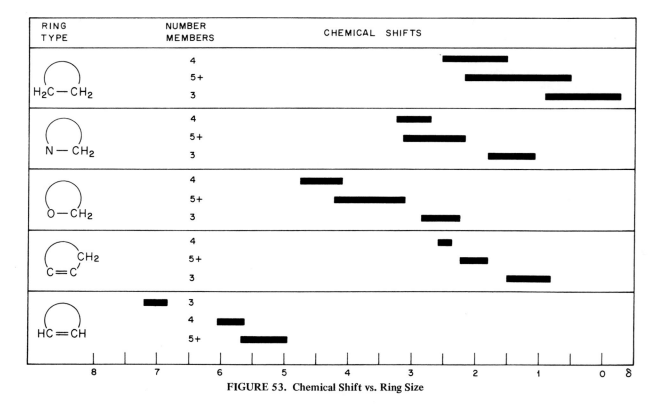

FIGURE 53. Chemical Shift vs. Ring Size

in Figure 53. For saturated systems the shift increases (higher delta values) in the order $3 < 5+ < 4$. For unsaturated systems the shifts of the alpha CH_2 groups increase in the order $3 < 5+ < 4$, but the shifts of the olefinic hydrogens increase in the order $5+ < 4 < 3$.

Although the examples given in Figure 53 are principally for groups alpha to a heteroatom, the trend seems to be general. It applies to groups alpha to external substituents and to groups beta and farther from the substituents or heteroatoms. More detailed data on these and other examples are shown in Charts S20 and S21.

In view of the many factors which can affect the shifts of hydrogens associated with nonaromatic ring systems, it may seem remarkable that there is an overriding correlation with ring size. It is fortunate, however, because it can be used to detect small rings.

F. STRONG INTRAMOLECULAR HYDROGEN BONDS

When an OH or NH hydrogen can form a semirigid five- or six-membered ring system with a double-bonded oxygen (such as in CO, SO_2, NO_2, etc.) or with various configurations of nitrogen, the ring system can be expected to close with a strong hydrogen bond. Such systems are shown in Charts S34 and S35.

The resonances of such strongly H-bonded hydrogens are at lower field than they would be if the hydrogen bond did not exist, and in the extreme cases they have the lowest field positions observed in this study. These hydrogens tend to exchange very slowly, and therefore produce sharp bands in many cases. The addition of acid tends to broaden the bands, probably by increasing the exchange to an intermediate rate. In at least the extreme cases, however, the H-bonded OH and acid OH bands do not coalesce. These OH bands behave like those of severely sterically hindered OH's. Because of these characteristics such bands are indicative of the types of structures shown in S34 and S35.

The ring system in which the OH or NH to be bonded is located must be made at least semirigid by two double bonds as shown in S34(1–4 and 22,23) and S35(2, 4, 22), or a ring and one double bond as in the remainder of the two charts. The systems are usually conjugated, so that the hydrogens of interest can be considered vinylogs of acidic or amide hydrogen, and it is assumed that this vinylogy contributes to the enhanced shift. Vinylogy does not account for all the shift or for the slow exchange rate in the case of the acid vinylogs, however.

CHAPTER 5

Coupling Constant Correlations

I. INTRODUCTION

The coupling constant seems to have been neglected somewhat in analytical work, partly because it is less useful than the chemical shift but partly because comprehensive correlations have not been available. It is hoped that this chapter will correct the latter deficiency, and will encourage the routine use of the coupling constant.

The coupling constant has not proved as generally useful as the chemical shift because couplings are less informative and can be measured less frequently than shifts. Nevertheless, coupling constants do provide important information about spatial positions and conformations, about the nature of the structural group in which the coupled atoms are located, and about some substituent atoms or groups. This information corroborates or supplements structure information provided by the shifts. It can resolve ambiguities in the shift correlations and can provide leads that are not readily obtainable from the shifts. Therefore, it is recommended that coupling constant data *always* be obtained when possible and used as a vital part of the structure determinations.

There are many factors that affect the coupling constant and their interaction is complex. These are mentioned in Chapter 1, Section IID, and are reviewed in detail by Bothner-By[T5] and Sternhell.[T40] These factors are fairly well understood so that theory is better at predicting coupling constants than chemical shifts, but the predicted values are still not good enough to use in structure determinations. Consequently, analytical work must depend on empirical correlations for coupling constants as it does for chemical shifts.

Coupling constants can be determined readily only for multiplets subject to first-order interpretation. This does not mean that the multiplets must *be* first order, only that it must be possible to impose a first-order interpretation on them. For multiplets not subject to first-order interpretation, the coupling constants must be calculated. The limitations and expense of such calculations make them justifiable only in special analytical problems, but when they are needed they can be decisive.

Coupling constants cannot be measured when the multiplicity is not observed. There can be a lack of observed multiplicity because there is no multiplicity, as with singlets, or because it is not resolved. Resolution of the multiplicity can sometimes be improved adequately by spin-decoupling some of the coupled nuclei. At other times computer techniques for enhancing the resolution will be effective.[T20]

The coupling constant data presented in this chapter include both those which were calculated and those which were measured as separations between adjacent peaks in first-order interpretations. The latter method of measurement may widen the ranges somewhat, but this is considered a small price to pay for the greater speed and lower cost of measurement. The resulting ranges are still useful for analytical work.

The data are presented in bar graphs to provide rapid access and to show the width of and the relationship between the ranges. These points were discussed in detail in the introduction to Chapter 4. The charts are presented at the condensed summary (index) and summary levels only. These two provide rapid access to the coupling constant which is characteristic of a given structural group or to the structural groups which will be characterized by a given coupling constant. Because coupling constants are not functions of compound types as are chemical shifts, detailed charts like those provided for chemical shifts are inappropriate.

The coupling constants for most organometallic compounds and for condensed ring saturated compounds have been omitted because the chemical shifts for these compounds were omitted. Some $H-C^{13}$ couplings in organometallics are shown when the metal atom materially affects this coupling, but the hydrogen-metal couplings are not shown. The couplings in condensed ring saturates are even more difficult to measure than are the chemical shifts. They would be of limited utility in general structure determinations. Those who need such information are referred to the data tabulated by Bothner-By.[T5]

II. DESCRIPTIONS OF THE CHARTS

A. THE INDEX CHART (I2)

Chart I2 presents an overall view of the coupling constants, to provide rapid orientation of the search and to serve as an index to the summary charts. It is divided into sections which present hydrogen–hydrogen, hydrogen–carbon-13, hydrogen–fluorine, hydrogen–nitrogen, and hydrogen–phosphorus couplings. Each section is further divided into subsections which present the data for aliphatic, aromatic, and hetero groups. Within each subsection, the data are divided by increasing number of bonds between coupled nuclei and by increasing unsaturation of the aliphatic groups. They are presented in numbered lines for easier reference.

Each line is numbered at each end to facilitate reading across the chart. At the extreme left, under the section and subsection names, is the line structure or word definition of the coupled system. This is followed by a bar showing the range of coupling constants characteristic of this system. The values covered by these bars are determined in hertz from the scales at top and bottom of each half of the chart.

The range bar is followed, in the third column from the right, by a statement of the number of compounds from which data were obtained. This serves as a measure

of the confidence that can be placed in the range. The next column shows the summary chart numbers and lines from which the data were taken, thus providing the index entry into the summary charts.

B. THE SUMMARY CHARTS (J1 TO J38)

The summary charts present the data in considerably greater detail than does the index chart. Access time will be significantly reduced, however, if the index chart is used to gain entry to the summary charts.

The summary charts are organized in the same way as the index chart. The major sections, which may occupy more than one chart, divide the couplings according to the nuclei involved (hydrogen–hydrogen, hydrogen–carbon-13, etc.) and the subsections further divide them by aliphatic, aromatic, and hetero groups. The aliphatic groups are further divided by increasing number of bonds between coupled nuclei and by increasing unsaturation. They are then divided by chains and rings, with the rings listed by increasing size.

The aromatic couplings are grouped by increasing degree of condensation, followed by heteroaromatic rings. Each category is further divided by ortho, meta, and para couplings and some couplings between different rings. Long-range couplings across four or more bonds in all the systems are listed last.

In the summary charts the signs of the coupling constants are taken into account when this is helpful in showing trends (and the signs are known). The scales on some of the charts range both above and below zero to take care of those cases where the constant changes sign (J1, J2, J6, J8, J22).

Structure diagrams are used on the summary charts to show exactly which nuclei are coupled in each correlation, and to show the structural environment in which they are located. Sometimes the hydrogens are indicated by letter and sometimes by short lines extending beyond the borders of a ring. When the use of a reference atom is helpful, this reference is also indicated on the diagram (Charts J1, J2, J24). The referencing system is similar to that used in the chemical shift charts.

The data are presented in lines numbered at the extreme right. In general, each line includes, from the left, the structure diagram, definitions of substituents if needed, the range bar, the numerical limits of the range, the sign, and the number of examples used in determining the range. The signs followed by question marks

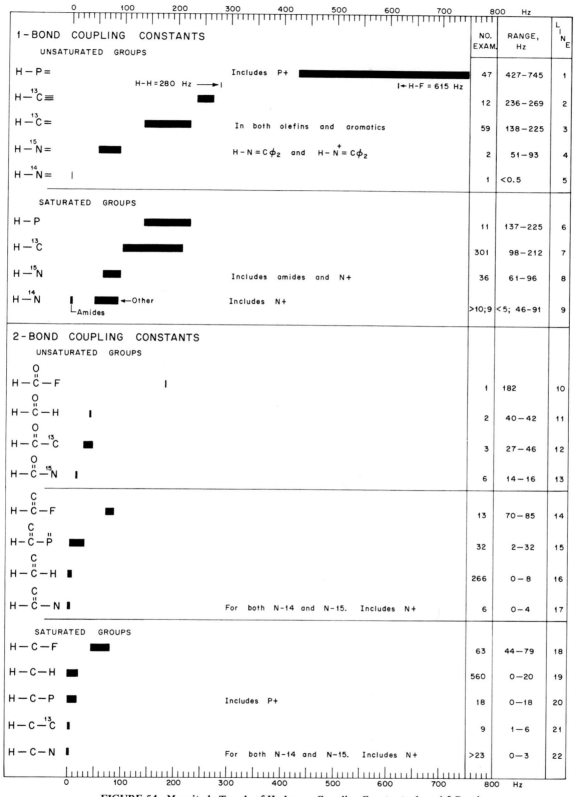

FIGURE 54. Magnitude Trends of Hydrogen Coupling Constants, 1- and 2-Bond

are estimated from trends for similar couplings. A question mark in place of a sign means the sign is not known and not estimated. In most cases hydrogen couplings are not significantly influenced by solvent, concentration, or temperature; therefore, no data on these factors are shown. The number of examples exceeds the number of compounds studied only where more than one constant of a given type was obtained from a single compound. This was not usually true except for some aliphatic chains. For this reason the number of examples shown usually is close to the number of compounds studied.

III. USEFUL GENERALIZATIONS

The separation between peaks in a first-order interpretation indicates the magnitude of the coupling constant but not the sign. The sign must usually be determined by calculation or at least by analogy to prior calculations. Consequently, it is the magnitude of the constant that is normally determined and used in analytical work. Trends in magnitude are not necessarily as clear and straightforward as trends which include the sign. Nevertheless, there are some magnitude trends which are useful analytically, and these will be discussed in the following paragraphs.

A. COUPLING CONSTANT VS. ISOTOPE

In Chapter 1, Section IID, it was pointed out that one of the factors which affect coupling constants is the nature of the isotope to which the hydrogen is coupled, and a preliminary indication of this effect was presented. It is now possible to refine this trend in the light of data presented in Figure 54 for one- and two-bond couplings and in Figure 55 for three-bond couplings. These figures have been organized to emphasize the isotope trend while showing other trends.

From the data of Figures 54 and 55 it appears that the magnitude of the constant decreases with isotope in the order $F > P > H > C^{13} > N^{15} > N^{14}$ for all the couplings. This improved order is the same as that given in Chapter 1 except for the interchange of F and P. It is also, except for hydrogen, the order of decreasing gyromagnetic (or magnetogyric) ratios, as indicated by the decreasing resonant frequencies shown in Table 10. This is in accord with coupling constant theory, which indicates that, other things being equal, the constant should vary as the product of the gyromagnetic ratios of the coupled nuclei [See Emsley et al.,[T19]].

Isotopes with spins of ½, like F, P, H, C^{13}, and N^{15}, tend to produce sharp well-defined multiplets from

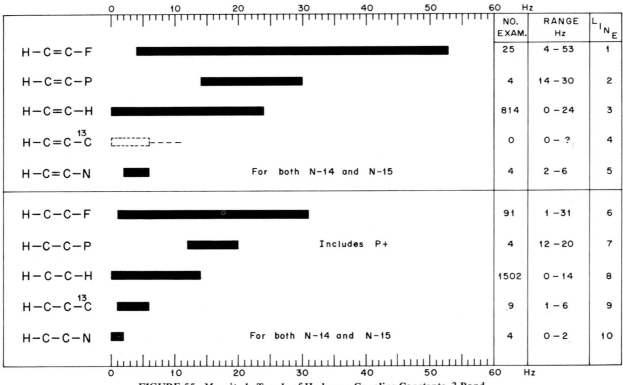

FIGURE 55. **Magnitude Trends of Hydrogen Coupling Constants, 3-Bond**

TABLE 10. Pertinent Isotope Characteristics

isotope	Resonant frequency at 23.5 kG	Spin, $h/2\pi$	Electric quadruple moment, 10^{-24} cm^2
F	94.1	½	0
P	40.5	½	0
H	100.0	½	0
C^{13}	25.1	½	0
N^{15}	10.1	½	0
N^{14}	7.2	1	7.1×10^{-2}

which the coupling constant can be measured with assurance. On the other hand, isotopes with spins greater than ½, such as N^{14} (Table 10), tend to give broad multiplets which may not be adequately resolved in many cases. Not only is the multiplicity increased by the higher spin, but the accompanying electric quadrupole moment tends to broaden the lines, making observation of the multiplicity more difficult. In fluctuating electric fields the electric quadrupole can cause complete or partial decoupling of the nitrogen spin from the hydrogen, giving an unrepresentative value for the coupling constant. All these factors combine to make the hydrogen–nitrogen-14 coupling constants difficult to measure and questionable in value in many cases. This may account for the difference in amide couplings between N-15 and N-14 [Figure 54(9)].

B. COUPLING CONSTANT VS. HYBRIDIZATION, BOND ORDER, OR BOND LENGTH

Another of the factors listed as affecting coupling constants in Chapter 1, Section IID, is bond order or length. This effect can be seen in Figure 55, lines 1–5 vs.

TABLE 11. Coupling Constant vs. Bond Order

Group	Magnitude, Hz	Number examined	Chart
H–C–C–H	4.2–7.8	279	J4(1,2)
H–C≡C–H	9.5–9.8	2	J9(4)
H–C–C^{13}	1.1–6.0	9	J22(1,2)
H–C=C^{13}	0.8–17.5	8	J22(5–7)
H–C≡C^{13}	41–61	11	J22(9–11)
H–C–N^{15}	0.0–1.4	6	J31(5,7)
H–C=N^{15}	2.4–16.3	24	J31(14,15)
H–C–C–N^{15}	0.7	1	J32(2)
H–C–C≡N^{15}	1.7	1	J32(4)

6–10. The coupling which takes place through higher-order bonds (double bonds) has larger constants than that which takes place through single bonds. Additional examples of this trend are listed in Table 11.

Figure 54 shows that the constant also increases when one of the coupled atoms or an intervening atom is hybridized, even if the higher-order bond is not in the direct path of the coupling. Compare the corresponding lines of sets 1–5 vs. 6–9 and 10–17 vs. 18–22. Other examples of this trend are shown in Table 12.

TABLE 12. Coupling Constant vs. Hybridization

Group	Magnitude, Hz	Number examined	Chart
H–C–^{13}C	1.1–6.0	9	J22(1,2)
H–C–^{13}C=	5.8–7.5	10	J22(3)
H–C(=)–^{13}C	26.6–46.3	3	J22(4)
H–C–^{13}C≡	10.0–11.4	5	J22(8)
H–C(=)–^{13}C≡	32.8	1	J22(12)
H–C–O–H	4.6–8.5	11	J15(2,3)
H–C(=)–O–H	11.7	1	J15(4)
H–C–O–C–H	0	36	J15(5)
H–C(=)–O–C–H	0.5	4	J15(6)
H–C(=)–O–C(=)–H	0.6	1	J15(7)
H–C–N–H	3.8–9.0	70	J16(2)
H–C(=)–N–H	1.7–13.6	10	J16(4)
H–C–S–H	5.8–8.9	43	J18(2)
H–C(=)–S–H	8.1–10.0	2	J18(5)

Long-range couplings over four or more bonds are usually the result of unsaturation of some of the intervening atoms, as indicated in Charts J9(5–17), J10(1–12), J14(1–8), J22(13–16), J23(9–11), J27(8–15), J28(14–16), and J38(6–21).

Exceptions to the trend of increasing constant with increasing unsaturation are exhibted by the H–C–C–H series, J4 and J5 vs. J6 and J9(1–3), and the H–C–C–F series, J26 vs. J27(1–4 and 7). In both of these series unsaturation of an intervening carbon with the double bond outside the direct line between the coupled atoms appears to have no significant effect on the coupling constant. It should also be noted that *cis-trans* isomerism modifies this trend by making the *cis* coupling constant less than the constant for the saturated system in most cases.

C. COUPLING CONSTANT VS. SUBSTITUENT ELECTRONEGATIVITY

Increasing electronegativity of an adjacent group should, as stated in Chapter 1, increase coupling constants in the positive direction. If the sign of the constant is positive, its magnitude (peak separation) will increase, but if the sign is negative its magnitude will decrease. Where the signs of the constants have been determined or inferred, this trend has been confirmed by the data presented in this chapter for one- and two-bond couplings but not for three-bond couplings.

1. One-Bond Coupling

Douglas[T18] found that $H-C^{13}$ coupling constants for monosubstituted methanes (positive sign) increased linearly with increasing electronegativity of the attached atoms. A different line was required for the atoms of each atomic period, however, indicating that other factors, such as bond length, were also active. This trend is shown roughly in Chart J19(2), where the increasing constants have ordered the metal atoms by increasing electronegativity for each period. Lines 7 to 17 also show a steady increase in constant with increasing electronegativity of the substituent, but the trend is somewhat obscured by the effects of hybridization of the substituents. Lines 3 and 17 show the constant increasing with the increasing electronegativity due to larger numbers of attached halogen atoms.

A similar trend for the H–P coupling constant, also positive, is indicated by Chart J33(7). The electronegativity of the substituent atoms increases in the order R = S < O, and the constant increases with increasing numbers of the more negative oxygen atoms.

2. Two-Bond Couplings

H–C–H (geminal) couplings have a negative sign and should therefore *decrease* in magnitude as the electronegativity of the substituent *increases.* This effect can be seen in Chart J1(1,2). Line 1 shows that the magnitude decreases for the more strongly electronegative halogens and oxygen compared to the less negative carbon and sulfur. It decreases still more when two chlorines are added to a methylene group. Line 2 shows that the addition of the more negative nitrogen decreases the magnitude of the constant compared to the less negative carbon and sulfur.

The same trend is shown for the small ring compounds in J1(5, 7, 11, and 12). It is shown even more clearly in the rearrangement of these data in Figure 58, in Section E of this chapter.

The H–C–F couplings are reported to be positive [Bovey,[T6] p. 228], and Chart J25(2–5) shows that their constants increase with increasing electronegativity as expected. Line 2 establishes the base case where attached atoms are carbon or hydrogen. Line 3 shows additional spread which is due to a number of factors, among which are the addition of Br or Cl. Line 4 shows the increase attributable to a second fluorine along with the additional increase due to the addition of oxygen. Line 5 shows the increase due to two additional fluorines.

The trend for the positive[T33] H–C–P couplings is not so clear. Chart J34(1, 3, and 5) shows the expected increase in magnitude when the more negative halogens, nitrogen, or oxygen are added. Lines 6, 7, and 9, however, are inconclusive.

Line 8 of J34 shows that the negative[T33] H–C–P+ coupling decreases in magnitude with increasing electronegativity, as expected.

Chart J31(11) indicates that the $H-C(=)-N^{15}$ coupling, of unknown sign, increases markedly with increasing electronegativity of the atom double bonded to the carbon. A similar trend for several constants is indicated in Figure 54(10–17).

3. Three-Bond Couplings

The effect of electronegativity on H–C–C–H (vicinal) couplings is too small to observe in the generalized data of Chart J4. Chart J9(1), however, indicates that the H–C(=)–C(=)–H coupling, which is thought to be positive, *decreases* in magnitude with increasing substituent electronegativity, the opposite of prediction. Similarly, Chart J26(1, 4, 6, 8, and 9) indicates that positive H–C–C–F couplings [Bovey,(T6) p. 228] decrease in magnitude with increasing electronegativity of the substituent,but line 2 indicates the opposite or is inconclusive. Chart J35(1) indicates that the positive H–C–C–P coupling (Bovey,(T6) p. 242] increases with electronegativity, as expected, but line 3 is inconclusive. It appears that the data in these charts are not adequate to confirm the trend of three-bond couplings with electronegativity.

D. EFFECT OF STERIC FACTORS ON THE COUPLING CONSTANT

1. Cis-Trans Isomerism

As was true for the chemical shifts, in many cases the coupling constant indicates with a high degree of confidence whether the compound under study is the *cis* or *trans* isomer, or whether the coupled atoms are *cis* or *trans* to each other. In other cases the difference in the constants is too small to be diagnostic. Illustrations of these two cases are found in the charts listed below:

Coupling Diagnostic for Isomer or Atom Positions

H—C=N[15]	J31(14,15)
H—C—C—H in rings	J4(10–12, 14–16), J5(7,8 in some cases)
H—C—C—F in rings	J26(10,11)
	In six-membered rings the coupling indicates the axial-equatorial relation of the coupled atoms
primarily	J5(7,8), J26(12)
H—C=C—H	J7(1–23), J8(1–7)
H—C=C—F	J27(5,6)
H—C=C—N	J32(5,6)
H—C=C—P	J35(5–8)
H—C=N(+)—H	J16(7)
H—C(=)—N—H	J16(4)
Long Range	J15(10), J17(2), J38(15,20)

Coupling not Diagnostic for Isomer

H—C—C—H in rings	J4(13,15,16), J5(1–9)
H—C(=)—C—H	J6(2)
Long range	J9(9,10), J10(1,2), J18(9), J27(11–14), J38(19)

It is important to note that while the coupling constant clearly indicates *cis-trans* relationships between hydrogens on three-membered rings, it does not indicate them clearly for hydrogens on larger rings.

2. Basal-Axial Couplings in SF₅ Compounds

The sulfur pentafluorides have a spatial structure consisting of four fluorines in a plane perpendicular to the nonfluorine substituent and a fifth fluorine in line with the nonfluorine substituent, as shown in Chart J29(1–8). The planar fluorines are called "basal" and the remaining one is called "axial." The data available on these compounds indicates that the substituent hydrogens couple to the basal fluorines but not to the axial one. This provides an unequivocal method of differentiating the axial and basal fluorines.

3. Nonequivalence

Nonequivalent methylene hydrogens (see Chapter 4, Section IIID-2) usually have different coupling constants to the same neighboring atom. In other words, geminal hydrogens which are not equivalent in chemical shift are also nonequivalent in coupling constants. As with nonequivalent shifts, however, the nonequivalence of the couplings may or may not be observed in a particular spectrum. It is helpful, therefore, to know the circumstances in which the nonequivalence is observed. The examples shown in these charts are listed below:

Observed Nonequivalent Couplings

H—C—C—H	J4(6–11), J5(7,8)
H—C—C—F	J26(8,9,12)
	In J26(9) it is the geminal fluorines which are nonequivalent.
H—C—O—P	J36(7,8)

In addition to the above examples, nonequivalent couplings are always observed for geminal hydrogens in saturated ring systems unless the rate of interconversion between conformers is so rapid the differences are averaged out. These different coupling constants are not always recorded, however, because they must usually be extracted by calculation rather than measured from peak separations.

One case of unobserved nonequivalent couplings is shown in J36(9). These couplings could be accidentally equivalent because of the preferred conformation imposed by the substituents.

The difference in coupling constants between *cis* and *trans* isomers of saturated ring systems are also examples of nonequivalent couplings of geminal hydrogens. They were listed under *cis-trans* isomerism because they differentiate between these isomers.

4. Steric Effects in Saturated Ring Systems

The discussion of this topic for chemical shifts in Chapter 4, Section IIID-4, applies to coupling constants, also. An indication of the long-range couplings made possible by ring systems is shown in Chart J9(6). Here the maximum H—C—C—C—H coupling is increased by a factor of 10 over the corresponding coupling for aliphatic chains (line 5). A less drastic increase in H—C—C—C—F couplings due to ring system steric factors is shown in Chart J27(8).

E. EFFECT OF NUMBER OF BONDS BETWEEN COUPLED NUCLEI

1. For Saturated Chains

For saturated chains the magnitude of the coupling constant falls off rapidly with increasing number of bonds between the coupled nuclei. This trend is shown in Figure 56. Lines 1–4 show that the constant for hydrogen–hydrogen couplings decreases by a factor of about 14 with the addition of a second single bond, then decreases by another factor of about 2 for the third bond and then by another factor of about 10 for the fourth bond. Lines 8–11 show similar decreases of factors of 10, 4, and 17 with the addition of the second, third, and fourth bonds between hydrogen and fluorine. These trends can be used to determine the relative positions of these pairs of coupled nuclei in saturated chains.

The trends are significantly different for hydrogen coupled to carbon-13, nitrogen-14, or phosphorus, however. The hydrogen–carbon-13 constant drops by a factor of about 40 between one and two bonds, but appears

to rise a bit between two and three bonds (lines 5–7). Similar sharp drops followed by rises are exhibted by the nitrogen-14 and phosphorus constants (lines 12–14 and 18–21). Addition of a fourth bond to the hydrogen–phosphorus system apparently results in complete elimination of the coupling. It may do the same for the carbon and nitrogen couplings, also.

The data are insufficient to establish the trend for nitrogen-15 with confidence, but the constant appears to decrease continuously with increasing number of bonds.

Magnetic asymmetry or restricted rotation, which introduce nonequivalence, broaden the hydrogen–hydrogen coupling ranges enough to interfere with the observation of the bond number trends. For this reason all constants for nonequivalent hydrogens have been omitted from the H—C—H data in Figure 56. Also, the H—N14 couplings for amides are out of line with the other couplings of this type [Figure 54(9)] and would interfere with observation of the bond number trend. Consequently, amide couplings have been omitted from Figure 56(12–14).

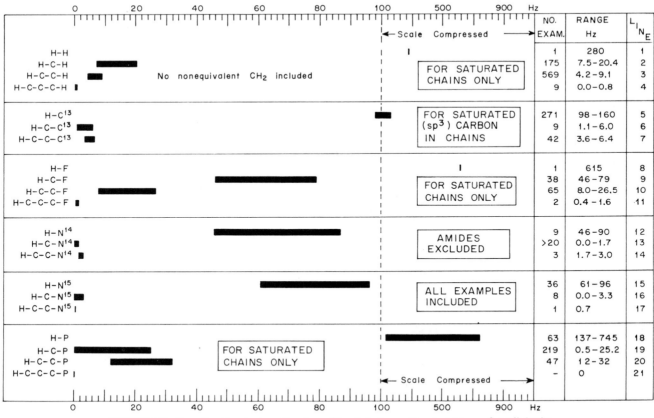

FIGURE 56. Coupling Constant vs. Number of Bonds between Coupled Atoms, for Aliphatics

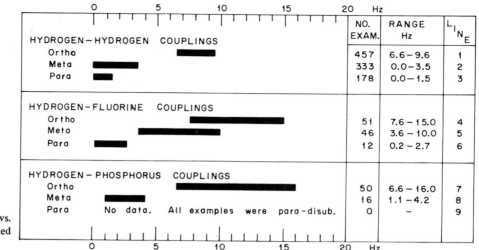

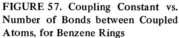

FIGURE 57. Coupling Constant vs. Number of Bonds between Coupled Atoms, for Benzene Rings

The data of Figure 56 show both the advantages and limitations of coupling constants in the structural analysis of saturated carbon–hydrogen chains. The following conclusions may be drawn:

a. Except for cases involving nonequivalent methylene hydrogens, the number of bonds between coupled hydrogens and between hydrogen and fluroine can be determined with considerable assurance from the coupling constants.

b. Hydrogen attached directly to C^{13}, N^{14}, N^{15}, or P^{31} is clearly distinguished by its coupling constant from hydrogen farther removed from these atoms. The reader should remember that the H–P coupling is so large that special effort must be made to show both peaks of the doublet on the spectrum. Offset scans will often be necessary. It must also be remembered that hydrogen is not usually coupled to N^{14} or N^{15} in amines (because of rapid exchange) and is not coupled to N^{14} in amides (reason unknown).

c. It is not always possible to determine from the coupling constants alone whether a hydrogen is two or three bonds removed from nitrogen or phosphorus. The coupling should usually indicate clearly when the hydrogen is less than four bonds removed from phosphorus, however.

2. For Aromatic Rings

For aromatic rings the coupling constant falls off somewhat less rapidly than for saturated chains, but still falls off rapidly enough to provide very helpful information about the relative positions of the coupled nuclei. The trends for hydrogen coupled to hydrogen, fluorine,

or phosphorus are shown in Figure 57. For the hydrogen and phosphorus couplings the ortho relation is unequivocally indicated by the magnitude of the coupling constant. The meta and para relations are distinguished most of the time by the hydrogen–hydrogen couplings, and are probably distinguished by the hydrogen phosphorus couplings, although lack of data prevents a check on the latter opinion.

The data of Figure 57 appear to indicate that it may not always be possible, from hydrogen–fluorine couplings, to differentiate the ortho from the meta relation if only one coupling constant is observable. The more detailed data of Chart J28, however, show that the two relations are clearly distinguishable if information about flanking substituents is available. The constant always distinguishes clearly between the meta and para relations for this pair of nuclei.

The ortho, meta, and para hydrogen couplings for condensed ring aromatics are about the same as the corresponding ones for benzene (J12). There is a clear distinction between the magnitudes of the ortho and meta constants, and a fair distinction between those of the meta and para constants. Couplings between hydrogens on different rings are so small they are not usually observed.

For the pyridine-type compounds there is also a clear distinction between ortho and meta constants, but not between meta and para (J13).

3. For Olefinic Chains

It was shown in Figure 55 that three-bond couplings through a double bond are larger than corresponding couplings through saturated systems. It is generally true that constants are larger when one or more double or triple bonds are in the pathway. The longer-range couplings in aliphatic chains are through unsaturated linkages [J9(5,9,10,16), J10(1,2,4–12), J27(8,11–15), and J38(6–13)]. In conjugated systems the coupling can be significant even across nine bonds [J10(12)].

F. COUPLINGS TRANSMITTED THROUGH HETERO-ATOMS

Replacement of a carbon in the coupling path by oxygen, nitrogen, sulfur, or even some metals does not significantly affect the magnitude of the coupling constant (J15–18, J29 and J36–38). When exchange is slowed sufficiently, alcoholic OH hydrogen will couple to hydrogens on the alpha group with the usual vicinal constant [J15(2,3)]. The NH hydrogen of amides also couples to the alpha group with the usual vicinal con-

stant [J16(2)], but thiolic SH hydrogen frequently couples to its alpha group with a slightly larger constant [J18(2,3)]. The four bonds between the alpha groups of aliphatic ethers effectively eliminate coupling of the hydrogens in those groups [J15(5)]. Thus, it appears that one can predict the coupling across a heteroatom from the corresponding coupling across carbon atoms.

G. COUPLING CONSTANT VS. RING SIZE

The geminal hydrogen coupling constant, H—C—H, exhibits a useful correlation with the size of the saturated ring in which the methylene is located. This correlation is shown in Figure 58. In this figure the data for carbocyclics and heterocyclics are presented along with the corresponding data for aliphatic chains for comparison.

The constants for methylenes alpha to heteroatoms (lines 4–6) show a decisive correlation with ring size for

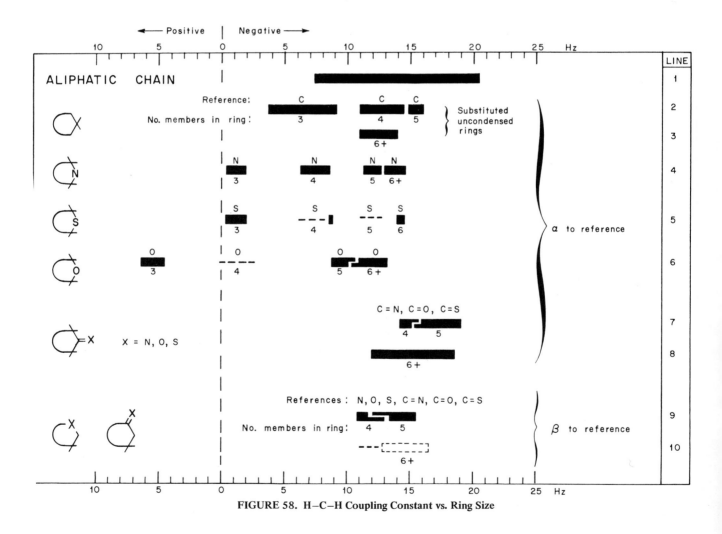

FIGURE 58. H—C—H Coupling Constant vs. Ring Size

rings smaller than six members. For six-membered and larger rings the constant reaches about the median value for aliphatic chains and does not change thereafter. Note that the positive sign of the constant for oxirane (three-membered with oxygen) must be taken into account to make the correlation accurate.

The constants for methylenes in saturated carbocyclic rings (line 2) differentiate between three-, four-, and five-membered rings but not between these and larger rings. Similarly, constants for carbocyclic rings containing an exocyclic C=X (lines 7 and 8) and for the beta methylenes in all the substituted rings (lines 9 and 10) differentiate between four- and five-membered rings but not between these and larger rings.

The added influence of heteroatom electronegativity is shown in the constants for the alpha methylenes (lines 4–6). The constants increase in the positive direction as the electronegativity increases in the order C < N < O. The values for sulfur fall on a different line, as was found for the $H-C^{13}$ couplings (see Section IIIC-1, this chapter). When the data of Douglas[T18] are combined with the data of Figure 58, it should be possible to predict the coupling constants for heterocyclic rings containing phosphorus, and perhaps for those containing silicon and other metals.

These correlations should be of value in the study of small rings.

H. VIRTUAL COUPLING

This book advocates the imposition of first-order interpretations on all spectra possible because this is usually the fastest way to obtain acceptable values for the spectral characteristics used in analysis. There are some cases, however, in which this practice can lead one astray unless virtual coupling is taken into account. The theoretical and some practical aspects of virtual coupling are discussed in detail by Anet[T1] and by Musher and Corey.[T36] For practical analytical work, however, the following simple discussion should serve to alert the chemist to the possible presence and effect of the phenomenon.

Virtual coupling is a special case of nuclear interaction in which the spectrum produced indicates a spin-coupling pattern which does not in fact exist. Like a virtual image, virtual coupling appears to be present but is not real. The simplest system in which the phenomenon can appear is a three-spin system such as the ABX system shown below:

$$\begin{array}{ccc} W & Z & D \\ | & | & | \\ R-C-C-C-D \\ | & | & | \\ H_A & H_B & H_X \end{array}$$

W and Z are strong shifting substituents, so that H_A and H_B will be shifted substantially to lower field away from H_X. The coupling between H_A and H_B, J_{AB}, will be the normal H—C—C—H constant of about 6 to 8 Hz. J_{BX} is also an H—C—C—H coupling and should have a constant of 6 to 8 Hz. J_{AX}, however, is an H—C—C—C—H coupling, should be essentially zero. The spectrum of H_X would therefore be expected to be a simple doublet because of coupling to H_B only. H_A would also be expected to produce a doublet because of coupling to H_B only, and H_B would be expected to produce a triplet (or double doublet) because of coupling to both the other hydrogens. No other couplings are considered here.

When the chemical shift between H_A and H_B is much larger than J_{AB} ($J/\Delta < 0.03$) the spectrum will have the expected doublet-triplet-doublet pattern. This pattern should still be recognizable as Δ_{AB} decreases, with changes in substituents, solvents, or frequency, until J/Δ approaches 2. When J/Δ becomes 2 or greater, however, the spectrum will begin to appear as if H_A and H_X were coupled with a substantial constant. This apparent coupling is not real because the calculated value of J_{AX} will still be zero. The coupling is therefore called *virtual*.

At the same time the apparent value of J_{BX} measured from the peak separations will decrease. In the limiting case, where $\Delta_{AB} = O$ (H_A and H_B have identical shifts), the spectrum will degenerate to a doublet for the H_A-H_B pair and a triplet for H_X, the characteristic spectrum of an A_2X system. The principal clue that this is an AA'X system instead of an A_2X one is that the peak separations will be $\frac{1}{2}(J_{AX} + J_{BX})$. In this example this is $\frac{1}{2}(7 + 0) = 3.5$, a value too small for the usual H—C—C—H coupling.

Whenever hydrogens with the same chemical shift are spin-coupled, with unequal constants, to additional hydrogens, virtual coupling will cause the observed peak separations to be the average of the coupling constants involved. Then when the peak separations are taken to be the coupling constants, the shorter-range constants will be too low and the longer-range ones will be too high.

A number of values known to be in error because of

virtual coupling are included in these charts because they are the values actually observed and are therefore useful for comparison and identification. In Chart J4(1) the apparent constant for the methyl—methylene coupling in chains longer than 6 carbons ($n = 3+$) is less than that in short chains ($n = 1,2$). As will be shown in Chapter 6, in the longer chains some of the methylenes beta and farther from the methyl have the same chemical shift as the methylene alpha to the methyl. This causes averaging of the H—C—C—H and longer-range couplings so that the H—C—C—H coupling as measured from the methyl peak separations is too low. In the shorter chains, however, the beta methylene has a sufficiently different shift from the alpha methylene so that the peak separations are not significantly affected and the constants measured therefrom are about right.

In Chart J13(15) virtual coupling has made an observed *ortho* coupling too low, and in line 17 it has made an observed *meta* coupling too high. In J13(8) virtual coupling may be responsible for the low values of the questioned couplings, but the high values in line 9 must be due to errors in line matching.

Chart J26(10) shows a clear example of coupling constant averaging through virtual coupling. The observed value is higher than the true *trans* coupling but lower than the true *cis* coupling.

As indicated by the asterisk note on Chart J34, P—C—C—P groups exhibit only one hydrogen—phosphorus coupling constant which must be the average of the H—C—P and H—C—C—P constants. Because these two constants are so near the same values, however, it is not known whether the resulting value for the H—C—P con-

stant is higher or lower than normal. It probably makes little practical difference.

Examples of spectra with virtual coupling are shown in Plates 47(c) and 50(k). Plate 47(c) shows an intermediate stage of virtual coupling where the upfield band is still recognizable as a doublet but it has a lot of fine structure on its base. The low-field band is too complex for first-order interpretation. Plate 50(k) shows a spectrum which could very easily be mistaken for that of an AB_2 system except that the coupling constants do not fit the other characteristics of the structure. Plate 22(a) and (d) show olefin spectra which have been simplified by virtual coupling.

How then does one avoid misinterpreting spectra involving virtual coupling? A major clue is that the data simply will not all fit together into a reasonable structure. When this happens, it is best to assume virtual coupling and see then if a reasonable structure can be devised. In order to obtain further evidence that virtual coupling exists, the sample should be run in several different solvents that might be expected to complex with it in different ways and thus change the relative chemical shifts of the virtually coupled groups. The resulting changes in the spectrum will not follow the corresponding changes in spectra of the wrong spin system. An $AA'B$ system may look like an A_2B system for one set of parameters, but will not look like it when the chemical shifts alone are changed. The comparison implied here may be with actual spectra or calculated theoretical spectra. Catalogs of both types, such as listed in the Data References, can save time and effort in resolving such problems.

CHAPTER 6

Typical NMR Spectra

I. INTRODUCTION

The definition and uses of band width and band shape were introduced in Chapter 1, Section IIE, and were illustrated briefly. However, full use of these two characteristics requires a much more extensive acquaintance with actual spectra. It is necessary to have a mental picture of the appearance of the various kinds of multiplets under different ratios of J/Δ, and it is desirable to have a picture of the overall appearance of the subspectra of various common structural groups. Such mental pictures lead to instant recognition of many structural features and save a great deal of time in interpreting spectra.

In some cases, however, the mental picture alone is inadequate. The identification of structural features from only *partial* subspectra, or the identification of virtual coupling, or the determination of the presence or absence of impurities requires a precise picture of the spectrum in question for comparison. This requirement is met by an extensive and detailed catalog of actual and theoretical spectra. The desirability of speed of access calls for a smaller collection of selected spectra which can fill the needs normally encountered, and a workable filing system for the major catalog.

It is the purpose of this chapter to present an introduction to a short collection of spectra or portions of spectra that the author has found to be helpful in everyday analysis and that are included in this volume. These spectra are grouped by similar molecular structures to permit the reader to see both the similarities and differences more easily. This grouping also expedites the searching. Unfortunately, it forces the reader to turn back and forth from text to figures during the initial familiarization period, and the author apologizes for this, but thinks the advantages of close grouping outweigh this disadvantage. Study of this collection will provide the mental picture one needs, and referral to it will provide many of the direct comparisons required.

A filing system that has proved to be workable and capable of providing rapid access to *all* the spectra in a large collection is described in the Appendix (p. 413). Although considerable time is required to set up this system, it is more than made up by the time saved in locating spectra when they are needed. Not only does it give rapid access to the desired spectrum if it is available, but it provides simultaneous access to all the spectra of a series of similar compounds. If the desired spectrum is not available, it may frequently be estimated from those of the similar compounds.

II. HYDROCARBON SPECTRA

A. PARAFFINS

The spectra of paraffinic hydrocarbons are characterized by overlapping, distorted and frequently unresolved multiplets, and by considerable virtual coupling. Coupling constants, when measurable, are unreliable. The interpretation of such spectra must rely heavily on band shapes and certain of the chemical shifts. A collection of typical spectra, such as presented in Plates 1–14, is essential to these interpretations.

Because these spectra vary significantly with frequency, they are shown for both 60 and 100 MHz. In the 60-MHz series (Plates 1–7) the methyl patterns are separated from the methylene and methine patterns to emphasize the characteristic changes that take place in each. In the 100-MHz series (Plates 7–14) the entire spectrum of each compound is shown as a unit to emphasize its overall appearance.

Plate 1 shows the three basic patterns produced by paraffinic methyl groups at 60 MHz—the distorted "triplet," the enhanced doublet, and the fairly sharp singlet. The chemical shifts of these three multiplets can be determined accurately, and are important in the interpretations (Charts 2–5). The appearance of these patterns is not very different at 100 MHz (Plates 8, 9, and 12).

Plate 2 shows the changes in the appearance of the "triplet" with increase in chain length. These patterns are helpful in determining whether a chain is 1, 2, 3, or 4 or more carbons long. The corresponding patterns at 100 MHz are shown in Plates 8 and 31.

Plate 3 shows the appearance of combinations of methyl doublets and triplets produced by the monoethyl paraffins. Again there is a usable change in appearance with chain length. The spectra of the 2-methyl paraffins consist of an intense doublet from the isopropyl methyls superimposed on a triplet from the other methyl. As the chain length increases the triplet pattern, indicated by its prominent central spike, moves downfield compared to the doublet. At 60 MHz the pattern stabilizes with the spike still showing on the upfield side of the low-field doublet peak, but at 100 MHz the stable pattern does not show the spike at all (Plate 9). The only indication of the triplet is then the shoulder on the low-field side of the low-field peak of the doublet.

The methyl subspectrum of the 3-methyl paraffins consists of a combination of one or two triplets from the end methyls and a doublet from the branch methyl. The relative shifts of these three seem to change with chain length, changing the appearance of the combination. This relative shifting is indicated a little better in the 100-MHz spectra (Plate 10).

The 4-methyl paraffin pattern is significantly different from that of the 3-methyl paraffins (Plate 3) and stabilizes quickly with increase in chain length to the same pattern as the initial 5-methyl paraffin (Plate 11).

Plate 4 shows the complex patterns which result from nonequivalent isopropyl methyls and the methyl parts of the structures which cause them to be nonequivalent. The 60-MHz spectra are generally uninter-

pretable, but the 100-MHz spectra can be interpreted to a somewhat surprising degree (Plate 13).

Plate 5 shows the effect of chain length on the subspectra of methylenes in paraffins. No matter what the end configuration, the band from the unsubstituted methylene chain narrows with increasing chain length, finally becoming a single sharp peak. For normal paraffins (Column 1) the methylene band becomes a single peak at a chain length of 4 to 5 methylenes (see Plate 8 for the corresponding 100-MHz spectra) and becomes sharper with increasing chain length thereafter. For alkyl benzenes (Column 2) the single peak is also reached at a chain length of 5 methylenes. The multiplet remaining on the low-field side of the single peak is from the beta methylene group. This multiplet is better separated at 100 MHz (Plate 31) and does not interfere with observation of the band from the remaining methylenes. This is a general characteristic of *n*-alkyl groups and is helpful in determining chain length. If spin coupling is eliminated at one end, the single peak is reached at a chain length of 3 methylenes (Column 3 and Plate 12). If spin coupling is eliminated at both ends a singlet is produced at all chain lengths.

Plate 6 shows that when methylenes and isopropyl methines are present in the same chain (Columns 1 and 2), the methylene band will collapse into a single peak at chain lengths of 4 and above, but the methine band does not collapse. This characteristic is even clearer in the 100-MHz spectra of Plates 9 and 12. When the methine is not part of an isopropyl group, however, (Column 3), its band is indistinguishable from that of the methylenes and it may broaden the base of the collapsing methylene band. This point is also shown better in the 100-MHz spectra of Plates 10 and 11.

Plate 7 shows the characteristic spectrum of the isolated ethyl group at both 60 and 100 MHz. This pattern or even parts of it can be used to identify this group. The methyl part of this group has the lowest δ value of all the methyl triplets (Chart 2).

Plate 14 shows the appearance of the 100-MHz spectra of paraffins with single long branches. Except for the ethyl or propyl branches (a) the methyl band is indistinguishable from that of a normal paraffin. The methylene band, however, has a shoulder or split peak which is more like the monoethyl paraffins. This combination of characteristics shows that the compound is neither normal nor monoethyl, but has a branch longer than methyl.

Spectra of mixtures of paraffins are shown in Figure

32, Chapter 3, Section IIIC. The characteristics of the spectra of the pure paraffins can be seen in the spectra of the mixtures.

The use of these spectra and the corresponding shift charts in determining the structures of paraffins is described in detail by Bartz and Chamberlain.[T4]

B. CYCLOPARAFFINS

The spectra of cycloparaffins are characterized by overlapping and generally unresolved multiplets. The bands are frequently almost featureless, so that the measurement of multiplicities and coupling constants is usually not possible and the chemical shifts are uncertain. Even band shapes are not as helpful as they are with acyclic paraffins. Hydrogen NMR below 300 MHz is at its greatest disadvantage with these compounds. Carbon-13 NMR, on the other hand, can resolve the peaks for the individual carbons even in complex condensed cycloparaffins and seems to be definitely the preferred method for studying these molecules. At present, hydrogen resonance at 60 and 100 MHz must be used on essentially a "fingerprint" basis with cycloparaffins, so that it is not reasonable to develope a set of typical spectra which are really useful. The spectra of Plate 15 are presented to help the reader recognize the presence of such molecules in a sample rather than to facilitate identification.

All unsubstituted single-ring cycloparaffins produce a sharp singlet band at room temperature, as shown in Plate 15(a). The chemical shift of this singlet helps identify the first three members of the series, but is of little help thereafter.

Any kind of substitution on a cyclopentane ring causes the spectrum to spread over a range of about 2 ppm, producing the broad and generally uninterpretable spectra of (b) to (e). The prominent band at about $\delta = 0.95$ in each spectrum represents the methyl groups, but the other bands have not been assigned. Methyl bands from methylated cyclopentanes are usually doublets, as shown.

Substitution of cyclohexanes also causes a spread in the spectra, but the extent and nature of the resulting pattern is strongly dependent on the degree and symmetry of substitution. Substitution with a single alkyl group leads to a two-band ring spectrum like those shown in (f) and (i). The assignment of these bands to specific ring hydrogens is uncertain, however.

Dimethylcyclohexanes can produce either a sharp ring spectrum with a doublet methyl band, or a broad ring spectrum with a singlet methyl band, depending on whether the substitution favors or hinders ring inversion.[T35] Similar behavior of the ring spectrum is also observed for other substituents,[T46] and can be an indication of the *cis-trans* relationship of the substituents if the positions of substitution are known.

When more than two substituents are attached to the cyclohexane ring the generalizations become more difficult, and will not be attempted here. Plate 15(h) shows the spectrum of trimethylcyclohexane to illustrate one of the drastic changes that can take place in the spectrum. Here part of the ring resonance (apparently a triplet) appears at a higher field than the methyl resonance (a doublet). Again the assignments are uncertain except for the methyl.

C. OLEFINS

1. Olefinic Hydrogens

The olefinic hydrogens in aliphatic olefins have distinctive spectra that give a considerable amount of information about the structure in which they are located. The band shape reveals at a glance whether the olefinic group is monosubstituted (vinyl group) or more highly substituted, and frequently reveals the degree and kind of higher substitution. It is emphasized that the correlations discussed in this section can be used with assurance only when the sample is run in CCl_4 solution at room temperature. Changes in solvent or temperature may alter some of the spectra enough to destroy the correlations. The 100-MHz spectra were run at 3% to 4% sample concentration, but many of the 60-MHz spectra were run on the neat (undiluted) sample. This concentration difference changed the olefinic hydrogen shifts significantly (see Chart 9) and is responsible for part of the difference in resolution at the two frequencies.

Important typical spectra at 60 MHz are presented in Plate 16. The characteristic broad but well-resolved spectra of the vinyl group with different alpha substituents are shown in (A) to (C). The band from the nonterminal hydrogen is frequently amenable to first-order interpretation as indicated. The markedly different spectra of the 1,1-disubstituted ethylenes are shown in (D) and (E). If neither substituent is branched at the alpha or beta carbon the olefinic hydrogens usually produce a single band broadened by spin coupling to the substituents (D). If one of the substituents is branched at the alpha or beta carbon, however, the olefinic hydrogens will have different shifts and will frequently give rise to two distinct bands (E). If both substituents are identical, no matter what the branching, only one olefinic band will be observed. In box (F) are shown the spectra characteristic of *cis* and *trans*-1,2-disubstituted ethylenes when the olefinic group is near the center of a long chain, as in the oleyl group. The isomer is clearly indicated by the spectrum.

At 100 MHz the olefinic hydrogen spectra show more detail and reveal more structural information. Plate 21 shows the vinyl group in full detail with first-order interpretations. In spectra (a) and (b) the entire vinyl group spectrum is shown to provide the mental image needed for recognition. In (a') and (b') the terminal hydrogen bands have been spread out to show detail and permit interpretation. In the remaining spectra the non-terminal band is shown on the left and the terminal bands are shown on the right.

In spectra (a) to (e) every line can be interpreted in first order. Spectrum (d) of an isolated vinyl is a simple 12-line ABC spectrum that is easy to interpret when it is realized that the *b* and *c* multiplets overlap and therefore have abnormal enhancements. Spectrum (e) shows the vinyl coupled to a methine, resulting in the doubling of every one of the 12 ABC lines. In (c) and (b) the vinyl is coupled to a methylene and the original lines are converted to triplets. In (a) spin coupling to a methyl has converted all the lines to quartets. These characteristics are useful in determining the degree of branching at the carbon alpha to the vinyl group.

There are slight changes in the line spacings in the terminal hydrogen bands between (b) and (c). This indicates some dependence of this pattern on substituent chain length, but the magnitude is probably too small to be of use analytically.

Plate 21 (f) to (h) show spectra that depart significantly from first order in number of lines. In each case the interpretations indicate a substantial reduction (f) and (h) or interchange (g) in the relative chemical shifts of *b* and *c*. These changes in overall relationships might be enough to introduce the extra lines. This point could be checked by calculation of a series of four- or five-spin spectra with appropriate parameters, but this is beyond the scope of this work.

The effects of changes in substituents on the spectra of the olefinic hydrogens in 1,1-disubstituted ethylenes are shown in Plate 22. Spectra (a) and (d) show the single band which results from symmetric substitution.

The identical chemical shifts for the two olefinic hydrogens lead to virtual coupling (Chapter 5, Section IIIH), which averages the coupling constants between the *a* and *b* hydrogens and simplifies the spectra. The methyl band which accompanies (a) is a triplet with the same averaged spacing and so is part of the methylene band which accompanies (d). These spectra would be very confusing if virtual coupling were not taken into account.

Spectra (b) to (i), except (d), show the effects of asymmetric substitution on the olefinic hydrogen pattern. Increasing the length of one of the chains while holding the other constant increases the difference in chemical shift between the olefinic hydrogens a small amount, producing the increasing multiplicity shown by the series (a) to (c) and the pair (d) and (e). Increasing the branchiness of one of the chains at the alpha or beta carbon, however, has a greater effect on the relative shift, as shown by the series (a), (g), (h), and (i).

Spectrum (f) does not seem to fit into these generalizations. Other spectra not reproduced here show that when the chain is longer than two carbons the effect of a methyl branch at the alpha carbon is reduced. This trend is seen by comparing spectra (g) and (f). Branches at the beta carbon, however, are still effective in separating the olefinic hydrogen bands, as shown by (h) vs. (f).

The olefinic hydrogen spectra of 1,2-disubstituted ethylenes are shown in Plates 23 and 24. For the symmetrically substituted molecules of Plates 23, the spectra are symmetric and the spectrum of the *trans* isomer has more peaks than that of the *cis* isomer. These features are useful in identifying these compounds.

For the asymmetrically substituted molecules of Plate 24, the spectra are asymmetric, and the *cis* and *trans* isomers are not so clearly distinguished. There is a noticeable effect of chain length on the spectra, however. Spectra (e) and (j) indicate an effect of chain branching which probably holds for other molecules as well.

Spectra of the olefinic hydrogen in trisubstituted ethylenes are shown in Plate 25. These spectra can be interpreted at least partially by first-order approximations, and the interpretations are shown for each. The spectra of compounds having geminal methyls (a to d) seem to have better resolved lines than the others. For the compounds *not* having geminal methyls, the *trans* isomer seems to have a much more complex-looking spectrum than the corresponding *cis* isomer (e to l). Note that in order to achieve this last correlation it has been necessary to reverse the names of two of the isomers (f, j and g, k). This seems justified in view of the clear correlation that results. The *trans* isomer has been chosen as the one with the most complex spectrum on the basis of the data in Plate 23.

2. Nonolefinic Hydrogens

Hydrogens in methyl and methylene groups alpha to olefin groups spin-couple to all the olefinic hydrogens, giving rise to some characteristic spectra that can be helpful in identifying such groups. Typical spectra of the alpha methyls at 60 MHz are shown in Plate 17 and typical spectra of alpha methylenes are shown in Plate 19. The observable multiplicity of these groups indicates the structural groups of which they are a part, and helps differentiate *cis* and *trans* isomers.

The higher resolution achieved at 100 MHz provides even more information from these spectra of the alpha groups. Plate 26 shows the changes in the alpha methyl band with changes in chain length and branchiness for 1,1-disubstituted ethylenes. Spectrum (a) shows the clean triplet which is due to coupling constant averaging because of virtual coupling. Spectra (b) to (d) show the effect of increasing chain length, indicating that some information about chain length can be obtained from these bands. Spectra (e) to (g) show the effect of chain branching, which is more marked than the effect of chain length. Spectrum (f) is almost like spectrum (a), indicating that the relative shift of the olefinic hydrogens is very small, as was observed for the olefin band itself in Plate 22(f). The olefin bands corresponding to the other methyl bands except (d) are also in Plate 22.

Plates 27 and 28 indicate that the spectra of methyl groups in linear and branched 1,2-disubstituted ethylene should be helpful in determining chain length and identifying the *cis* or *trans* isomer. The spectra of Plate 27 become more complex with increasing chain length, but there does not seem to be a clear generalization for identifying the isomer. Similarly, there are no clear generalizations from Plate 28. The patterns must be used by direct comparison of compound pairs.

Plate 28 (e) and (f) show the separate bands produced by geminal methyls in trisubstituted ethylenes. These bands should be helpful in identifying such structures.

Typical 60-MHz spectra of methyl groups beta or farther from an olefin group are shown in Plate 18. These band shapes are discussed in the next paragraph.

The spectra of the saturated parts of linear alpha olefins are shown at 60 MHz in Plate 20 and at 100 MHz in Plates 29 and 30. The pattern of the alpha methylene band does not change with chain length except to lose a little resolution. The bands produced by the remaining methylenes ($\beta+$) and by the methyls, however, vary with chain length very much like those of the paraffins. In the olefin spectra the methylene bands are a little wider than in the paraffin spectra, and they do not collapse to a single peak until the length of the $\beta+$ methylene chain has reached 5 or 6 as compared to 4 for the paraffins. The methyl band patterns also stabilize at a chain length of 5 or 6 instead of 4. These differences help to differentiate short-chain paraffins from short-chain olefins.

D. AROMATICS

The spectra of the alkyl chains in *n*-alkyl benzenes are shown in Plates 5 and 31. Plate 5 shows the methylene bands at 60 MHz, and Plate 31 shows the entire alkyl spectrum at 100 MHz. Because of the greater chemical shift of both alpha and beta methylenes, these spectra have broader methylene bands and require longer chain lengths to stabilize the patterns than the spectra of either paraffins or olefins. Also, the beta methylene band is still observed after the $\gamma+$ methylene band has collapsed to a singlet. The beta methylene band is especially prominent at 100 MHz.

The aromatic ring spectra for aromatic hydrocarbons have been scattered throughout the Plates showing all the other aromatic spectra because they are needed for reference in those plates. A separate discussion at this point may be helpful, however.

The spectrum of benzene is a very sharp singlet, shown in Plate 44(a). This peak is sharp enough for use as a resolution adjustment standard and as a reference for internal lock systems. Attachment of a single alkyl chain to the ring will produce significant broadening of the ring spectrum because of minor chemical shifting of the ortho and para hydrogens and spin coupling of the alpha alkyl hydrogens with the ring hydrogens. A typical spectrum of a monoalkylbenzene is shown in Plate 44(b). The more branched the alkyl chain is at the alpha carbon, the greater the width of the ring spectrum.

Ortho-disubstitution with identical unbranched groups leads to a narrow ring spectrum [Plate 45(b)] but if one or both of the groups are branched at the alpha carbon the ring spectrum may be appreciably broadened [Plate 45(f)] .

Meta-disubstitution with either branched or unbranched groups leads to a characteristic ring spectrum like that of Plate 47(a). Although the details of this spectrum depend on the branchiness of the substituents and on whether they are alike or not, the overall pattern is recognizable as that of meta dialkyl substitution.

Para-disubstitution with identical unbranched groups leads to a narrow ring which shows the effects of spin coupling between ring and chain hydrogens, as in Plate 48(b). If one of the groups is branched, the ring spectrum may look more like Plate 48(c) or (d).

Ring spectra of other methylated benzenes are shown in Plate 49(a), Plates 50(a) and 51(b), (f), and (g). These all show the effects of spin coupling of ring and methyl hydrogens, modified at times by crowding or virtual coupling.

The spectra of some important condensed ring aromatics are shown in Plate 53. These are divided into those that resemble AA′BB′ systems (a) to (c) and others (d) to (f). True AA′BB′ spectra are symmetric, as shown in Plate 45 (c) to (e), but the spectra of Plate 53 (a) to (c) are not quite symmetric. It is this asymmetry, which is due to weak coupling between hydrogens on different rings, that indicates that these latter spectra are from condensed rings rather than from single rings.

Note that in the spectra of phenanthrene (d) and chrysene (e) the lowest-field bands are produced by the "angular" hydrogens, *a* in (d) and *a,b* in (e).

III. SUBSTITUTED HYDROCARBONS

A. MONOSUBSTITUTED SATURATED ALIPHATIC CHAINS

Determination of the structure of saturated chains in nonhydrocarbons is made much easier if there is available a set of typical spectra of the simpler alkyl radicals having substantial chemical shift for the alpha and beta groups. From such a set one can identify the small radicals by direct comparison. The differences in chemical shift normally encountered will not often distort the spectra beyond recognition.

A set of the spectra of the aliphatic alcohols will serve very well as the desired comparison set. These are presented in Plates 32 to 39. Plates 32 to 34 show the 60-MHz spectra of all the aliphatic alcohols through C_5, and Plate 35 shows the linear alcohols from C_6 to C_{12}. Plates 36 to 39 show the spectra of the same compounds at 100 MHz. The spectra are interpreted as completely as

possible to first order, and all band assignments are indicated. Chemical shifts can be read from the scales at the top and bottom of each figure, and coupling constants are stated beside the interpretations.

These are excellent spectra for a beginner to study. They show how the multiplets change in appearance with changes in configuration. They show when first-order interpretations can be imposed on distorted multiplets and when they cannot. They afford examples of overlapped multiplets which have been sorted out and labeled. Finally, they provide the mental images of the overall appearance of the spectra of the lower aliphatic radicals.

Although the spectra of alkyl radicals encountered in most analytical work can be recognized from the alcohol spectra, there are two extreme cases that can cause trouble. An overall view of the problem is shown by the various spectra of the normal butyl group in Plate 40. Spectra (b) through (e) can be recognized as butyl groups if allowance is made for differences in the shift of the beta methylene, but spectra (a) and (f) would not be recognized. The absence of the third band in (a) indicates that it is a hydrocarbon, and comparison with the paraffin spectra quickly shows it to be *n*-butane. The unusually large beta hydrogen shift in (f), however, would remain a problem without the example shown.

Spectra of the linear ketones and acids resemble those of the linear alpha olefins more than those of the linear alcohols, except that the alpha methylene produces a triplet rather than a quartet. The beta hydrogen band is a little more prominent in the carbonyl compound spetra, also, because it is shifted to a little lower field.

B. THE X-CH_2-CH_2-Y GROUP

This group is commonly encountered in organic chemistry, and it has characteristic spectra which can be illustrated with a small number of examples. The spectra are of two spin classes, A_2B_2 and AA′BB′. Both classes are illustrated in Plate 41. Both sets are presented in order of increasing chemical shift difference between the methylenes. The A_2B_2 set (a) to (d) progresses through symmetric patterns from the singlet at zero relative shift to two triplets at maximum relative shift.

The AA′BB′ set (e) to (i) also has symmetric patterns, but the patterns depend on differences in coupling constants as well as differences in shift. The differences in coupling constants are attributed to restricted rotation, which produces unequal rotamer populations. It is possible, however, that these systems are actually ABCD in which A and B have nearly equal shifts and C and D also have nearly equal shifts. Judging from the theoretical spectra of Wiberg and Nist,[T43] (g) and (h) are AA'BB' systems, but (f) and (i), and therefore (e), may be ABCD systems. The latter three appear to have three different coupling constants, and may or may not have more than two shifts.

It is important to note that no matter what the spin system, when there is no chemical shift difference there is no multiplicity [(a) and (e)]. It also makes no difference whether the lack of relative shift is normal because of identical substituents (a) or accidental (e). A singlet will be produced in both cases. If accidental equivalence is suspected, changing to a strongly anisotropic solvent may reveal the full pattern, as in (f). The spectrum of the truly equivalent system will remain a singlet in any solvent.

C. THE SUBSTITUTED VINYL GROUP

Because the vinyl group can take part in chemical resonance as well as induction, its NMR spectrum varies drastically with the nature of the substituent. A set of the typical spectra of this group showing the range of this variation is very helpful in identifying the group itself and its probable substituent. This set of spectra is presented in Plates 42 and 43.

The isolated vinyl group (no spin coupling to outside groups) will give a 12-line spectrum when the relative chemical shifts among its three hydrogens are significantly larger than the coupling constants. These spectra are easily interpreted to first order, as shown in Plates 42(e) and 43(b) to (e).

When the relative shifts are of the same magnitude as the coupling constants, however, combination lines are produced and it is necessary to resort to calculation to arrive at a satisfactory interpretation, as in Plate 42(a) to (d). Note that some of the combination lines can be as intense as the first-order lines, making the first-order interpretation uncertain.

When the vinyl group is spin-coupled to an outside group *all* the lines are multiplied by the coupling, as in Plate 43(a). In this case each of the 12 original lines has been converted to a triplet by coupling to the methylene.

The chemical shifts are shown for each band to provide an idea of the location and width of these patterns. In some cases, such as Plate 43(e) and perhaps (c), the separation between parts of the vinyl spectrum is so large that it may be difficult to realize that they all belong to the same group.

Note that in the spectra of Plate 42, the relative order of the shifts changes with the substituent; that is, the order is *a*, *b*, *c* in (a) and (b), it is *c*, *b*, *a* in (c) and is *b*, *a*, *c* in (d) and (e). In Plate 43 the relative order stays the same but the relative positions within that order change.

D. SUBSTITUTED BENZENE RINGS

The various forms of substitution of the benzene ring can lead to spectra produced by one to five hydrogens with a range of 3.5 ppm in chemical shifts and a range of 16 Hz in coupling constants. Furthermore, for spin systems like ABCD, small changes in one or two of the parameters can make significant changes in the appearance of the spectra. As a result, the number of significantly different spectra which can be produced by the substituted benzene ring is so large that it is impractical to cover it with a set of theoretical spectra. Many thousands of spectra would be required.

On the other hand, this class of compounds is so important in organic chemistry that its members are encountered almost every day in analytical work. Identification of the number, kinds, and positions of substituents is greatly facilitated by a set of reference spectra, but a means must be found to gain rapid access to the relevant section of these spectra for each problem. One means of gaining this rapid access is through a small set of typical spectra that show the types of patterns produced by the benzene ring with types of substitutions generally encountered. Plates 44 to 53 are the result of an attempt to provide such a set of spectra. It is expected that this set will lead the searcher to the proper spin system, which indicates the number and positions of the substituents. The shift and spin-coupling charts should then help identify the substituents. Finally, the proposed structure can be checked in detail from a large catalog of aromatic spectra.

First-order interpretations have been shown where possible, and assignments of bands to specific hydrogens have been made when the bands could be separately identified. In making first-order interpretations of the more complex spectra, it is usually necessary to look at the envelopes formed by individual lines rather than at the lines themselves. This is usually easier to do when the spectrum is not spread out or when the resolution has been slightly reduced. For example, the first-order multiplicity is easier to see in a 1000-Hz scan than in a 100-Hz scan of the spectrum of a monosubstituted benzene [Plate 44(k) and (n)].

In all spin system designations only the ring hydrogens are considered. Spin couplings to nonring hydrogens are discussed as perturbations on the ring spectrum. This simplifies the grouping of the spectra, and can be tolerated because the couplings to nonring hydrogens are much smaller than to most ring hydrogens.

1. Monosubstituted Benzenes

Plate 44 presents typical spectra of the monosubstituted benzene ring. The coupling constants in this system vary very little with substituent, so the appearance of a spectrum is determined primarily by the relative shifts induced by the substituent and by spin coupling between ring and substituent hydrogens. A relatively small number of spectra can cover the range of patterns normally encountered.

Spectrum (a) of Plate 44 shows the sharp peak produced by unsubstituted benzene. The addition of substituents with successively stronger shifting power causes this peak to change to a successively wider but roughly symmetric band, as in (b) to (d).

As the shifting power of the substituent increases further, the spectrum usually breaks into two distinct bands, the lower-field band representing two hydrogens and the upfield band representing three hydrogens. The reason for this consistent behavior is that upfield shifting substituents (OR, NR$_2$) usually move both the *ortho* and *para* hydrogen bands upfield away from the meta hydrogen band, but downfield shifting substituents (CO—R, SO—R, SO$_2$—R, P(Z)R$_2$) move only the *ortho* hydrogen band downfield away from the other two. Because the *meta* hydrogens produce a triplet while the *ortho* hydrogens produce a doublet, it is sometimes possible to determine the direction of the major shift from the multiplicity of the low-field band. For example, the low-field bands in spectra (f), (g), and (k) to (n) are rough triplets, indicating that the major shift is upfield. The low-field bands of spectra (h) and (i) are rough doublets, indicating that the major shift is downfield. This information, of course, helps identify the substituent.

The low-field band of spectrum (j) is a double doublet because of spin coupling of the *ortho* hydrogens to

phosphorus. This characteristic identifies phosphorus in some form as the substituent.

The multiplicity of the low-field bands of spectra (e) and (o) is not clear. Comparison with the other spectra leads to the conclusion that the low-field band in (e) is a doublet, as it should be, but in (o) the triplet which should be observed at low field is obscured by overlap with the triplet of the *para* hydrogen. In this case the *para* hydrogen band has not been shifted upfield far enough to uncover the *meta* band. In these and other cases where additional chemical shift is needed to clarify the pattern, a change in solvent or the addition of a shift reagent (Chapter 2, Section IIB-5a) may help.

2. Disubstituted Benzenes

a. Ortho-Disubstituted

Typical spectra of *ortho*-disubstituted benzene rings are presented in Plate 45. When the substituents are alike the spin system is AA′BB′. The coupling constants remain reasonably constant with changes in substituent, so that the spectral pattern depends primarily on the relative chemical shift of the two sets of hydrogens. Spectra (a) to (e) of Plate 45 shows how the pattern changes with increasing relative shift. The patterns are symmetric and are quite distinctive. They are clearly different from those of the X-CH$_2$-CH$_2$-Y systems of Plate 41.

When the substituents are unlike, the system is ABCD, and the number of possible patterns becomes very large. Examples of the types encountered in practice are given in Plate 45(f) to (m), with first-order interpretations and band assignments where possible. The reader should note the band broadening caused by spin coupling between the ring and methyl hydrogens in spectra (j) and (k). This coupling occurs with the ring hydrogens both *ortho* and *para* to the methyl. Some clearer examples will be shown later.

These ABCD patterns are sometimes similar to the patterns from monosubstituted benzenes. For example, spectrum (j) of Plate 45 resembles spectrum (1) of Plate 44. Close inspection reveals intensity differences between the low-field bands, however.

The first-order interpretations shown in Plate 45 consist of two major doublets from the "outside" hydrogens and two major triplets from the "inside" hydrogens. Each is split further by additional coupling but this is ignored for the present. In some cases, however, the coupling constants are modified a little by one substituent so that the first-order pattern is two major doublets and two major double doublets, again with additional splitting, as shown in Plate 46. The resolution is exceptionally good and there is no spin coupling to outside hydrogens in this case. If the resolution were substantially poorer these bands could very easily be considered two doublets and two triplets. It is likely that most of the other patterns are actually two doublets and two double doublets, also.

b. Meta-Disubstituted

Typical patterns produced by meta-disubstituted benzene rings are shown in Plate 47. When the substituents are alike the system is AB_2C and the patterns are fairly distinctive. Spectra (a), (b), (d), and (e) show the pattern from a ring with upfield shifting substituents. The low-field band represents the hydrogen meta to both substituents, while the narrowest upfield band represents the hydrogen between the substituents.

If the substituents are downfield shifting, the pattern changes to that of spectrum (f). Here the low-field band represents the hydrogen between the substituents, and the upfield band represents the hydrogen meta to both substituents.

Spectrum (c) is a special case where the three adjacent hydrogens are equivalent or nearly so. The resultant virtual coupling averages the coupling constants so that the three adjacent hydrogens produce a doublet and the isolated hydrogen produces a rough quartet. This is obviously not an AB_3 system, however.

When the substituents are unlike, the system is ABCD. The patterns produced by these systems are not always easy to distinguish from those produced by the ortho-disubstituted ABCD systems. For example, spectrum (h) of Plate 47 is very much like spectrum (j) of Plate 45. Spin decoupling of the methyls in each case might clear up the ambiguity.

Spin coupling of ring and methyl hydrogens is seen clearly in spectrum (i) of Figure 47 as the fine structure on bands b, c, and d. It does not show clearly on band a, although it may be present.

A complete interpretation of this type of ABCD system is shown in spectrum (j).

c. Para-Disubstituted

Para-disubstituted benzene rings have the most distinctive patterns of all the benzene series. These patterns are shown in Plate 48. When the substituents are alike,

the system is A_2A_2' with zero chemical shift. The resulting spectrum is always a sharp singlet (a) unless there is spin coupling between the ring hydrogens and the substituent (b). In the latter case, virtual coupling should average the coupling constants and reduce the splitting, as has apparently happened with the methyl coupling in (b). Para-difluorobenzene (not shown) produces a triplet with a constant of 6.1 Hz, which is just about the average of the lower values given for the ortho and meta couplings in Chart J28 (2) and (4).

When the substituents are unlike, the system is AA'BB' and produces a pattern which is clearly distinguishable from the AA'BB' of the ortho-disubstituted system ring. The pattern from isolated ring hydrogens of a para-disubstituted system consists of two sets of mirror-image doublets that have little satellite lines on each side of each principal line, as shown in Plate 48(h). The presence of the satellite lines positively identifies the para-disubstituted ring.

The pattern of the para-disubstituted ring depends only on the relative shift between the pairs of hydrogens and on spin coupling to outside groups. Plates 48(c), (d), (e), and (h) show how the pattern changes with increasing relative shift when little or no outside coupling is present. Spectra (f) and (g) show the effect of spin coupling to a methyl group. Even when this coupling is not resolved, the reduction in height of the affected peaks indicates its presence and shows which band is closest to the coupled group. For example, the reduced height of the low-field band in spectrum (e) is due to spin coupling with the CH_2 group attached to the ring.

Spectrum (h) should be symmetric. The asymmetry shown appears, from inspection of other spectra of this compound, to be due to saturation. This emphasizes that care must be taken to avoid saturation of very sharp peaks such as these. It may be necessary to run such a spectrum several times at different rf (H_1) settings to make sure saturation is reduced to an acceptable level.

3. Trisubstituted Benzenes

a. 1,2,3-Trisubstituted

Some patterns typical of 1,2,3-trisubstituted benzene rings are presented in Plate 49. When the substitution is symmetric, the system is AB_2 and the spectra are like those in (a) to (e). The methyl group shifts the para hydrogen about the same as the ortho hydrogen, so that the ring hydrogens on 1,2,3-trimethyl benzene have essentially no relative shift. The resulting spectrum is a singlet broadened by spin coupling to the methyl hydrogens. Even this broadening is probably reduced by the averaging action of virtual coupling.

Spectra (b) and (c) are straightforward AB_2 patterns. (See Wiberg and Nist,[T43] pp. 11–19). In spectrum (d) the pattern is reversed, indicating that the relative shifts are in the opposite direction from those in (b) and (c). Spectrum (e) is a reversed pattern modified by spin coupling to the methyls. This coupling is seen clearly on the *b* band but is much reduced on the *a* band.

When the substitution is asymmetric the system is ABC, and the spectra may sometimes resemble those of the substituted vinyl group [compare Plates 49(f) and 42(a)]. The chemical shift should remove any uncertainty about these two, however.

Spectra (g) and (h) are fairly simple patterns which seem to be typical of these aromatic systems. Spectra of this type may be a strong indication of 1,2,3-trisubstitution. Note that these are 250-Hz scans rather than 100-Hz scans like the rest.

Spin coupling between methyl and ring hydrogens is much more prominent in the wider scan of (i) than in (g) and (h). Note that in (i) band *a* is upfield of *b* and *c*.

b. 1,2,4-Trisubstituted

1,2,4-Trisubstituted benzene rings are inherently asymmetric and the systems are ABC. Some typical spectra are shown in Plate 50. Spectra (a) through (e) seem to be unique to the 1,2,4-trisubstituted ring. Spectrum (i) might be mistaken at first glance for Plate 49(i), but the differences should be clear on inspection. Spectra (h) and (j) resemble Plate 42(e) and 43(c) and (d), but the two sets are actually more like mirror images. Because of the differences in allowed arrangements of shifts and coupling constants, it appears that the ABC spectra of vinyl groups, 1,2,3-trisubstituted benzene rings, and 1,2,4-trisubstituted benzene rings are recognizably different.

Plate 50(k) shows a spectrum in which two of the ring hydrogens accidentally have the same shift. The resulting virtual coupling and coupling constant averaging produce a spectrum that looks like an AB_2 spectrum. The averaged coupling constant is about the size of a *meta* constant, so that this spectrum is very much like that of the true AB_2 system of Plate 51(e). The differences between the two are that the ABB′ spectrum has broader lines than the AB_2 spectrum and, although not shown here, the chemical shifts are quite different.

The broader lines should lead one to suspect virtual coupling in the ABB′ case, but the principal clue will probably come from the chemical shifts. It may not be possible to devise a true AB_2 system with the same shifts as the ABB′ system.

Once virtual coupling is suspected, the positions of substitution on the ring should be checked by infrared spectroscopy. This should show a clear difference between the ABB′ and AB_2 systems.

c. 1,3,5-Trisubstituted

Some representative spectra of 1,3,5-trisubstituted benzene rings are shown in Plate 51(a) to (e). When the substituents are identical the system is A_3 and the pattern is a singlet modified only by spin coupling to outside hydrogens. Spectrum (a) shows the singlet and spectrum (b) shows the fine structure due to spin coupling to methyl groups.

When two different substituents are attached to the ring, the system is AB_2. Because all the couplings between ring hydrogens are *meta* couplings, the peak separations are small and the bands are narrow. A relatively small chemical shift then gives a nearly first-order spectrum as in Plate 51(e). Spectrum (d) shows a similar chemical shift but with the bands made more complex by spin coupling to the methyl hydrogens. Spectrum (c) shows a pattern produced by a very small chemical shift plus spin coupling to methyls.

If all three substituents are different, the system will be ABC with small coupling constants. The resulting patterns (not shown) should be distinguishable from the other ABC systems by the narrower bands and smaller peak separations.

4. Tetrasubstituted Benzenes

The tetrasubstituted benzene ring has only two hydrogens, which can be *ortho*, *meta*, or *para* to each other. They can form an A_2 or an AB system depending on whether the substitution is symmetric or asymmetric about them. The A_2 systems produce a singlet as in Plate 52(a), or a singlet broadened and modified by spin coupling to outside hydrogens as in Plates 51(f) and (g) and 52(b) to (d).

The AB systems produce two doublets with *ortho* couplings [Plate 51(h)] or *meta* couplings [Plate 52(e)] or *para* couplings [Plates 51(i) to (k) and 52(f)]. Any of these may be modified by spin coupling to outside hydrogens.

Plate 52(g) and (h) show the spectra produced when

two of the substituents are fluorine. The coupling between the ring hydrogens and fluorine is so large that the effect can no longer be treated as a perturbation on the ring spectrum. These systems are therefore labeled ABX_2 for the symmetric one (h) and ABXX' for the asymmetric one (g). The ABX_2 system produces a pattern of two triplets which are further split by the *para* coupling between the ring hydrogens. The ABXX' system produces a pair of double doublets because each ring hydrogen couples unequally to the two fluorines. Each line of this pattern is further split by the *para* coupling between the ring hydrogens.

5. Penta- and Hexasubstituted Benzenes

Pentasubstituted benzene rings have only one hydrogen and can therefore produce only a singlet modified at times by coupling to outside hydrogens, as shown in Plate 52(i) to (k).

Hexasubstituted benzene has no ring hydrogens and produces no ring spectrum. This can mislead one into thinking the compound is not aromatic, except that the chemical shifts of the groups alpha and beta to the ring will indicate that a ring can be present. This uncertainty can be quickly resolved by infrared, ultraviolet, or mass spectroscopy, or by carbon-13 NMR.

6. Condensed Ring Aromatics

The spectra of important condensed ring aromatics are shown in Plate 53 and are discussed in Section IID of this chapter. No attempt is made to present spectra of the substituted condensed ring aromatics because the spectra necessary to show reliable trends are not available.

E. Phosphorus Compounds

Some important peculiarities of the spectra of phosphorus compounds can be shown conveniently by direct comparison with similar spectra from nonphosphorus compounds.

In general, the spectra of alkyl phosphines closely resemble those of similar paraffins. Plate 54 shows the striking similarity between the spectrum of tributyl phosphine and normal pentane. The ratio of methyl to methylene hydrogen is twice as high in the pentane as in the phosphine, but one could easily be led astray if the intensities were not available.

The similarity between neat tri-*n*-octyl phosphine

and 11-butyldocosane is shown in Plate 55(c) and (d). Here the methyl-to-nonmethyl ratio is almost the same for both compounds. The uncertainty in intensity measurement due to band overlap could make these two compounds difficult to distinguish if there were no other method. Fortunately, the spectrum of trioctyl phosphine changes significantly with the addition of a solvent [(a) and (b)] whereas the spectrum of the paraffin does not. If the solvent is trifluoroacetic acid, the change is dramatic, as shown in Plate 56. Conversion of the phosphine to a phosphonium salt by this method causes a marked downfield shift of the bands produced by the alpha groups, *d*, and thereby reveals the presence of phosphorus.

Oxidation of an aromatic phosphine to the corresponding phosphine oxide increases the chemical shift of the *ortho* hydrogen and increases its coupling to the phosphorus. This produces the marked change in the spectrum shown in Plate 57(a) and (b). Oxidation of an alkyl phosphine increases the H—C—P and H—C—C—P coupling but does not make a sizable change in chemical shift. The result is a marked broadening of the base of the methylene band as shown in Plate 57(c) and (d).

Alkyl esters of phosphorus acids are characterized by sizable coupling constants between phosphorus and the hydrogen group closest to it. This additional coupling adds distinctive multiplicity to the spectra, as shown in Plate 58. The low-field triplet characteristic of carboxylic esters of longer alkyl chains [band *d* in spectrum (a)] is converted to a quartet in the phosphorus ester [band *d* in spectrum (b)]. Similarly, the quartet of an ethyl ester is converted to a quintet [spectra (c) and (d)]. Both these changes are indicative of phosphorus esters. The detailed shapes of the alpha alkyl group bands are shown more clearly in Plate 59.

The dramatic doubling of the *ortho* hydrogen band in aromatic ring spectra by spin coupling to phosphorus is shown in Plate 60. This is also indicative of the presence of phosphorus in the molecule.

Plate 61 illustrates the problem of locating the widely separated peaks of the H—P doublet. Although the chemical shift of the hydrogen attached to phosphorus ($\delta = 7.15$) in the subject compound is almost the same as that of the aromatic ring hydrogens, the actual spectral peaks are so far removed from this position that they might be overlooked or might be obscured by other bands. Care must be exercised to locate such peaks in more complex spectra.

Chemical Shift Correlation Charts

CHART I 1 (top)

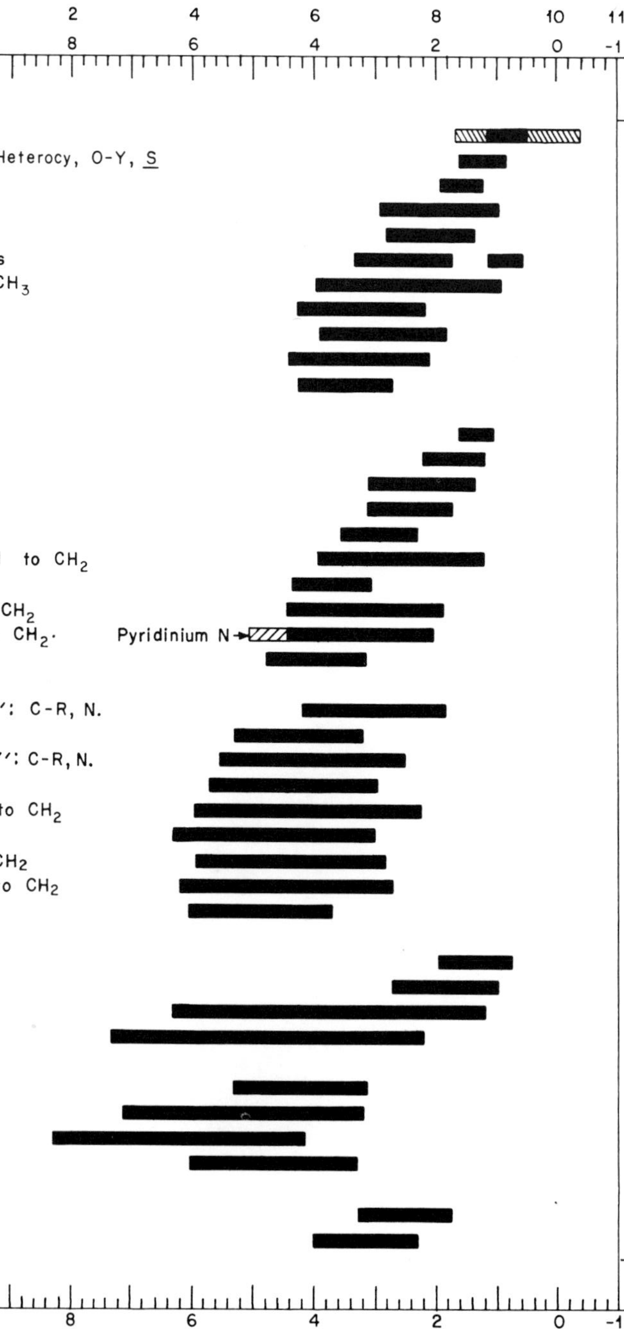

X: Any group
X: C=Y, C≡Y, φ, NR₂, NR₃⁺, N=Y, P(Z)Y₂, Heterocy, O-Y, S
X: F, Cl, Br, I, NO₂, -N⁺(Pyridinium), PR₃⁺
X: Combinations of the above groups.
X: C-Y, N-Y, O, S
Includes 5 & 6-membered heterocyclic aromatics
P: Any phosphorus group with P bonded to CH₃
S: Any sulfur group with S bonded to CH₃
N: Any nitrogen group with N bonded to CH₃

X: C-Z, N-Z, O, S.   Y: C-Z, N.
Includes CH₂ on pyridinium ring
P: Any phosphoras group with P bonded to CH₂
S: Any sulfur group with S bonded to CH₂
N: Any nitrogen group with N bonded to CH₂.

Y: C-Z, O.   Y′: C-R, N.
Y: C-Z, N-Z, O.   Y′: C-R, N.
P: Any phosphorous group with P bonded to CH₂
S: Any sulfur group with S bonded to CH₂
N: Any nitrogen group with N bonded to CH₂

Pyridinium N→

LINE	Structure	SUMMARY CHART (LINE)	NO. CPDS.
1	$CH_3-C-C-X$	S1(2-4)	1292
2	CH_3-C-X	S1(5,6,11,14,16,19,20,22,24,25)	1177
3	CH_3-C-X	S1(21,23,26-29)	46
4	$CH_3-C(X)_{2,3}$	S1(7-10,12,13,15,-18,24,27-29)	101
5	$CH_3-(C≡X, C≡C-Y)$	S2(1-6,11-17)	596
6	$CH_3-Aromatic$	S2(7-10,18)	786
7	$CH_3-\underline{P}$	S2(19-24)	55
8	CH_3-CF,Cl,Br,I	S2(25)	4
9	$CH_3-\underline{S}$	S3(1-13)	208
10	$CH_3-\underline{N}$	S3(14-32)	306
11	CH_3-O-X	S3(33,34)	1139
12	$R-C-CH_2-C-R$	S4(1)	790
13	$R-C-CH_2-C-X$	S4(2-16)	920
14	$R-C-CH_2-C(X)-Y$ or $Y-C-CH_2-C-X$	S5	163
15	$R-C-CH_2-(C≡X, C≡Y)$	S6(1-11)	422
16	$R-C-CH_2-Aromatic$	S6(12-14)	114
17	$R-C-CH_2-\underline{P}$	S6(15-17)	46
18	$R-C-CH_2-(F,Cl,Br,I)$	S6(21 23)	35
19	$R-C-CH_2-\underline{S}$	S7(1-11)	189
20	$R-C-CH_2-\underline{N}$	S7(12-27)	229
21	$R-C-CH_2-O-X$	S6(18-20)	525
22	$X-C-CH_2-(C=Y, C≡Y', Arom., \underline{P}, \underline{S}, \underline{N})$	S8, S9	620
23	$X-C-CH_2-(O-Y,F,Cl,Br,I)$	S10	636
24	$X-CH_2-(C=Y, C≡Y')$	S11	489
25	$X-CH_2-Aromatic$	S12(1-19)	356
26	$X-CH_2-\underline{P}$	S12(20-39)	78
27	$X-CH_2-(F,Cl,Br,I)$	S12(40-54)	141
28	$X-CH_2-\underline{S}$	S13(1-33)	232
29	$X-CH_2-\underline{N}$	S14(1-34)	246
30	$X-CH_2-O-Y$	S13(34-50)	228
31	$(R-C)_2CH-C-R$	S14(35-37)	156
32	$(R-C)_2CH-C-X$	S14(38)	—
33	$(R-C)_2CH-C-X$	S14(39)	—
34	$R-C-CH(X)-Y$	S14(40)	—
35	$(X-C)_2CH-Y$	S15(1-3)	17
36	X_nC-CHY_2	S15(4-20)	60
37	$X-CHY_2$	S15(21-48)	100
38	$X-CH(Y)-Z$	S15(49-59)	52
39	$X-C-C≡CH$	S18(36-40)	35
40	$X-C≡CH$	S18(41-45)	23

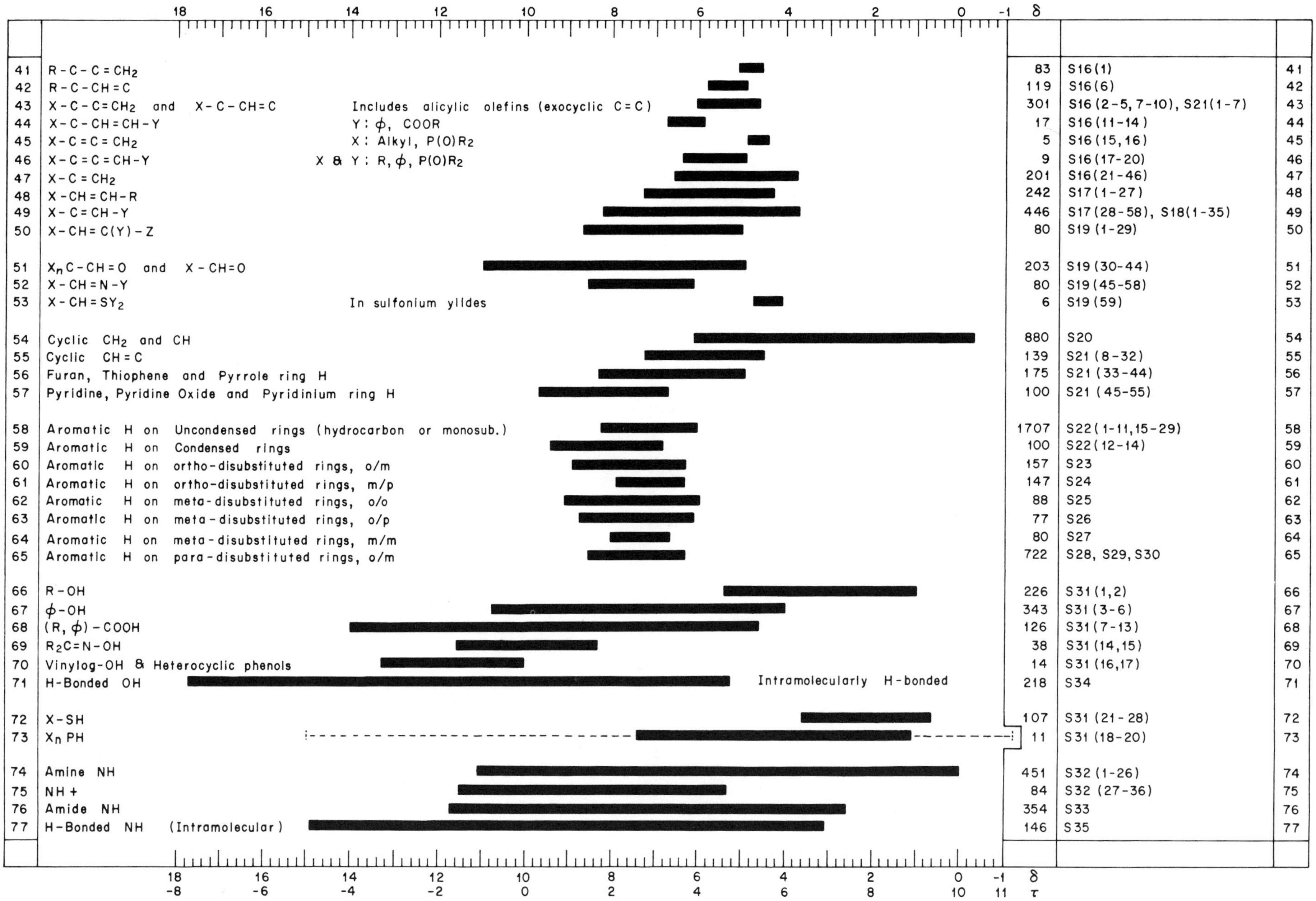

CHART I1. Index to and Condensed Summary of Chemical Shifts

		δ
41	S16(1)	83
42	S16(6)	119
43	S16(2-5,7-10), S21(1-7)	301
44	S16(11-14)	17
45	S16(15,16)	5
46	S16(17-20)	9
47	S16(21-46)	201
48	S17(1-27)	242
49	S17(28-58), S18(1-35)	446
50	S19(1-29)	80
51	S19(30-44)	203
52	S19(45-58)	80
53	S19(59)	6
54	S20	880
55	S21(8-32)	139
56	S21(33-44)	175
57	S21(45-55)	100
58	S22(1-11,15-29)	1707
59	S22(12-14)	100
60	S23	157
61	S24	147
62	S25	88
63	S26	77
64	S27	80
65	S28, S29, S30	722
66	S31(1,2)	226
67	S31(3-6)	343
68	S31(7-13)	126
69	S31(14,15)	38
70	S31(16,17)	14
71	S34	218
72	S31(21-28)	107
73	S31(18-20)	11
74	S32(1-26)	451
75	S32(27-36)	84
76	S33	354
77	S35	146

41	R-C-C=CH2
42	R-C-CH=C
43	X-C-C=CH2 and X-C-CH=C
44	X-C-CH=CH-Y
45	X-C=C=CH2
46	X-C=C=CH-Y
47	X-C=CH2
48	X-CH=CH-R
49	X-C=CH-Y
50	X-CH=C(Y)-Z
51	XnC-CH=O and X-CH=O
52	X-CH=N-Y
53	X-CH=SY2
54	Cyclic CH2 and CH
55	Cyclic CH=C
56	Furan, Thiophene and Pyrrole ring H
57	Pyridine, Pyridine Oxide and Pyridinium ring H
58	Aromatic H on Uncondensed rings (hydrocarbon or monosub.)
59	Aromatic H on Condensed rings
60	Aromatic H on ortho-disubstituted rings, o/m
61	Aromatic H on ortho-disubstituted rings, m/p
62	Aromatic H on meta-disubstituted rings, o/o
63	Aromatic H on meta-disubstituted rings, o/p
64	Aromatic H on meta-disubstituted rings, m/m
65	Aromatic H on para-disubstituted rings, o/m
66	R-OH
67	φ-OH
68	(R,φ)-COOH
69	R2C=N-OH
70	Vinylog-OH & Heterocyclic phenols
71	H-Bonded OH
72	X-SH
73	XnPH
74	Amine NH
75	NH+
76	Amide NH
77	H-Bonded NH (Intramolecular)

Includes alicyclic olefins (exocyclic C=C)

Y: φ, COOR
X: Alkyl, P(O)R2
X & Y: R, φ, P(O)R2

In sulfonium ylides

Intramolecularly H-bonded

CHART I1

Cross Index between Summary and Detailed Charts

Summary Chart (Line)	Detailed Charts (Lines) from which Data Were Taken
S1(1)	1(1)
S1(2)	13(1,8), 14(0,1)
S1(3)	Almost all charts
S1(4)	12(15)
S1(5)	1(16), 8(3–8), 12(1), 16(2), 17(2), 18(1,15), 19(2,3), 20(3), 21(2,3), 22(2), 23(7), 30(1,4), 40(2), 41(2), 48(5), 50(1), 51(5,6) 53(1), 55(1,14), 56(1,6,13,17), 57(1,6), 64(1), 69(2), 70(2), 71(2), 83(1,8), 86(1)
S1(6)	27(12), 64(14)
S1(7)	1(5), 12(1)
S1(8)	16(1), 18(1,15), 19(2), 22(2), 27(9), 31(1,7), 50(9), 83(1)
S1(9)	69(2), 71(2), 83(8)
S1(10)	40(2), 41(2)
S1(11)	1(20), 24(1,9), 25(1), 58(1), 84(1,7,11)
S1(12)	24(1)
S1(13)	25(1), 84(1,7,11)
S1(14)	29(1,8), 60(1,8), 61(1,9), 62(9,13), 79(1,6)
S1(15)	79(1)
S1(16)	13(1,8), 14(1), 15(4), 16(1), 17(1), 30(7), 31(1,7), 32(1), 33(1), 34(1,6), 35(1,7), 37(1), 42(1,11), 43(1), 46(5,10), 52(1,7), 58(4,8), 59(4), 73(1), 77(3,8,12), 78(1), 79(14), 85(1,7), 89(5)
S1(16a)	14(1)
S1(17)	16(1), 30(7), 31(1,7), 32(1), 46(5), 50(9,14), 58(4), 85(1)
S1(18)	73(1)
S1(19)	45(1), 46(1), 47(1), 48(1), 50(5,14), 51(1), 53(1), 54(1,6,11,12,15,16), 56(5,10), 63(1,6,9), 86(5), 87(11)
S1(20)	45(5), 46.1(1,2,3), 55(5,9), 63(13,16)
S1(21)	38(4), 41(3), 43(2), 62(13)
S1(22)	80(1), 81(1), 82(8), 85(1), 88(10)
S1(23)	89(1,6)
S1(24)	18(5,8,11), 22(2), 23(1), 25(2), 27(1,3,6), 28(1), 29(1), 30(4), 36(2), 57(11), 64(7), 66(7), 67(4), 68(4), 82(1), 83(1)
S1(25)	64(4,10), 65(1,5,8,11), 66(1,4,11), 67(1,8,11), 68(1,7), 69(2), 70(2), 71(2), 73(1), 77(1)
S1(26)	38(10)
S1(27)	38(1), 40(3), 42(2)
S1(28)	38(4), 41(3), 43(2)
S1(29)	38(7)

FUNCTIONAL GROUP		RANGE, δ-VALUES	NO. CPDS.	NO. EXAM.	LINE
CH_4		0.23	1	1	1
$CH_3-\overset{\mid}{\underset{\mid}{C}}-\overset{\mid}{\underset{\mid}{C}}-X$	X = Alkyl aromatics, aromatic naphthenes	1.05 (-)0.41	120	164	2
	X = All other compounds	1.15 – 0.65	1170	1397	3
	X = Condensed cyclic olefins	1.63 – 0.84	2	4	4
$CH_3-\overset{\mid}{\underset{\mid}{C}}-C=X$	X = CR_2, O, N-R, N-OH, N-NR_2	1.27 – 0.80	204	275	5
	X = S, or C=X = Quinone ring	1.32 – 1.15	8	8	6
$CH_3-\overset{\mid}{\underset{Y}{C}}-C=X$	Y = C=C< X = CR_2	1.13 – 1.03	5	5	7
	Y = ϕ, CO-R, O-Y X = CR_2, O	1.42 – 1.15	13	17	8
	Y = S-Y X = CR_2	1.60 – 1.15	27	28	9
	Y = Cl, Br X = CR_2	2.3 – 1.6	0	0	10
$CH_3-\overset{\mid}{\underset{\mid}{C}}-C\equiv X$	X = C-R, N	1.37-1.13-0.93	7	7	11
$CH_3-\overset{\mid}{\underset{X}{C}}-C\equiv C-$	X = CO-O-R	1.4 – 0.93	1	1	12
	X = O-R, (O,S)-P(O,S)(OR)$_2$, P(O,S)Y$_2$	1.70 – 1.40	13	13	13
$CH_3-\overset{\mid}{\underset{\mid}{C}}-$Heterocy.	X=O, S, N. $CH_3-\overset{\mid}{\underset{\mid}{C}}-$ at any position except on N+.	1.35 – 1.20	11	13	14
$CH_3-\overset{\mid}{\underset{\mid}{C}}\left(\langle\!\!\rangle_S\right)_2$		1.80 – 1.70	2	2	15
$CH_3-C\left(\langle\!\bigcirc\!\rangle\right)_n$	Normal n = 3 2 1	2.15, 1.67	1, 4	1, 4	16
		1.50 – 1.04	160	201	
	At edge of single arom. ring in rigid systems	1.80 – 1.30	6	6	16a
$CH_3-\overset{\mid}{\underset{X}{C}}-\langle\!\bigcirc\!\rangle$	X = C=C<, CO-Y, CN, O-Y, N(R)Y, PO(OR)$_2$	1.65 – 1.20	8	8	17
	X = S-R	1.9 – 1.5	0	0	18
$CH_3-\overset{\mid}{\underset{\mid}{C}}-\underline{N}$	$\underline{N}$ = N<^X_Y	1.58-1.40-0.82	117	159	19
	$\underline{N}$ = $\overset{+}{N}R_3$, N=CR_2, N=C=O, N=C=S	1.48 – 1.20	21	21	20
	$\underline{N}$ = NO_2, $-\overset{+}{N}\langle\!\bigcirc\!\rangle$	1.73 – 1.48	9	10	21
$CH_3-\overset{\mid}{\underset{\mid}{C}}-\underline{P}$	$\underline{P}$ = PY_2, P(O)Y$_2$, P(S)Y$_2$	1.43 – 1.00	39	42	22
	$\underline{P}$ = $\overset{+}{P}R_3$	1.70 – 1.20	6	7	23
$CH_3-C(O-X)_{1,2}$		1.63 – 0.94	461	554	24
$CH_3-\overset{\mid}{\underset{\mid}{C}}-\underline{S}$	$\underline{S}$ = S-Y, SR_2, S=CR_2, SO-Y, SO_2-Y	1.58 – 1.15	158	175	25
$CH_3-\overset{\mid}{\underset{\mid}{C}}-F$		1.7 – 1.4	0	0	26
CH_3-CCl_n	n = 3 2 1	2.70	1	1	27
		2.50 – 2.18	4	4	
		1.75 – 1.43	8	8	
CH_3-CBr_n	n = 3 2 1	2.45	1	1	28
		1.90 – 1.60	12	12	
CH_3-CI_n	n = 3 2 1	1.90 – 1.60	5	5	29

CHART S1. Summary of Chemical Shifts of $CH_3-C-C-X$, CH_3-C-X, and CH_3-CX_n

Cross Index between Summary and Detailed Charts

Summary Chart (Line)	Detailed Charts (Lines) from which Data Were Taken
S2(1)	1(7), 8(12–17), 12(1), 16(2), 17(2,3), 19(1), 22(1), 23(7), 39(5,10), 40(1), 41(1), 50(1), 69(1), 70(1), 71(1), 83(1,8)
S2(2)	20(1,2), 21(1), 27(12), 50(1)
S2(3)	1(6), 12(2), 20(3), 21(1), 23(7), 27(12), 50(1), 70(1), 71(9)
S2(4)	17(2,3)
S2(5)	1(20), 24(1), 25(1)
S2(6)	84(1,7,11)
S2(7)	14(15)
S2(8)	13(1,8), 14(1,15), 17(1), 30(7), 32(1), 33(1), 34(1,6,7), 35(1,7), 36(1), 37(1), 42(1), 43(1), 44(1–6), 46(5,10), 52(7), 54(6), 58(4,8), 59(5), 61(1), 62(9), 77(3,8,12), 78(1), 85(7), 89(4)
S2(9)	27(16), 33(1), 34(1,7), 52(1), 58(8), 59(4), 61(1), 62(9,13), 78(1), 79(14), 85(7)
S2(10)	15(4,5,6)
S2(11)	55(1,14), 56(1), 57(1)
S2(12)	56(6,13)
S2(13)	56(17), 57(6)
S2(14)	18(1,15), 21(2), 48(5)
S2(15)	51(5,6)
S2(16)	X, 30(1), 64(1), 86(1)
S2(17)	64(14)
S2(18)	29(1,8), 60(1,8), 79(1,6,14)
S2(19)	80(1), 88(1,4,7)
S2(20)	81(1), 82(8), 88(2,3,5,6,8,9)
S2(21)	89(1)
S2(22)	89(1)
S2(23)	88(1–10)
S2(24)	89(5)
S2(25)	38(1,4,7,10)

X indicates data which are not included in detailed charts.

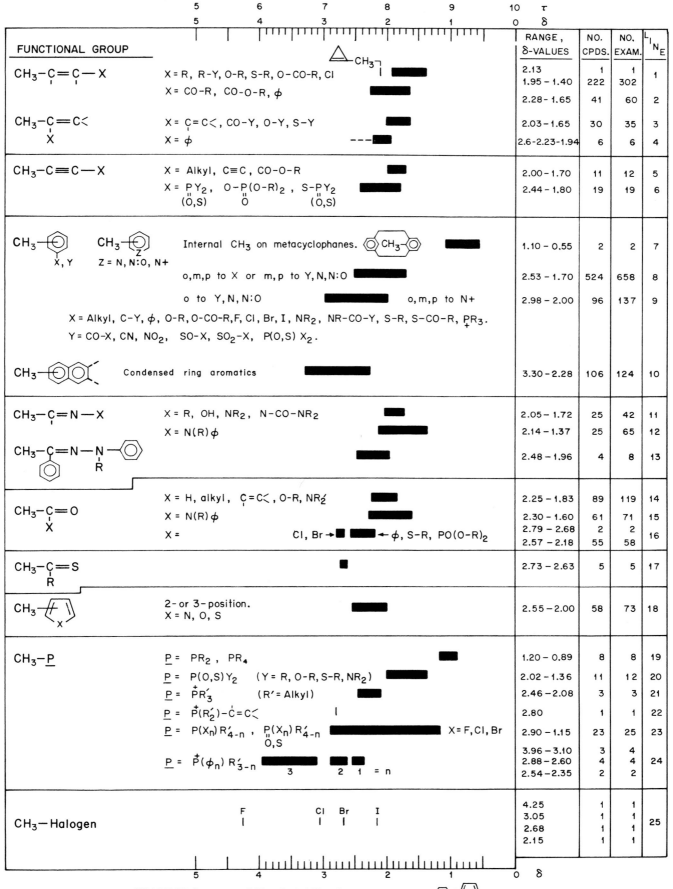

CHART S2. Summary of Chemical Shifts of CH₃ –(C=X, C≡X, ⌬, ⟨ₓ⟩, P, Hal.)

CHART S2

Cross Index between Summary and Detailed Charts

Summary Chart (Line)	Detailed Charts (Lines) from which Data Were Taken
S3(1)	65(1), 69(2)
S3(2)	64(4), 68(7), 70(2), 82(5)
S3(3)	64(10), 65(5), 67(7), 68(1)
S3(4)	65(8)
S3(5)	77(1)
S3(6)	66(10)
S3(7)	66(1)
S3(8)	66(4)
S3(9)	66(4)
S3(10)	66(11), 67(1,7,11)
S3(11)	67(8)
S3(12)	65(11)
S3(13)	66(10)
S3(14)	45(1), 46(1), 47(1)
S3(15)	46(1), 87(11)
S3(16)	48(1), 50(5), 51(1), 52(8), 53(1), 54(1,6,11,15,16), 87(5)
S3(17)	54(1)
S3(18)	54(12)
S3(19)	56(5)
S3(20)	56(10), 87(11)
S3(21)	63(1)
S3(22)	63(6)
S3(23)	63(9)
S3(24)	55(5)
S3(25)	55(9)
S3(26)	63(6)
S3(27)	63(9)
S3(28)	60(1,8)
S3(29)	45(5)
S3(30)	46.1(1–3)
S3(31)	62(13)
S3(32)	59(1)
S3(33)	18(11), 22(2), 23(1), 66(7)
S3(34)	18(5,8), 30(4), 57(11), 64(7), 67(4), 68(4), 82(1)

CHART S3. Summary of Chemical Shifts of CH₃−(S, N, O)

FUNCTIONAL GROUP		RANGE, δ-VALUES	NO. CPDS.	NO. EXAM.	LINE
CH_3-S-X	X = Alkyl	2.15 – 1.80	39	49	1
	X = C=C<, ϕ, CO-Y, P(O,S)Y₂	2.47 – 2.00	33	42	2
	X = CS-Y, CN, Sn-R, SO₂-R	2.74 – 2.30	33	40	3
$CH_3-S=C<^{Y_1}_{Y_2}$ (with X below S)	X = Alkyl	3.10 – 2.70	10	10	4
	X = ϕ	3.42 – 3.15	5	5	5
$(CH_3)_2 S=C<^{Y_1}_{Y_2}$ (with O below S)	Y₁ = H, Alkyl, Y₂ / Y₂ = CO-Z, CN	3.61 – 3.28	7	7	6
CH_3-S-X (O below S)	X = Alkyl, ϕ	2.52 – 2.24	6	10	7
	X = O-R, NR₂	2.75 – 2.44	9	10	8
	X = Cl	3.20	1	1	9
CH_3-S-X (O, O)	X = Alkyl, ϕ, O-R, S-R, NX₂	3.33 – 2.75	37	44	10
	X = Cl	3.52	1	1	11
$CH_3-S<^X_Y$ (+)	X = H, R, X₂ Y = CO-Y, CN	3.57 – 2.88	26	28	12
$(CH_3)_3 S=O$ (+)		3.88	1	1	13
$CH_3-N<^X_Y$	X = Y = Alkyl	2.38 – 2.10	19	19	14
	X = ϕ, PR₂, P(Z)Y₂ Y = Alkyl	3.00 – 2.32	92	100	15
	X = CO-Z, S-R, SO₂-R Y = Alkyl, ϕ	3.37 – 2.57	76	89	16
	X = CS-NH₂ Y = Alkyl	3.20 – 3.03	6	6	17
	X = CS-(R, S-R) Y = Alkyl	3.62 – 3.30	6	6	18
	X = N=CR₂ Y = Alkyl	2.67 – 2.32	9	18	19
	X = N=CR₂, P(Z)Y₂ Y = ϕ	3.28 – 2.76	17	20	20
	X = CH=N-R Y = Alkyl	2.97 – 2.80	4	4	21
	X = N=O Y = Alkyl, aralkyl	3.82 – 2.85	10	31	22
	X = N=O Y = ϕ	4.12 – –3.32	2	4	23
$CH_3-N=C<^X_R$	X = Alkyl R = Alkyl	3.25 – 2.84	11	16	24
	X = ϕ R = Alkyl	3.7 – 2.3	0	0	25
$CH_3-N=C=X$	X = O	3.3 – 2.7	0	0	26
	X = S	3.6 – 3.37–3.2	1	1	27
$CH_3-N\langle\text{pyrrole}\rangle$		3.93 3.05	2	2	28
$CH_3-N<^R_{R'}-X$ (+)	X = Alkyl R = Alkyl	3.50 – 2.63	25	27	29
	X = ϕ R = Alkyl	3.70 – 3.35	4	4	30
$CH_3-N\langle\text{pyridinium}\rangle$ (+)		4.41 – 4.23	2	3	31
CH_3-NO_2		4.30	1	1	32
CH_3-O-X	X = R(ex. ϕ), SO-Y	3.55 – 2.98 – 2.70	35	37	33
	X = ϕ, CO-Y, CS-Y, N=CR₂, SO₂-Y, SY₂, PY₂ (Z)	4.26 – 3.37	1104	1160	34

Cross Index between Summary and Detailed Charts

Summary Chart (Line)	Detailed Charts (Lines) from which Data Were Taken
S4(1)	Almost all charts
S4(2)	1(9), 8(10), 16(5), 18(16), 19(7), 22(4), 39(11), 40(5), 41(5), 69(4), 70(4), 71(4,10), 83(3,10)
S4(3)	20(4), 21(5), 27(14), 55(2,15), 56(2,7,14), 57(2,7), 64(15)
S4(4)	1(21), 24(3,10), 25(4), 84(2,8,12)
S4(5)	58(2)
S4(6)	29(2,9), 60(2,9), 61(3,9), 62(14), 79(2,7)
S4(7)	13(2,9), 14(2), 15(7), 16(4), 30(9), 31(3,9), 32(4), 33(2), 34(2,9), 35(2,8), 37(2), 42(4), 43(4), 46(7,11), 52(2,9), 58(5,9), 59(5), 73(4), 77(4,9,13), 78(2), 85(2,8), 89(8)
S4(8)	45(2), 46(2), 47(2), 48(2), 50(6), 51(2), 52(10), 53(2), 54(2,7,13,17), 60(2,9), 63(2,7,10), 87(12)
S4(9)	45(6), 46.1(4,5), 55(6,10), 63(14,17)
S4(10)	59(2), 62(14)
S4(11)	80(2), 81(2), 89(2,9)
S4(12)	18(12), 22(4), 23(2), 25(5), 30(5)
S4(13)	18(6,9), 23(8), 64(8), 66(8), 67(5), 68(5), 82(3)
S4(14)	64(5,11), 65(2), 68(8), 69(4), 70(4), 71(4), 73(5)
S4(15)	65(6,9,12), 66(2,5,12), 67(2,9,12), 68(2), 77(2), 82(6), 83(10)
S4(16)	38(2,5,8,11), 39(11), 40(6), 41(6), 42(5), 43(5)

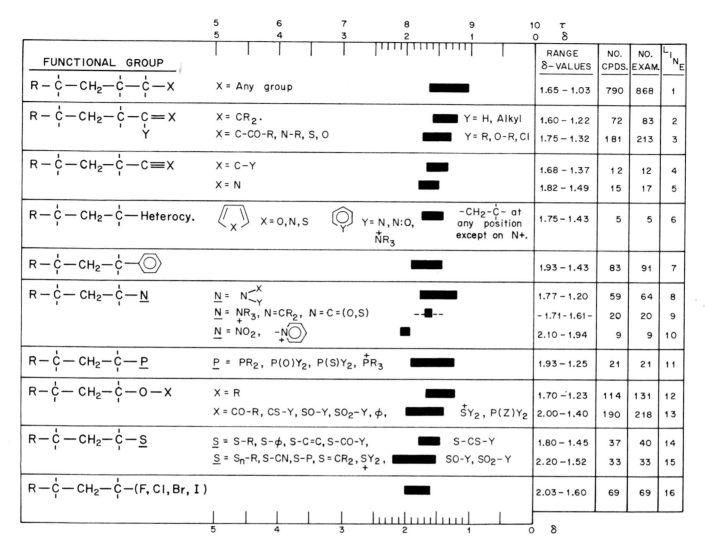

CHART S4. Summary of Chemical Shifts of R–C–CH₂–C–C–X and R–C–CH₂–C–X

R = H, Alkyl, C–Z.   Y, Z = Any reasonable group unless specifically identified.

Cross Index between Summary and Detailed Charts

Summary Chart (Line)	Detailed Charts (Lines) from which Data Were Taken
S5(1)	1(8), 16(5)
S5(2)	16(4), 19(6), 22(4), 69(4), 71(4), 83(3,10)
S5(3)	40(5), 41(5), X
S5(4)	24(3), 25(4), 58(2), X, 84(2,8,12, estimated from CH_3 data)
S5(5)	13(2)
S5(6)	16(4), 31(3,9), 32(4), 46(7), 50(10), 57(15), 58(5), 73(4), 85(2), 89(8)
S5(7)	61(3), 79(2)
S5(8)	18(2,16), 19(7), X
S5(9)	18(12), 23(2), X
S5(10)	45(2)
S5(11)	45(6), 55(10)
S5(12)	59(2)
S5(13)	64(5), 65(2)
S5(14)	66(12), 82(6)
S5(15)	66(12)
S5(16)	38(2,5,8,11), 40(6), 41(6), 42(5), X
S5(17)	40(5), 41(5), 42(4), 43(4), X

X indicates data not included in detailed charts.

FUNCTIONAL GROUP		RANGE δ-VALUES	NO. CPDS.	NO. EXAM.	LINE
R—C—CH₂—C—C=C— X = C=C		1.52 – 1.43	4	4	1
OR X	X = φ, CO-R, COOR, O-Y, S-Y	1.75 – 1.43	10	10	2
X—C—CH₂—C—C=C— X = Cl, Br, I		2.4-2.00-1.7	2	2	3
R—C—CH₂—C—C≡X OR Y—C—CH₂—C—C≡X X = Y = N		2.25 – 1.77	11	11	4
Y = COOR, (O,S)-P(Z)Y₂, P(Z)Y₂ X = C-R		1.75 – 1.53	10	10	5
(⬡)ₙC—CH₂—C(⬡)ₘ n+m = 6 5 4 3 2		4.25, 3.65 / 3.05, 2.50 / 1.97 – 1.94	1,0 / 0,2 / 2	1,0 / 0,2 / 2	6
R—C—CH₂—C—⬡ OR X—C—CH₂—C—⬡ X	X = C=C, CO-Y, CN, O-Y, NR₂, S-R, P(Z)Y₂, P⁺R₃	2.07 – 1.67	4	5	7
Het.—C—CH₂—C—Het. Het. = (thiophene) (pyridine)		2.13 – 1.95	2	3	8
(X—CO)ₙ C—CH₂—C (CO-X)ₘ X = R, O-R, Cl, Br n+m= 2,3		2.14 – 1.68	23	27	9
R—C—CH₂—C(O-R)ₙ OR (R-O)ₙC—CH₂—C(O-R')ₘ n+m= 2,3,4		2.00 – 1.37	17	17	10
N—C—CH₂—C—N N = NR₂		1.65 – 1.50	3	3	11
N = NR₃⁺, N=C—φ		2.10 – 2.05	2	2	12
N = NO₂		2.50	1	1	13
S—C—CH₂—C—S S = S-R, S-CO-R		1.94 – 1.65	15	15	14
S = SO₂-R, S-P(Z)Y₂		2.37 – 2.00	4	5	15
S = (SO₂-R)₂		3.10	1	1	16
R—C—CH₂—CXₙ X = F, Cl, Br, I n = 2		2.36 – 2.20	8	8	17
Y—C—CH₂—C—X X = F, Cl, Br, I Y = C=C, φ, CO-R, NR₂, SO₂ O-R, O-CO-R, P⁺R₃		2.50 – 1.86	42	42	18

R = H, Alkyl, C–Z.   Y, Z = Any reasonable group unless specifically identified.

CHART S5. Summary of Chemical Shifts of R–C–CH₂–CXₙ and YₙC–CH₂–CXₙ

Cross Index between Summary and Detailed Charts

Summary Chart (Line)	Detailed Charts (Lines) from which Data Were Taken
S6(1)	1(10), 16(5), 19(5), 22(5), 23(8), 39(6,11), 40(4), 41(4), 69(3), 70(3), 71(3,10), 72(1,7), 83(2,9)
S6(2)	20(4), 21(4)
S6(3)	18(2,16), 19(7), 48(6)
S6(4)	21(5), 30(2), 64(2), 86(2), X
S6(5)	56(7,14)
S6(6)	55(2,15), 56(2,7,14), 57(2)
S6(7)	57(7)
S6(8)	64(15)
S6(9)	1(21), 24(2,10), 25(3), 84(8,12)
S6(10)	84(2,8,12)
S6(11)	58(2)
S6(12)	13(2,9), 14(2), 16(3), 30(8), 31(2,8), 32(2), 33(2), 34(2,9), 35(2,8), 37(2), 42(3), 43(3), X, 46(6,11), 52(2,9), 54(7), 58(5,9), 59(5), 61(3,10), 73(3), 77(4,9,13), 78(2), 85(2,8), 89(7)
S6(13)	33(2), 34(2,9), 52(2), 58(9), 59(5), 61(3,10), 62(14), 77(13), 78(2), 85(8)
S6(14)	15(7)
S6(15)	80(2), 81(2), 82(9), 85(3)
S6(16)	88(11), 89(2,9)
S6(17)	89(8)
S6(18)	18(12), 22(4), 23(2), 66(8), 74(7,8)
S6(19)	18(6,9), 23(8), 30(5), 36(3), 57(12), 64(8), 67(5), 68(5), 82(3), X
S6(20)	X
S6(21)	38(8), 39(11)
S6(22)	38(2,5), 40(6), 41(6), 42(5), 43(5)
S6(23)	38(11)

X indicates data not included in detailed charts.

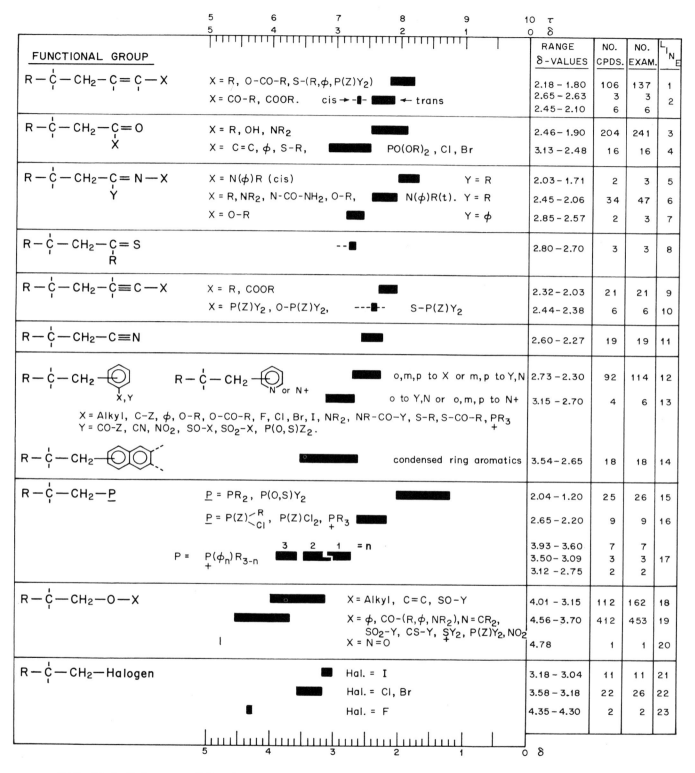

CHART S6. Summary of Chemical Shifts of R−C−CH₂−(C=X, C≡X, ⬡, P, O, Hal.)

Cross Index between Summary and Detailed Charts

Summary Chart (Line)	Detailed Charts (Lines) from which Data Were Taken
S7(1)	65(2), 69(4), 70(4), 73(5)
S7(2)	64(5), 65(6), 68(2,8), 71(4)
S7(3)	64(11)
S7(4)	65(9), 77(2)
S7(5)	66(5)
S7(6)	66(2,5)
S7(7)	67(2)
S7(8)	66(12), 67(12)
S7(9)	67(9)
S7(10)	65(12)
S7(11)	82(6), X
S7(12)	45(2)
S7(13)	46(2), 48(2), 50(6), 51(2), 54(2,17), 87(12), also see 53(2)
S7(14)	54(2), 63(2), 86(6)
S7(15)	54(13)
S7(16)	63(7)
S7(17)	63(10)
S7(18)	52(10), 54(7)
S7(19)	55(6)
S7(20)	55(10)
S7(21)	63(14)
S7(22)	63(17)
S7(23)	53(2), 60(2)
S7(24)	45(6)
S7(25)	46.1(4,5)
S7(26)	62(14)
S7(27)	59(2)

X indicates data not included in detailed charts.

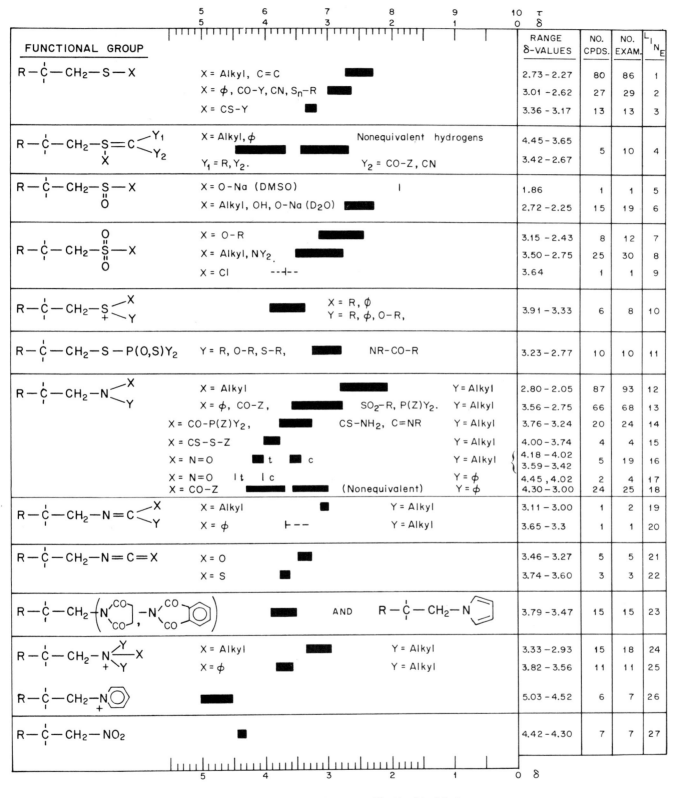

FUNCTIONAL GROUP		RANGE δ-VALUES	NO. CPDS.	NO. EXAM.	LINE
R–C–CH₂–S–X	X = Alkyl, C=C	2.73 – 2.27	80	86	1
	X = φ, CO-Y, CN, Sₙ–R	3.01 – 2.62	27	29	2
	X = CS-Y	3.36 – 3.17	13	13	3
R–C–CH₂–S=C‹Y₁,Y₂ (X)	X = Alkyl, φ Nonequivalent hydrogens	4.45 – 3.65	5	10	4
	Y₁ = R, Y₂. Y₂ = CO-Z, CN	3.42 – 2.67			
R–C–CH₂–S–X (O)	X = O–Na (DMSO)	1.86	1	1	5
	X = Alkyl, OH, O–Na (D₂O)	2.72 – 2.25	15	19	6
R–C–CH₂–S–X (O,O)	X = O–R	3.15 – 2.43	8	12	7
	X = Alkyl, NY₂.	3.50 – 2.75	25	30	8
	X = Cl	3.64	1	1	9
R–C–CH₂–S⁺‹X,Y	X = R, φ Y = R, φ, O–R,	3.91 – 3.33	6	8	10
R–C–CH₂–S–P(O,S)Y₂	Y = R, O–R, S–R, NR–CO–R	3.23 – 2.77	10	10	11
R–C–CH₂–N‹X,Y	X = Alkyl Y = Alkyl	2.80 – 2.05	87	93	12
	X = φ, CO-Z, SO₂–R, P(Z)Y₂. Y = Alkyl	3.56 – 2.75	66	68	13
	X = CO-P(Z)Y₂, CS-NH₂, C=NR Y = Alkyl	3.76 – 3.24	20	24	14
	X = CS-S-Z Y = Alkyl	4.00 – 3.74	4	4	15
	X = N=O t c Y = Alkyl	4.18 – 4.02 / 3.59 – 3.42	5	19	16
	X = N=O I t I c Y = φ	4.45, 4.02	2	4	17
	X = CO-Z (Nonequivalent) Y = φ	4.30 – 3.00	24	25	18
R–C–CH₂–N=C‹X,Y	X = Alkyl Y = Alkyl	3.11 – 3.00	1	2	19
	X = φ Y = Alkyl	3.65 – 3.3	1	1	20
R–C–CH₂–N=C=X	X = O	3.46 – 3.27	5	5	21
	X = S	3.74 – 3.60	3	3	22
R–C–CH₂–(N‹CO,CO , –N‹CO,CO) AND R–C–CH₂–N⟨pyrrole⟩		3.79 – 3.47	15	15	23
R–C–CH₂–N⁺‹Y,Y–X	X = Alkyl Y = Alkyl	3.33 – 2.93	15	18	24
	X = φ Y = Alkyl	3.82 – 3.56	11	11	25
R–C–CH₂–N⁺⟨pyridinium⟩		5.03 – 4.52	6	7	26
R–C–CH₂–NO₂		4.42 – 4.30	7	7	27

R = H, Alkyl, C–Z.   Y, Z = Any reasonable group unless specifically identified.

CHART S7. Summary of Chemical Shifts of R–C–CH₂–(S, N)

Cross Index between Summary and Detailed Charts

Summary Chart (Lines)	Detailed Charts (Lines) from which Data Were Taken
S8(1)	1(9), 19(5), 22(3)
S8(2)	16(4), 19(5), 20(4), 21(4), 22(3), 69(3), 70(3), 71(3,10), 72(2,7), 83(2,9)
S8(3)	19(5), X, 39(6,11), 40(4), 41(4)
S8(4)	21(5)
S8(5)	18(16), 31(9), X
S8(6)	31(9), X
S8(7)	31(9)
S8(8–10)	X
S8(11)	19(6), 24(3)
S8(12)	18(2), 31(3), X
S8(13)	X
S8(14)	31(3)
S8(15)	48(6), X
S8(16)	X
S8(17)	1(21), 24(2), 25(3)
S8(18)	58(5), X
S8(19)	X
S8(20)	16(3)
S8(21)	13(2), 31(2,8), 46(6), 50(16), 58(5), 73(3), 85(2)
S8(22)	30(8), 32(2), 42(3), 43(3), 57(15), X
S8(23)	13(2), 32(3)
S8(24)	32(3)
S8(25)	80(2), 82(9), 84(3), 85(2)
S8(26)	81(2)

X indicates data not included in detailed charts.

Functional Group	Substituents	Range δ-values	No. Cpds.	No. Exam.	Line
R—C—CH₂—C=C— OR X—C—CH₂—C=C— (X)	X = C=C	2.18 - 1.85	11	12	1
	X = φ, CO-R, COOR, O-R, O-P(O)(OR)₂, S-R, S-φ, S-CO-R, S-P(O,S)(OR)₂	2.45 - 2.05	25	28	2
	X = (COOR)₂ Quinone ring, SO-R, F, Cl, Br, I	2.65 - 2.42	5	6	3
X—C—CH₂—C=O (Y)	X = C=C Y = R	2.65 - 2.40	3	3	4
	X = φ, CO-R, COOR, CN, NRφ, S-R, O-R Y = R	2.88 - 2.40	20	24	5
	X = φ₂, CO-φ, ⁺NH₃, P(O)(OR)₂ Y = R	3.41 - 2.91	8	9	6
	X = φ₃ Y = R	4.0 - 3.5	0	0	7
	X = φ, CO-R, COOφ, Cl Y = φ	3.52 - 2.87	11	13	8
	X = COOR Y = Cl	2.62	1	1	9
	X = φF₅, Br Y = Cl	3.47 - 3.18	2	2	10
	X = C=C, C≡C Y = O-R	2.35 - 2.09	6	8	11
	X = φ, CO-R, COOR, CN, NR₂, N-CO-R, N-CO-N, N-CS-N, S-R, Sₙ-R, S-SO₂-R, SO-R, SO₂-R, SO₂-OH, O-R, O-φ Y = O-R	2.88 - 2.30	61	66	12
	X = Cl, Br, I Y = O-R	3.07 - 2.70	4	4	13
	X = φₙ (n=3) Y = O-R	4.00 - 3.55	4	4	14
	X = φₙ (n=2) Y = O-R	3.30 - 3.06	7	7	14
	X = φ, CO-NR₂, S-R, S-CS-NR₂, O-R, O-φ Y = NR₂	2.74 - 2.29	15	15	15
	X = ⁺NR₃, Cl Y = NR₂	3.25 - 2.72	3	3	16
X—C—CH₂—C≡C—R	X = C≡CR, COOR, O-R	2.42 - 2.25	4	6	17
X—C—CH₂—C≡N	X = φ, NR₂, NRφ, S-R, O-R	2.69 - 2.36	29	29	18
	X = SO₂-O-Na, Cl, Br	2.97 - 2.80	3	3	19
X—C—CH₂—⟨◯⟩	X = C=C	2.64 - 2.60	2	2	20
	X = φ, CO-(R,φ), COOR, CN, NR₂, NR-CO-(R,φ), S-R, P(O)(OR)₂	3.02 - 2.55	42	44	21
	X = NR₃, O-R, O-CO-R, ⁺O-N=CR₂, Cl, Br	3.22 - 2.69	31	34	22
X—C⟨◯⟩—CH₂—⟨◯⟩	X = φ, OH	3.23 - 2.75	4	5	23
X—C⟨◯⟩₂—CH₂—⟨◯⟩	X = OH	3.50 - 3.43	2	2	24
X—C—CH₂—P	P = PR₂, Pφ₂, P(O)R₂, P(O)(OR)₂ X = C≡C, φ, P	2.25 - 1.99	8	8	25
	P = P(O)φ X = P(O)Rφ	2.86 - 2.37	4	6	26

R = H, Alkyl, C-Z.

CHART S8. Summary of Chemical Shifts of X—C—CH₂—(C=Y, C≡Y, ⟨◯⟩, P)

Cross Index between Summary and Detailed Charts

Summary Chart (Lines)	Detailed Charts (Lines) from which Data Were Taken
S9(1)	69(4), X
S9(2)	65(2), 73(4), X
S9(3)	X
S9(4)	71(4)
S9(5)	73(4), X
S9(6)	X
S9(7)	72(1), X
S9(8,9)	X
S9(10)	X
S9(11)	66(12), X
S9(12)	45(2), X
S9(13)	46(7), X
S9(14–16)	X
S9(17)	X, 53(2)
S9(18–23)	X
S9(24)	55(6)
S9(25)	55(10)
S9(26)	X
S9(27)	45(6), X
S9(28)	X

X indicates data not in detailed charts.

FUNCTIONAL GROUP		RANGE δ-VALUES	NO. CPDS.	NO. EXAM.	LINE
$X-\overset{\mid}{\underset{\mid}{C}}-CH_2-S-Y$	X = C=C, C≡C Y = R	2.7-2.5-2.3	1	1	1
	X = φ, COOR, CO-NR₂, CN, NR₂, NRφ, NH-COOR, S-R, O-R, O-CO-R Y = R	2.90 – 2.54	32	33	2
	X = $\overset{+}{NR_3}$ Y = R	3.20 – 2.85	2	2	3
	X = C=C Y = φ	3.0 – 2.7	0	0	4
	X = φ, NR₂ Y = φ	3.23 – 3.07	3	3	5
	X = COOR, $\overset{+}{NH_3}$ Y = Sₙ-R	3.12 – 2.85	4	4	6
	X = C=C, CO-Z, $NH-\overset{+}{C(NH_2)_2}$ Y = CO-Z	3.12 – 2.88	7	8	7
	X = $\overset{+}{NR_3}$ Y = CO-Z	3.8-3.5-3.3	1	1	8
	X = CO-NH₂ Y = CS-(NR₂, NRφ)	3.48 – 3.40	2	2	9
$X-\overset{\mid}{\underset{\mid}{C}}-CH_2-\overset{O}{\underset{O}{S}}-Y$	X = φ, CH, O-R Y = R, O-R, O-Na	3.47 – 3.15	4	4	10
	X = COOR, SO₂-R, Cl Y = R, O-R, O-Na	3.80 – 3.35	10	11	11
$X-\overset{\mid}{\underset{\mid}{C}}-CH_2-\overset{\mid}{\underset{R}{N}}-Y$	X = NR, O-R, O-φ Y = R	2.90 – 2.30	24	24	12
	X = φ, COOR, CN, S-R, O-CO-R, Cl Y = R	2.97 – 2.52	88	88	13
	X = NR₂, O-R Y = φ	3.51 – 2.87	8	8	14
	X = φ, CO-R, CN Y = φ	3.66 – 3.38	5	5	15
	X = Cl Y = φ	3.90 – 3.69	5	5	16
	X = NR-COOR, NR-CO-R Y = CO-(R, O-R, NR₂)	3.30 – 3.20	2	2	17
	X = φ, COOR, O-CO-R Y = CO-(R, O-R, NR₂)	3.80 – 3.41	11	11	13
	X = O-R, $\overset{+}{NR_3}$ Y = CO-(R, O-R, NR₂)	3.95 – 3.24	6	6	19
	X = COOR Y = CS-(NRφ, CS-NR₂)	3.72 – 3.32	7	8	20
	X = O-R Y = CS-CS-R	3.88	1	1	21
	X = S-CO-R Y = $C\overset{NH_2}{\underset{NH_2}{<}}$ ⊕	3.59 – 3.49	3	3	22
	X = O-φ Y = $C\overset{NH_2}{\underset{NH_2}{<}}$ ⊕	3.81 – 3.70	7	7	23
$X-\overset{\mid}{\underset{\mid}{C}}-CH_2-N=\overset{\mid}{\underset{R}{C}}-Y$	X = N=CR₂ Y = R	3.5-3.42-3.1	1	1	24
	X = N=CRφ Y = φ	3.97 – 3.72	4	4	25
$X-\overset{\mid}{\underset{\mid}{C}}-CH_2-\overset{+}{\underset{R_2}{N}}-Y$	X = φ, S-R, S-CO-φ, S-S-R Y = R	3.50 – 3.10	12	12	26
	X = CO-Nφ, $\overset{+}{NR_3}$, O-R, O-φ, O-CO-(φ,R), O-CS-φ, Cl, Br Y = R	4.03 – 3.02	57	58	27
	X = $\overset{+}{NR_2}φ$ Y = φ	4.19	1	1	28

R = H, Alkyl, C–Z .   Z = Any reasonable group.

CHART S9. Summary of Chemical Shifts of X–C–CH₂–(S, N)

Cross Index between Summary and Detailed Charts

Summary Chart (Lines)	Detailed Charts (Lines) from which Data Were Taken
S10(1)	22(4), 25(4), X
S10(2)	32(4), X
S10(3)	18(13), X
S10(4—8)	X
S10(9)	18(6), X
S10(10,11)	X
S10(12)	18(9), X
S10(13)	X
S10(14)	30(8), X
S10(15)	X
S10(16)	74(7)
S10(17)	X
S10(18)	57(15)
S10(19)	39(6), X
S10(20)	38(2,5), 40(5), 41(5), 42(4), X
S10(21)	38(2), X

X indicates data not included in detailed charts.

CHART S10. Summary of Chemical Shifts of X–C–CH₂–(O,F,Cl,Br,I)

FUNCTIONAL GROUP	X =	Y =	RANGE δ-VALUES	NO. CPDS.	NO. EXAM.	LINE
$X-\overset{|}{\underset{|}{C}}-CH_2-O-Y$	C=C, C–C	Y = R	3.75 – 3.36	9	9	1
	φ, CO-Z, CN, S-R, N-CO-R, N-SO₂-Z, SO₂-O-Na, F, Cl, Br	Y = R	3.99 – 3.40	65	65	2
	NR₂, NRφ, O-R, O-φ, O-CO-Z	Y = R	4.05 – 3.20	180	220	3
	$\overset{+}{N}R_3$, NO₂, φ₂, F₂, Cl₂, Br₂	Y = R	4.32 – 3.72	23	23	4
	NR₂, O-R	Y = φ	4.43 – 3.71	54	54	5
	COOR, CO-NH₂, Br	Y = φ	4.42 – 4.21	5	5	6
	$\overset{+}{N}R_3$, O-CO-R	Y = φ	4.73 – 4.06	51	51	7
	φ, NR₂, S-R	Y = CO-R	4.32 – 4.00	15	15	8
	O-R, O-φ, O-CO-R, Cl	Y = CO-R	4.44 – 3.95	37	40	9
	$\overset{+}{N}R_3$	Y = CO-R	4.71 – 4.41	11	11	10
	φ, O-R	Y = CO-φ	4.57 – 4.35	12	12	11
	$\overset{+}{N}R_3$, O-φ, O-CO-φ, Cl	Y = CO-φ	4.88 – 4.60	43	43	12
	Cl, Cl₃	Y = CO-Cl	4.87 – 4.54	2	2	13
	φ, $\overset{+}{N}R_3$, O-φ, O-CO-NRφ, Cl, Br	Y = CO-(NR₂, NRφ)	4.54 – 4.22	11	11	14
	$\overset{+}{N}R_3$	Y = CS-φ	5.30 – 5.28	2	2	15
	φ Nonequivalent	Y = SO-φ	3.93, 3.30	1	2	16
	O-φ, Cl	Y = SO₂-φ	4.34 – 4.16	5	7	17
	φ	Y = N = CR₂	4.37 – 4.30	3	3	18
$X-\overset{|}{\underset{|}{C}}-CH_2-F$	C=C, OH --|--		4.6-4.37-4.1	1	1	19
$X-\overset{|}{\underset{|}{C}}-CH_2-(Cl, Br, I)$	C=C, φ, CO-Z, CN, O-R, O-φ, O-CO-Z, O-SO₂-R, S-R, S-φ, O-P(O)R(O-R), Cl, Br, I		4.10 – 3.30	75	81	20
	NRφ, N-CO-R, $\overset{+}{N}R_3$, N(N=O)-CO-NRφ, Cl₂		4.30 – 3.55	24	24	21

R = H, Alkyl C-Z.   Z = Any reasonable group.

CHART S10. Summary of Chemical Shifts of X–C–CH₂ –(O,F,Cl,Br,I)

Cross Index between Summary and Detailed Charts

Summary Chart (Lines)	Detailed Charts (Lines) from which Data Were Taken
S11(1)	1(10, 19(4), 22(3), X
S11(2)	16(3), 19(5), 69(3), X
S11(3)	71(3), 72(1), 83(9), X
S11(4)	39(11), 41(4), 72(6,10)
S11(5)	22(3), 40(4), X
S11(6)	23(8), 83(2)
S11(7)	X
S11(8)	18(16), 31(8), X
S11(9–19)	X
S11(20)	19(5), 24(2), X
S11(21)	18(2), 31(2), X
S11(22–35)	X
S11(36)	84(3)
S11(37)	24(2), X
S11(38)	84(12)
S11(39)	X
S11(40)	25(3), X
S11(41)	X, 84(8)
S11(42–44)	X
S11(45)	58(5), X
S11(46,47)	X

X indicates data not included in detailed charts.

FUNCTIONAL GROUP	X =	RANGE δ-VALUES	NO. CPDS.	NO. EXAM.	LINE
X—CH₂—C=C—	X = C=C, P(O)(OR)₂	2.83 – 2.48	21	22	1
	X = φ, CN, COOR, S–R, NR₂	3.39 – 2.82	40	41	2
	X = S–φ, S–CO–R, S–P(O,S)(OR)₂, S–CN	3.62 – 3.35	43	43	3
	X = SO₂–R, S–(Benzothiazole-2), Br, I	4.04 – 3.77	13	15	4
	X = NR₃(q), O–R, N=C=S, Cl	4.25 – 3.83	20	23	5
	X = ⁺O–CO–R, O–P(O)(OR)₂, F	4.68 – 4.35	5	5	6
X—CH₂—C=O (R)	X = COOH, NR₂, P(O)(OR)₂	3.42 – 2.91	8	9	7
	X = φ, CO–R, S–R, S–φ	3.70 – 3.42	11	11	8
	X = SO₂–R	4.56 – 4.37	2	2	9
X—CH₂—C=O (φ)	X = S–R	3.78 – 3.73	3	3	10
	X = φ, S–φ	4.37 – 3.95	14	14	11
	X = CN, Br	4.52 – 4.32	10	11	12
	X = SO–(R, φ) Nonequivalent	4.35 – 4.19	8	8	13
		4.60 – 4.35	8	8	14
	X = S–CN, SO₂–R, SO₂–φ, N–CO–R, ⁺NR₃ (pri. & sec.)	5.05–4.82–4.53	13	14	15
	X = O–φ, N in phthalimide	5.18 – 5.13	2	2	16
	X = ⁺SR₂ in TFA	5.17 – 4.90	2	2	17
	in CDCl₃	5.55 – 5.39	4	4	18
	X = O–CO–R	5.53 – 5.22	11	11	19
X—CH₂—C=O (OR)	X = C=C, C≡C, NR₂ (in esters)	3.17 – 3.00	5	5	20
	X = φ, CN, CO–R, COOR, S–R	3.62 – 3.10	21	24	21
	X = COOφ, CO–CF₃, Sₙ–R, NR₂(acids), N=C=O, Br, I	4.03 – 3.47	25	28	22
	X = S–CS–NR₂, SO–R, NR–CO–φ, NH–SO₂–φ, O–R, Cl	4.31 – 3.80	16	19	23
	X = NR₂ (pri., tert. & quat.)	4.46 – 3.59	7	7	24
	X = ⁺SO₂–R, NH–CS–CS–NR₂, N–CS–S–C–CO	4.61 – 4.24	6	8	25
	X = O–φ	4.81 – 4.55	10	11	26
	X = ⁺SR₂	5.25	1	1	27
X—CH₂—C=O (Y)	X = φ, φF₅ Y = Cl	4.26 – 4.00	2	4	28
	X = Br Y = CO–R	4.30	1	1	29
X—CH₂—C=O (NR₂)	X = NR₂, P(O)(OR)₂	3.20 – 2.53	13	14	30
	X = φ, CO–R, CO–NR₂, CN, SR, S–CO–R	3.64 – 3.00	27	27	31
	X = NH–CO–φ, O–R, Cl, I	4.38 – 3.89	12	12	32
	X = CO–φ, O–φ, F, ⁺Pφ₃	5.11 – 4.64	8	8	33
X—CH₂—C=N—Y	X = φ Y = NR–CS–NR₂	3.52	1	1	34
	X = NO₂ Y = OH	5.40 – 5.22	2	2	35
X—CH₂—C≡C—R	X = P(O)(OR)₂	2.9–2.70–2.6	1	1	36
	X = COOR, NR₂	3.40 – 3.10	18	18	37
	X = S–P(O,S)(OR)₂	3.66 – 3.55	3	3	38
	X = NR–CO–R, Br	4.01 – 3.83	3	3	39
	X = O–R	4.28 – 4.05	17	17	40
	X = N in phthalimide, O–φ, O–CO–R, O–P(O)(OR)₂	4.71 – 4.44	7	8	41
	X = O–(C in s-triazine)	5.09	1	1	42
X—CH₂—C≡C—⬡	X = ⁺NR₃(quat.)	5.28 – 5.19	2	2	43
X—CH₂—C≡N	X = C=C, COOR	3.48 – 2.15	4	4	44
	X = φ, NR₂, NRφ	3.92 – 3.45	24	24	45
	X = CO–φ, CO–NH–COOR, O–R, Cl	4.52 – 4.11	5	5	46
	X = O–φ, O–CO–R	5.05 – 4.67	9	9	47

R = H, Alkyl, C–Z.   Z = Any reasonable group.

CHART S11. Summary of Chemical Shifts of X—CH₂—(C=Y, C≡Y)

Cross Index between Summary and Detailed Charts

Summary Chart (Lines)	Detailed Charts (Lines) from which Data Were Taken
S12(1)	16(3), 85(2)
S12(2)	31(2), X
S12(3)	31(8), 46(6), 50(10), 58(5), 73(3), 74(1,3)
S12(4)	13(2), X, 73(2), 74(1,6)
S12(5)	X, 74(12), 89(7)
S12(6)	32(2), 43(3), 44(6), 74(6), 89(7), X
S12(7–9)	X
S12(10)	42(3), 44(5,7), 74(7,8,11)
S12(11)	74(9)
S12(12,13)	X
S12(14)	X, 32(3), 57(15)
S12(16,17)	63(10)
S12(18)	30(8), 89(7), X
S12(19)	74(10)
S12(20)	80(2)
S12(21)	80(8), 88(11)
S12(22)	X
S12(23)	81(2)
S12(24,25)	X
S12(26)	81(2)
S12(27)	82(9), X
S12(28)	86(8,9), 88(13)
S12(29)	X, 88(13)
S12(30)	88(11,14)
S12(31)	89(2), X
S12(32)	89(7), X
S12(33)	X
S12(34)	89(7)
S12(35–37)	X
S12(38)	89(7), X
S12(39)	X
S12(40)	39(6)
S12(41)	X
S12(42)	38(11)
S12(43)	80(8), 88(11,13)
S12(44)	88(14)
S12(45)	40(4), X
S12(46)	88(11,14), X
S12(47)	42(3), 44(5,7), X
S12(48)	38(2), X
S12(49)	X
S12(50)	88(13)
S12(51)	38(8), 39(11), 41(4), X
S12(52)	43(3), X
S12(53)	38(5), X
S12(54)	X

X indicates data not included in detailed charts.

CHART S12. Summary of Chemical Shifts of X–CH$_2$–(⬡, P,F,Cl,Br,I)

FUNCTIONAL GROUP		RANGE δ-VALUES	NO. CPDS.	NO. EXAM.	LINE
X–CH$_2$–⬡	X = C=C, P(O)(OR)$_2$	3.35–2.96	9	9	1
	X = COOR	3.60–3.25	6	6	2
	X = CO-R, CO-NR$_2$, CN, S-R, S-S-R, NR$_2$	3.94–3.30	70	77	3
	X = φ, CO-Cl, S-φ, S$_n$-(R,φ), S-CO-C	4.26–3.65	56	60	4
	X = CO-φ, S-CN, SO$_2$-Y, NH-COOR, $\overset{+}{P}$R$_3$, $\overset{+}{P}$R$_2$φ	4.52–3.84	25	30	5
	X = S-CS-SR, NRφ, N≡C, N(imide), O-R, Br, I, $\overset{+}{P}$φ$_2$R	4.74–4.12	50	63	6
	X = Nφ-CO-(R,φ) Nonequivalent	4.87–3.53	7	7	7
		5.70–4.73	7	7	8
	X = NR-CO-(R,φ) Includes cis & trans	4.80–4.00	15	19	9
	X = S(C)=CY$_2$, O-SO-(φ,Oφ), Cl	4.93–4.05	36	49	10
	X = $\overset{+}{S}$R$_2$	5.02–4.68	6	9	11
	X = $\overset{+}{N}$R$_3$ Secondary	4.39–3.65	7	8	12
	Quaternary	5.15–4.49	12	12	13
	X = NH-CS-Y, O-φ, O-N=CR$_2$	5.26–4.80	17	17	14
	X = NR-N=O cis	4.80–4.55	4	5	16
	trans	5.30–5.13	4	5	17
	X = O-CO-(R,NR$_2$), O-P(O)(OR)$_2$, $\overset{+}{P}$φ$_3$, F	5.36–4.82	38	45	18
	X = S(φ)=CY$_2$ Nonequivalent	5.80, 4.60	1	2	19
X–CH$_2$–P(R,φ)$_2$	X = P(R,φ)$_2$	2.80	1	1	20
	X = Cl	3.78–3.76	2	2	21
	X = O-R	4.28	1	1	22
X–CH$_2$–PY$_2$ ‖ O	X = φ Y = R	3.15–3.00	2	2	23
	X = CO-φ, Cl, Br Y = R	3.96–3.59	3	3	24
	X = COONa Y = φ	3.59	1	1	25
	X = P(O)φ$_2$ Y = φ	4.83–3.56	8	4	26
X–CH$_2$–P(OR)$_2$ (‖ O,S)	X = P(O,S)(OR)$_2$	2.61–2.25	8	8	27
	X = CO-R, CO-NR$_2$, Br, I	3.24–2.53	16	17	28
	X = O-R, Cl	3.75–3.49	8	9	29
X–CH$_2$–(PF$_2$, PF$_4$, P(O)F$_2$, P(O)Cl$_2$) X = Cl		4.40–3.35	7	7	30
X–CH$_2$–$\overset{+}{P}$Y$_3$	X = C=C, COOH Y$_3$ = R$_3$	3.88–3.36	4	4	31
	X = φ, CN Y$_3$ = R$_3$	4.65–3.95	5	7	32
	X = CO-φ, Br Y$_3$ = R$_3$	5.04–4.81	2	2	33
	X = φ Y$_3$ = φ$_2$R	4.50	1	1	34
	X = O-R Y$_3$ = φ$_2$R	5.60–5.26	2	2	35
	X = Cl Y$_3$ = φ$_2$R	5.93	1	1	36
	X = C=CHφ, C=CR$_2$ Y$_3$ = φ$_3$	5.00–4.43	2	3	37
	X = φ, CO-NH$_2$ Y$_3$ = φ$_3$	5.36–4.82	5	7	38
	X = COOR, O-R Y$_3$ = φ$_3$	5.82–5.42	4	5	39
X'–CH$_2$–F	X = C=C	4.68	1	1	40
	X = φ, CO-NR$_2$	5.27–4.87	4	4	41
	X = F	5.5–5.1	0	0	42
X–CH$_2$–Cl	X = P(R,φ)$_2$, P(O)R$_2$, P(O,S)(OR)$_2$	3.96–3.49	10	11	43
	X = P(Z)F$_2$	4.15–3.35	3	3	44
	X = C=C, C≡C, COOR, CN	4.17–3.83	17	24	45
	X = CO-NR$_2$, SO$_2$-NR$_2$, P(O)Cl$_2$, PF$_4$	4.48–3.98	16	16	46
	X = φ, CO-φ, S-R	4.93–4.05	36	43	47
	X = Cl, Br	5.30–5.18	3	3	48
	X = $\overset{+}{P}$φ$_2$R, $\overset{+}{P}$φ$_3$	6.32–5.93	4	4	49
X–CH$_2$–(Br, I)	X = P(O,S)(OR)$_2$	3.24–3.00	3	3	50
	X = C=C, C≡C, COOR, P(O)R$_2$, I	3.99–3.59	10	12	51
	X = φ, CO-CO-R, CO-φ, CO-NR$_2$	4.69–4.12	28	35	52
	X = $\overset{+}{P}$R$_3$, Br, Cl	5.18–4.81	3	3	53
	X = $\overset{+}{P}$φ$_3$	6.32–6.00	3	3	54

R = H, Alkyl, C–Z.   Y, Z = Any group unless specifically identified.

CHART S12. Summary of Chemical Shifts of X–CH$_2$–(⬡, P,F,Cl,Br,I)

Cross Index between Summary and Detailed Charts

Summary Chart (Line)	Detailed Charts (Lines) from which Data Were Taken
S13(1)	69(3), X
S13(2)	65(2), X
S13(3)	73(3), X
S13(4)	X
S13(5)	71(3), X
S13(6)	73(2), X
S13(7,8)	X
S13(9)	83(9), 84(12)
S13(10–13)	X
S13(14)	74(11)
S13(15)	74(10)
S13(16)	X
S13(17)	68(2)
S13(18–21)	X
S13(22)	76(6), X
S13(23)	74(12)
S13(24)	X
S13(25)	66(12), X
S13(26–29)	X
S13(30)	74(9)
S13(31–34)	X
S13(35)	22(3), 25(3), X
S13(36)	32(2)
S13(37)	18(12), X
S13(38,39)	X
S13(40)	32(3), X
S13(41)	X
S13(42)	83(2), 84(8)
S13(43)	X
S13(44)	74(7,8)
S13(45)	X
S13(46)	23(8), X
S13(47)	30(8), X
S13(48,49)	X
S13(50)	30(8), 57(15), X

X indicates data not included in detailed charts.

FUNCTIONAL GROUP		RANGE δ-VALUES	NO. CPDS.	NO. EXAM.	LINE
X—CH₂—S—R	X = C=C, CO-R	3.15 - 2.82	24	24	1
	X = COOR, CO-NHφ, S-R	3.62 - 3.10	18	20	2
	X = φ, CO-φ	3.86 - 3.45	26	32	3
	X = O-CO-R, Cl	4.86 - 4.70	3	3	4
X—CH₂—S—⬡	X = C=C, CO-R	3.48 - 3.35	8	8	5
	X = φ, CO-φ	4.24-3.94-3.68	20	20	6
	X = NRφ, Cl	4.89 - 4.72	6	6	7
	X = Nφ₂	5.20	1	1	8
X—CH₂—S—Y	X = C=C, C=C Y = P(O,S)Z₂	3.66 - 3.35	25	25	9
	X = CO-NHφ Y = CO-R	3.64	1	1	10
	X = C=C Y = C=NH / NH₂	3.90	1	1	11
	X = COOH Y = CS-NR₂	4.21 - 4.20	2	2	12
	X = φ Y = C(NH₂)₂+	4.46	1	1	13
X—CH₂—S=C—C—⬡ (Y, Z, O)	X = φ Y = CH₃. Z = H. Nonequivalent	4.85, 4.35	1	2	14
	X,Y = φ Z = CO-N-φ. Nonequiv.	5.80, 4.60	1	2	15
X—CH₂—S—C≡N	X = C=C-φ	3.58	1	1	16
	X = S-CN	4.40	1	1	17
	X = CO-φ	4.78 - 4.67	2	2	18
X—CH₂—S(=O)—R	X = COOR. Nonequivalent	4.0 , 3.8	1	2	19
	X = CO-φ. Nonequivalent	4.35 - 4.19	8	8	20
		4.60 - 4.35	8	8	21
X—CH₂—S(=O)(=O)—Y	X = C=C Y = R, φ	4.04 - 3.77	8	10	22
	X = φ Y = R	4.29 - 3.84	6	8	23
	X = CO-NR-NR₂ Y = R	4.48 - 3.94	2	2	24
	X = CO-R, COOR, SO₂-R, NH-COOR, O-R. Y = R, φ	4.61 - 4.24	14	16	25
	X = CO-φ, NR-CO-(R,φ), NH-SO₂-(R,φ), Cl, Br, I Y = R, φ	5.18 - 4.48	20	21	26
	X = NH-CO-NHR in TFA Y = R	5.95	1	1	27
	X = C=C Y = O-R	3.88 - 3.87	2	2	28
	X = φ, Cl Y = O-R, NR₂	4.52 - 4.34	7	7	29
X—CH₂—SR +	X = φ	5.02 - 4.68	6	9	30
	X = COOR	5.25	1	1	31
	X = CO-φ in TFA	5.17 - 4.90	2	2	32
	in CDCl₃	5.55 - 5.39	4	4	33
X—CH₂—O—R	X = P(O)(OR)₂	3.69	1	1	34
	X = C=C, C≡C, COOR, CO-NR₂, CN, PR₂	4.31 - 3.83	34	37	35
	X = φ, O-R, SO₂-φ	4.83 - 4.15	28	34	36
	X = O-φ	5.36 - 4.55	13	13	37
	X = Pφ₃ +	5.82 - 5.51	2	3	38
X—CH₂—O—⬡	X = C≡C, COOR, CO-NR₂, CN	5.09 - 4.55	15	16	39
	X = φ, CO-φ	5.25 - 4.80	14	14	40
	X = O-CO-R, O-CO-φ	5.70 - 5.50	3	3	41
X—CH₂—O—Y	X = C=C, C≡C Y = P(O)(OR)₂	4.64 - 4.35	4	4	42
	X = φ Y = P(O)(OR)₂	5.20 - 4.93	3	3	43
	X = φ Y = SO-(φ,O-φ)	4.93 - 4.32	6	12	44
	X = CN Y = SO₂-φ	4.72	1	1	45
	X = C=C, C≡C, CN, CCl₃ S-R Y = CO-(R,φ)	5.05 - 4.67	14	14	46
	X = φ Y = CO-(R,φ)	5.38 - 4.83	30	35	47
	X = CO-φ Y = CO-(R,φ)	5.53 - 5.22	11	11	48
	X = O-φ, Cl Y = CO-(R,φ)	6.05 - 5.50	5	5	49
	X = φ Y = CO-NR₂, N=CR₂, Cl	5.41 - 5.00	47	47	50

R = H, Alkyl, C-Z.   Z = Any reasonable group unless specifically identified.

CHART S13. Summary of Chemical Shifts of X—CH₂—(S,O)

Cross Index Between Summary and Detailed Charts

Summary Chart (Line)	Detailed Charts (Lines) from which Data Were Taken
S14(1)	45(2)
S14(2)	X
S14(3)	46(6), X
S14(4–10)	X
S14(11)	50(17), X
S14(12–16)	X
S14(17)	46(6)
S14(18,19)	X
S14(20,21)	50(15)
S14(22–27)	X
S14(28)	59(2)
S14(29–34)	X
S14(35–37)	Most charts
S14(38)	1(12), 8(11), 13(3), 14(3), 18(3,7,10,13,17), 23(3), 38(3,6), 45(3,7), 54(3), 57(3), 65(3), 66(3,13), 68(9), 82(4,10), 87(13), 88(12), 89(12).
S14(39)	1(12,22), 13(3), 18(3,7,10,13,17), 30(3), 37(3), 38(3,6,9), 45(3,7), 48(7), 54(3), 55(3), 56(3,8), 57(3,13), 58(3), 59(3), 63(8,11,15,18), 64(6,9,16), 65(3), 66(13), 67(3,6), 68(3), 72(2), 78(3), 82(4,7), 86(7), 87(13).
S14(40)	1(11), 13(3), 18(3,13,17), 30(3), 37(3), 38(3,6), 48(6), 59(3), 65(3), 66(13)

X indicates data not included in detailed charts.

FUNCTIONAL GROUP		RANGE δ-VALUES	NO. CPDS.	NO. EXAM.	LINE
X—CH₂—NR₂	X = NR₂	2.99-2.72	4	4	1
	X = C=C, C≡C, CO-R, COOR (not COOH)	3.40-2.98	31	32	2
	X = φ, COOH, CN	3.96-3.30	34	34	3
	X = N-CO-N-CO-C	4.02	1	1	4
	X = CO-φ, Nφ-CO-R, N-CO-C-C-CO	4.63-4.42	17	17	5
	X = N-CO-C-CO-N-CO	4.89	1	1	6
X—CH₂—N—R (—Y)	X = COOR (Incl. COOH), CN Y = φ	3.93-3.77	3	3	7
	X = φ Y = φ	4.52-4.20	7	8	8
	X = S-φ Y = φ	4.89-4.72	5	5	9
	X = C≡C, COOR(Incl. COOH) Y = CO-(R,φ)	4.06-3.67	12	13	10
	X = φ, CO-φ Y = CO-(R,φ)	4.80-4.00	16	20	11
	X = SO₂-(R,φ) Y = CO-(R,φ)	5.18-4.65	8	8	12
	X = φ, COO(R,φ) Y = COOR, SO₂-(R,φ)	4.33-4.11	4	4	13
	X = SO₂-R Y = COOR, SO₂-(R,φ)	4.83-4.44	6	6	14
	X = SO₂-R in TFA Y = CO-NH-R	5.95	1	1	15
	X = φ Y = CS-(S-Zn, N-N-COOR, CS-NR₂)	5.09-4.80	5	5	16
X—CH₂—N—Y (φ)	X = φ Y = R	4.52-4.20	7	8	17
	X = S-φ Y = φ	5.20	1	1	18
	X = NR₂, CN Y = CO-(R,φ)	4.69-4.49	3	3	19
	X = φ. Nonequiv. } Y = CO-(R,φ)	4.87-3.53	7	7	20
	X = φ. Nonequiv. } Y = CO-(R,φ)	5.70-4.73	7	7	21
X—CH₂—N=C=S	X = C=C	4.15	1	1	22
X—CH₂—N≡C	X = φ	4.59	1	1	23
X—CH₂—NR₃ +	X = φ, COOR Pri., sec. & tert.	3.99-3.59	6	7	24
	X = C=C, COOR Quaternary	4.46-3.89	5	5	25
	X = φ, CO-φ Quaternary	5.15-4.49	13	13	26
	X = C≡C-φ Quaternary	5.28-5.19	2	2	29
X—CH₂—NO₂	X = NO₂	6.20	1	1	28
X—CH₂—N—Y	X = φ, NR₂ Y = -CO-C-C-CO-	4.63-4.42	16	16	29
	X = NRφ Y = -CO-C-C-CO-	5.06-4.88	16	16	30
	X = C≡C Y = -CO-φ-CO-	4.60-4.44	2	3	31
	X = CO-φ, CO-NR₂ Y = -CO-φ-CO-	5.15-5.13	2	2	32
	X = NR₂ Y = -CO-C-CO-N-CO-	4.89	1	1	33
	X = COOH Y = -CO-C-S-CS-	4.56	1	1	34

Methine Chemical Shifts Correlations

Because CH bands are usually broad and low and frequently overlap CH₂ bands, chemical shift data for CH's are much less abundant and less reliable than those for CH₂'s. It is possible, however, to estimate the CH shift ranges as functions of the corresponding CH₂ ranges. These estimates are given below. Ranges for paraffinic CH's are plotted below, and additional CH correlations are plotted in CHART S15.

	RANGE	NO. CPDS.	NO. EXAM.	STERIC ENVIRON. (See CHARTS 1 & 7)	
R—C—CH(C—R)₂	1.18-0.75			Shielded	35
	1.72-1.18	156	161	Normal	36
	1.94-1.60			Crowded	37
(R—C)₂ CH—C—X	Extend corresponding CH₂ range by 0.5 higher δ-value.				38
(R—C)₂ CH—X	Extend corresponding CH₂ range by 1.0 higher δ-value.				39
X—CH—C—R (Y)	Extend corresponding CH₂ range by 1.0 higher δ-value.				40

R = H, Alkyl, C-Z.   Y, Z = Any reasonable group unless specifically identified.

CHART S14. Summary of Chemical Shifts of X—CH₂—N and Some CH

Cross Index between Summary and Detailed Charts

Summary Chart (Line)	Detailed Charts (Lines) from which Data Were Taken
S15(1)	X
S15(2)	X
S15(3)	X
S15(4)	19(9), X
S15(5)	18(3), X
S15(6)	X
S15(7)	X
S15(8)	18(13), X
S15(9)	18(7)
S15(10)	13(3), X
S15(11)	31(4,10), X
S15(12)	X
S15(13)	X
S15(14)	38(12)
S15(15)	X
S15(16)	X, 38(3)
S15(17)	X, 38(3)
S15(18)	X
S15(19)	38(6), X
S15(20)	X
S15(21)	X, 18(3)
S15(22)	X
S15(23)	X
S15(24)	X
S15(25)	X, 18(13)
S15(26)	22(6)
S15(27)	85(4)
S15(28)	31(4,10), 58(6), X, 74(2)
S15(29)	X, 73(6), 74(4)
S15(30)	13(3), 37(3), 46(12), X
S15(31)	X, 42(6), 43(6)
S15(32)	32(5)
S15(33)	89(10)
S15(34)	65(3)
S15(35)	68(9)
S15(36)	X
S15(37)	X
S15(38)	88(13)
S15(39)	X
S15(40)	88(12)
S15(41)	42(6), X
S15(42)	38(3), X
S15(43)	X
S15(44)	X, 38(6), 43(6)
S15(45)	38(9), X
S15(46)	23(9), X
S15(47)	X
S15(48)	59(3)
S15(49)	X
S15(50)	X
S15(51)	X
S15(52)	X
S15(53)	X
S15(54)	X
S15(55)	X
S15(56)	X
S15(57)	X
S15(58)	X
S15(59)	X

X indicates data which are not in detailed charts.

CHART S15. Summary of Chemical Shifts of X–C–CHY₂ and X–CH(Z)–Y

Scale (top): τ 1 2 3 4 5 6 7 / δ 9 8 7 6 5 4 3

FUNCTIONAL GROUP			RANGE δ-VALUES	NO. CPDS.	NO. EXAM.	LINE
$(X-C)_2CH-Y$	X = O-R	Y = OH	4.02 – 3.15	6	6	1
	X = O-CO-R	Y = OH	4.19 – 4.13	2	2	2
	X = O-CO-R	Y = O-CO-R	5.30 – 5.13	9	9	3
$X-\overset{\mid}{\underset{\mid}{C}}-CHY_2$	X = C=C	Y = COOR	3.31 – 3.22	2	3	4
	X = φ, COOR	Y = COOR	3.77 – 3.55	4	4	5
	X = φ, COOR, NR₂, Br, Cl₂	Y = O-R	4.57 – 4.38	3	7	6
	X = Cl	Y = O-R	4.63 – 3.96	2	3	7
	X = O-R	Y = O-R	5.17 – 4.50	3	3	8
	X = O-CO-R	Y = O-CO-R	6.80	1	1	9
X_nC-CHY_2	X_n = φ, φ₂	Y = φ	4.38 – 4.10	3	3	10
	X_n = CO-R, COOR	Y = φ	4.60 – 4.48	7	7	11
	X_n = NR₃	Y = φ	4.82 – 4.72	2	2	12
	X_n = ⁺Cl₂, Cl₃	Y = φ	5.17 – 4.99	2	2	13
	X_n = F₂	Y = F	6.02 – 5.85	6	6	14
	X_n = OH, (O-R)₂	Y = Cl	5.76 – 5.55	2	2	15
	X_n = SH, Cl	Y = Cl	5.90 – 5.73	5	5	16
	X_n = φ₂, Cl₂, Cl₃	Y = Cl	6.35 – 5.93	10	10	17
	X_n = Br	Y = Br	5.69	1	1	18
	X_n = Br₂, Br₃	Y = Br	6.33 – 6.04	2	2	19
	X_n = φ	Y = Br	7.08	1	1	20
$X-CHY_2$	X = φ, COOR	Y = COOR	4.42 – 4.17	2	2	21
	X = OH	Y = COOR	5.01	1	1	22
	X = NH-CO-R	Y = COOR	5.45 – 5.28	3	3	23
	X = NR₃	Y = COOR	6.29	1	1	24
	X = ⁺C≡C, O-R	Y = O-R	5.18 – 5.00	7	7	25
	X = C=C	Y = O-R	6.71	1	1	26
	X = P(O)(OR)₂	Y = φ	4.24 – 4.20	2	2	27
	X = CO-R, COOR, CN, S-R, S-S-R	Y = φ	5.29 – 4.80	8	8	28
	X = NH-P(O)(OR)₂, S-φ, S₃-(R,φ)	Y = φ	5.69 – 5.28	8	9	29
	X = φ	Y = φ	6.22 – 5.23	3	4	30
	X = O-NH₂, Cl, Br	Y = φ	6.45 – 6.10	3	3	31
	X = O-R	Y = φ	6.74 – 5.35	18	22	32
	X = Pφ₃	Y = φ	8.27	1	1	33
	X = ⁺S-R	Y = S-R	4.18	1	1	34
	X = S-φ	Y = S-φ	5.28	1	1	35
	X = COOR	Y = F	6.15 – 5.91	2	2	36
	X = S-R, SO₂-R	Y = F	6.72 – 6.62	2	2	37
	X = P(O)(OR)₂	Y = Cl	5.90 – 5.55	5	5	38
	X = CO-(R,O-R,Sn)	Y = Cl	6.11 – 5.76	6	6	39
	X = P(O)Cl₂	Y = Cl	6.25	1	1	40
	X = φ, CO-(φ,NR₂)	Y = Cl	6.70 – 6.31	10	10	41
	X = F, Cl, Br	Y = Cl	7.45 – 7.20	3	50+	42
	X = CO-R, CN	Y = Br	6.31 – 5.85	2	2	43
	X = CO-φ, Cl, Br	Y = Br	7.08 – 6.70	3	4	44
	X = I	Y = I	5.20 – 4.93	1	2	45
	X = C=C	Y = O-CO-R	7.13 – 7.04	3	3	46
	X = φ	Y = O-CO-R	7.71	1	1	47
	X = NO₂	Y = NO₂	7.52	1	1	48
$X-\overset{}{\underset{Z}{CH}}-Y$	X = NR₂ Y = C-OH Z = COOR		3.60 – 3.29	3	3	49
	X = NR₂ Y = C-φ Z = COOR		3.73	1	1	50
	X = NR₂ Y = C-SH Z = COOR		4.48	1	1	51
	X = NR₃ Y = C-S-R, C-SO₂-OH Z = COOR		4.52 – 4.26	3	3	52
	X = ⁺NR₃ Y = C-φ Z = COOR		4.62 – 4.39	4	4	53
	X = NRφ Y = C-NO₂, COOR Z = φ		5.19 – 4.98	2	2	54
	X = NR-CO-(R,O-R) Y = C-(COOφ, CO-NR₂, OH, S-R) Z = COO(R,φ,NR₂)		4.67 – 4.29	9	9	55
	X = NR-CO-(R,O-R) Y = C-φ Z = COO(R,φ,NR₂)		5.18 – 4.51	12	12	56
	X = O-(R,φ) Y = CO-φ Z = φ		5.41 – 5.35	2	2	57
	X = OH Y = CO-R Z = φ		5.19 – 4.98	2	2	58
	X = OH Y = CO-(φ, O-R, Cl) Z = φ		6.04 – 5.13	13	13	59

Scale (bottom): 9 8 7 6 5 4 3 δ

R = H, Alkyl, C-Z.   Z = Any reasonable group unless specifically identified.

Cross Index between Summary and Detailed Charts

Summary Chart (Line)	Detailed Charts (Lines) from which Data Were Taken
S16(1)	1(14), 9(1,4), 16(9), 19(11), 22(8), 39(13), 40(10), 41(10), 69(7), 71(7), see also 72(3), 83(6,13)
S16(2)	16(9), 19(11), 69(7), 71(7), 72(3)
S16(3)	22(8), 23(12), X, 39(13), 40(10), 41(10), 83(6,13)
S16(4)	39(8)
S16(5)	23(12), 72(8), X
S16(6)	1(18), 9(1–3), 16(10), 19(12), 22(9), 39(14), 40(11), 41(11), 69(8), 71(8), 72(5), 83(6,14)
S16(7)	71(8), 72(4)
S16(8)	X
S16(9)	39(9,14), 40(11), 41(11)
S16(10)	16(10), 19(12,14), 22(9,11), 23(13), 69(8), 71(8), 72(4,9)
S16(11)	76(9,10), X
S16(12–14)	X
S16(15)	1(15)
S16(16)	81(4)
S16(17)	1(19)
S16(18–20)	81(5)
S16(21,22)	23(4,5)
S16(23,24)	23(10), 83(7)
S16(25)	1(13)
S16(26,27)	17(6–9)
S16(28)	70(7), 71(12)
S16(29)	72(15)
S16(30)	72(17)
S16(31)	72(16)
S16(32,33)	72(14)
S16(34,35)	20(7), 21(10,13,16,17)
S16(36)	21(16,17)
S16(37)	X
S16(38)	21(13)
S16(39)	X
S16(40,41)	61(6)
S16(42,43)	80(4)
S16(44)	39(8), X
S16(45)	40(10), 41(10)
S16(46)	39(13), X

X indicates data not included in detailed charts.

Functional Group	X / Y substituents	Range δ-Values	No. Cpds.	No. Exam.	Line
X—C—C=CH₂	X = Alkyl, C-Y	5.10 – 4.55	145	240	1
	X = φ, COOR, S-R, S-φ, S-CO-R	5.13 – 4.62	18	24	2
	X = O-Y, N-Y, N=CR₂, S-P(O,S)(OR)₂, I, Cl, Br	5.40 – 4.77	64	78	3
	X = F, F₂, F₃	5.16 – 5.12	6	6	4
	X = (O-CO-R)₂, SO₂-R, NR₃⁺	6.00 – 5.40	7	10	5
X—C—CH=CR₂	X = Alkyl, C-Y	5.82 – 4.90	211	255	6
	X = S-φ, S-CO-R	5.50 – 5.05	24	38	7
	X = NR₂, NRφ	5.75 – 5.52	14	14	8
	X = F, Cl, Br, I	5.91 – 4.68	21	21	9
	X = φ, COOR, O-Y, S-Y, SO₂-R	6.05 – 5.00	122	136	10
X—C—C=CH—Y	X = OH, O-CO-R, S-CN, SO₂-R (all cisoid) Y = φ	6.72 – 6.40	7	7	11
	X = P(O)(OR)₂ (c, t ?) Y = COOR	5.90	1	1	12
X—C—CH=C—Y	X = OH, O-CO-R, S-CN, SO₂-R Y = cis-φ	6.38 – 6.10	8	8	13
	X = P(O)(OR) Y = cis-COOR	6.74	1	1	14
X—C=C=CH₂	X = Alkyl	4.57 – 4.42	3	3	15
	X = P(O)R₂	4.90 – 4.56	2	2	16
X—C=C=CH—Y	X = Alkyl, Y = Alkyl	5.10 – 4.92	4	4	17
	X = Alkyl, Y = P(O)R₂	5.88 – 5.70	2	2	18
	X = P(O)R₂, Y = φ	6.22	1	1	19
	X = φ, φ₂, Y = P(O)R₂	6.40 – 6.28	2	2	20
C=C (H trans / H cis)	X = O-R, X = O-C=C, trans	4.05 – 3.74	15	20	21
	X = O-R, cis	4.18 – 4.00	15	20	22
	X = O-CO-R, O-P(O)(OR)₂, trans	4.70 – 4.47	4	6	23
	X = O-CO-R, O-P(O)(OR)₂, cis	4.70 – 4.65	4	6	24
	X = C=C, c & t	5.15 – 4.84	6	16	25
	X = φ, trans	5.50 – 4.90	20	23	26
	X = φ, cis	5.72 – 4.76	20	23	27
	X = S-R, S-φ, c & t	5.20 – 4.52	5	14	28
	X = S-benzothiazole, c & t	5.74, 5.67	1	2	29
	X = SF₅, c & t	5.94, 5.65	1	2	30
	X = SR₂⁺, c & t	6.43, 6.38	1	2	31
	X = SO₂-(R, OR, Oφ), trans	6.31 – 5.95	6	7	32
	X = SO₂-(R, OR, Oφ), cis	6.58 – 6.13	6	7	33
	X = CO-R, COOR, trans	5.92 – 5.39	35	48	34
	X = CO-R, COOR, cis (Includes acrolein)	6.50 – 6.02	36	55	35
	X = CO-H, trans (Acrolein only)	6.57 – 6.20	1	6	36
	X = CO-NR₂, c & t	6.28 – 5.27	7	14	37
	X = -C=C-CO-H, cc & tt	6.05, 5.50	1	2	38
	X = CN, c & t	6.05, 5.94	1	2	39
	X = (N ring), trans	5.48 – 5.17	3	9	40
	X = (N ring), cis	6.23 – 5.60	3	9	41
	X = PR₂, trans	5.70 – 5.48	2	2	42
	X = PR₂, cis	6.23 – 6.20	2	2	43
	X = F, c & t	4.78 – 4.13	2	5	44
	X = Cl, Br, c & t	6.00 – 5.30	2	4	45
	X = I, c & t	6.58 – 5.52	2	4	46

R = H, Alkyl, C-Z.   Y, Z = Any reasonable group unless specifically identified.
cis and trans indicate relative steric positions of X (or Y) and observed H. c = cis. t = trans.

CHART S16. Summary of Chemical Shifts of X–C–CH=CH–Y, X–C=C=CH–Y, X–C=CH₂

Cross Index between Summary and Detailed Charts

Summary Chart (Line)	Detailed Charts (Lines) from which Data Were Taken
S17(1)	23(6)
S17(2)	23(11)
S17(3)	1(16), 19(14), 22(11)
S17(4)	70(8)
S17(5)	71(13), 72(13), 83(15)
S17(6)	72(17)
S17(7)	39(9,14), 40(11), 41(11), X
S17(8,9)	21(12), X
S17(10,11)	20(8)
S17(12)	80(4)
S17(13–15)	X
S17(16)	20(8), X
S17(17)	21(11), X
S17(18)	1(17)
S17(19)	17(12,13)
S17(20)	61(7)
S17(21)	70(8), 71(13), 72(13), 83(15)
S17(22)	72(16,17)
S17(23)	72(14,15), X
S17(24)	40(11), 41(11), X
S17(25)	39(9)
S17(26)	23(6), 83(7)
S17(27)	23(11)
S17(28,29)	20(12,14), 21(14)
S17(30)	17(18)
S17(31)	20(14)
S17(32)	17(18)
S17(33)	17(17)
S17(34)	17(15,16), X
S17(35)	X
S17(36)	75(2,7)
S17(37)	75(5)
S17(38,39)	76(2,6)
S17(40)	X
S17(41)	75(8,9)
S17(42–58)	X

X indicates data not in detailed charts.

FUNCTIONAL GROUP

X–C=CH–R′ (R′=Alkyl, C-Z)

		RANGE δ-VALUES	NO. CPDS.	NO. EXAM.	LINE
X = O-R	c & t	5.2 – 4.3	0	0	1
X = O-CO-R	c & t	6.0 – 5.2	0	0	2
X = C=C	c & t	5.85 – 5.30	7	7	3
X = S-R	c & t	5.55 – 4.45	8	8	4
X = S-φ, S-CO-R, S-P(O,S)(OR)$_2$	c & t	6.07 – 5.75	3	3	5
X = SF$_5$	c & t	6.37	1	1	6
X = F, Cl, Br, I	c & t	5.91 – 4.68	21	21	7
X = CO-R, CO-NR$_2$	trans	6.5 – 5.8	0	0	8
X = CO-R, CO-NR$_2$	cis	6.95 – 6.35	4	6	9
X = COOH	trans	6.43 – 5.77	16	20	10
X = COOH	cis	7.10 – 6.70	24	28	11

X–CH=CR$_2$

		RANGE δ-VALUES	NO. CPDS.	NO. EXAM.	LINE
X = PR$_2$		5.56 – 5.38	2	2	12
X = $_+$NR-Y	Y = R, φ, CO-Z, CS-Z, CN, SO$_2$-φ	5.95 – 5.60	20	20	13
X = NR$_3$		6.2 – 6.0	3	3	14
X = CO-NR , CO-O-O-R,	CO-O-CO-C=C	5.84 – 5.38	10	10	15
X = COOR, COO-C=C,		6.15 – 5.58	23	38	16
X = CO-R, CO-C=C		6.92 – 6.02	10	13	17
X = C=C		6.67 – 5.60	19	27	18
X = φ		6.97 – 6.20	18	21	17
X = Pyridyl-(2,3,4)		6.76 – 6.55	3	9	20
X = S-R, S-φ, S-CO-R,	S-P(O,S)(OR)$_2$	6.50 – 5.67	12	13	21
X = SF$_5$, SR$_2$		6.73 – 6.37	3	3	22
X = S-benzothiazole,	SO$_2$-(R,OR)	7.05 – 6.59	7	8	23
X = Cl, Br		6.42 – 5.70	8	8	24
X = F, I		6.96 – 6.35	5	5	25
X = O-R, O-P(O)(OR)$_2$		6.60 – 6.22	17	22	26
X = O-CO-R		7.27 – 7.22	1	2	27

X–C=CH–C=C–Y

		RANGE δ-VALUES	NO. CPDS.	NO. EXAM.	LINE
X = cis-(CO-R,COOR)	Y = R	6.65 – 6.05	5	5	28
X = trans-(CO-R,COOR)	Y = R	7.35 – 7.15	3	3	29
X = cis-φ	Y = φ	6.60	1	1	30
X = c & t-COOR	Y = COOR	8.40 – 6.68	3	5	31

X–C=CH–⟨phenyl⟩

	RANGE δ-VALUES	NO. CPDS.	NO. EXAM.	LINE
X = cis–C=C-φ	6.90	1	1	32
X = trans–φ	6.55	1	1	33
X = cis-φ	7.12 – 6.80	5	5	34
X = cis-φ, o,p-dinitrophenyl.	7.64	1	1	35
X = c&t-(S-R,S-CO-R)	6.40 – 6.00	10	16	36
X = c&t-S-φ	6.78 – 6.12	11	11	37
X = trans-SO$_2$-(R,φ)	7.13 – 6.88	3	3	38
X = cis-SO$_2$-(R,φ,Cl)	8.08 – 7.58	14	14	39
X = cis-SO$_2$-C=C	7.40	1	1	40
X = c&t-SR$_2$	8.00 – 7.52	6	9	41
X = cis-CN$^+$	7.36 – 7.24	3	3	42
X = cis-NR$_2$. o,p-dinitrophenyl.	7.31	1	1	43
X = NO$_2$ (c or t ?)	8.14 – 7.91	4	4	44
X = cis-CO-R	7.72 – 7.30	12	12	45
X = cis-CO-R. o-nitrophenyl.	8.09	1	1	46
X = c & t-CO-φ	8.20 – 7.00	12	12	47
X = cis-COOR	8.06 – 7.47	31	31	48
X = cis-COOR In TFA solvent.	8.18	1	1	49

X–C=CH–C(=O)–Y

		RANGE δ-VALUES	NO. CPDS.	NO. EXAM.	LINE
X = cis-S-C=C	Y = R	5.68	1	1	50
X = cis-φ	Y = R, Cl	6.86 – 6.42	14	14	51
X = cis-φ	Y = C=C	7.10 – 7.02	3	3	52
X = cis(?)-NR$_2$	Y = φ	5.69	1	1	53
X = c&t-φ	Y = φ	7.39 – 6.73	6	6	54
X = cis-pyridyl-2	Y = φ	7.81 – 7.75	2	2	55
X = cis-CO-φ	Y = φ	7.95	1	1	56
X = cis-COOR	Y = φ	8.00 – 7.75	9	9	57
X = cis(?)-φ	Y = 4-pyr.	7.45	1	1	58

R = H, Alkyl, C-Z.   Y, Z = Any reasonable group unless specifically identified.
cis and trans indicate relative steric positions of X and observed H. c = cis. t = trans.

CHART S17. Summary of Chemical Shifts of X–CH=CR$_2$ and X–C=CH–Y(A)

Cross Index between Summary and Detailed Charts

Summary Chart (Line)	Detailed Charts (Lines) from which Data Were Taken
S18(1)	X
S18(2)	20(12)
S18(3)	20(14)
S18(4,5)	X
S18(6,7)	20(12), X
S18(8–15)	X
S18(16,17)	75(1)
S18(18,19)	75(4)
S18(20,21)	76(1,5)
S18(22,23)	75(8)
S18(24,25)	75(7)
S18(26)	40(11)
S18(27)	41(11)
S18(28)	X
S18(29)	80(4)
S18(30)	81(7)
S18(31)	61(7)
S18(32–35)	X
S18(36)	1(23), 24(8), 25(10), 84(5,10,14)
S18(37)	25(10)
S18(38)	84(5)
S18(39)	84(14)
S18(40)	84(10)
S18(41)	1(23)
S18(42)	80(5)
S18(43)	81(6)
S18(44)	84(6)
S18(45)	84(5)

X indicates data not in detailed charts.

FUNCTIONAL GROUP			RANGE δ-VALUES	NO. CPDS.	NO. EXAM.	LINE
X—C=CH—C—OR (O)	X = trans-(NR₂, NRφ)		4.72 – 4.30	4	4	1
	X = c & t – C = CR₂		5.82 – 5.47	3	3	2
	X = c & t – C=C–COOR		6.17 – 5.96	4	6	3
	X = cis–φ		6.75 – 6.12	29	29	4
	X = cis– Cl		6.28	1	1	5
	X = trans–COOR		6.27 – 5.99	12	12	6
	X = cis–COOR		6.94 – 6.63	12	12	7
	X = trans–CO-N(R₂,Rφ)		6.60 – 6.25	8	8	8
	X = cis–CO–φ		6.90 – 6.69	9	9	9
	X = cis–pyridinium–4		7.09 – 6.57	2	2	10
X—C=CH—C—Z (O)	X = cis–φ	Z = O-CO-C=C	6.49	1	1	11
	X = cis – COOR	Z = N(R₂, Rφ)	6.60 – 6.25	7	7	12
X—C=CH—C≡N	X = NR₂ (c or t ?)		3.71	1	1	13
	X = cis–φ		5.84 – 5.53	3	3	14
	X = cis–CN		6.26	1	1	15
φ—C=CH—S	φ = trans	S = S-R	6.08 – 5.88	5	8	16
	φ = cis	S = S-R	6.75 – 6.47	5	8	17
	φ = trans	S = S-φ	6.52 – 6.25	6	6	18
	φ = cis	S = S-φ	6.88 – 6.66	5	5	19
	φ = trans	S = SO₂-(R,φ)	6.59 – 6.35	4	4	20
	φ = cis	S = SO₂-(R,φ)	7.20 – 6.70	10	10	21
	φ = trans	S = SR₂⁺	6.94 – 6.30	3	6	22
	φ = cis	S = +"	7.48 – 6.93	4	5	23
	φ = trans	S = S-CO-R	7.02	1	1	24
	φ = cis	S = S-CO-R	7.30	1	1	25
X—C=CH—Hal.	X = c & t–Cl Hal. = Cl		6.30 – 6.18	2	2	26
	X = c & t–Br Hal. = Br		7.04 – 6.60	2	4	27
	X = cis–COOR Hal. = Cl		7.50	1	1	28
X—C=CH—P	X = P = PR₂		7.18, 6.78	2	2	29
	X = P = P(O)φ₂		7.40, 6.90	2	2	30
X—C=CH—(pyridyl) (2,3,4)	X = cis(?)-pyridyl-(2,4)		7.80 – 7.65	2	2	31
	X = cis(?)–COOR		8.05 – 7.61	3	3	32
	X = cis(?)–CO–φ		8.23 – 8.12	2	2	33
X—C=CH—Y	X = trans-CO-NR₂ Y = OH		5.04	1	1	34
	X = trans-CO-R Y = NH-CO-R		5.46 – 5.33	3	3	35
X—C—C≡CH	X = C-Y		2.12 – 1.74	14	15	36
	X = OH		2.63 – 2.28	15	15	37
	X = P(O)(OR)₂		2.76	1	1	38
	X = S-P(O,S)(OR,SR)₂		2.80 – 2.48	3	3	39
	X = O-P(O)(OR)₂		3.28 – 3.18	2	2	40
X—C≡CH	X = C≡C-R		2.30	1	1	41
	X = PR₂		3.07 – 2.63	4	5	42
	X = P(O)R₂		3.33 – 2.95	4	4	43
	X = P(Z)Y₂, except PR₂, P(O)R₂ & P(O)(OR)₂		3.35 – 2.65	13	14	44
	X = P(O)(OR)₂		4.00	1	1	45

R = H, Alkyl, C–Z.   Y, Z = Any reasonable group unless specifically identified.
cis and trans indicate relative steric positions of X and observed H. c = cis. t = trans.

CHART S18. Summary of Chemical Shifts of X—C=CH—Y(B), X—C-C≡CH, and X—C≡CH

Cross Index between Summary and Detailed Charts

Summary Chart (Line)	Detailed Charts (Lines) from which Data Were Taken
S19(1)	17(10,11)
S19(2)	17(18)
S19(3)	17(14)
S19(4)	20(10)
S19(5)	X
S19(6)	40(11)
S19(7)	X
S19(8)	17(18)
S19(9)	X
S19(10)	17(17)
S19(11)	76(7)
S19(12–26)	X
S19(27)	40(11)
S19(28,29)	X
S19(30)	18(17), X
S19(31–39)	X
S19(40)	21(15,17), X
S19(41–43)	X
S19(44)	34(4,5), X
S19(45)	56(4)
S19(46,47)	56(9,16)
S19(48)	55(17)
S19(49,50)	57(4)
S19(51)	56(9)
S19(52,53)	55(4)
S19(54)	56(18)
S19(55)	55(8)
S19(56,57)	57(9)
S19(58,59)	X, 63(3)

X indicates data not in detailed charts.

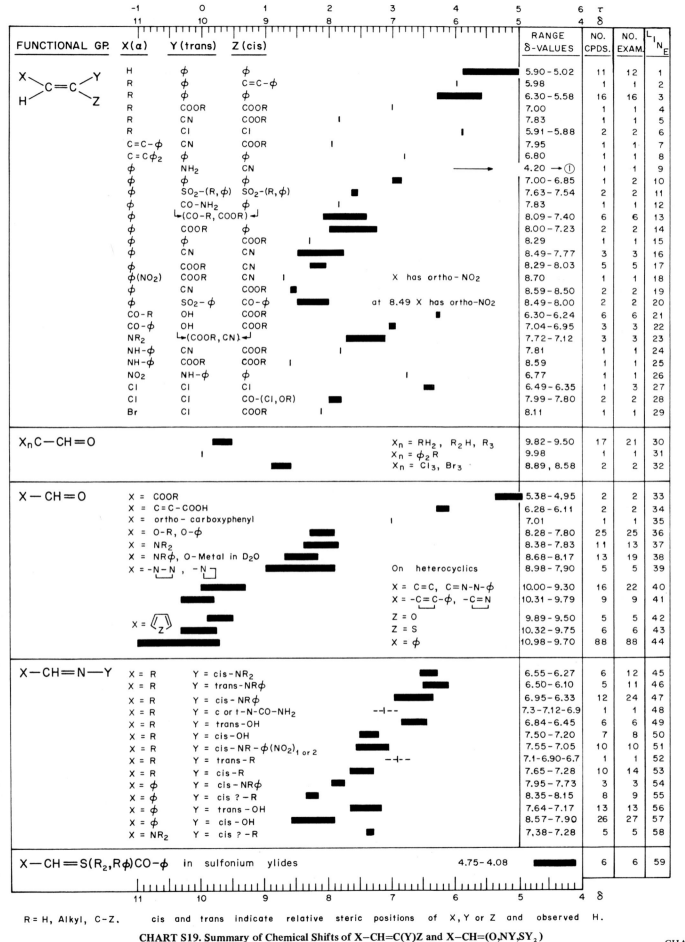

CHART S19. Summary of Chemical Shifts of X–CH=C(Y)Z and X–CH=(O,NY,SY₂)

$R = H$, Alkyl, $C-Z$.   cis and trans indicate relative steric positions of X, Y or Z and observed H.

Cross Index between Summary and Detailed Charts

Summary Chart (Line)	Detailed Charts (Lines) from which Data Were Taken
S20(1)	10(1)
S20(2)	14(4)
S20(3)	26(1,3,8)
S20(4)	10(7), 12(15)
S20(5)	14(4), 47(5)
S20(6)	26(1,3)
S20(7)	28(4), 78(8)
S20(8)	78(13)
S20(9)	26(2)
S20(10)	26(2,3)
S20(11)	47(6)
S20(12)	10(6)
S20(13)	10(10), 26(11), 27(10), 47(14,15)
S20(14)	14(7), 26(6)
S20(15)	10(10), 12(5), 27(10), 53(5)
S20(16)	14(7,10), 26(6), 47(14,15)
S20(17)	26(11), 27(2), 28(7)
S20(18)	26(6)
S20(19)	10(10)
S20(20)	10(2–5), 11(5–7)
S20(21)	10(8,9,11), 12(6,7,16), 26(9), 27(11,12), 28(5,6,9), 47(7,8,10–13), 53(6,7,8,10)
S20(22)	14(5,6,8,9), 15(1,2), 87(5,6)
S20(23)	26(4,5), 27(4,5,7), 79(9,10,14,15,17)
S20(24)	10(8,9,11), 12(5), 26(4,5), 27(4,5,7,8,11,12), 53(6–10)
S20(25)	14(5,6,8,9), 47(7,8,10–13), 78(9,10)
S20(26)	15(1,2), 78(12,14)
S20(27)	26(9,10), 28(5,6)
S20(28)	27(4,5,7,8), 78(15)
S20(29)⎱ S20(30)⎰	78(16,17), 87(5)
S20(31)	12(4)
S20(32)	14(8), 26(4,5), 27(12), 78(9,10)
S20(34)	15(1,2)
S20(35)	12(3)
S20(36)	27(12)
S20(37)	14(8,9)
S20(38)	78(9,10)
S20(39)	28(5,6,9)
S20(40)	14(8)
S20(41)	28(12)

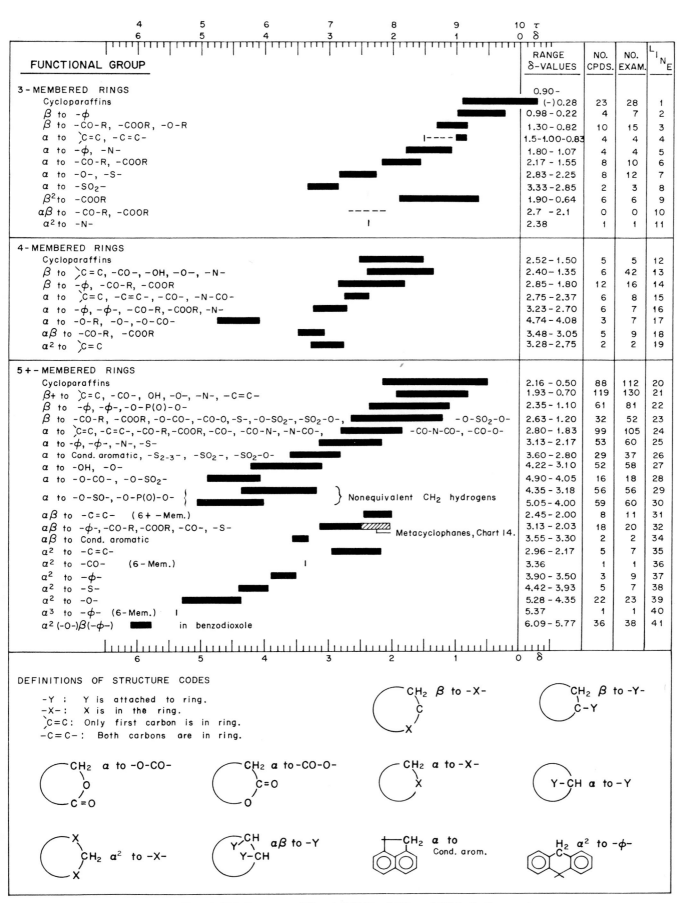

FUNCTIONAL GROUP	RANGE δ-VALUES	NO. CPDS.	NO. EXAM.	LINE
3-MEMBERED RINGS				
Cycloparaffins	0.90 – (-)0.28	23	28	1
β to -φ	0.98 – 0.22	4	7	2
β to -CO-R, -COOR, -O-R	1.30 – 0.82	10	15	3
α to >C=C, -C≡C-	1.5-1.00-0.83	4	4	4
α to -φ, -N-	1.80 – 1.07	4	4	5
α to -CO-R, -COOR	2.17 – 1.55	8	10	6
α to -O-, -S-	2.83 – 2.25	8	12	7
α to -SO₂-	3.33 – 2.85	2	3	8
β² to -COOR	1.90 – 0.64	6	6	9
αβ to -CO-R, -COOR	2.7 – 2.1	0	0	10
α² to -N-	2.38	1	1	11
4-MEMBERED RINGS				
Cycloparaffins	2.52 – 1.50	5	5	12
β to >C=C, -CO-, -OH, -O-, -N-	2.40 – 1.35	6	42	13
β to -φ, -CO-R, -COOR	2.85 – 1.80	12	16	14
α to >C=C, -C≡C-, -CO-, -N-CO-	2.75 – 2.37	6	8	15
α to -φ, -φ-, -CO-R, -COOR, -N-	3.23 – 2.70	6	7	16
α to -O-R, -O-, -O-CO-	4.74 – 4.08	3	7	17
αβ to -CO-R, -COOR	3.48 – 3.05	5	9	18
α² to >C=C	3.28 – 2.75	2	2	19
5+-MEMBERED RINGS				
Cycloparaffins	2.16 – 0.50	88	112	20
β+ to >C=C, -CO-, OH, -O-, -N-, -C≡C-	1.93 – 0.70	119	130	21
β to -φ, -φ-, -O-P(O)-O-	2.35 – 1.10	61	81	22
β to -CO-R, -COOR, -O-CO-, -CO-O-, -S-, -O-SO₂-, -SO₂-O-, -O-SO₂-O-	2.63 – 1.20	32	52	23
α to >C=C, -C≡C-, -CO-R,-COOR, -CO-, -CO-N-, -N-CO-, -CO-N-CO-, -CO-O-	2.80 – 1.83	99	105	24
α to -φ, -φ-, -N-, -S-	3.13 – 2.17	53	60	25
α to Cond. aromatic, -S₂₋₃-, -SO₂-, -SO₂-O-	3.60 – 2.80	29	37	26
α to -OH, -O-	4.22 – 3.10	52	58	27
α to -O-CO-, -O-SO₂-	4.90 – 4.05	16	18	28
α to -O-SO-, -O-P(O)-O- } Nonequivalent CH₂ hydrogens	4.35 – 3.18	56	56	29
	5.05 – 4.00	59	60	30
αβ to -C≡C- (6+-Mem.)	2.45 – 2.00	8	11	31
αβ to -φ-, -CO-R, -COOR, -CO-, -S- — Metacyclophanes, Chart 14.	3.13 – 2.03	18	20	32
αβ to Cond. aromatic	3.55 – 3.30	2	2	34
α² to -C≡C-	2.96 – 2.17	5	7	35
α² to -CO- (6-Mem.)	3.36	1	1	36
α² to -φ-	3.90 – 3.50	3	9	37
α² to -S-	4.42 – 3.93	5	7	38
α² to -O-	5.28 – 4.35	22	23	39
α³ to -φ- (6-Mem.)	5.37	1	1	40
α² (-O-)β(-φ-) in benzodioxole	6.09 – 5.77	36	38	41

DEFINITIONS OF STRUCTURE CODES

-Y : Y is attached to ring.
-X- : X is in the ring.
>C=C : Only first carbon is in ring.
-C=C- : Both carbons are in ring.

CH₂ β to -X-
CH₂ β to -Y- C-Y
CH₂ α to -O-CO-
CH₂ α to -CO-O-
CH₂ α to -X-
Y-CH α to -Y
CH₂ α² to -X-
Y-CH / Y-CH αβ to -Y
-CH₂ α to Cond. arom.
H₂ α² to -φ-

CHART S20. Summary of Chemical Shifts of CH₂ and CH in Cyclics

Cross Index between Summary and Detailed Charts

Summary Chart (Line)	Detailed Charts (Lines) from which Data Were Taken
S21(1,2)	10(12)
S21(3,4)	10(14,16)
S21(5)	10(13)
S21(6,7)	10(15)
S21(8–10)	12(8,9,13)
S21(11)	12(12)
S21(12)	28(11)
S21(13)	87(9,10)
S21(14)	12(11,14)
S21(15)	12(11)
S21(16)	16(13)
S21(17)	28(10)
S21(18)	87(7,8)
S21(19)	28(10)
S21(20,21)	12(12)
S21(22,23)	16(13)
S21(24)	83(16)
S21(25)	87(7,8)
S21(26)	28(10)
S21(27,28)	12(10)
S21(29)	27(16,18)
S21(30)	16(13)
S21(31)	12(12)
S21(32)	13(17)
S21(33,34)	60(4,5)
S21(35,36)	29(4–7)
S21(37,38)	79(3–5)
S21(39,40)	60(10,11)
S21(41,42)	29(10–18)
S21(43,44)	79(8–13)
S21(45–47)	61(5,8,12,14)
S21(48–50)	62(1–8)
S21(51,52)	62(11,12)
S21(53–55)	62(16–18)

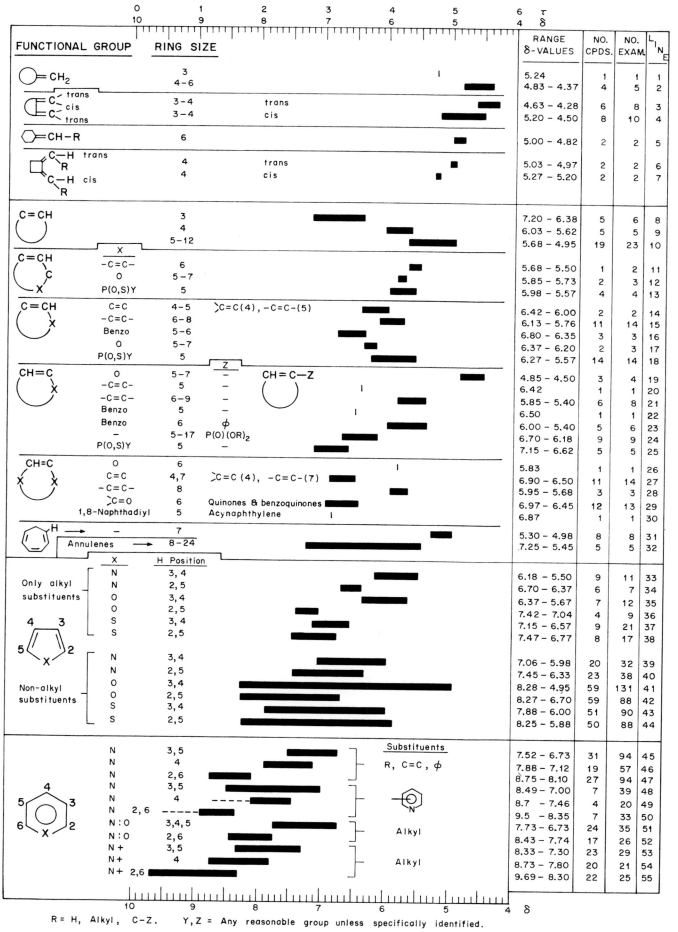

CHART S21. Summary of Chemical Shifts of =CH₂ and =CH in Cyclics

R = H, Alkyl, C-Z.   Y, Z = Any reasonable group unless specifically identified.

CHART S21

Cross Index between Summary and Detailed Charts

Summary Chart (Line)	Detailed Charts (Lines) from which Data Are Taken
S22(1)	13(4), 16(11)
S22(2)	27(15), 30(11), 31(5,11), 32(6), 42(9,10), 43(9), 44(5,6,8,9), 46(9), 50(12,20), 56(15), 57(16), 58(7), 61(11), 73(10, 11), 74(5,13), 85(6)
S22(3)	14(17)
S22(4)	14(13)
S22(5)	14(11,12)
S22(6)	14(14,17)
S22(7)	16(14), 17(20), 76(9,10)
S22(8)	16(15), 17(19)
S22(9)	75(3,6), 76(4,8)
S22(10)	13(14,15)
S22(11)	13(12,13)
S22(12)	15(3,12–14)
S22(13)	15(3,9–11)
S22(14)	15(3,9)
S22(15)	44(1), 46(13)
S22(16)	35(6), 54(19), 63(12)
S22(17)	42(11), 52(5), 54(8), 56(11), 58(11), 77(5,10)
S22(18)	46.1(8), 80(6)
S22(19)	81(8)
S22(20)	77(11), 89(13)
S22(21)	37(4,5,7), 46(14,15)
S22(22)	35(4,10), 36(4,5), 55(13), X
S22(23)	X, 34(12), 77(6)
S22(24)	33(6), X, 43(11), 44(4), 52(13), 55(7), 56(19), 63(5), 77(14)
S22(25)	34(4,11), X, 52(4), 57(10), 61(13), 63(4), 78(5), 85(10)
S22(26)	33(4,5), X, 59(9), 77(7), 78(4), 86(4)
S22(27)	33(6), 34(12,13), 35(11), 36(4), 37(6,8), 43(11), 44(3,4), 46(16), 55(13), 77(6), X
S22(28)	33(4,5), 34(4,11), 35(5), 52(4,13), 55(7), 56(19), 57(10), 59(7,8), 61(13), 63(4,5), 77(7,14), 85(10), 86(4), X
S22(29)	78(4–7)

X indicates data which are not included in detailed charts.

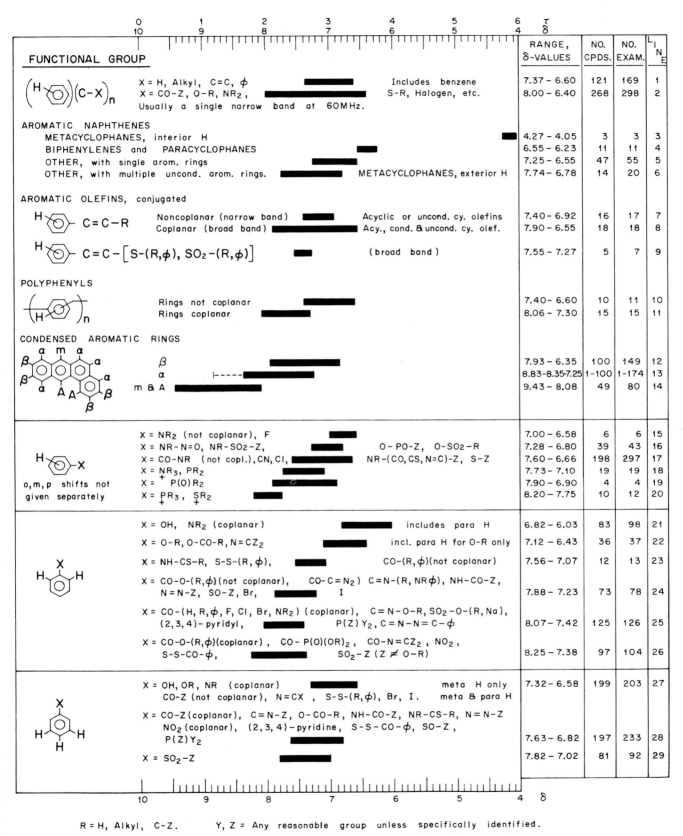

R = H, Alkyl, C-Z.   Y, Z = Any reasonable group unless specifically identified.

CHART S22. Summary of Chemical Shifts of Aromatic H in (⬡)ₙ, ⬡(C–X)ₙ, ⬡–X

Cross Index between Summary and Detailed Charts

Summary Chart (Lines)	Detailed Charts (Lines) Showing Similar Data
S23(1–3)	X
S23(4)	42(11), 44(2), X
S23(5–8)	X
S23(9)	33(7)
S23(10–18)	X
S23(19)	36(4), 37(11)
S23(20–23)	X
S23(24)	46(18)
S23(25–35)	X
S23(36)	59(11), X
S23(37–44)	X

X indicates data not included in detailed charts.

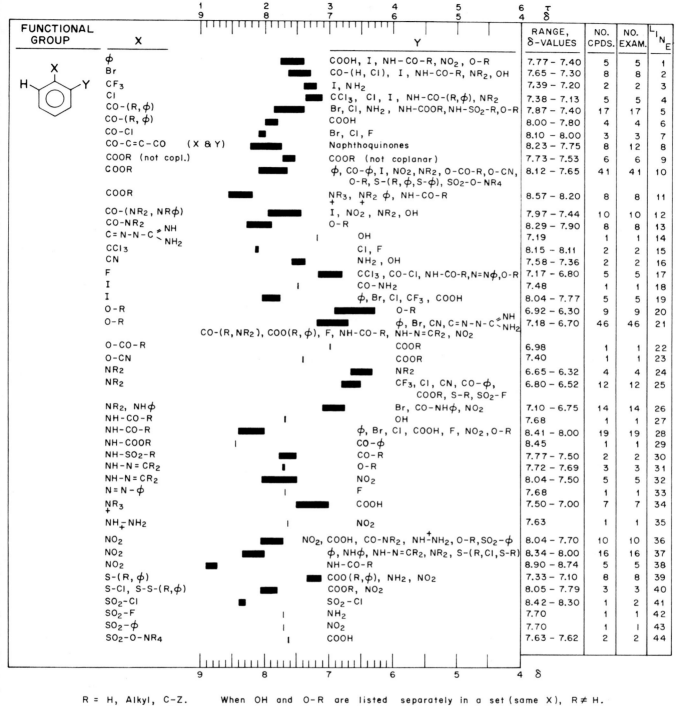

FUNCTIONAL GROUP X	Y	RANGE, δ-VALUES	NO. CPDS.	NO. EXAM.	LINE
φ	COOH, I, NH–CO–R, NO₂, O–R	7.77 – 7.40	5	5	1
Br	CO–(H, Cl), I, NH–CO–R, NR₂, OH	7.65 – 7.30	8	8	2
CF₃	I, NH₂	7.39 – 7.20	2	2	3
Cl	CCl₃, Cl, I, NH–CO–(R,φ), NR₂	7.38 – 7.13	5	5	4
CO–(R,φ)	Br, Cl, NH₂, NH–COOR, NH–SO₂–R, O–R	7.87 – 7.40	17	17	5
CO–(R,φ)	COOH	8.00 – 7.80	4	4	6
CO–Cl	Br, Cl, F	8.10 – 8.00	3	3	7
CO–C=C–CO (X & Y)	Naphthoquinones	8.23 – 7.75	8	12	8
COOR (not copl.)	COOR (not coplanar)	7.73 – 7.53	6	6	9
COOR	φ, CO–φ, I, NO₂, NR₂, O–CO–R, O–CN, O–R, S–(R,φ,S–φ), SO₂–O–NR₄	8.12 – 7.65	41	41	10
COOR	NR₃, NR₂φ, NH–CO–R	8.57 – 8.20	8	8	11
CO–(NR₂, NRφ)	I, NO₂, NR₂, OH	7.97 – 7.44	10	10	12
CO–NR₂	O–R	8.29 – 7.90	8	8	13
C=N–N–C⟨NH,NH₂	OH	7.19	1	1	14
CCl₃	Cl, F	8.15 – 8.11	2	2	15
CN	NH₂, OH	7.58 – 7.36	2	2	16
F	CCl₃, CO–Cl, NH–CO–R, N=Nφ, O–R	7.17 – 6.80	5	5	17
I	CO–NH₂	7.48	1	1	18
I	φ, Br, Cl, CF₃, COOH	8.04 – 7.77	5	5	19
O–R	O–R	6.92 – 6.30	9	9	20
O–R	φ, Br, CN, C=N–N–C⟨NH,NH₂, CO–(R,NR₂), COO(R,φ), F, NH–CO–R, NH–N=CR₂, NO₂	7.18 – 6.70	46	46	21
O–CO–R	COOR	6.98	1	1	22
O–CN	COOR	7.40	1	1	23
NR₂	NR₂	6.65 – 6.32	4	4	24
NR₂	CF₃, Cl, CN, CO–φ, COOR, S–R, SO₂–F	6.80 – 6.52	12	12	25
NR₂, NHφ	Br, CO–NHφ, NO₂	7.10 – 6.75	14	14	26
NH–CO–R	OH	7.68	1	1	27
NH–CO–R	φ, Br, Cl, COOH, F, NO₂, O–R	8.41 – 8.00	19	19	28
NH–COOR	CO–φ	8.45	1	1	29
NH–SO₂–R	CO–R	7.77 – 7.50	2	2	30
NH–N=CR₂	O–R	7.72 – 7.69	3	3	31
NH–N=CR₂	NO₂	8.04 – 7.50	5	5	32
N=N–φ	F	7.68	1	1	33
NR₃⁺	COOH	7.50 – 7.00	7	7	34
NH⁺–NH₂	NO₂	7.63	1	1	35
NO₂	NO₂, COOH, CO–NR₂, NH–⁺NH₂, O–R, SO₂–φ	8.04 – 7.70	10	10	36
NO₂	φ, NHφ, NH–N=CR₂, NR₂, S–(R,Cl,S–R)	8.34 – 8.00	16	16	37
NO₂	NH–CO–R	8.90 – 8.74	5	5	38
S–(R,φ)	COO(R,φ), NH₂, NO₂	7.33 – 7.10	8	8	39
S–Cl, S–S–(R,φ)	COOR, NO₂	8.05 – 7.79	3	3	40
SO₂–Cl	SO₂–Cl	8.42 – 8.30	1	2	41
SO₂–F	NH₂	7.70	1	1	42
SO₂–φ	NO₂	7.70	1	1	43
SO₂–O–NR₄	COOH	7.63 – 7.62	2	2	44

R = H, Alkyl, C–Z.   When OH and O–R are listed separately in a set (same X), R ≠ H.
Z = Any reasonable group.

CHART S23. Summary of Chemical Shifts of ⟨benzene ring⟩–X, Y (H)

Cross Index Between Summary and Detailed Charts

Summary Chart (Lines)	Detailed Charts (Lines) Showing Similar Data
S24(1–4)	X
S24(5)	42(11), 44(2), X
S24(6–10)	X
S24(11)	33(8), X
S24(12–18)	X
S24(19)	37(12)
S24(21,22)	X
S24(23)	46(18)
S24(24–33)	X
S24(34)	59(11), X
S24(35–46)	X

X indicates data not included in detailed charts.

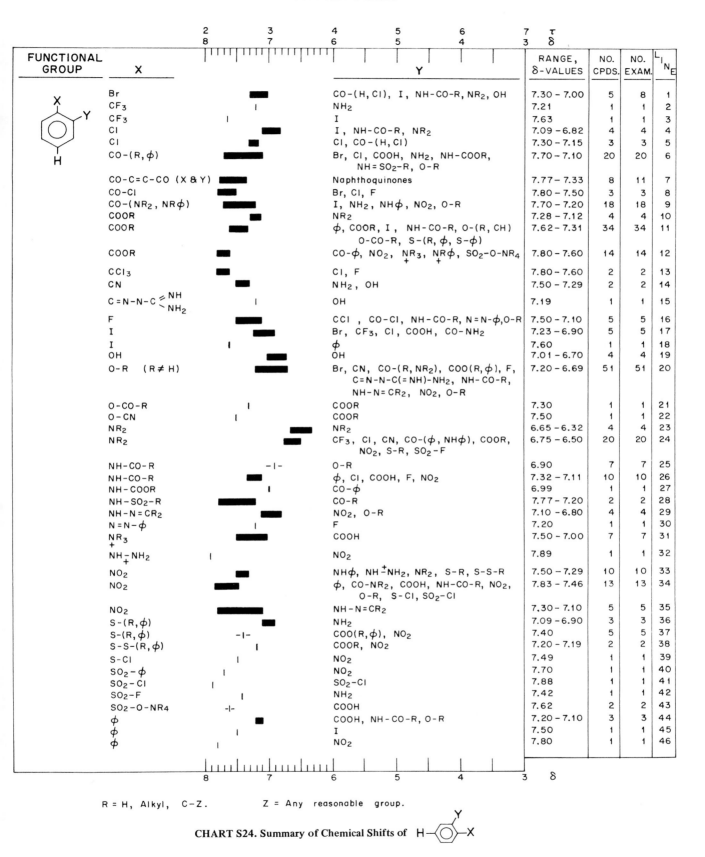

FUNCTIONAL GROUP	X	Y	RANGE, δ-VALUES	NO. CPDS.	NO. EXAM.	LINE
	Br	CO-(H, Cl), I, NH-CO-R, NR₂, OH	7.30 – 7.00	5	8	1
	CF₃	NH₂	7.21	1	1	2
	CF₃	I	7.63	1	1	3
	Cl	I, NH-CO-R, NR₂	7.09 – 6.82	4	4	4
	Cl	Cl, CO-(H, Cl)	7.30 – 7.15	3	3	5
	CO-(R, φ)	Br, Cl, COOH, NH₂, NH-COOR, NH=SO₂-R, O-R	7.70 – 7.10	20	20	6
	CO-C=C-CO (X & Y)	Naphthoquinones	7.77 – 7.33	8	11	7
	CO-Cl	Br, Cl, F	7.80 – 7.50	3	3	8
	CO-(NR₂, NRφ)	I, NH₂, NHφ, NO₂, O-R	7.70 – 7.20	18	18	9
	COOR	NR₂	7.28 – 7.12	4	4	10
	COOR	φ, COOR, I, NH-CO-R, O-(R, CH) O-CO-R, S-(R, φ, S-φ)	7.62 – 7.31	34	34	11
	COOR	CO-φ, NO₂, NR₃⁺, NRφ⁺, SO₂-O-NR₄	7.80 – 7.60	14	14	12
	CCl₃	Cl, F	7.80 – 7.60	2	2	13
	CN	NH₂, OH	7.50 – 7.29	2	2	14
	C=N-N-C⟨NH⟩⟨NH₂⟩	OH	7.19	1	1	15
	F	CCl, CO-Cl, NH-CO-R, N=N-φ, O-R	7.50 – 7.10	5	5	16
	I	Br, CF₃, Cl, COOH, CO-NH₂	7.23 – 6.90	5	5	17
	I	φ	7.60	1	1	18
	OH	OH	7.01 – 6.70	4	4	19
	O-R (R≠H)	Br, CN, CO-(R, NR₂), COO(R, φ), F, C=N-N-C(=NH)-NH₂, NH-CO-R, NH-N=CR₂, NO₂, O-R	7.20 – 6.69	51	51	20
	O-CO-R	COOR	7.30	1	1	21
	O-CN	COOR	7.50	1	1	22
	NR₂	NR₂	6.65 – 6.32	4	4	23
	NR₂	CF₃, Cl, CN, CO-(φ, NHφ), COOR, NO₂, S-R, SO₂-F	6.75 – 6.50	20	20	24
	NH-CO-R	O-R	6.90	7	7	25
	NH-CO-R	φ, Cl, COOH, F, NO₂	7.32 – 7.11	10	10	26
	NH-COOR	CO-φ	6.99	1	1	27
	NH-SO₂-R	CO-R	7.77 – 7.20	2	2	28
	NH-N=CR₂	NO₂, O-R	7.10 – 6.80	4	4	29
	N=N-φ	F	7.20	1	1	30
	NR₃⁺	COOH	7.50 – 7.00	7	7	31
	NH-NH₂⁺	NO₂	7.89	1	1	32
	NO₂	NHφ, NH-NH₂⁺, NR₂, S-R, S-S-R	7.50 – 7.29	10	10	33
	NO₂	φ, CO-NR₂, COOH, NH-CO-R, NO₂, O-R, S-Cl, SO₂-Cl	7.83 – 7.46	13	13	34
	NO₂	NH-N=CR₂	7.30 – 7.10	5	5	35
	S-(R, φ)	NH₂	7.09 – 6.90	3	3	36
	S-(R, φ)	COO(R, φ), NO₂	7.40	5	5	37
	S-S-(R, φ)	COOR, NO₂	7.20 – 7.19	2	2	38
	S-Cl	NO₂	7.49	1	1	39
	SO₂-φ	NO₂	7.70	1	1	40
	SO₂-Cl	SO₂-Cl	7.88	1	1	41
	SO₂-F	NH₂	7.42	1	1	42
	SO₂-O-NR₄	COOH	7.62	2	2	43
	φ	COOH, NH-CO-R, O-R	7.20 – 7.10	3	3	44
	φ	I	7.50	1	1	45
	φ	NO₂	7.80	1	1	46

R = H, Alkyl, C–Z.   Z = Any reasonable group.

CHART S24. Summary of Chemical Shifts of H—⟨◯⟩—X (with Y)

Cross Index Between Summary and Detailed Charts

Summary Chart (Lines)	Detailed Charts (Lines) Showing Similar Data
S25(1)	X
S25(2)	43(13), 44(3)
S25(3–7)	X
S25(8)	42(12), X
S25(9–19)	X
S25(20)	33(8)
S25(21)	X
S25(22)	33(7)
S25(23,24)	X
S25(25)	52(6)
S25(26–30)	X
S25(31)	44(4), X
S25(32)	46(17)
S25(33–42)	X
S25(43)	52(14)
S25(44–48)	X
S25(49)	X, 59(10)
S25(50)	35(10), 36(4), 37(10)
S25(51–61)	X

X indicates data not included in detailed charts.

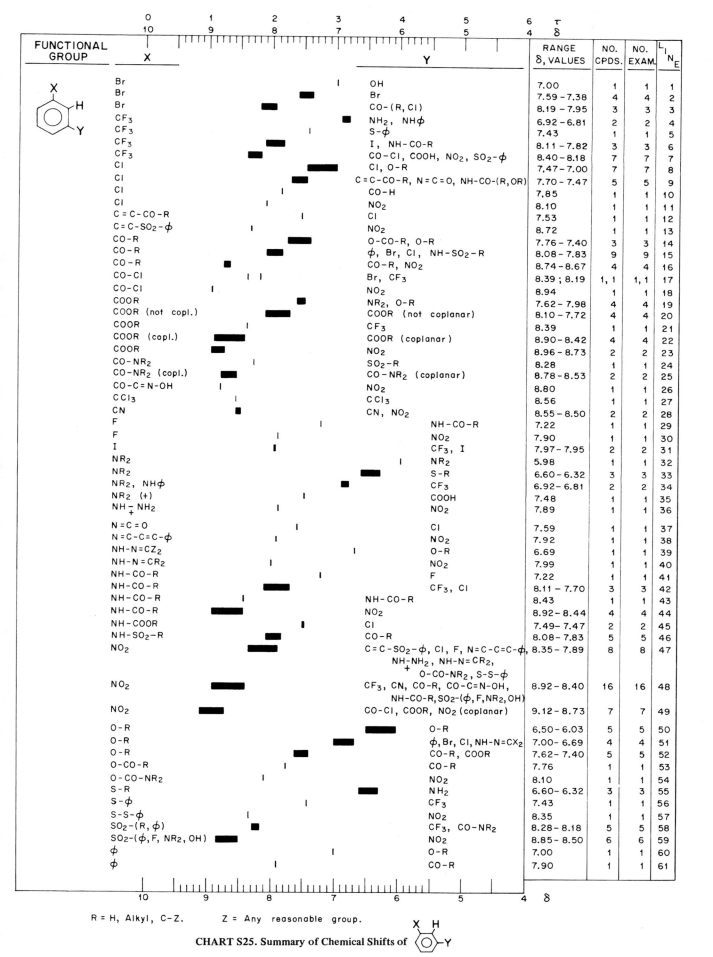

FUNCTIONAL GROUP X	Y	RANGE δ, VALUES	NO. CPDS.	NO. EXAM.	LINE
Br	OH	7.00	1	1	1
Br	Br	7.59 – 7.38	4	4	2
Br	CO-(R, Cl)	8.19 – 7.95	3	3	3
CF_3	NH_2, NHφ	6.92 – 6.81	2	2	4
CF_3	S-φ	7.43	1	1	5
CF_3	I, NH-CO-R	8.11 – 7.82	3	3	6
CF_3	CO-Cl, COOH, NO_2, SO_2-φ	8.40 – 8.18	7	7	7
Cl	Cl, O-R	7.47 – 7.00	7	7	8
Cl	C=C-CO-R, N=C=O, NH-CO-(R,OR)	7.70 – 7.47	5	5	9
Cl	CO-H	7.85	1	1	10
Cl	NO_2	8.10	1	1	11
C=C-CO-R	Cl	7.53	1	1	12
C=C-SO_2-φ	NO_2	8.72	1	1	13
CO-R	O-CO-R, O-R	7.76 – 7.40	3	3	14
CO-R	φ, Br, Cl, NH-SO_2-R	8.08 – 7.83	9	9	15
CO-R	CO-R, NO_2	8.74 – 8.67	4	4	16
CO-Cl	Br, CF_3	8.39 ; 8.19	1, 1	1, 1	17
CO-Cl	NO_2	8.94	1	1	18
COOR	NR_2, O-R	7.62 – 7.98	4	4	19
COOR (not copl.)	COOR (not coplanar)	8.10 – 7.72	4	4	20
COOR	CF_3	8.39	1	1	21
COOR (copl.)	COOR (coplanar)	8.90 – 8.42	4	4	22
COOR	NO_2	8.96 – 8.73	2	2	23
CO-NR_2	SO_2-R	8.28	1	1	24
CO-NR_2 (copl.)	CO-NR_2 (coplanar)	8.78 – 8.53	2	2	25
CO-C=N-OH	NO_2	8.80	1	1	26
CCl_3	CCl_3	8.56	1	1	27
CN	CN, NO_2	8.55 – 8.50	2	2	28
F	NH-CO-R	7.22	1	1	29
F	NO_2	7.90	1	1	30
I	CF_3, I	7.97 – 7.95	2	2	31
NR_2	NR_2	5.98	1	1	32
NR_2	S-R	6.60 – 6.32	3	3	33
NR_2, NHφ	CF_3	6.92 – 6.81	2	2	34
NR_2 (+)	COOH	7.48	1	1	35
NH-$\overset{+}{N}H_2$	NO_2	7.89	1	1	36
N=C=O	Cl	7.59	1	1	37
N=C-C=C-φ	NO_2	7.92	1	1	38
NH-N=CZ_2	O-R	6.69	1	1	39
NH-N=CR_2	NO_2	7.99	1	1	40
NH-CO-R	F	7.22	1	1	41
NH-CO-R	CF_3, Cl	8.11 – 7.70	3	3	42
NH-CO-R	NH-CO-R	8.43	1	1	43
NH-CO-R	NO_2	8.92 – 8.44	4	4	44
NH-COOR	Cl	7.49 – 7.47	2	2	45
NH-SO_2-R	CO-R	8.08 – 7.83	5	5	46
NO_2	C=C-SO_2-φ, Cl, F, N=C-C=C-φ, NH-NH_2, NH-N=CR_2, O-CO-$\overset{+}{N}R_2$, S-S-φ	8.35 – 7.89	8	8	47
NO_2	CF_3, CN, CO-R, CO-C=N-OH, NH-CO-R, SO_2-(φ,F,NR_2,OH)	8.92 – 8.40	16	16	48
NO_2	CO-Cl, COOR, NO_2 (coplanar)	9.12 – 8.73	7	7	49
O-R	O-R	6.50 – 6.03	5	5	50
O-R	φ, Br, Cl, NH-N=CX_2	7.00 – 6.69	4	4	51
O-R	CO-R, COOR	7.62 – 7.40	5	5	52
O-CO-R	CO-R	7.76	1	1	53
O-CO-NR_2	NO_2	8.10	1	1	54
S-R	NH_2	6.60 – 6.32	3	3	55
S-φ	CF_3	7.43	1	1	56
S-S-φ	NO_2	8.35	1	1	57
SO_2-(R,φ)	CF_3, CO-NR_2	8.28 – 8.18	5	5	58
SO_2-(φ,F,NR_2,OH)	NO_2	8.85 – 8.50	6	6	59
φ	O-R	7.00	1	1	60
φ	CO-R	7.90	1	1	61

R = H, Alkyl, C–Z.   Z = Any reasonable group.

CHART S25. Summary of Chemical Shifts of

CHART S25

Cross Index Between Summary and Detailed Charts

Summary Chart (Lines)	Detailed Charts (Lines) Showing Similar Data
S26(1)	43(11), 44(3), X
S26(2–4)	X
S26(5)	42(11), 44(2), X
S26(6–9)	X
S26(10)	34(5), X
S26(11,12)	X
S26(13)	33(7)
S26(14)	X
S26(15)	52(14), X
S26(16–21)	X
S26(22)	44(4)
S26(23)	X
S26(24)	46(17), X
S26(25–36)	X
S26(37)	59(10), X
S26(38)	35(10), 36(4), 37(11)
S26(39–49)	X

X indicates data not included in detailed charts.

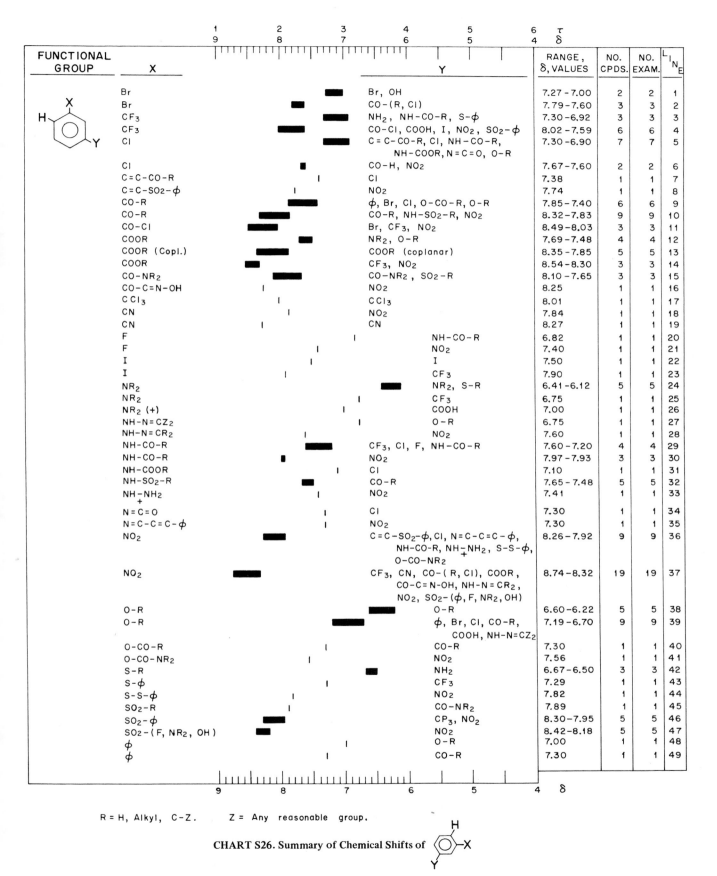

FUNCTIONAL GROUP X	Y	RANGE, δ, VALUES	NO. CPDS.	NO. EXAM.	LINE
Br	Br, OH	7.27 - 7.00	2	2	1
Br	CO-(R, Cl)	7.79 - 7.60	3	3	2
CF₃	NH₂, NH-CO-R, S-φ	7.30 - 6.92	3	3	3
CF₃	CO-Cl, COOH, I, NO₂, SO₂-φ	8.02 - 7.59	6	6	4
Cl	C=C-CO-R, Cl, NH-CO-R, NH-COOR, N=C=O, O-R	7.30 - 6.90	7	7	5
Cl	CO-H, NO₂	7.67 - 7.60	2	2	6
C=C-CO-R	Cl	7.38	1	1	7
C=C-SO₂-φ	NO₂	7.74	1	1	8
CO-R	φ, Br, Cl, O-CO-R, O-R	7.85 - 7.40	6	6	9
CO-R	CO-R, NH-SO₂-R, NO₂	8.32 - 7.83	9	9	10
CO-Cl	Br, CF₃, NO₂	8.49 - 8.03	3	3	11
COOR	NR₂, O-R	7.69 - 7.48	4	4	12
COOR (Copl.)	COOR (coplanar)	8.35 - 7.85	5	5	13
COOR	CF₃, NO₂	8.54 - 8.30	3	3	14
CO-NR₂	CO-NR₂, SO₂-R	8.10 - 7.65	3	3	15
CO-C=N-OH	NO₂	8.25	1	1	16
CCl₃	CCl₃	8.01	1	1	17
CN	NO₂	7.84	1	1	18
CN	CN	8.27	1	1	19
F	NH-CO-R	6.82	1	1	20
F	NO₂	7.40	1	1	21
I	I	7.50	1	1	22
I	CF₃	7.90	1	1	23
NR₂	NR₂, S-R	6.41 - 6.12	5	5	24
NR₂	CF₃	6.75	1	1	25
NR₂ (+)	COOH	7.00	1	1	26
NH-N=CZ₂	O-R	6.75	1	1	27
NH-N=CR₂	NO₂	7.60	1	1	28
NH-CO-R	CF₃, Cl, F, NH-CO-R	7.60 - 7.20	4	4	29
NH-CO-R	NO₂	7.97 - 7.93	3	3	30
NH-COOR	Cl	7.10	1	1	31
NH-SO₂-R	CO-R	7.65 - 7.48	5	5	32
NH-NH₂ (+)	NO₂	7.41	1	1	33
N=C=O	Cl	7.30	1	1	34
N=C-C=C-φ	NO₂	7.30	1	1	35
NO₂	C=C-SO₂-φ, Cl, N=C-C=C-φ, NH-CO-R, NH-NH₂(+), S-S-φ, O-CO-NR₂	8.26 - 7.92	9	9	36
NO₂	CF₃, CN, CO-(R, Cl), COOR, CO-C=N-OH, NH-N=CR₂, NO₂, SO₂-(φ, F, NR₂, OH)	8.74 - 8.32	19	19	37
O-R	O-R	6.60 - 6.22	5	5	38
O-R	φ, Br, Cl, CO-R, COOH, NH-N=CZ₂	7.19 - 6.70	9	9	39
O-CO-R	CO-R	7.30	1	1	40
O-CO-NR₂	NO₂	7.56	1	1	41
S-R	NH₂	6.67 - 6.50	3	3	42
S-φ	CF₃	7.29	1	1	43
S-S-φ	NO₂	7.82	1	1	44
SO₂-R	CO-NR₂	7.89	1	1	45
SO₂-φ	CP₃, NO₂	8.30 - 7.95	5	5	46
SO₂-(F, NR₂, OH)	NO₂	8.42 - 8.18	5	5	47
φ	O-R	7.00	1	1	48
φ	CO-R	7.30	1	1	49

R = H, Alkyl, C-Z.   Z = Any reasonable group.

CHART S26. Summary of Chemical Shifts of [ring]-X

CHART S26

Cross Index between Summary and Detailed Charts

Summary Chart (Lines)	Detailed Charts (Lines) Showing Similar Data
S27(1–4)	X
S27(5)	42(11), X
S27(6–9)	X
S27(10)	34(5), X
S27(11–13)	X
S27(14)	33(8), X
S27(15)	33(7), X
S27(16)	52(14), X
S27(17–23)	X
S27(24)	46(19)
S27(25–35)	X
S27(36)	59(10), X
S27(37)	36(4), 37(12), X
S27(38–46)	X

X indicates data not included in detailed charts.

FUNCTIONAL GROUP X	Y	RANGE, δ-VALUES	NO. CPDS	NO. EXAM.	LINE
Br	Br, OH	7.00 – 6.95	2	2	1
Br	CO–(R, Cl)	7.41 – 7.30	3	3	2
CF_3	I, NH_2, S–φ	7.29 – 7.20	3	3	3
CF_3	CO–Cl, COOH, NH–CO–R, NO_2, SO_2–φ	7.80 – 7.59	6	6	4
Cl	C=C–CO–R, C=N=O, NH–CO–R, NH–COOR, O–R, Cl	7.38 – 7.00	7	7	5
Cl	CO–H, NO_2	7.60 – 7.53	3	3	6
C=C–CO–R	Cl	7.38 – 7.30	2	2	7
C=C–SO_2–φ	NO_2	7.74	1	1	8
CO–R	φ, Br, Cl, NH–SO_2–R, O–CO–R, O–R	7.65 – 7.30	12	12	9
CO–R	CO–R, NO_2	7.85 – 7.67	4	5	10
CO–Cl	Br	7.39	1	1	11
CO–Cl	CF_3	7.72	1	1	12
CO–Cl	NO_2	7.85	1	1	13
COOR	COOR (not coplanar), NR_2, O–R	7.34 – 7.20	5	5	14
COOR	COOR (coplanar), CF_3, NO_2	7.77 – 7.38	8	8	15
CO–NR_2	CO–NR_2, SO_2–R	7.51 – 7.40	2	2	16
CO–C=N–OH	NO_2	7.80	1	1	17
CCl_3	CCl_3	7.51	1	1	18
CN	CN, NO_2	7.84 – 7.74	2	2	19
F	NH–CO–R	7.22	1	1	20
F	NO_2	7.53	1	1	21
I	I	6.67	1	1	22
I	CF_3	7.20	1	1	23
NR_2	NR_2, S–R	7.01 – 6.82	5	5	24
NH_2	CF_3	7.20	1	1	25
NR_2 (+)	COOH	7.33	1	1	26
NH–N=CZ_2	NO_2, O–R	7.30 – 7.28	2	2	27
NH–CO–R	Cl, F, NH–CO–R	7.42 – 7.20	3	3	28
NH–CO–R	CF_3, NO_2	7.67 – 7.40	4	4	29
NH–COOR	Cl	7.20 – 7.10	2	2	30
NH–SO_2–R	CO–R	7.65 – 7.48	5	5	31
NH–NH_2 (+)	NO_2	7.63	1	1	32
N=C–C=C–φ	NO_2	7.30	1	1	33
N=C=O	Cl	7.30	1	1	34
NO_2	Cl, F, N=C–C=C–φ, NH–CO–R, NH–N=CR_2, NH–NH_2 (+), O–CO–NR_2, S–S–φ	7.67 – 7.30	10	10	35
NO_2	C=C–SO_2–φ, CF_3, CO–(R,Cl), COOR, CO–C=N–OH, CN, NO_2, SO_2–(φ,F,NR_2,OH)	8.05 – 7.60	23	3	36
O–R	φ, Br, Cl, CO–R, COOR, O–R, NH–N=CZ_2	7.34 – 6.92	14	14	37
O–CO–R	CO–R	7.48	1	1	38
O–CO–NR_2	NO_2	7.56	1	1	39
S–R	NH_2	7.01 – 6.94	3	3	40
S–φ	CF_3	7.29	1	1	41
S–S–φ	NO_2	7.51	1	1	42
SO_2–(R,φ)	CF_3, CO–NR_2, NO_2	7.77 – 7.51	4	4	43
SO_2–(F,NR_2,OH)	NO_2	8.05 – 7.79	5	5	44
φ	O–R	7.18	1	1	45
φ	CO–R	7.30	1	1	46

R = H, Alkyl, C–Z.   Z = Any reasonable group.

CHART S27. Summary of Chemical Shifts of

Cross Index Between Summary and Detailed Charts

Summary Chart (Line)	Detailed Charts (Lines) Showing Similar Data
S28(1)	43(11), 44(3), X
S28(2,3)	X
S28(4)	42(11), 44(2), X
S28(5–9)	X
S28(10)	34(5), X
S28(11–17)	X
S28(18)	33(7), X
S28(19–51)	X

X indicates data not included in detailed charts.

FUNCTIONAL GROUP X	Y	RANGE, δ-VALUES	NO. CPDS.	NO. EXAM.	LINE
Br	Br, Cl, CN, F, Z, NH−C(O,S)−R, NH−COOR, NH−N=C−φ, NR₂, O−(R,φ), O−C(O,S)−NR₂, S−CF₃, S−S−φ	7.53−7.18	34	34	1
Br	CF₃, CO−(R, φ, C=N-, NH₂,OH), NH₃⁺, SO−(φ, Cl, O−Na)	7.82−7.50	25	25	2
CF₃	Br, Cl, CO−R, F, I, NH−CO−R	7.79−7.49	6	6	3
Cl	R, Br, Cl, I, NH−C=C, NR₂, NH−N=CR₂, NH−P(Z)Y₂	7.29−7.00	20	20	4
Cl	C=CRZ, O−(R, CF₃, CN, C=NH), O−(COOφ, SO₂−C=C)	7.43−7.10	29	29	5
Cl	NH−(CO−R, COOR, CS−R, SO₂−φ), N(CN)−C=NH, N=N−φ, S−(R, CS−R, CS−S−R, S−φ)	7.54−7.17	39	39	6
Cl	CO−(R,φ, Cl, C=N−OH, C=N₂), CO−(NR₂, O−O−CO−φ), CZₙ(R)₃₋ₙ	7.64−7.14	54	54	7
Cl	SO₂−(φ, Cl, F, NR₂, S−R, S−φ), SO₂−thienyl (2,3)	7.72−7.42	18	18	8
CO−(R, φ)	φ, Br, CF₃, Cl, F, I, NR₂, S−R, O−(R,φ)	8.08−7.56	78	78	9
CO−(R, φ)	CN, CO−(R,φ)(coplanar), NH₃⁺, NH−(CO,SO₂)−R, NO₂, SO₂−F	8.32−7.86	28	28	10
CO−N(R₂, Rφ, NR₂)	SO₂−R	8.03−7.52	13	13	11
CO−N(R₂, Rφ, NR₂)	φ, Cl, CO−NR₂, F, NO₂, O−R	8.10−7.64	13	13	12
CO−Nφ₂	O−R	7.27	1	1	13
CO−C=CR₂	NO₂, O−R	8.01−7.70	3	3	14
CO−Cl	Cl, CO−Cl, NO₂, O−R	8.38−8.02	4	4	15
COOR	NR₂	7.99−7.73	12	12	16
COOR	O−(R, Na, CF₃,CO−φ)	8.19−7.77	20	20	17
COOR	φ, COOR, F, I, NH−(CO, SO₂)−R, S−(R, CF₃)	8.20−7.98	16	16	18
COOR	C=N−R, CN, CO−φ(TFA), NR₃⁺, N=N−φ, SO₂−(R, NR₂)	8.45−8.11	10	10	19
COOR	NO₂	8.33−8.15	36	36	20
CO−O−O−CO−φ	Cl	7.99	1	1	21
CO−C=N−OH	Cl	7.99	1	1	22
CO−C=N−OH	NO₂	8.19	1	1	23
CO−C=N−N−CO−R	S−φ	8.07	1	1	24
CO−C=N₂	Cl	7.67	1	1	25
CO−C=N₂	NO₂	7.84	1	1	26
CO−N₃	NO₂	8.25	1	1	27
CO−S−R	O−R	7.95	1	1	28
CS−φ	O−R	7.73	1	1	29
CS−O−R	O−R	8.23−8.19	2	2	30
C=CR₂	Cl, O−R	7.32−7.06	5	5	31
C=C−φ, C(φ)=C−φ	O−CO−R	7.14−7.12	2	2	32
C(φ)=CCl₂	Cl	7.29−7.25	1	2	33
C=C−CO−(R,φ) trans	NR₂	7.24	1	1	34
C=C−CO−(R,φ) trans	Cl, NO₂, O−R	7.68−7.42	4	5	35
C=C−CO−(R,φ) trans	NR₃⁺	7.97	1	1	36
C=C−COOH trans	O−CO−R, O−R	7.73−7.61	2	2	37
C=C−CN trans	NR₂	7.30	1	1	38
C=C(CN)₂	O−R	8.12−7.99	2	2	39
C(φ)=C(CN)₂	O−R	7.40	1	1	40
C(OH)=C(CN)₂	NO₂	8.02	1	1	41
C=C−SO₂−(R,φ) trans	Cl, O−R	7.42−7.30	3	3	42
C=C(SO₂−R)₂	NO₂	7.77	1	1	43
C=C−NO₂	F, NR₂	7.42−7.37	2	2	44
C=C(CN)−φ	O−R	7.81	1	1	45
C=C(CN)−COOR	NR₂	7.95−7.90	2	2	46
C=N−R	COOR, NH−CO−R	7.79−7.69	2	2	47
C=N−φ	NR₂, OH	7.38−7.20	3	3	48
C=N−C=C−φ	O−R	6.80	1	1	49
C=N−N−φ	NR₂	7.53−7.45	3	3	50
C=N−N=CR₂	O−R	7.78−7.73	7	7	51

R = H, Alkyl, C−Z.   Z = Any reasonable group.

CHART S28. Summary of Chemical Shifts of Y−⟨C₆H₄⟩−X (A)

Cross Index Between Summary and Detailed Charts

Summary Chart (Lines)	Detailed Charts (Lines) Showing Similar Data
S29(1,2)	X
S29(3)	42(10), X
S29(4)	44(1), X
S29(5)	X
S29(6)	44(4), X
S29(7)	46(18)
S29(8–31)	X
S29(32)	59(11), X
S29(33,34)	X
S29(35)	59(10), X
S29(36–45)	X
S29(46)	37(11)
S29(47,48)	X

X indicates data not recorded in detailed charts.

CHART S29

FUNCTIONAL GROUP X	Y	RANGE, δ-VALUES	NO. CPDS.	NO. EXAM.	LINE
CN	Br, F, NH₂, O-(R, CF₃)	7.69 – 7.47	7	7	1
CN	CO-R, COOR	7.93 – 7.72	3	3	2
CCl₃	CCl₃, Cl, F	7.95 – 7.88	3	3	3
F	R-Z, φ, Br, CCl₃, CF₃, C=C-NO₂, F, NH₂, NH-CO-R, N=C=S, N=N-φ, O-CO-R, O-R, S-R	7.13 – 6.75	21	21	4
F	CN, CO-(R, C=C, φ, NHφ), COOR, NO₂, SO₂-(φ, Cl, NH₂)	7.37 – 7.00	27	27	5
I	Br, CF₃, Cl, CO-R, COOH, I, NH-(CO, SO₂)-R	7.71 – 7.30	7	9	6
NR₂	NR₂, NHφ	6.57 – 6.32	4	4	7
NR₂	Br, Cl, F	6.72 – 6.36	9	9	8
NR₂	φ, C=C-(CN, CO-H, NO₂), C=CZ₂, C=N-(φ, N-φ), CN, NO₂, SH, NH-C(O,S)-R, N=(CZ₂, N-φ,O)	6.79 – 6.49	33	33	9
NR₂	CO-(R, φ), COOR, O-(R, SO₂-O-K)	6.88 – 6.48	24	24	10
NR₂	SO₂-(φ, F, NR₂)	6.91 – 6.56	19	19	11
NHφ	NR₂	6.81 – 6.80	2	2	12
NHφ	NO₂	7.09	1	1	13
NR₃⁺	Br, OH, S-(R, φ)	7.60 – 7.22	7	7	14
NR₃⁺	φ, C=C-COOH, CO-φ, COOH, NR̲φ⁺, SO₂-(φ, NR₂)	7.89 – 7.62	25	25	15
NR₃⁺ TFA→I I←DMSO	CO-R	7.83, 7.20	2	2	16
NH-N=CR₂⁺ I TFA	NO₂	7.53	1	1	17
NH-CO-R	Br, CF₃, Cl, F, I, NR₂, O-CO-R, O-R, SO-OH	7.73 – 7.31	34	34	18
NH-CO-R	C=N-R, CO-H, COOR, NH-CO-R, NO₂, S-CO-R, SO₂-(R, F, NR₂)	8.01 – 7.68	30	30	19
NH-COOR	Br, O-R	7.42 – 7.40	2	2	20
NR-CS-R	Br, Cl, NR₂	7.28 – 7.02	3	3	21
NH-CS-NR₂	O-R	7.32 – 7.09	8	8	22
NH-CS-NRZ	SO₂-(φ, NR₂)	7.81 – 7.52	8	8	23
NH-CS-CS-NH₂	O-R	7.93	1	1	24
NH-CS-O-R	SO₂-(φ, NH₂)	7.80 – 7.67	3	3	25
NH-SO₂-R	Cl, I, NH-SO,-R	7.17 – 7.01	5	5	26
NH-SO₂-R	CO-R, COOH	7.44 – 7.26	9	9	27
NH-NH₂	NO₂, SO₂-R	7.30 – 6.88	3	3	28
NH-N=C-φ	Br	6.95	1	1	29
NH-N=C-φ	SO₂-NH₂	7.51 – 7.50	1	2	30
NH-N=C(R₂,Z₂)	Cl, NO₂, O-R	7.29 – 7.02	21	21	31
NO₂	NR₂, NH-NH₂, NH-N=CR₂, N₃, NO₂(not coplanar), S-φ, S-SO₂-φ, S-CO-NR₂	8.35 – 7.95	16	16	32
NO₂	φ, C=CZ₂, F, NH-CO-R, NH-N=CR₂, O-(R, φ), O-CO-(R, φ, CF₃), O-CS-NR₂, O-P(O,S)(O-R)₂, P(O)(OH)₂, SO₂-(φ, NR₂)	8.43 – 8.12	37	37	32
NO₂	COOR	8.33 – 8.15	36	36	34
NO₂	CO-[C=C, C=N-OH, C(=N₂)-CO-R, Cl, NR₂, N₃, R, φ], NO₂(copl.)	8.55 – 8.22	17	17	35
N=CZ₂	NR₂	7.60	1	1	36
N=CH-φ	O-R	7.76	1	1	37
N=C=(O,S)	F, O-CF₃	7.15 – 7.10	2	2	38
N=N-φ	Cl, F, NR₂	7.89 – 7.72	12	12	39
N=N-φ	COOR, O-R	8.52 – 7.80	5	5	40
N=NO-φ or NO=N-φ	O-R	8.31 – 7.82	3	6	41
N=O	NR₂, OH	7.79 – 7.67	3	3	42
NH-P(O)(OR)₂	Cl	6.96	1	1	43
N(CN)-C=NH	Cl	7.46	1	1	44
N₃ (triazo)	NO₂	7.18	1	1	45
OH	OH	6.63	1	1	46
OH	Br, N=O, O-(R, φ)	6.81 – 6.59	9	9	47
OH	C=N-φ, NR₃⁺, SR₂⁺, SO₂-O-(Na, NH₄)	7.34 – 6.88	9	9	48

R = H, Alkyl, C-Z.   Z = Any reasonable group.

CHART S29. Summary of Chemical Shifts of Y–◯–X (B)

Cross Index between Summary and Detailed Charts

Summary Chart (Lines)	Detailed Charts (Lines) Showing Similar Data
S30(1)	36(4)
S30(2–46)	X

X indicates data not included in detailed charts.

| | | τ | | | |
| | | δ | | | |
FUNCTIONAL GROUP	X	Y	RANGE, δ-VALUES	NO. CPDS.	NO. EXAM.	LINE
![structure: benzene ring with X top, Y bottom, H,H sides] O-(R,φ)	O-(R,φ)	C=N-(C=C-φ, N=CR₂), N=CH-φ, NR₂, NH-CO-(R,φ,O-R), NH-CS-(CS-NH₂, NR₂), NH-N=CZ₂, O-CO-(R,NR₂),OH, O-R,S-(R,φ,C=C),S-CS-S-R	6.99-6.59	53	53	1
	O-(R,φ)	φ, Br, Cl, CO-(R,φ,C=C), COO(R,φ), CO-(Cl, Nφ₂, NH-N=CR₂), CN, CS-(φ, O-R), F, NO₂, P(O,S)φ₂, SO-R, SO₂-(φ, NR₂)	7.18-6.61	118	118	2
	O-R	C=CH-(R, SO₂-φ), C=CH-CO-(R,φ,OH), C=C(SO₂-φ)-CO-φ	6.95-6.71	9	9	3
	O-R	C=C(φ)-CN, C(φ) = C(CN)₂	6.94-6.88	2	2	4
	O-R	C=C(CN)₂	7.20-7.19	2	2	5
	O-R	N=N-φ, & NO=N-φ or N=NO-φ	7.38-6.93	7	10	6
	O-CF₃	Cl, CN, COOH, N=C=O	7.37-7.10	4	4	7
	O-CO-(R,φ)	C(R,φ) = C(φ)R, C=C-COOH, COOR, F, NH-CO-R, NO₂, O-R	7.32-6.98	13	13	8
	O-CO-CF₃	NO₂	7.11	1	1	9
	O-CO-NR₂	Br, O-R	7.06-7.04	2	2	10
	O-CO-O-φ	Cl	7.20-7.19	2	2	11
	O-CS-NR₂	Br	6.70	1	1	12
	O-CS-NR₂	NO₂	7.26	1	1	13
	O-(CN, C=NH, SO₂-φ)	Cl	7.25-7.13	4	4	14
	O-SO₂-O-K	NH₂	7.15	1	1	15
	O-P(O,S)(O-R)₂	NO₂	7.58-7.30	8	8	16
	O-C-NH-(R, NR₂, SO₂-φ) (‖ NH)	CH₃	7.11-6.87	5	5	17
	S-(R,φ)	CO-(R, C=N-N-Z), COOH, NH₂, O-R	7.39-7.08	9	9	18
	S-(R,φ)	Cl, F, ⁺NH₃, NO₂	7.62-7.13	22	26	19
	S-CF₃	Br	7.50	1	1	20
	S-CF₃	COOH	7.78	1	1	21
	S-CO-R	NH-CO-R	7.33	1	1	22
	S-CO-NR₂	NO₂	7.75	1	1	23
	S-CS-(R,S-R)	Cl, O-R	7.43-7.37	6	6	24
	S-S-φ	Br, Cl	7.39-7.38	2	2	25
	S-SO₂-φ	NO₂	7.59	1	1	26
	⁺SR₂	OH	7.87	1	1	27
	SO-R	O-R	7.62	1	1	28
	SO-OH	NH-CO-R	7.79	1	1	29
	SO₂-(R,φ)	NR₂, NH-NH₂	7.73-7.44	5	5	30
	SO₂-(R,φ)	Br, Cl, C=C-φ, F, NH-CO-R, NH-CS-(NR₂, O-R), O-R	8.03-7.72	17	17	31
	SO₂-R	CO-NH-(R, φ, NR₂)	8.19-7.72	11	11	32
	SO₂-(R,φ)	COOH, ⁺NR₃, NO₂	8.26-8.07	4	4	33
	SO₂-(F,Cl)	Br, Cl, F, NH₂, NH-CO-R	8.06-7.73	6	6	34
	SO₂-F	CO-R	8.21	1	1	35
	SO₂-NR₂	NH₂	7.91-7.40	14	14	36
	SO₂-NR₂	Cl, F, NH-CO-R, NH-CS-(NH₂, O-R), NH-N=CR₂, O-R	8.08-7.65	60	60	37
	SO₂-NR₂	COOH, ⁺NH₃, NO₂	8.36-7.90	22	22	38
	SO₂-O-(H, Na)	φ, Br, NHφ, OH	7.91-7.75	5	5	39
	SO₂-S-(R,φ)	Cl	7.93-7.48	2	2	40
	SO₂-thienyl-(2,3)	Cl	7.94-7.90	2	2	41
	P(O)(OH)₂	NO₂	7.99	1	1	42
	P(O,S)φ₂	O-R	7.63-7.58	2	2	43
	φ	NH₂, O-R	7.47-7.04	4	4	44
	φ	CO-(R, NH₂), COOR	7.80-7.64	5	5	45
	φ	⁺NH₃, NO₂, SO₂-OH	7.92-7.50	7	7	46

R = H, Alkyl, C-Z. Z = Any reasonable group.

CHART S30. Summary of Chemical Shifts of Y⟨○⟩X (C)

Cross Index Between Summary and Detailed Charts

Summary Chart (Lines)	Detailed Charts (Lines) from which Data Were Taken
S31(1,2)	18(4), 22(12), 25(9), 26(12), 32(7)
S31(3–5)	37(13), X
S31(6)	37(14)
S31(7)	18(4)
S31(8)	20(15), 31(6)
S31(9)	18(4), 20(15), 24(8), 26(7), 31(6)
S31(10–13)	33(9)
S31(14,15)	57(5,14)
S31(16,17)	X
S31(18,19)	80(7)
S31(20)	82(11)
S31(21)	68(10)
S31(22)	65(4), 68(11,17), 69(9), 73(12)
S31(23)	65(4), 68(18)
S31(24)	65(4), 68(19)
S31(25)	68(12), 73(12)
S31(26)	68(13)
S31(27)	68(14), 70(9)
S31(28)	68(15,16)

X indicates data not included in detailed charts.

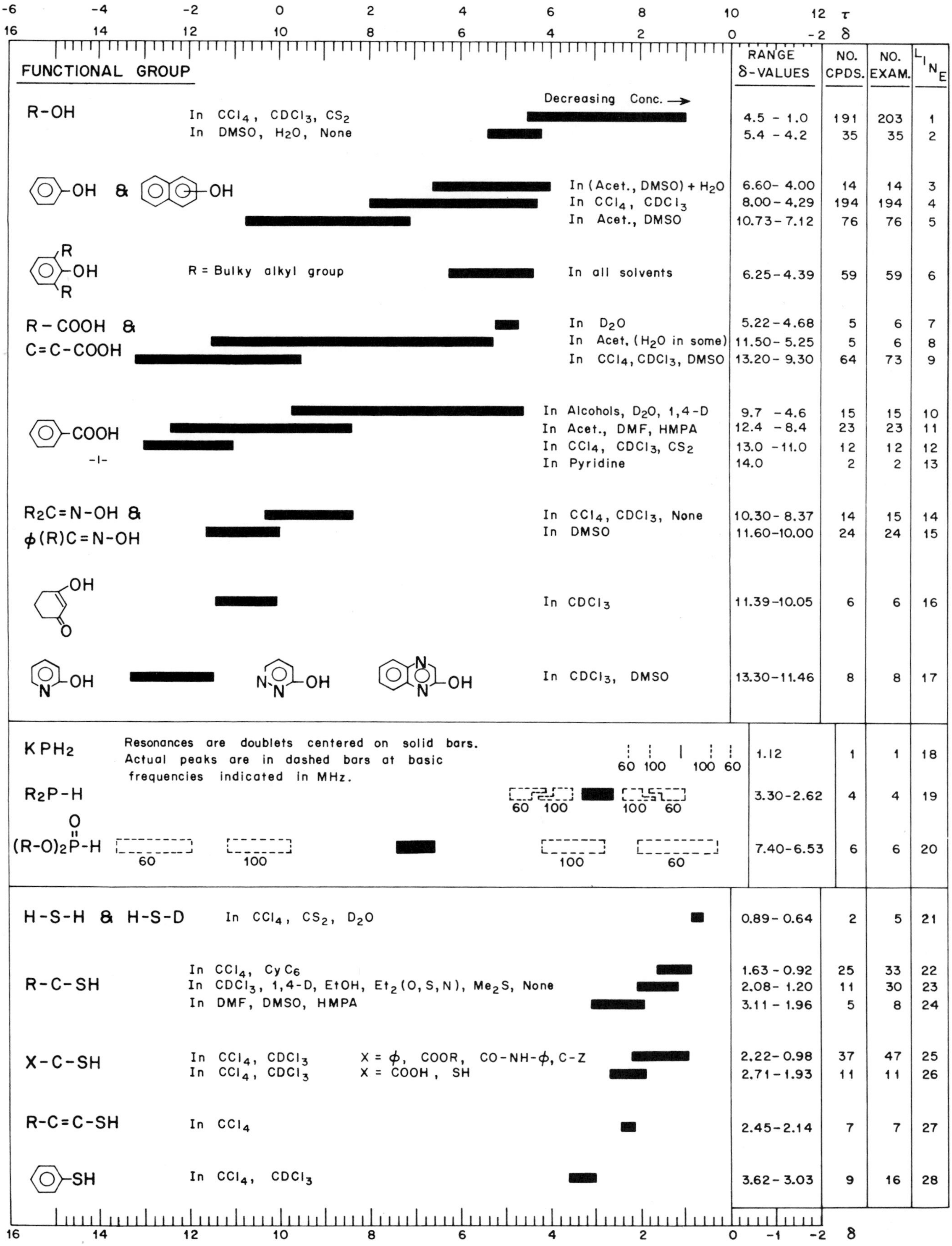

FUNCTIONAL GROUP

Functional Group	Conditions	RANGE δ-VALUES	NO. CPDS.	NO. EXAM.	LINE
R–OH	In CCl$_4$, CDCl$_3$, CS$_2$ (Decreasing Conc. →)	4.5 – 1.0	191	203	1
	In DMSO, H$_2$O, None	5.4 – 4.2	35	35	2
⬡–OH & ⬡⬡–OH	In (Acet., DMSO) + H$_2$O	6.60 – 4.00	14	14	3
	In CCl$_4$, CDCl$_3$	8.00 – 4.29	194	194	4
	In Acet., DMSO	10.73 – 7.12	76	76	5
⬡(R)(R)–OH	R = Bulky alkyl group, In all solvents	6.25 – 4.39	59	59	6
R–COOH & C=C–COOH	In D$_2$O	5.22 – 4.68	5	6	7
	In Acet. (H$_2$O in some)	11.50 – 5.25	5	6	8
	In CCl$_4$, CDCl$_3$, DMSO	13.20 – 9.30	64	73	9
⬡–COOH –I–	In Alcohols, D$_2$O, 1,4-D	9.7 – 4.6	15	15	10
	In Acet., DMF, HMPA	12.4 – 8.4	23	23	11
	In CCl$_4$, CDCl$_3$, CS$_2$	13.0 – 11.0	12	12	12
	In Pyridine	14.0	2	2	13
R$_2$C=N–OH & φ(R)C=N–OH	In CCl$_4$, CDCl$_3$, None	10.30 – 8.37	14	15	14
	In DMSO	11.60 – 10.00	24	24	15
(cyclohexenone –OH)	In CDCl$_3$	11.39 – 10.05	6	6	16
(pyridone/N–OH systems)	In CDCl$_3$, DMSO	13.30 – 11.46	8	8	17
KPH$_2$	Resonances are doublets centered on solid bars. Actual peaks are in dashed bars at basic frequencies indicated in MHz. (60 100 | 100 60)	1.12	1	1	18
R$_2$P–H	(60 100 | 100 60)	3.30 – 2.62	4	4	19
(R–O)$_2$P(=O)–H	(60 100 | 100 60)	7.40 – 6.53	6	6	20
H–S–H & H–S–D	In CCl$_4$, CS$_2$, D$_2$O	0.89 – 0.64	2	5	21
R–C–SH	In CCl$_4$, Cy C$_6$	1.63 – 0.92	25	33	22
	In CDCl$_3$, 1,4-D, EtOH, Et$_2$(O,S,N), Me$_2$S, None	2.08 – 1.20	11	30	23
	In DMF, DMSO, HMPA	3.11 – 1.96	5	8	24
X–C–SH	In CCl$_4$, CDCl$_3$ X = φ, COOR, CO–NH–φ, C–Z	2.22 – 0.98	37	47	25
	In CCl$_4$, CDCl$_3$ X = COOH, SH	2.71 – 1.93	11	11	26
R–C≡C–SH	In CCl$_4$	2.45 – 2.14	7	7	27
⬡–SH	In CCl$_4$, CDCl$_3$	3.62 – 3.03	9	16	28

R = H, Alkyl, C–Z.   Z = Any reasonable group.

CHART S31. Summary of Chemical Shifts of OH, PH, and SH

Cross Index between Summary and Detailed Charts

Summary Chart (Lines)	Detailed Charts (Lines) from which Data Were Taken
S32(1,2)	47(4)
S32(3)	45(4), 46(4), X
S32(4)	45(4)
S32(5)	46(20), X
S32(6,7)	X
S32(8)	46(20), X
S32(9)	X
S32(10–18)	X
S32(19)	60(7,12,15)
S32(20)	60(7,15)
S32(21)	X
S32(22)	55(12)
S32(23)	56(5)
S32(24)	56(12), X
S32(26)	X
S32(27–29)	45(8), X
S32(30,31)	46.1(7), X
S32(32,33)	45(9), 46.1(7), X
S32(34,35)	45(10), 46.1(7), X
S32(36)	62(19)

X indicates data not included in detailed charts.

FUNCTIONAL GROUP	Conditions	Notes	RANGE, δ-VALUES	NO. CPDS.	NO. EXAM.	LINE
NH (ring)	In CDCl₃	3-Membered Ring	0.03	1	1	1
	In CCl₄, CDCl₃	5- to 7-Membered Rings	2.45–1.15	14	15	2
Alkyl–NH–R	In CCl₄, CDCl₃, DMSO, None		3.27–0.53	157	157	3
	In H₂O		4.7 –4.3	1	2	4
⬡–NH–R	In CCl₄, CDCl₃		4.70–2.89	155	155	5
	In CDCl₃	m- or p-SO₂ (F,Cl)	5.36–4.16	4	4	6
	In Acet., DMSO		6.81–3.86	44	44	7
⬡–NH–⬡	In CDCl₃		5.53–5.00	10	10	8
	In Acet.	p-NO₂	8.20	1	1	9
pyridinyl–NH–R	In Acet., CCl₄, CDCl₃	R = H	4.97–4.73	6	6	10
	In DMSO	R = H	6.11–5.60	2	2	11
	In CCl₄, CDCl₃. m- or p-NO₂.	R = Alkyl	6.19–5.18	5	5	12
pyrazolyl–NH–R	In CDCl₃	R = Alkyl	5.20–5.00	3	3	13
pyrimidinyl–NH–R	In CDCl₃	R = H	5.53–4.45	11	11	14
	In DMSO	R = H	7.93–6.04	3	3	15
triazinyl–NH–R	In CDCl₃	R = H	5.86–5.48	2	2	16
	In DMSO	R = H	8.14–6.69	6	6	17
pyrimidinyl–NH–pyrimidinyl	In CDCl₃		8.81	1	1	18
pyrrole, indole	In CCl₄, CDCl₃		8.30–7.50	5	5	19
	In Acet.		10.12–9.25	2	2	20
pyrrole–CO-(H,OR)	In CDCl₃	+ corresponding indoles.	11.08–9.45	4	4	21
⬡–C(X)=NH	X = C–CN. NH may be H-bonded to CN. In CDCl₃		5.02	1	1	22
R₂C=N–NH₂	In CDCl₃		5.39–5.35	2	2	23
R₂C=N–NH–⬡	In Acet., CCl₄, CDCl₃, CH₂Br₂, MeOH, None		9.16–6.57	12	35	24
	In DMSO		11.03–10.98	2	2	26
R–NH₃⁺	In TFA		8.08–6.60	44	44	27
	In DMSO		8.35–6.21	3	3	28
	In CDCl₃		8.79–7.70	4	4	29
⬡–NH₃⁺	In TFA		8.82–8.00	7	7	30
	In DMSO		10.27–8.55	5	5	31
(R,⬡)₂NH₂⁺	In TFA		10.15–5.95	5	5	32
	In CDCl₃		11.50–8.42	2	2	33
(R,⬡)₃NH⁺	In TFA		8.25–7.09	8	8	34
	In CDCl₃		11.41–6.00	5	5	35
pyridinium⁺NH	In HAc, HCl, TFA		10.47–5.35	1	3	36

R = H, Alkyl, C–Z.   Z = Any reasonable group.

CHART S32. Summary of Chemical Shifts of NH in Amines and N⁺ Salts

CHART S32

Cross Index between Summary and Detailed Charts

Summary Chart (Lines)	Detailed Charts (Lines) from which Data Were Taken
S33(1)	48(4)
S33(2)	48(4), 50(13,21), X
S33(3)	48(4), 50(21), X
S33(4,5)	50(8)
S33(6,7)	51(4)
S33(8,10)	52(16), X
S33(9)	X
S33(11,12)	53(4)
S33(13,14)	54(10)
S33(15,16)	54(4,5)
S33(17)	54(9)
S33(18)	X
S33(19–21)	55(18)
S33(22)	87(14)
S33(23–29)	X

X indicates data not included in detailed charts.

FUNCTIONAL GROUP		RANGE, δ-VALUES	NO. CPDS.	NO. EXAM.	LINE
R-CO-NH-R	In Acet., D₂O, DMSO	9.60-6.25	8	11	1
	In CCl₄, CDCl₃	8.15-5.40	50	58	2
	In TFA, None	8.53-7.35	16	16	3
C=C-CO-NH-R	In CDCl₃	6.38-5.71	5	5	4
	In Acet., DMSO	8.65-7.20	2	2	5
⬡-CO-NH-R	In Acet., CCl₄, CDCl₃, DMSO	8.63-6.03	12	12	6
	In TFA	11.70-8.18	2	2	7
(R, ⬡)-CO-NH-⬡	In CCl₄, CDCl₃, CD₃CN	9.60-6.83	74	78	8
	In TFA	9.26-8.91	9	9	9
	In Acet., DMSO	11.10-9.67	19	19	10
⟨CO NH⟩	In CCl₄, CDCl₃	8.10-7.58	4	4	11
⟨CO NH CO⟩	In CCl₄, CDCl₃	9.20-8.47	9	9	12
R-CO-N-NH₂		4.51-3.66	10	10	13
R-CO-NH-N<	In CDCl₃, DMSO	10.54-7.93	11	11	14
R₂N-CO-NH-R	In CCl₄, CDCl₃, DMSO	6.09-4.13	15	15	15
R₂N-CS-NH-R	In CCl₄, CDCl₃, DMSO	7.52-5.55	13	14	16
R₂N-C(O,S)-NH-⬡	In Acet., CDCl₃, DMSO	9.58-7.08	8	8	17
R-NH-CS-CS-NH-R	In Acet., CDCl₃, DMSO	10.70-10.35	6	6	18
R₂C=N-N-CO-NH₂	In CDCl₃, DMSO	6.62-5.61	8	8	19
R₂C=N-NH-CO-N<	In CDCl₃	9.58-7.91	5	5	20
	In DMSO	10.84-8.88	5	5	21
Y₂P(Z)-NH-R In CCl₄, CDCl₃. Y = R, O-R, NR₂, φ Z = O, S, Nothing		4.75-2.60	18	18	22
(R, ⬡)-SO₂-NH-R	In CCl₄, CDCl₃	6.03-4.87	18	18	23
	In Acet.	7.28-6.51	5	5	24
	In DMSO	7.71-7.21	3	3	25
(R, ⬡)-SO₂-NH-⬡	In CDCl₃	7.23-6.49	2	2	26
	In TFA	8.16	1	1	27
	In Acet.	9.62-8.40	5	5	28
(R, ⬡)-SO₂-NH-CO-(R,NR₂)	In CDCl₃	9.25-6.94	11	11	29

R = H, Alkyl, C-Z.   Z = Any reasonable group unless specifically identified.

CHART S33. Summary of Chemical Shifts of NH in Amides and Imides

Cross Index between Summary and Detailed Charts

Summary Chart (Lines)	Detailed Charts (Lines) from Which Data Were Taken
S34(1–23)	X. None of these data are included in detailed charts.

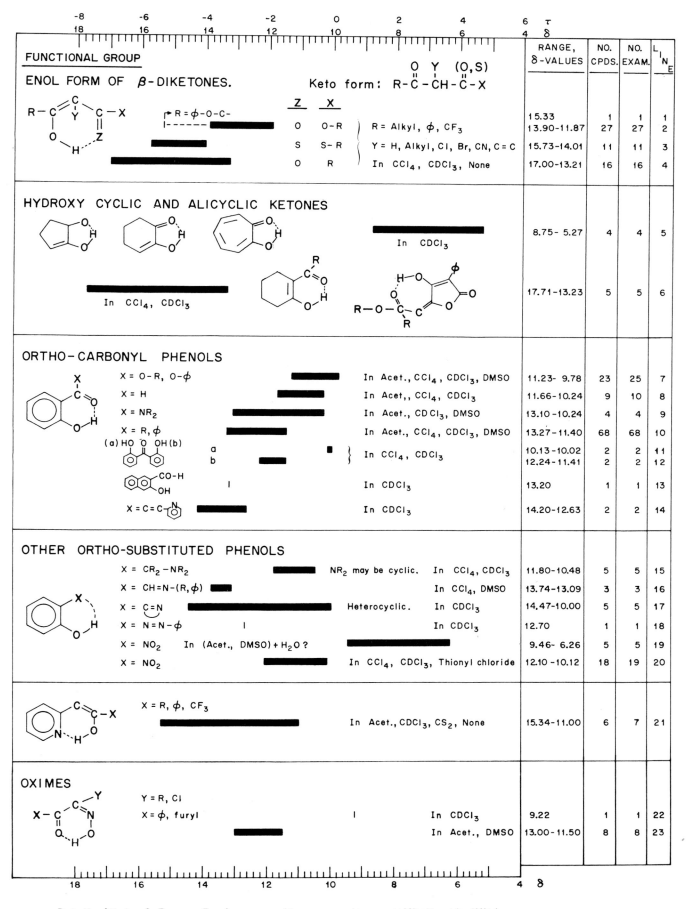

CHART S34. Summary of Chemical Shifts of OH in Intramolecular H-Bonds

R = H, Alkyl, C-Z.   Z = Any reasonable group unless specifically identified.

Cross Index between Summary and Detailed Charts

Summary Chart (Lines)	Detailed Charts (Lines) from which Data Were Taken
S35(1–18)	X
S35(19–21)	52(17), X
S35(22)	56(12), X

X indicates data not included in detailed charts.

CHART S35. Summary of Chemical Shifts of NH in Intramolecular H-Bonds

Scale (top): τ / δ — marks at -6/16, -4/14, -2/12, 0/10, 2/8, 4/6, 6/4, 8/2

FUNCTIONAL GROUP		RANGE, δ-VALUES	NO. CPDS.	NO. EXAM.	LINE
AMINE NH					
R–N(C)–N(C=O) ... O (succinimide-type)	In CDCl₃	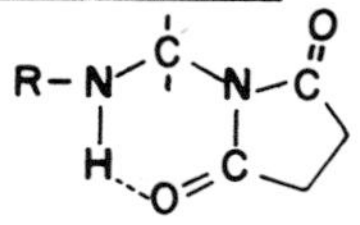 5.30–4.60	17	17	1
R–N(C=C)–C(=O)–O–R ... O	In CCl₄, CDCl₃	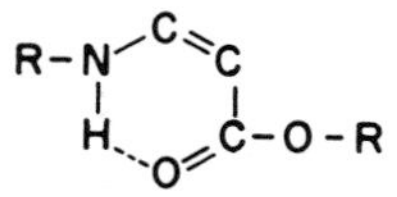 8.48–6.00	4	5	2
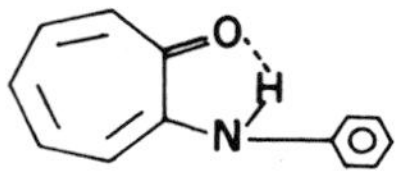 (tropolone ... O–H ... N–phenyl)	In DMSO	9.90	1	1	3
(N–H ... N=O diphenyl)	In CDCl₃	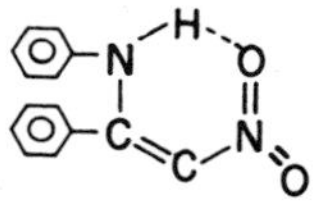 11.47	1	1	4
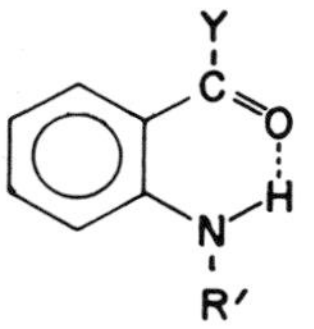 Y = O–R, R′ = H	In CCl₄, CDCl₃	5.77–5.55	6	7	5
Y = O–R, R′ 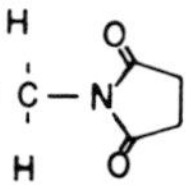 = C–N(succinimide)	In CDCl₃	8.55	1	1	6
Y = R,φ, R′ = H	In CCl₄, CDCl₃	7.30–5.91	4	4	7
Y = R,φ, R′ = CH₃	In CDCl₃	10.46–9.57	2	2	8
Y = NR₂, R′ = φ	In CDCl₃	11.24–8.99	9	9	9
X = NO₂, R′ = H	In (Acet., DMSO) + H₂O	4.16–3.08	10	10	10
X = NO₂, R′ = H	In CDCl₃	6.55–5.22	4	4	11
X = NO₂, R′ = H	In Acet.	7.73–7.08	3	3	12
X = NO₂, R′ = Alk.,φ,C–Z	In CCl₄, CDCl₃, DMSO, Thionyl-Cl	10.18–7.84	12	12	13
X = NO₂, R′: N=CR₂	In CDCl₃	11.77–10.80	6	6	14
X = SO₂–(F,R), R′ = H	In CDCl₃	5.19–5.11	2	2	15
X = C=N–R, R′ = H	In CDCl₃	10.09–7.70	2	2	16
AMIDE NH					
R–C(=O)–C=C–N–CO–(R,φ) ... O–H	In CDCl₃	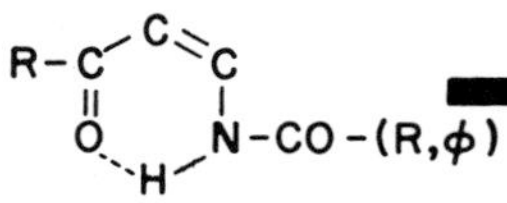 13.88–12.32	4	4	17
(tropolone)–N–CO–(R,O–R)	In CDCl₃	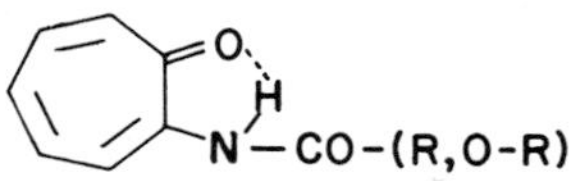 9.35–8.91	2	2	18
X = SO₂–(R, NR₂)	In CDCl₃, CD₃CN	9.90–9.07	3	6	19
X = NO₂	In CDCl₃, CD₃CN	11.35–9.25	7	11	20
X = CO–Y, Y = R,φ,O–R	In CDCl₃, CD₃CN	12.50–10.80	5	7	21
PHENYL HYDRAZONE NH–N=					
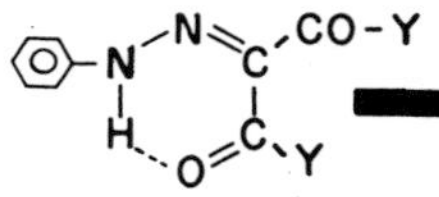 φ–N(H)–N=C(CO–Y)–C(=O)–Y, Y = R,φ,O–R	In Acet., CDCl₃	14.90–12.51	41	41	22

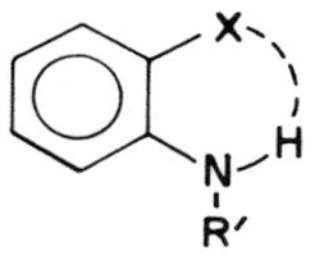

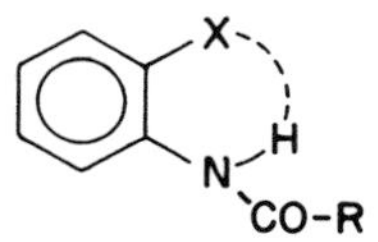

Scale (bottom): δ — marks at 16, 14, 12, 10, 8, 6, 4, 2

R = H, Alkyl, C–Z.   Z = Any reasonable group.

CHART S35. Summary of Chemical Shifts of NH in Intramolecular H-Bonds

190

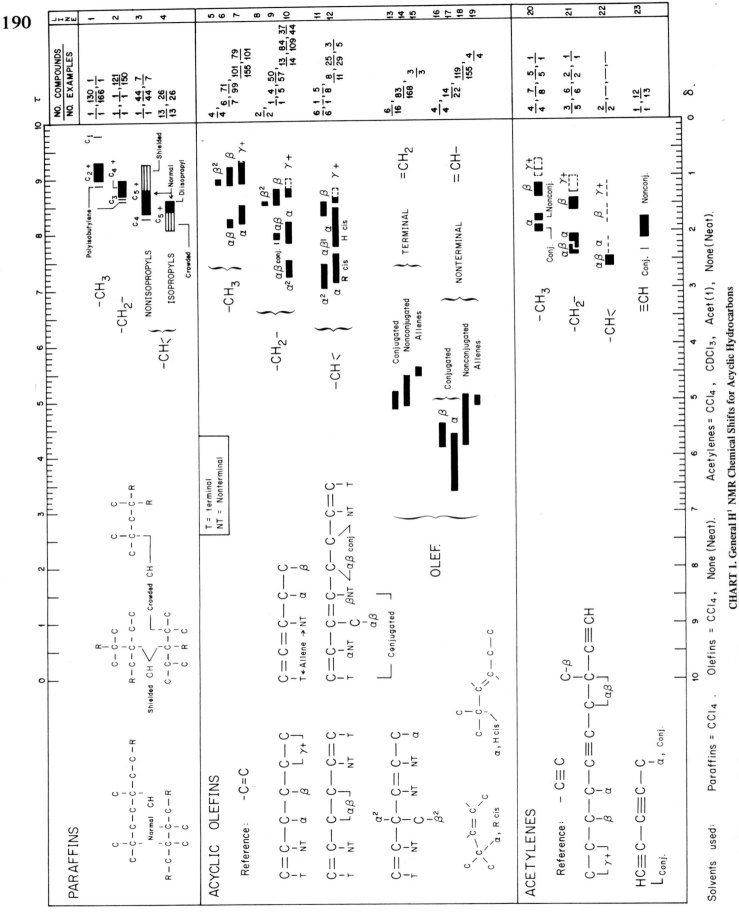

The following table appears at the top of the chart:

LINE	NO. COMPOUNDS / NO. EXAMPLES
1	$\frac{1}{1}$, $\frac{130}{166}$, $\frac{1}{1}$
2	$\frac{1}{1}$, $\frac{1}{1}$, $\frac{121}{150}$
3	$\frac{1}{1}$, $\frac{44}{44}$, $\frac{7}{7}$
4	$\frac{13}{13}$, $\frac{26}{26}$
5	$\frac{4}{4}$
6	$\frac{6}{7}$, $\frac{71}{99}$, $\frac{101}{155}$, $\frac{79}{101}$
7	
8	$\frac{2}{2}$, $\frac{1}{1}$, $\frac{4}{5}$, $\frac{50}{57}$, $\frac{13}{14}$, $\frac{84}{109}$, $\frac{37}{44}$
9	
10	
11	$\frac{6}{6}$, $\frac{1}{1}$, $\frac{5}{8}$, $\frac{8}{11}$, $\frac{25}{29}$, $\frac{3}{5}$
12	
13	$\frac{6}{16}$, $\frac{83}{168}$, $\frac{3}{3}$
14	
15	
16	$\frac{4}{4}$, $\frac{14}{22}$, $\frac{119}{155}$, $\frac{4}{4}$
17	
18	
19	
20	$\frac{4}{4}$, $\frac{7}{8}$, $\frac{5}{5}$, $\frac{1}{1}$
21	$\frac{3}{5}$, $\frac{6}{6}$, $\frac{2}{2}$, $\frac{1}{1}$
22	$\frac{2}{2}$, —, —, —
23	$\frac{1}{1}$, $\frac{12}{13}$

CHART 1. General H¹ NMR Chemical Shifts for Acyclic Hydrocarbons

Solvents used: Paraffins = CCl₄. Olefins = CCl₄, None (Neat). Acetylenes = CCl₄, CDCl₃, Acet(1), None(Neat).

CHART 1

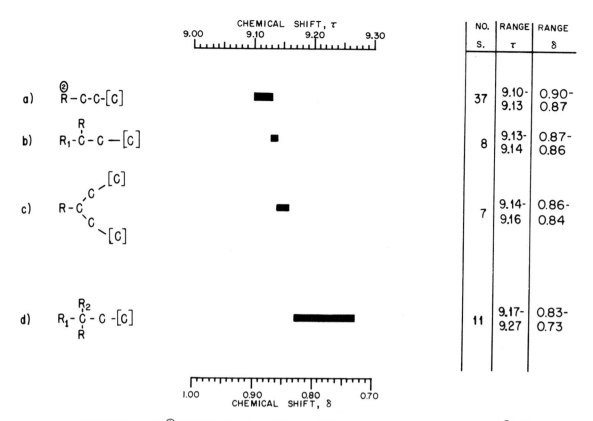

CHART 2. Precise[3] H[1] NMR Chemical Shifts for $[CH_3]-CH_2$. Paraffinic Methyl Triplets[1](A1)

1. Chemical shifts are measured from the middle peak of the triplet.
2. R, R_1, R_2, = Alkyl radicals, C_mH_{2m+1}, with $m > 0$.
3. Samples run at 50 vol. % concentration in CCl_4, using tetramethylsilane as internal reference. Data satisfactory for 0–50% concentration range in CCl_4.

CHART 2

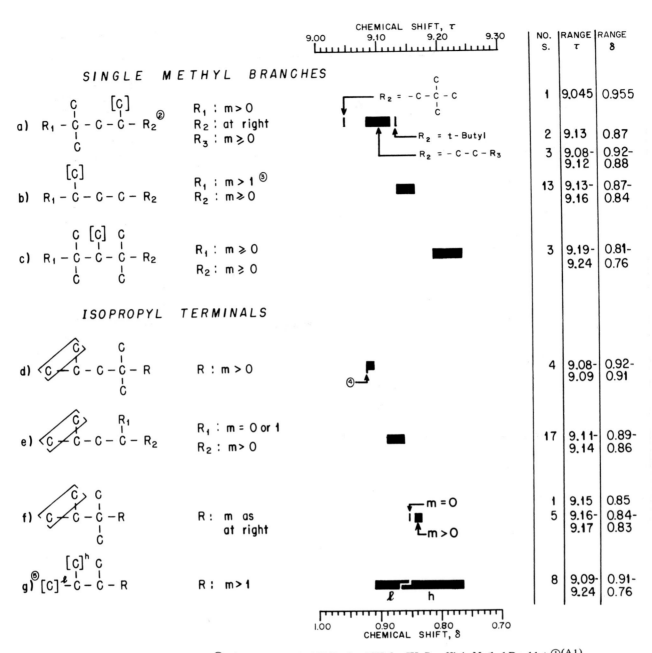

CHART 3. Precise[6] H[1] NMR Chemical Shifts for [CH$_3$]—CH. Paraffinic Methyl Doublets[1](A1)

1. Chemical shifts are measured from the doublet center.
2. R, R$_1$, R$_2$, R$_3$ = alkyl radicals, C$_m$H$_{2m+1}$, with values of m as shown. When m = O, R = H.
3. R$_1$ includes t-butyl and internal gem methyl groups.
4. Chemical shift of CH$_3$ doublet produced by triisopropylmethane.
5. Nonequivalent methyls in isopropyl groups. They produce two separate doublets, one in the ℓ block and one in the h block.
6. Samples run at 50 vol. % concentration in CCl$_4$, using tetramethylsilane as internal reference. Data satisfactory for 0–50% concentration range in CCl$_4$.

CHART 3

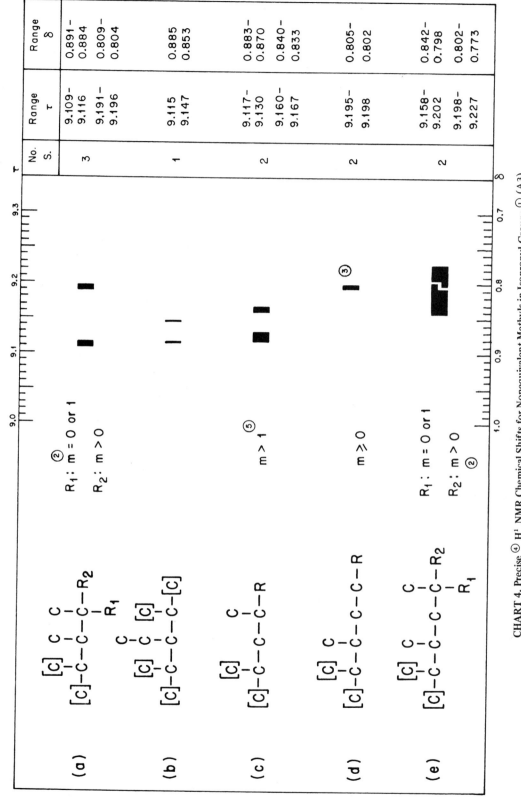

	No. S. τ	Range τ	Range δ
(a)	3	9.109–9.116	0.891–0.884
		9.191–9.196	0.809–0.804
(b)	1	9.115 9.147	0.885 0.853
(c)	2	9.117–9.130	0.883–0.870
		9.160–9.167	0.840–0.833
(d)	2	9.195–9.198	0.805–0.802
(e)	2	9.158–9.202	0.842–0.798
		9.198–9.227	0.802–0.773

CHART 4. Precise ④ H¹ NMR Chemical Shifts for Nonequivalent Methyls in Isopropyl Groups ① (A3)

1. These groups normally produce two separate doublets, one in each of the two blocks shown for a single structure. Chemical shifts are measured from the doublet centers.
2. R and R₁ and R₂ = alkyl radicals, $C_m H_{2m+1}$, with values of m as shown. When m = O, R = H.
3. This structure produces only one doublet, indicating a condition in which the nonequivalence is not observed.
4. Samples were run at 20 vol. % concentration in CCl₄ at 32°C., using tetramethylsilane as internal reference. Shift due to concentration <±0.001 ppm between 5% and 20%. Data satisfactory for 0–20% and possibly 0–50% range in CCl₄.
5. The group β to the isopropyl must be asymmetric (R>C).

CHART 4

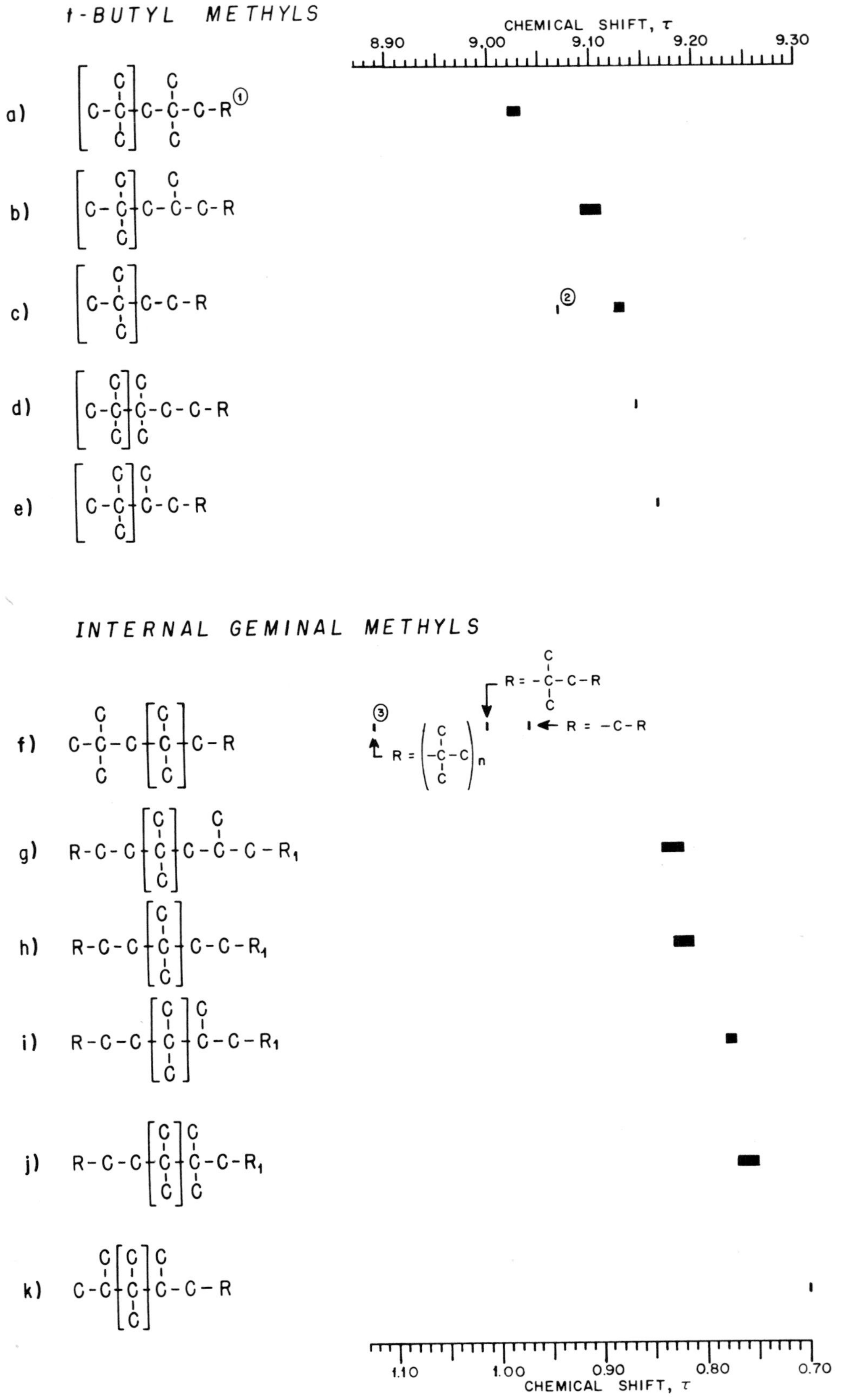

t-BUTYL METHYLS

CHEMICAL SHIFT, τ

	Structure	NO. S.	RANGE τ	RANGE δ
a)	[C–C(C)(C)–C–C(C)(C)–C–R①]	3	9.02–9.03	0.98–0.97
b)	[C–C(C)(C)–C–C(C)–C–R]	8	9.09–9.11	0.91–0.89
c)	[C–C(C)(C)–C–C–R] ②	10	9.12–9.13	0.88–0.87
d)	[C–C(C)(C)–C(C)–C–C–R]	4	9.14	0.86
e)	[C–C(C)(C)–C(C)–C–R]	3	9.16	0.84

INTERNAL GEMINAL METHYLS

R = –C(C)(C)–C–R

③ R = (–C(C)(C)–C–)n

R = –C–R

	Structure	NO. S.	RANGE τ	RANGE δ
f)	C–C(C)(C)–C–[C(C)(C)]–C–R	1	8.99	1.01
		2	9.03	0.97
		1	8.88	1.12
g)	R–C–C–[C(C)(C)(C)]–C–C(C)–C–R₁	2	9.16–9.18	0.84–0.82
h)	R–C–C–[C(C)(C)]–C–C–R₁	3	9.17–9.19	0.83–0.81
i)	R–C–C–[C(C)]–C(C)–C–R₁	4	9.22–9.23	0.78–(0.783) 0.77
j)	R–C–C–[C(C)(C)]–C(C)–C–R₁	3	9.23–9.25	0.77–0.75
k)	C–C(C)–[C(C)(C)]–C(C)(C)–C–R	2	9.30	0.70

CHEMICAL SHIFT, τ

CHART 5. Precise ④ H¹ NMR Chemical Shifts for [CH₃]–C≤. Paraffinic Methyl Singlets[A1]

1. R, R₁, = alkyl radicals, C_mH_{2m+1}, with $m \geqslant 0$, but ≠ t-butyl or internal gem methyl radicals except where indicated.
2. 2,2-dimethylpropane (neopentane), 9.065 τ.
3. Polyisobutylene.
4. Samples run at 50 vol. % concentration in CCl_4, using tetramethylsilane as internal reference. Data satisfactory for 0–50% concentration range in CCl_4.

CHART 5

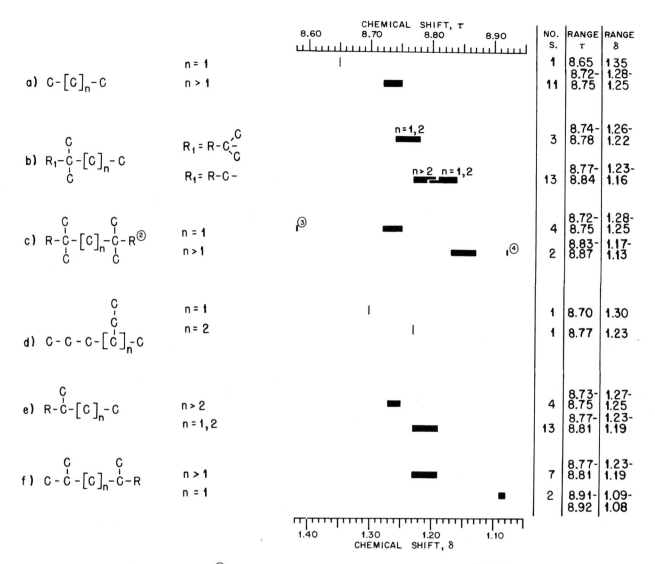

CHART 6. Precise [5] H[1] NMR Chemical Shifts for Paraffinic Methylenes [1] (A1)

1. Chemical shifts were measured from the centers of the methylene resonance patterns.

2. R = alkyl radical, C_mH_{2m+1}, with $m>0$.

3. Polyisobutylene, $\left[\sim C - \underset{C}{\overset{C}{C}} \sim \right]_n$, 8.58τ.

4. Poly-3-methylbutene-1, $\left[\sim C - C - \underset{C}{\overset{C}{C}} \sim \right]_n$, 8.92τ.

5. Samples run at 50 vol. % concentration in CCl_4, using tetramethylsilane as internal reference. Data satisfactory for 0–50% concentration range in CCl_4.

CHART 6

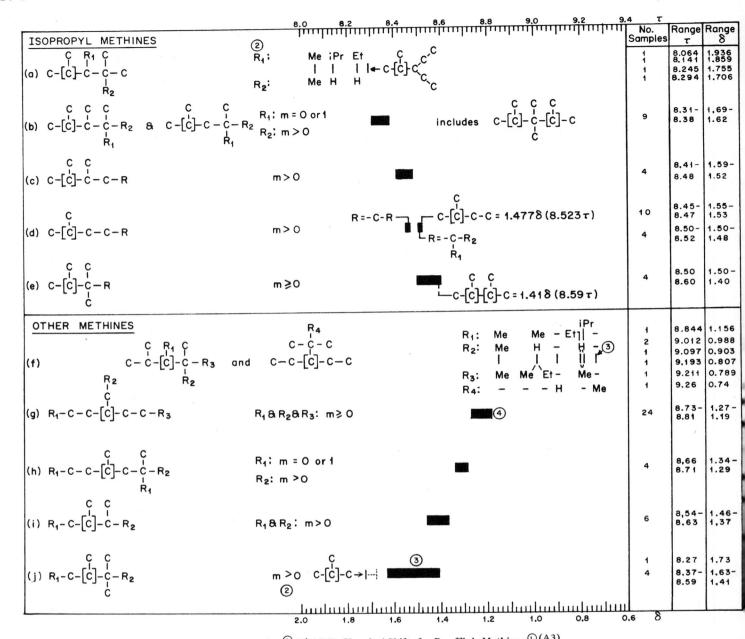

CHART 7. Precise Ⓢ H¹ NMR Chemical Shifts for Paraffinic Methines ① (A3)

1. Chemical shift was measured from the center of the multiplet pattern when this was identifiable. Otherwise it was measured from the center of area of the complex methine band.
2. R, R_1, R_2, R_3, R_4 = alkyl radicals, C_mH_{2m+1}, with values of m as shown.
3. Chemical shift is uncertain.
4. Coincident with and indistinguishable from the methylene band of the same compound. Located by area measurements and spin decoupling.
5. Samples run at 20 vol. % concentration in CCl_4 at 32° C, using tetramethylsilane as internal reference. Shift due to concentration <0.001 ppm between 5% and 20%. Data satisfactory for 0–20% and perhaps 0–50% range in CCl_4.

CHART 7

NO. OF EXAMPLES		LINE
25	γ -CH$_3$	1
11	-(CH$_2$)$_n$CH$_3$ $n \geqslant 2$	2
31	β CH$_3$ = C-C-CH$_3$	3
11	all -CH$_2$CH$_3$ and -CH(CH$_3$)R }With cis R	4
10	}With no cis R	5
2	CH$_2$ = CH C(CH$_3$)$_2$R	6
5	-C(CH$_3$)$_2$ R IN di & tri SUBSTITUTED ETHYLENES }With no cis R	7
3	}With cis R	8
4	γ -CH$_2$-	9
6	β -CH$_2$-	10
3	β -CH<	11
41	α -CH$_3$	12
10	CH$_2$ = C(CH$_3$)R	13
7	trans R CH = CH CH$_3$	14
6	cis R CH = CH CH$_3$ if R is $-\overset{C}{\underset{C}{C}}-C$	15
12	R$_2$C = C(CH$_3$)R, cis & trans R CH = C(CH$_3$)R	16
6	R$_2$C = CH CH$_3$	17
30	α -CH$_2$-	18
11	all -CH$_2$ CH$_3$	19
6,1	CH$_2$ = CH – CH$_2$ (CH$_2$)$_n$ CH$_3$ → $n \geqslant 1$ ← CR$_2$ = CH – CH$_2$ (CH$_2$)$_n$ CH$_3$	20
3,4	R CH = CH – CH$_2$ (CH$_2$)$_n$ CH$_3$ cis trans $n \geqslant 1$	21
6,1	all -CH$_2$ CR$_3$ ←and -CH$_2$CHR$_2$	22
5,8	α -CH< With cis R No cis R	23

7.0τ .1 .2 .3 .4 .5 .6 .7 .8 .9 8.0 .1 .2 .3 .4 .5 .6 .7 .8 .9 9.0τ .1 .2 .3

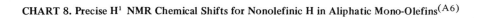

δ 3.0 .9 .8 .7 .6 .5 .4 .3 .2 .1 2.0 .9 .8 .7 .6 .5 .4 .3 .2 .1 1.0 .9 .8 .7
CHEMICAL SHIFT, PPM.

CHART 8. Precise H^1 NMR Chemical Shifts for Nonolefinic H in Aliphatic Mono-Olefins[A6]

Data applicable to samples run at 5–100 vol. % concentration in CCl$_4$, using tetramethylsilane as internal reference.

CHART 8

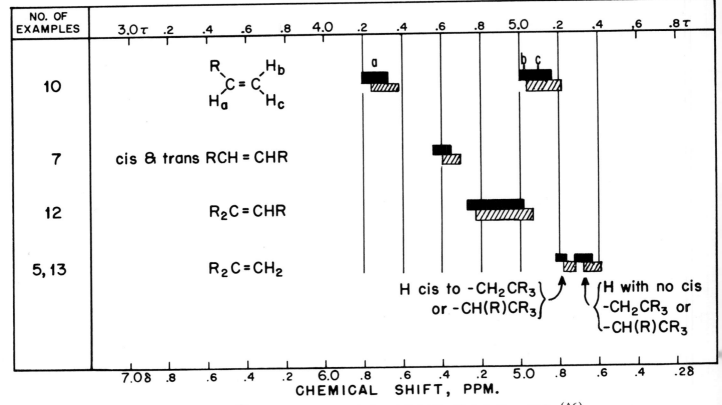

CHART 9. Precise ① H¹ NMR Chemical Shifts for Olefinic H in Aliphatic Mono-Olefins[A6]

1. Solid bars are for neat liquids (no solvent). Cross hatched bars are for 50% solutions in CCl_4. Internal reference: tetramethylsilane.

CHART 9

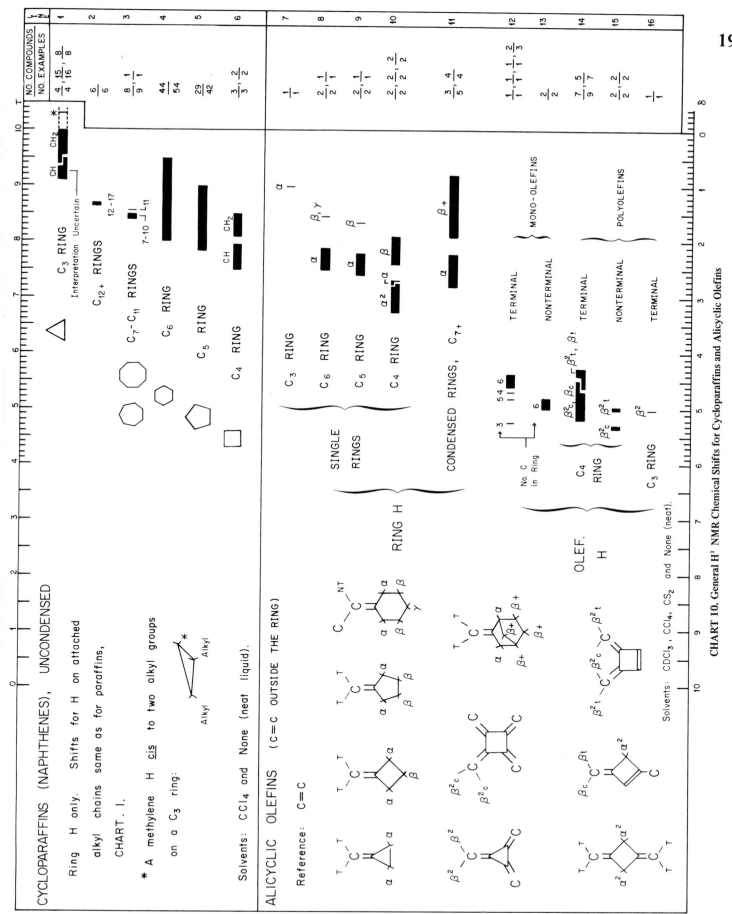

CHART 10. General H¹ NMR Chemical Shifts for Cycloparaffins and Alicyclic Olefins

CHART 10

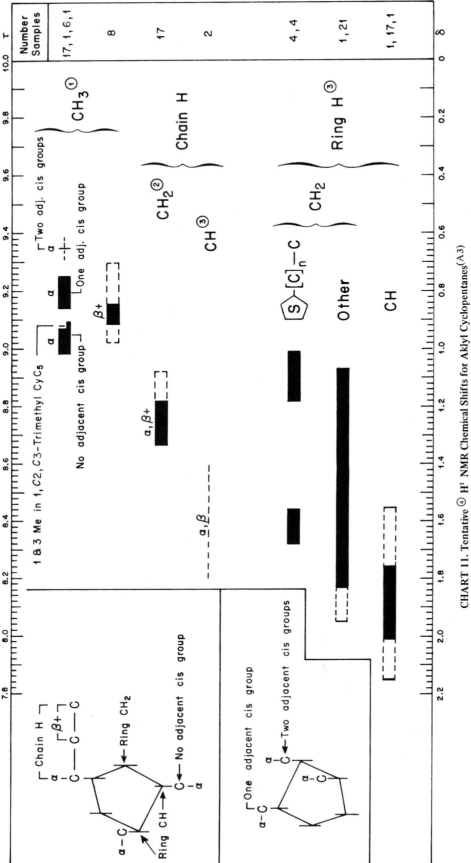

CHART 11. Tentative [4] H[1] NMR Chemical Shifts for Alkyl Cyclopentanes[(A3)]

1. Methyl bands could be identified and assigned positively, and their shifts could be measured accurately.
2. Band identifications reasonably positive but locations of band centers sometimes uncertain.
3. Both band assignments and locations of band centers were frequently uncertain.
4. Composite of data from samples run as neat liquids at 60 MHz and as 20 vol. % or less in CCl_4 at 100 MHz.

CHART 11

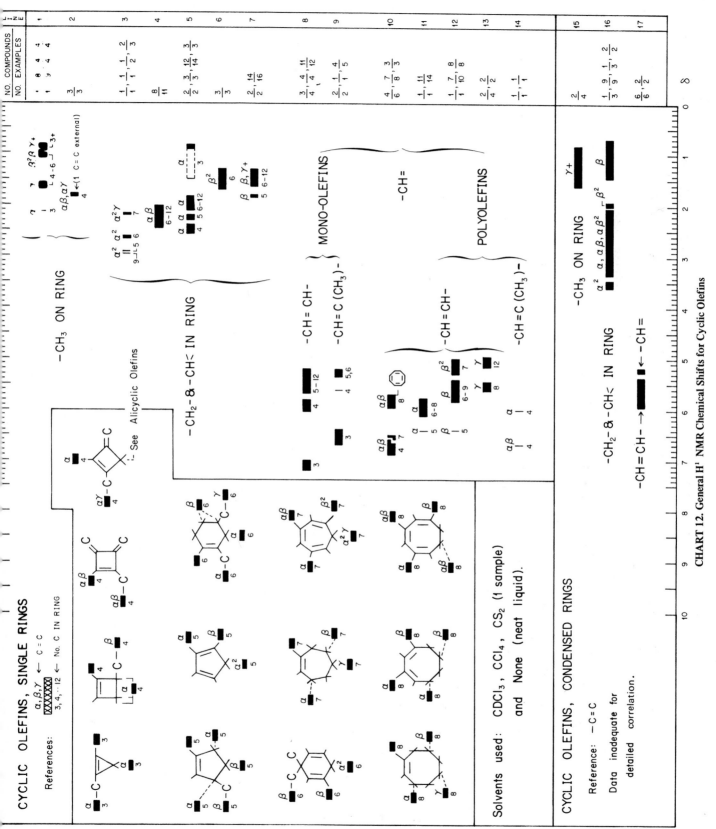

CHART 12. General H¹ NMR Chemical Shifts for Cyclic Olefins

CHART 12

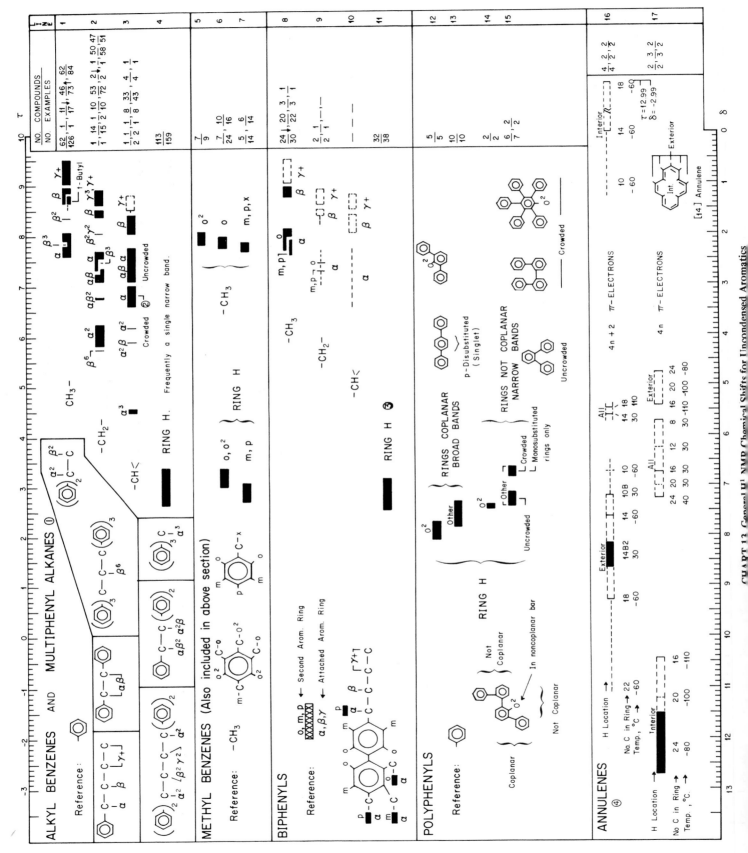

CHART 13 General H¹ NMR Chemical Shifts for Uncondensed Aromatics

CHART 13

Notes for Chart 13 – Uncondensed Aromatics

1. Solvents used: CCl_4, $CDCl_3$, CS_2 and none (neat liquids). The observed solvent effects are reported as $\Delta\delta$ (solution-neat) for $CDCl_3$ and CCl_4 solvents in the following table:

H Type	Position	Solvent	$\Delta\delta$ Range	$\Delta\delta$ Avg.	Compounds/Examples
Alkyl	α	$CDCl_3$	0.00 – 0.25	0.11	37/49
"	β	and	(−0.01) − 0.15	0.06	24/28
"	γ^+	CCl_4	(−0.02) − 0.10	0.02	8/8
Arom.	Ring	$CDCl_3$	0.05 − 0.17	0.11	11/11
Arom.	Ring	CCl_4	(−0.10) − 0.05	−0.02	23/25

2. When biphenyl is not substituted in an ortho position the aromatic band is broad, presumably because of interaction of the coplanar rings. When at least one substituent is placed in an ortho position, the aromatic bands resemble those of correspondingly substituted single rings, presumably because ring interaction has been reduced to a negligible level.

3. At low temperatures annulenes exhibit separate resonance bands for hydrogens held outside the ring (exterior) and those held inside the ring (interior). At high temperatures these bands collapse to about the weighted average position, presumably because of rapid interconversion of hydrogen types through rapid conformational changes in the ring. This interconversion is prevented when the ring is bridged, as in 10B and 14B2.

4. Annulenes with $4n + 2$ π-electrons exhibit downfield shifts for exterior hydrogens and upfield shifts for interior hydrogens. This is consistent with a ring current similar to that of aromatic rings. Annulenes with $4n$ π-electrons exhibit shifts in the opposite direction, consistent with a ring current in the direction opposite to that of aromatics. For detailed discussion see Pople and Untch.(53)

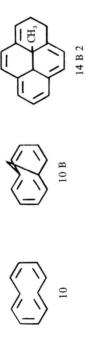

10

10 B

14 B 2

for CH_3:
$\tau = 14.23$
$\delta = -4.23$

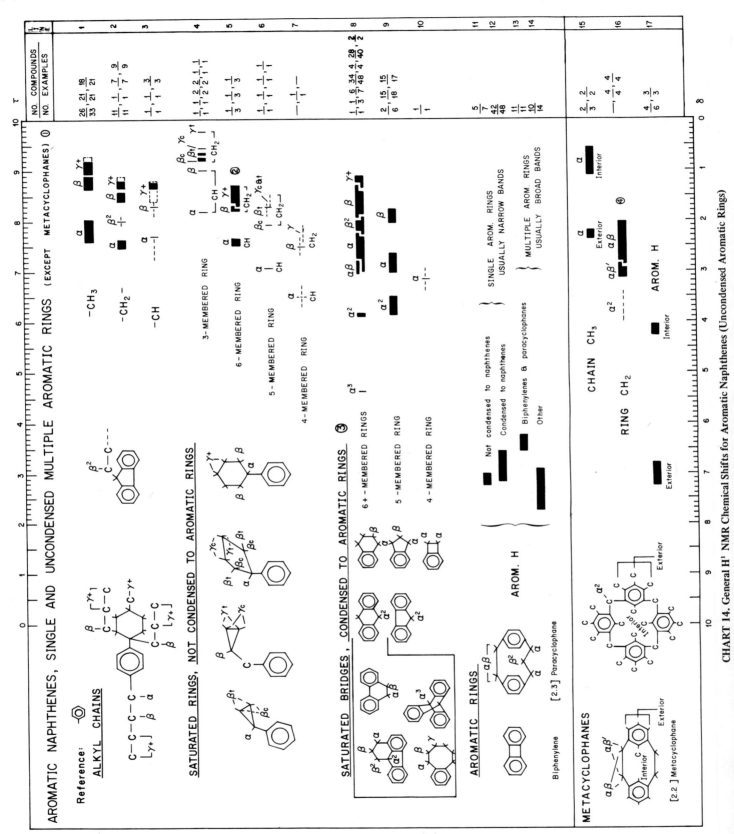

CHART 14. General H¹ NMR Chemical Shifts for Aromatic Naphthenes (Uncondensed Aromatic Rings)

CHART 14

Notes for Chart 14–Aromatic Naphthenes

1. Solvents used: CCl_4, $CDCl_3$, CS_2, and none (neat liquids).
2. The interpretation of the β and $\gamma+$ portions of the spectra of 6- membered rings is uncertain.
3. Bridge methylenes α or β to the aromatic ring occasionally have nonequivalent hydrogens. Only the center points of the resulting multiplets are recorded on the chart. The observed chemical shift differences between bands from the individual nonequivalent hydrogens are as follows:

5-Membered Rings			6-Membered Rings		
CH_2 Posit.	Shift Diff.	No. Examples	CH_2 Posit.	Shift Diff.	No. Examples
α	0.12	1	α	0.30–0.68	5
β	0.70	1	β	0.20–0.30	4

4. In [2.2] metacyclophanes the hydrogens in each bridge methylene are nonequivalent. One chemical shift falls in the $\alpha\beta$ bar and the other in the $\alpha\beta'$ bar.

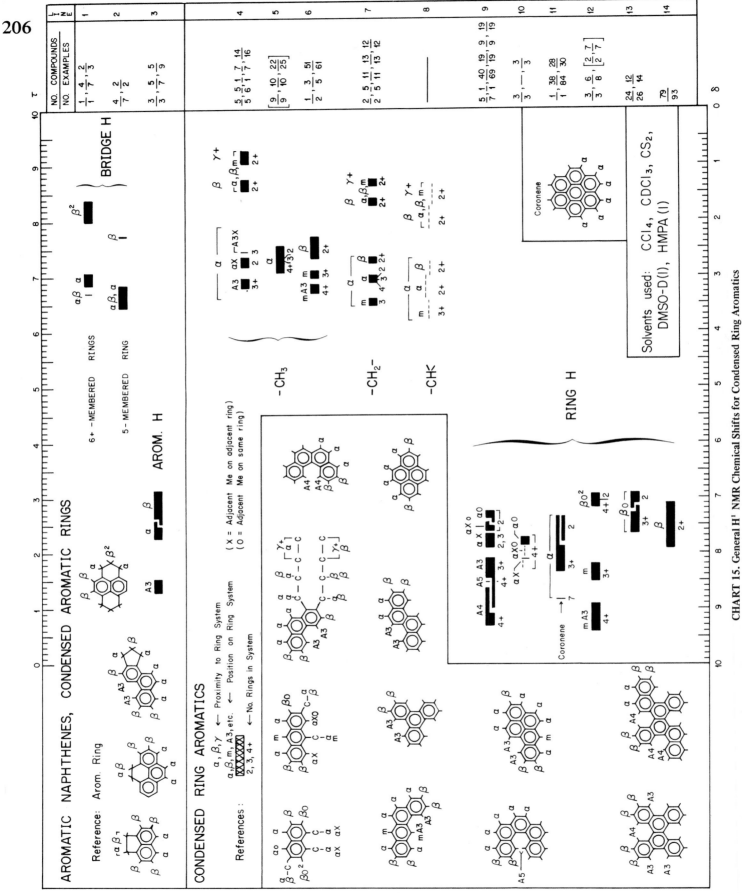

CHART 15. General H¹ NMR Chemical Shifts for Condensed Ring Aromatics

CHART 15

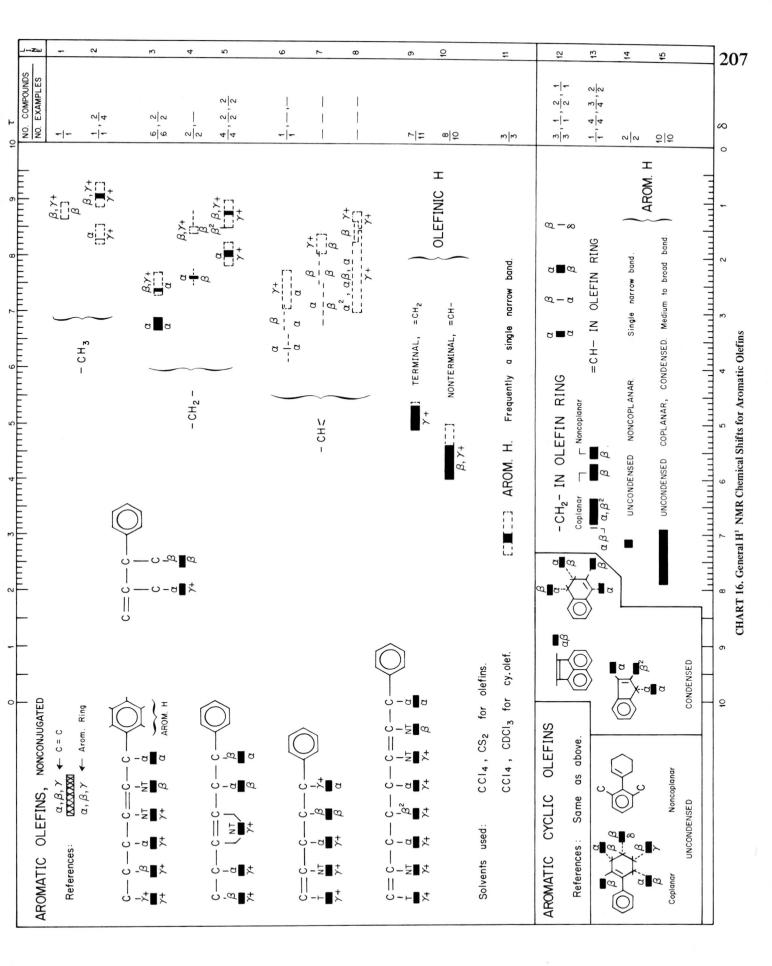

CHART 16. General H¹ NMR Chemical Shifts for Aromatic Olefins

CHART 16

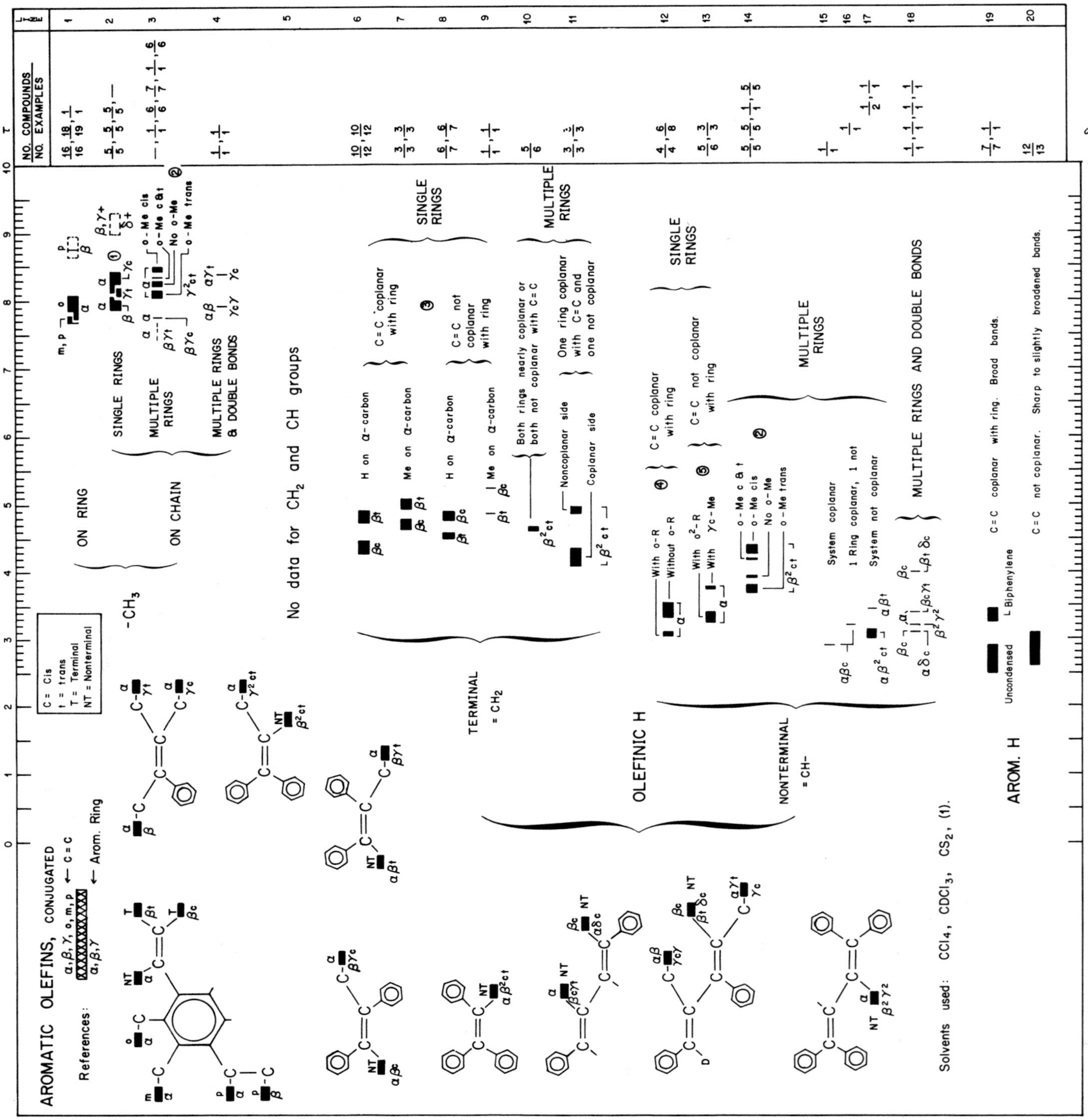

CHART 17. General H¹ NMR Chemical Shifts for Aromatic Olefins, Conjugated

CHART 17

Notes for Chart 17 – Aromatic Olefins, Conjugated

1. When the methyl is γc to the ring the system is presumed to be not coplanar, and the methyl is held in the upfield shifting region of the ring field. When the methyl is γt to the ring the system is presumed to be coplanar, and the methyl is in the downfield shifting region of the ring field.

2. o-Me means C=C ortho to CH₃. The *cis-trans* relationship is between the ortho methyl on the ring and the methyl on the olefin (line 3) or the hydrogen on the olefin (line 14). The various chemical shifts reflect the effects of conformational and perhaps electronic changes caused by addition of ortho methyl groups to 1,1-diphenylpropene (for methyl and =CH shifts) and to 1,1-diphenyl-2-methylpropene (for methyl shifts).

3. Note that the βc and βt shifts for the noncoplanar case are reversed from those of the coplanar case.

4. R is alkyl or CH=CH₂. o-R means C=C ortho to R.

5. R is alkyl or CH=CH₂. o²-R means C=C ortho to two R's, one on each side. The C=C group can be forced out of the plane of the aromatic ring by two ortho substituents on the ring or by a methyl in the gamma cis position on the C=C.

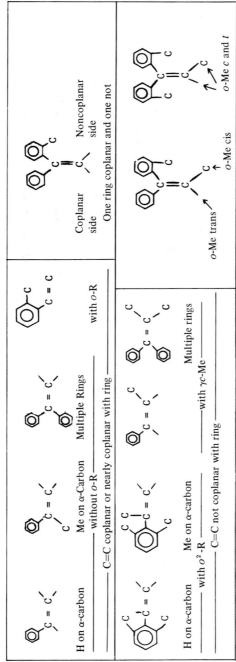

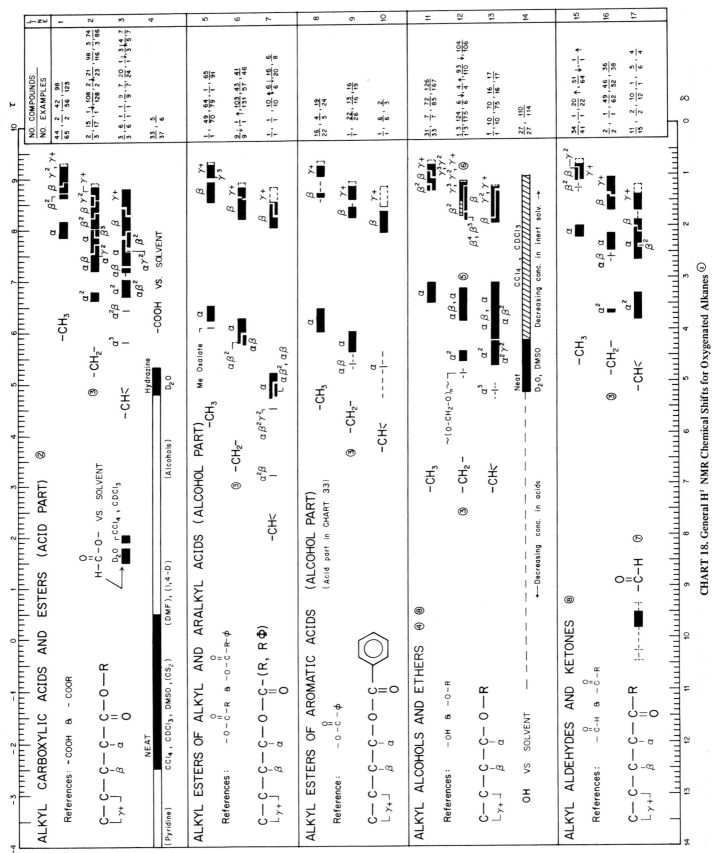

CHART 18. General H¹ NMR Chemical Shifts for Oxygenated Alkanes ①

CHART 18

Notes for Chart 18—Oxygenated Alkanes

1. Solvents used: none (neat liquid), $CDCl_3$, and CCl_4. Some acids, salts, and alcohols also run in D_2O and DMSO-D.

2. TFA solvent shifted bands downfield (higher δ-values) as follows: 0.4–0.6 ppm for α groups, 0.1–0.3 ppm for β groups, negligible for $\gamma+$ groups.

3. Data plotted for only the band centers of CH_2 groups, even if the two hydrogens were nonequivalent, because the CH_2 band center is more reliably measured. Observed shift differences for nonequivalent hydrogens (22 examples in complete chart) varied from 0.09 to 0.55 ppm. No correlation with structure was feasible.

4. In TFA solvent, alcohols frequently exhibit the ester spectrum instead of the alcohol spectrum. TFA should not be used in obtaining alcohol shifts.

5. In $HO-CH_2-CH_2-OR$ groups, shift of $HO-CH_2$ is 0.0–0.2 ppm to lower field (higher δ value) than CH_2-OR. In general, though, for alcohols and ethers, $\alpha\beta$ positions are not distinguishable from α positions by chemical shift alone.

6. For groups in a single chain, the β-CH_2 band is almost always at 0.2–0.3 ppm lower field (higher δ-valve) than the $\gamma+$-CH_2 band.

7. The dotted vertical lines represent questionable observations.

8. These sections include data for alkyl-cycloalkyl and alkyl-aralkyl compounds.

9. Additional illustrative structures:

W = COOH, COOR, O—CO—R, O—CO—ϕ, OH, OR, CHO, CO—R, alone or in combinations indicated by titles of chart sections.

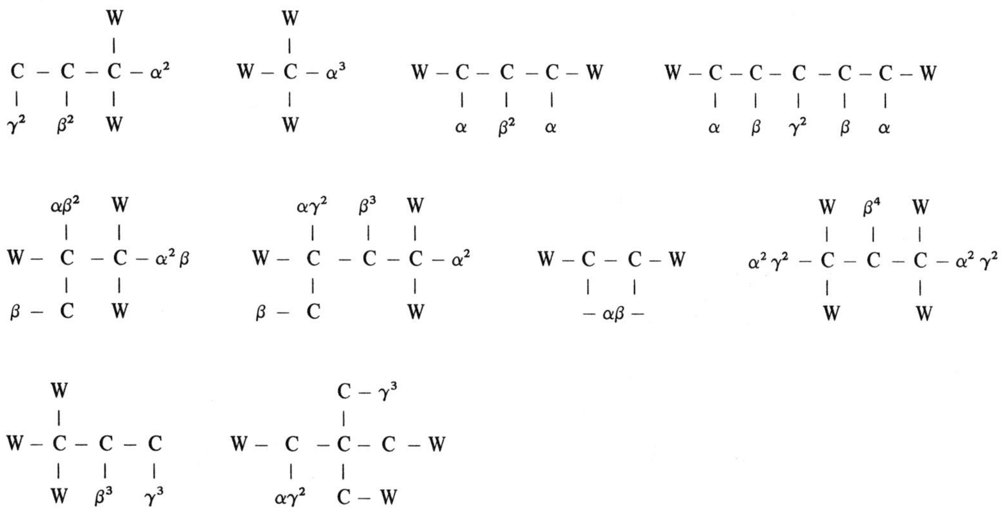

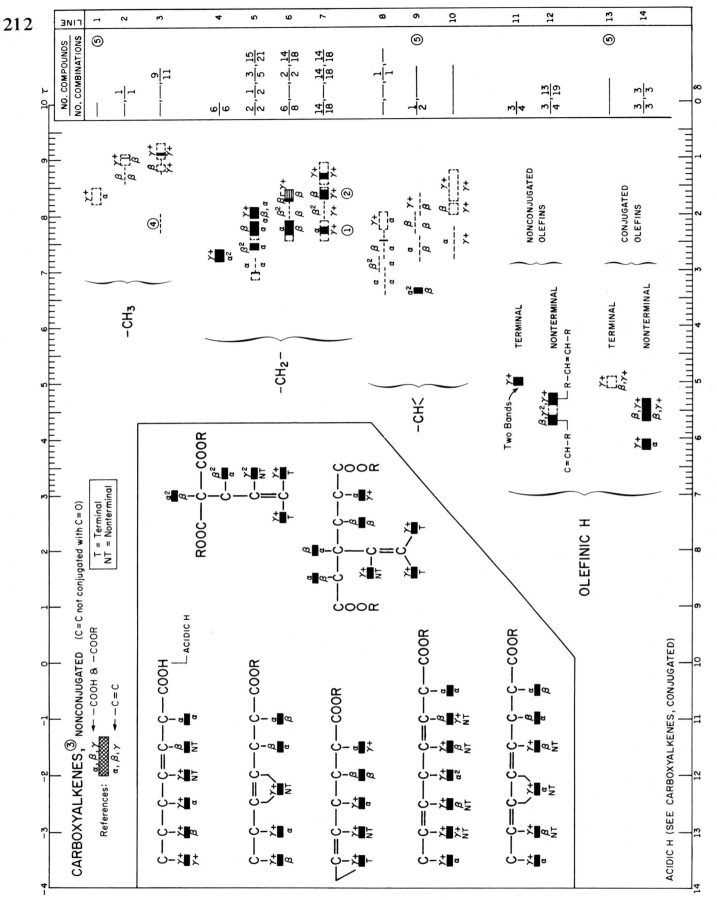

CHART 19. General H¹ NMR Chemical Shifts for Unsaturated Aliphatic Acids and Their Esters. ③ (A4)

CHART 19

Notes for Chart 19—Unsaturated Aliphatic Acids and Esters, Nonconjugated

1. CH$_2$ groups β/α, α/β, and $\alpha/\gamma+$ have indistinguishable chemical shifts at 60 MHz.

2. The chemical shifts of CH$_2$ groups $\beta/\gamma+$ and $\gamma+/\beta$ are both indistinguishable and uncertain at 60 MHz.

3. This chart has been extended where necessary to cover unsaturated aldehydes and ketones which are nonconjugated.

4. This indicates the shift range of ketone methyls α/γ^2 and $\alpha/\gamma+$. It does not apply to acids, esters, or aldehydes.

5. Because of the interactions through space in these molecules, interpolation and extrapolation of shifts are not always linear. The dotted boxes and lines must therefore be used with caution.

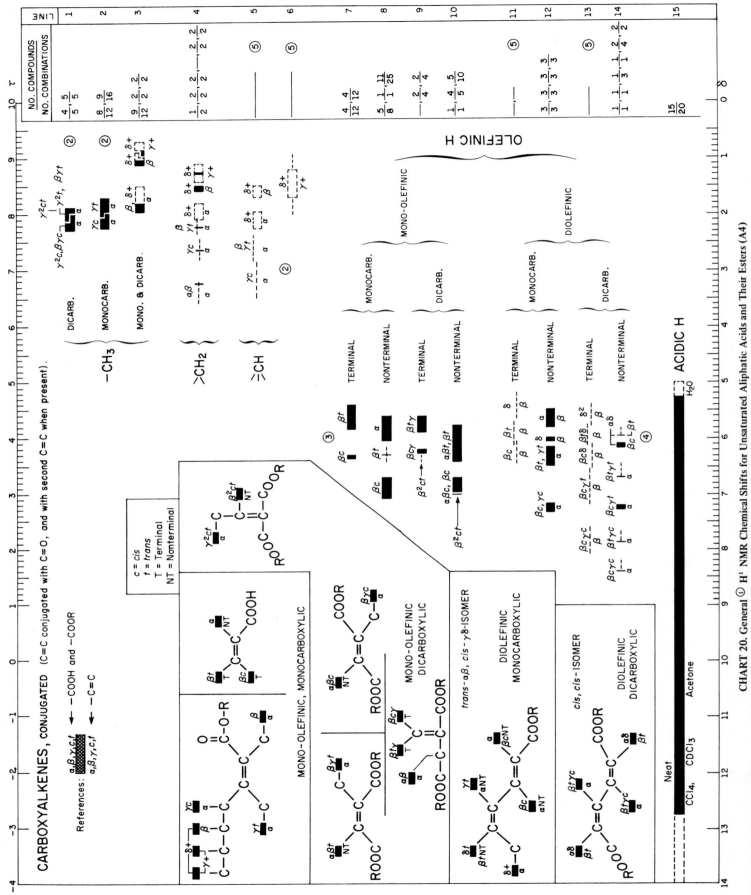

CHART 20. General ① H¹ NMR Chemical Shifts for Unsaturated Aliphatic Acids and Their Esters (A4)

CHART 20

Notes for Chart 20–Unsaturated Aliphatic Acids and Esters, Conjugated

1. Solvents used in obtaining these data were: CCl_4, $CDCl_3$, acetone, D_2O, tetrahydrofuran, and none (neat liquid). Chemical shifts of groups closely associated with carboxyl changed markedly in some other solvents. Benzene shifted these values 0.1 to 0.5 ppm upfield. CF_3COOH shifted these values 0.1 to 0.4 ppm downfield. Salts of these acids in D_2O exhibited an upfield shift of 0 to 0.5 ppm.

2. and 3. The relative shift between groups *cis* and *trans* to the carboxyl is enhanced enough to be diagnostic of the steric position. Note also that the shifts of groups β and γt to carboxyl are indistinguishable, and that nonterminal olefinic hydrogens β to carboxyl resonate at lower field than those α to carboxyl.

4. Steric position with respect to the additional C=C in diolefinic acids also appears to contribute to chemical shift differences. This effect is not shown in other applicable sections of this chart because it is too small to be reliable.

5. Because of the interactions through space in these molecules, interpolation and extrapolation of shifts are not always linear. The dotted boxes and lines must therefore be used with caution.

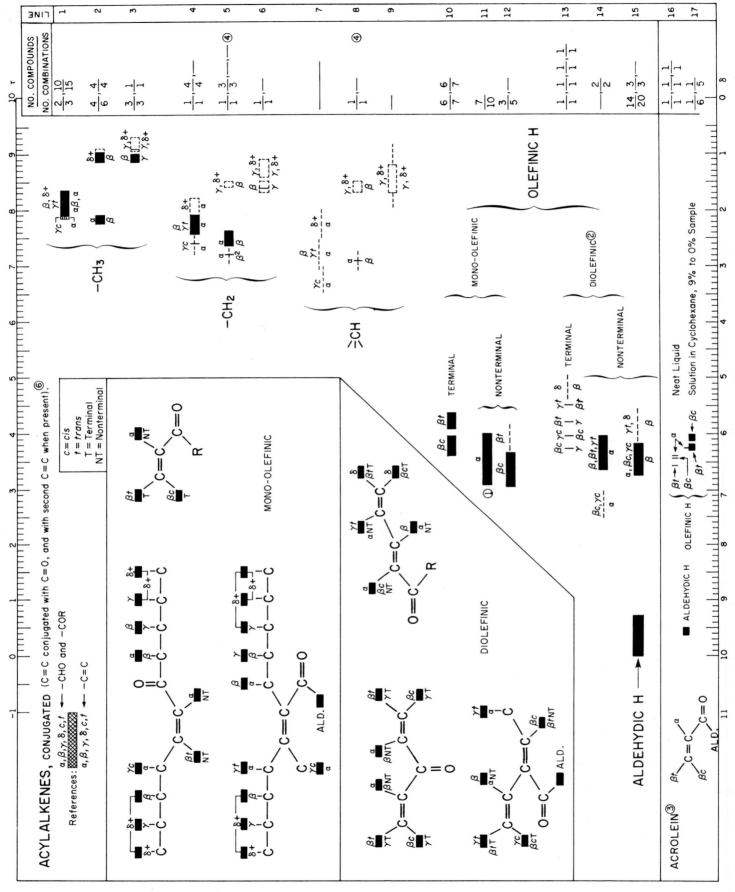

CHART 21 General H¹ Chemical Shifts for Unsaturated Aliphatic Aldehydes and Ketones ⑤ (A4)

CHART 21

Notes for Chart 21—Unsaturated Aliphatic Aldehydes and Ketones

1. When the resonance of the nonterminal α hydrogen arises from the same olefinic group as that of a βc hydrogen, the α resonance is always at lower field than that of the terminal βc but is usually at higher field than that of the nonterminal βc.

2. Data for olefinic hydrogens of diolefinic ketones and aldehydes are too scarce and too scattered for good correlation.

3. The published data for acrolein do not fit the correlations for the remainder of the unsaturated aldehydes and ketones because of reversal of βc and βt assignments. Charting these data separately shows this reversal more clearly. It also shows the marked solvent effects on acrolein hydrogen shifts.

4. Because of the interactions through space in these molecules, interpolation and extrapolation of shifts are not always linear. The dotted boxes and lines must therefore be used with caution.

5. Data for unsaturated compounds containing two carbonyl groups were not available for this chart. Chemical shifts for such compounds may be estimated from Chart 20, Carboxyalkenes, conjugated. The shifts of olefinic hydrogens α to ketone or aldehyde carbonyl are about the same as those β to these carbonyls, but the shifts of other groups are about as indicated in Chart 20.

6. The chemical shifts for nonconjugated aldehydes and ketones are included on Chart 19, Carboxyalkenes, nonconjugated.

218

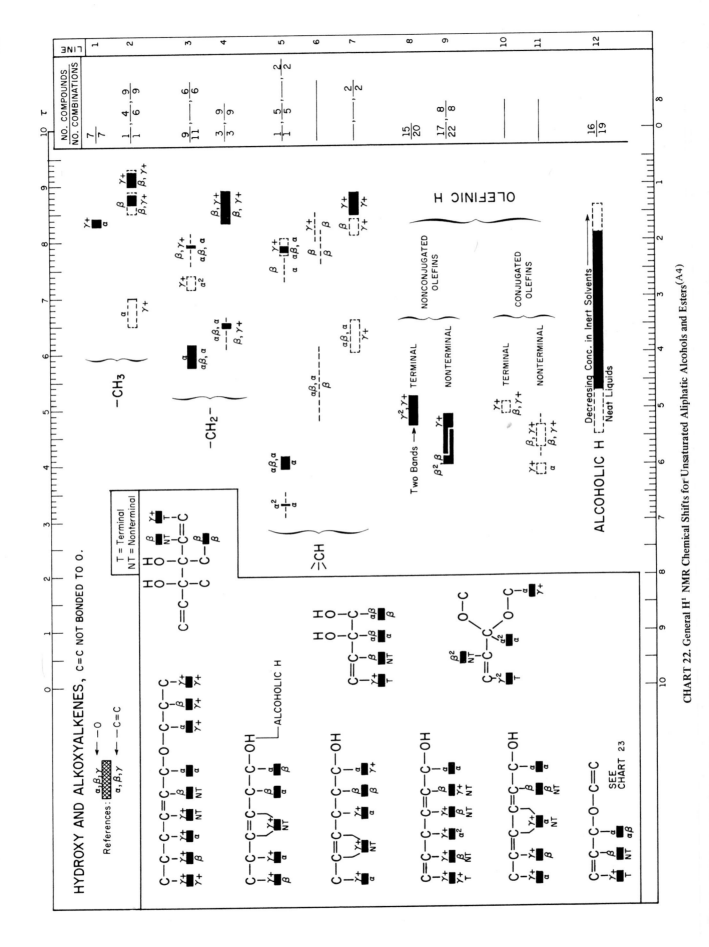

CHART 22. General H[1] NMR Chemical Shifts for Unsaturated Aliphatic Alcohols and Esters(A4)

CHART 22

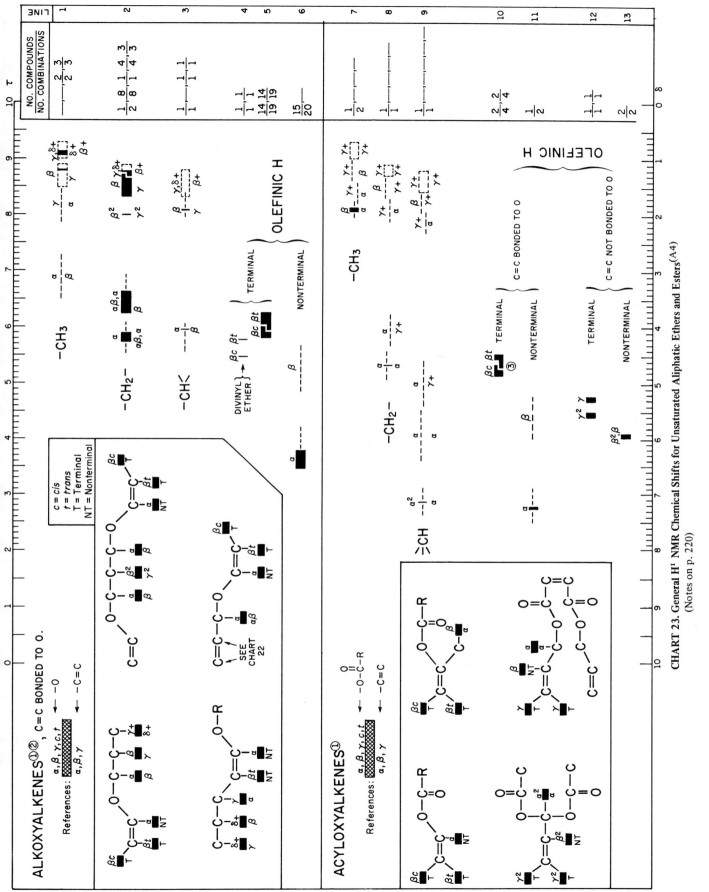

219

CHART 23. General H[1] NMR Chemical Shifts for Unsaturated Aliphatic Ethers and Esters[(A4)]

(Notes on p. 220)

CHART 23

Notes for Chart 23—Unsaturated Aliphatic Ethers and Esters

1. This chart section simply presents the few data available for these compounds. A more detailed correlation is not feasible.

2. Data shown in this section for alkoxyalkenes (ethers) should apply also to similar hydroxyalkenes (alcohols), although direct confirmation of this assumption is lacking.

3. The chemical shifts of the βc and βt terminal olefinic hydrogens are clearly distinguishable for vinyl acetate but not for α-methylvinyl acetate.

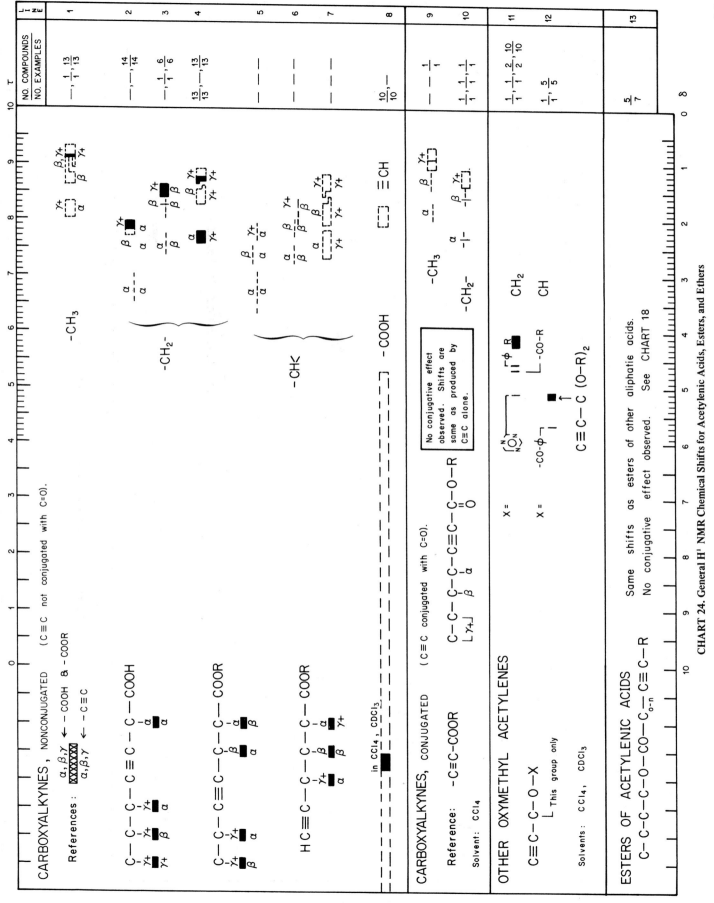

CHART 24. General H¹ NMR Chemical Shifts for Acetylenic Acids, Esters, and Ethers

CHART 24

222

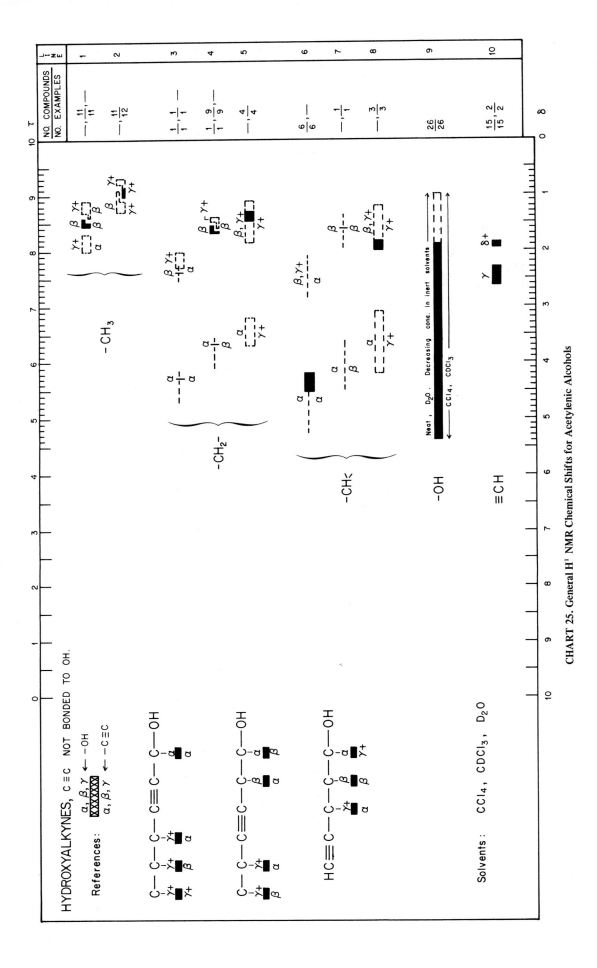

CHART 25. General H¹ NMR Chemical Shifts for Acetylenic Alcohols

CHART 25

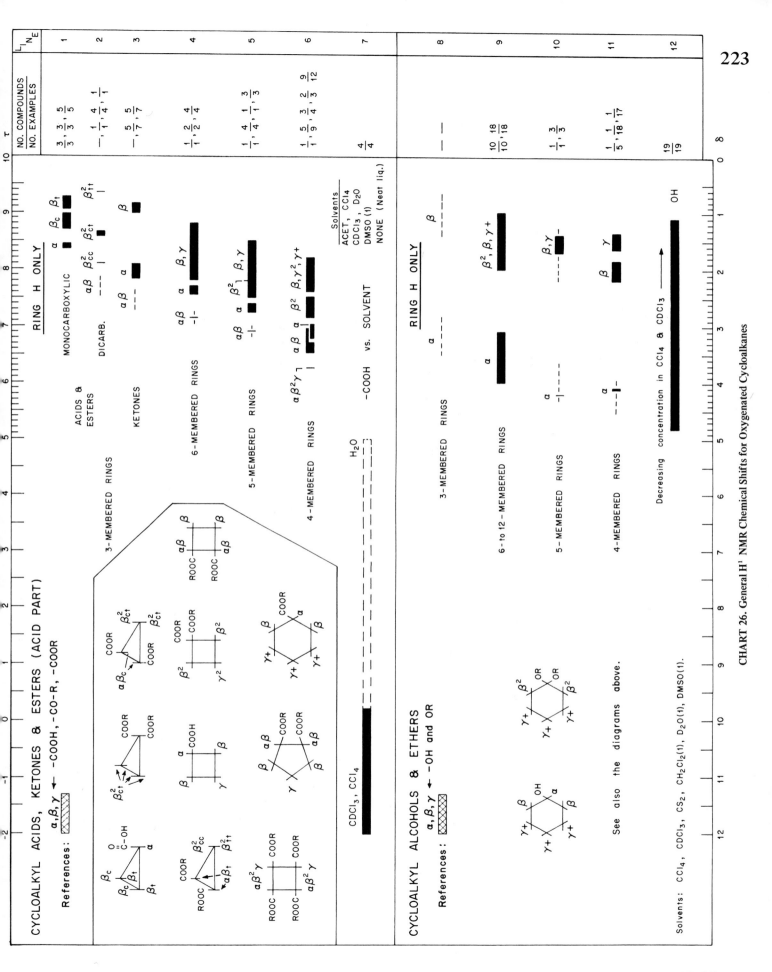

CHART 26. General H[1] NMR Chemical Shifts for Oxygenated Cycloalkanes

CHART 26

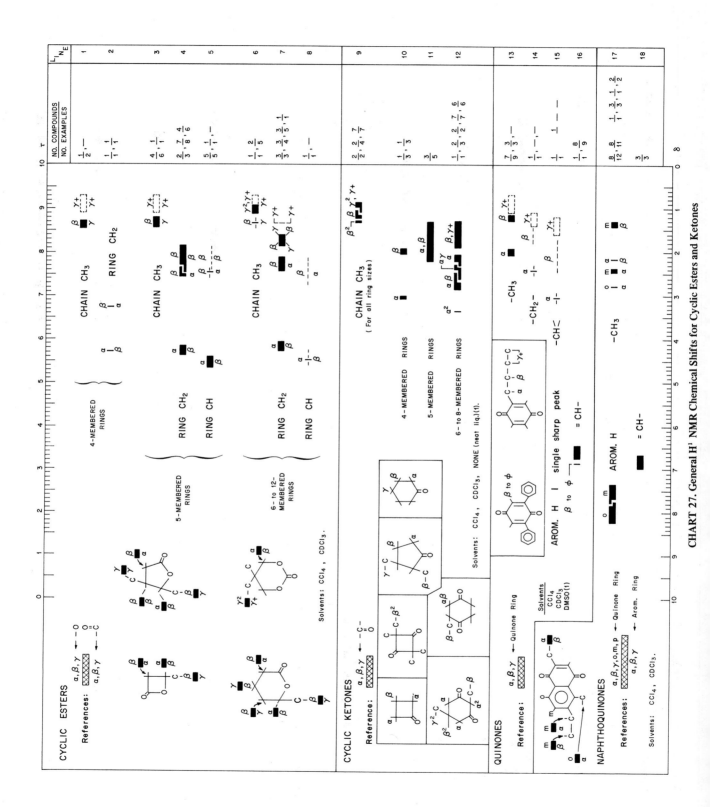

CHART 27. General H¹ NMR Chemical Shifts for Cyclic Esters and Ketones

CHART 27

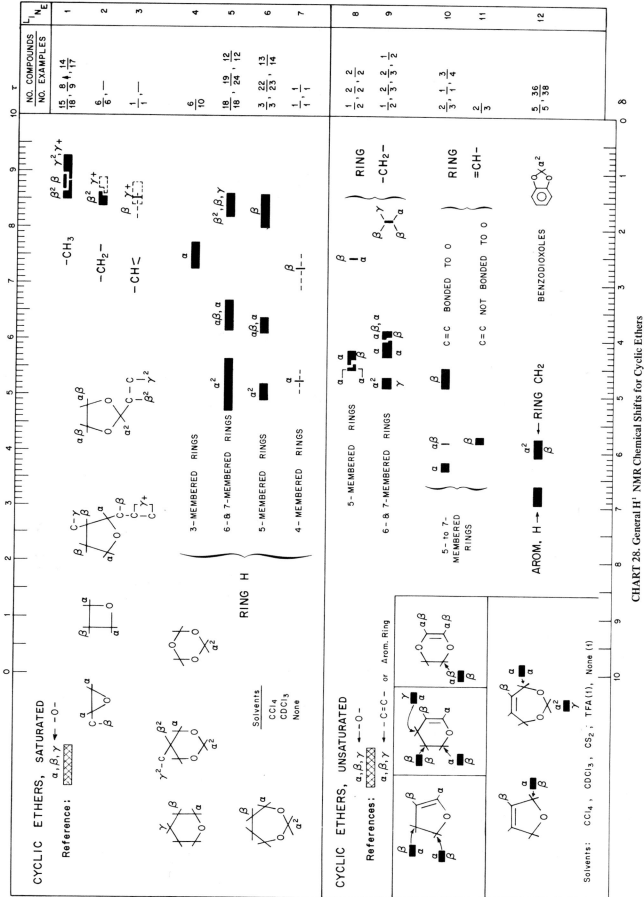

CHART 28. General H[1] NMR Chemical Shifts for Cyclic Ethers

CHART 28

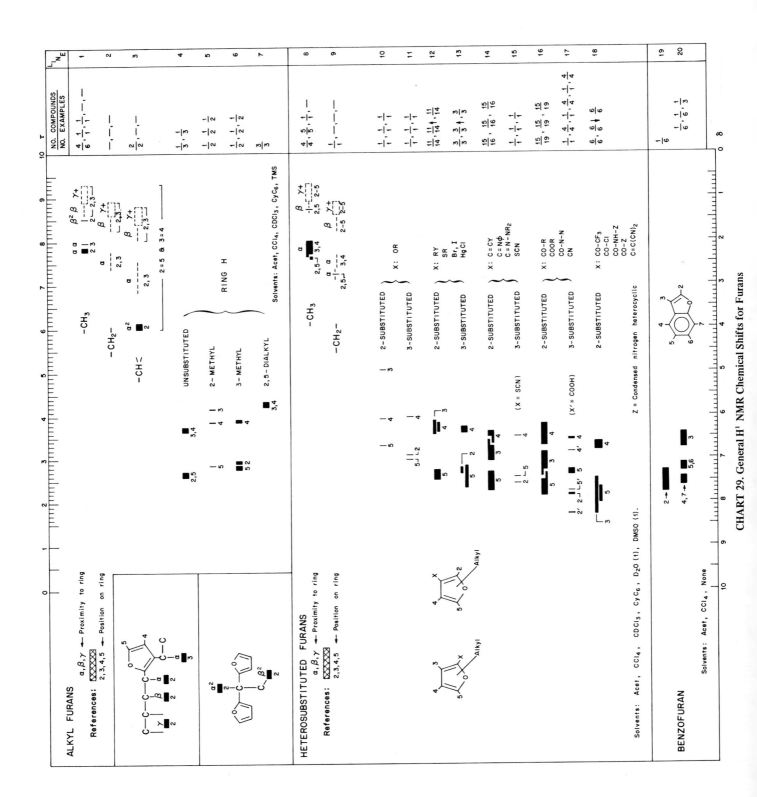

CHART 29. General H¹ NMR Chemical Shifts for Furans

CHART 29

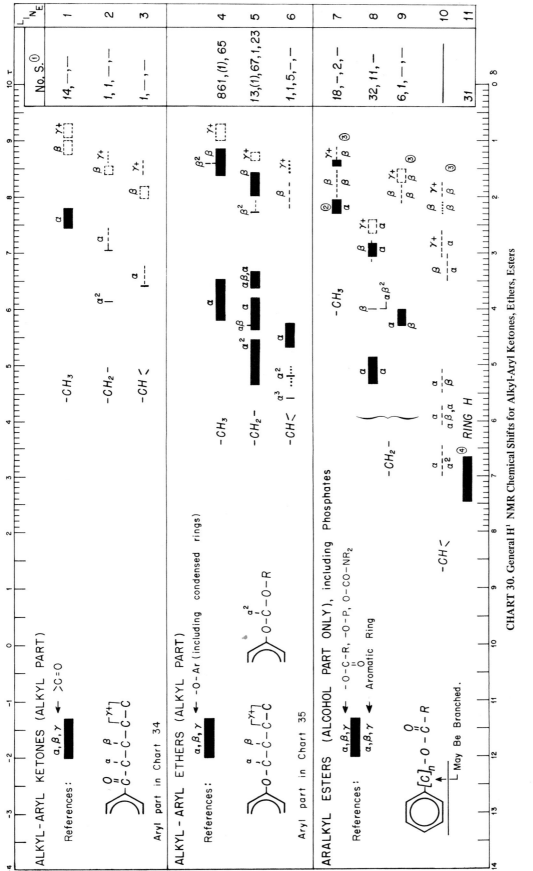

CHART 30. General H¹ NMR Chemical Shifts for Alkyl-Aryl Ketones, Ethers, Esters

1. Total number of compound–solvent combinations included in each solid bar.
2. When present on the same ring, methyls ortho to $-CH_2OCOR$ resonate 0.05 to 0.1 ppm downfield from methyls meta or para to $-CH_2OCOR$ (benzyl esters only).
3. Chemical shifts of groups γ+ to the aromatic ring are the same as those for aliphatic esters. (Chart 18).
4. Frequently produces a single narrow band. When resonance positions differ, hydrogens ortho to $-COOCR$ resonate at lowest field.

CHART 30

228

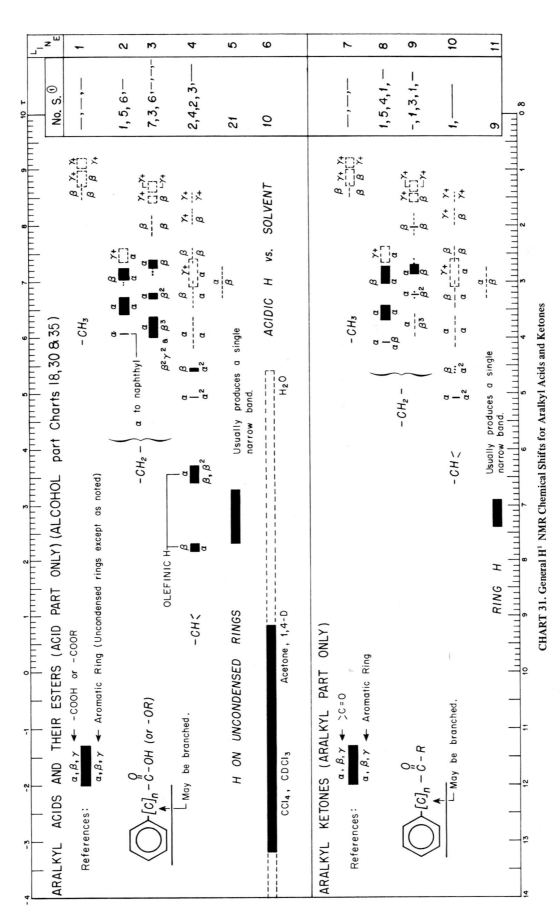

CHART 31. General H¹ NMR Chemical Shifts for Aralkyl Acids and Ketones

1. Total number of compound-solvent combinations included in each solid bar.

CHART 31

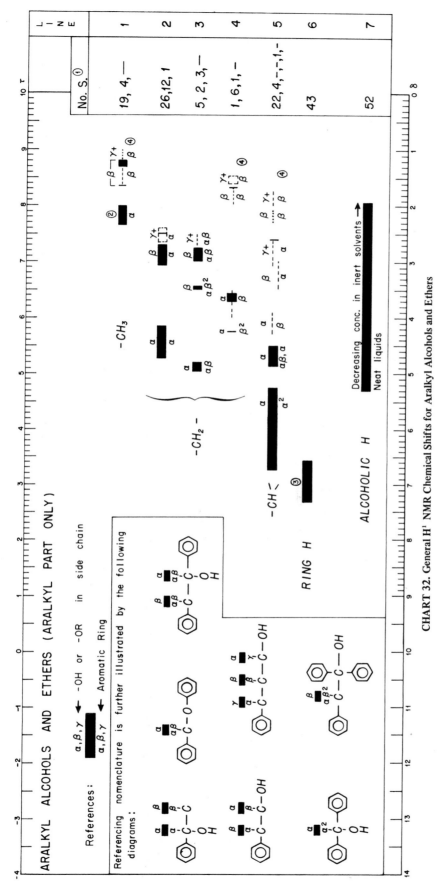

CHART 32. General H¹ NMR Chemical Shifts for Aralkyl Alcohols and Ethers

1. Total number of compound-solvent combinations included in each solid bar.
2. When present on the same ring, methyls ortho to $-CH_2O-$ resonate 0 to 0.1 ppm downfield from methyls meta or para to $-CH_2O-$. In diphenylmethanols this relationship may be reversed.
3. Ring H's may produce either a single narrow band or a pair of bands, depending on

solvent and type of ring substitution. When two bands are produced, hydrogens ortho to $-CH_2O-$ resonate 0.1 to 0.5 ppm downfield from hydrogens meta or para to $-CH_2O-$.
4. Chemical shifts of groups $\gamma+$ to the aromatic ring are the same as those for aliphatic alcohols and ethers (Chart 18).

CHART 32

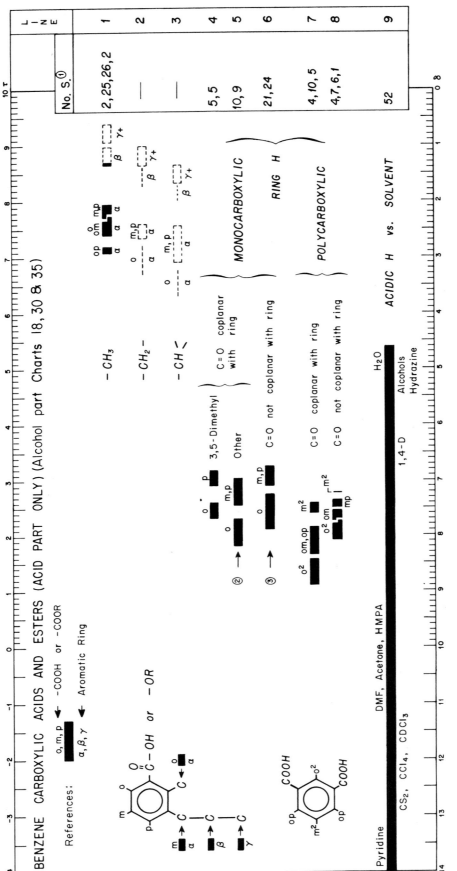

CHART 33. General H¹ NMR Chemical Shifts for Aromatic Acids and Esters

1. Total number of compound–solvent combinations included in each solid bar.
2. In a single compound, separation between *o* and *m* or *p* bands varies from 0.54 to 0.80 ppm.
3. In a single compound, separation between *o* and *m* or *p* bands varies from 0.66 to 1.50 ppm.

CHART 33

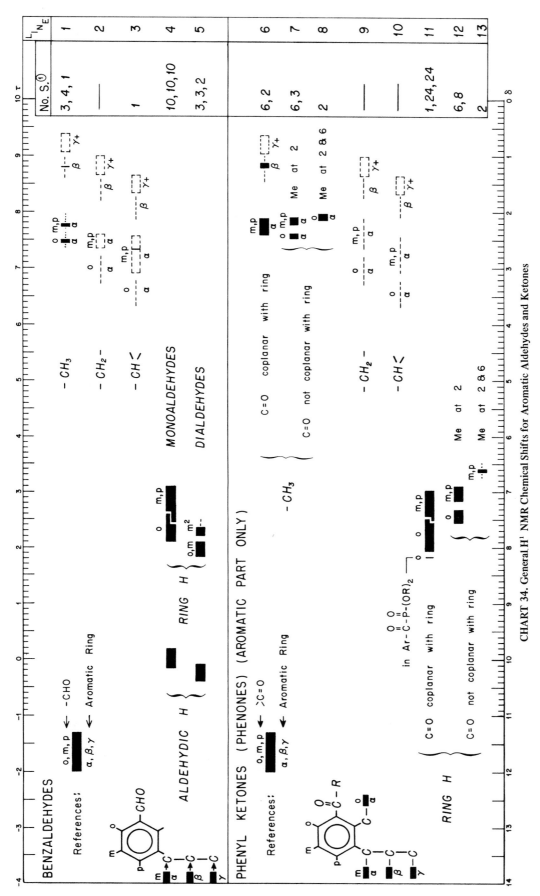

CHART 34. General H¹ NMR Chemical Shifts for Aromatic Aldehydes and Ketones

1. Total number of compound–solvent combinations included in each solid bar.

CHART 34

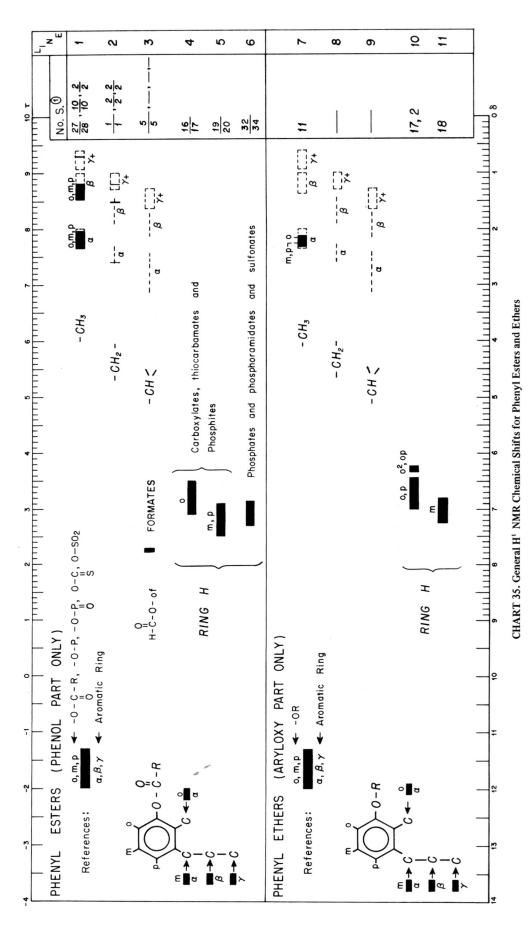

CHART 35. General H¹ NMR Chemical Shifts for Phenyl Esters and Ethers

1. Total number of compound–solvent combinations included in each solid bar.

CHART 35

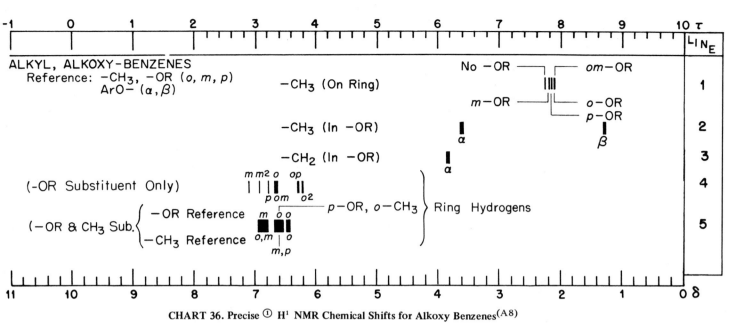

CHART 36. Precise ① H¹ NMR Chemical Shifts for Alkoxy Benzenes[A8]

1. Samples run at <10% concentration in CCl_4, using TMS internal reference.

CHART 36

234

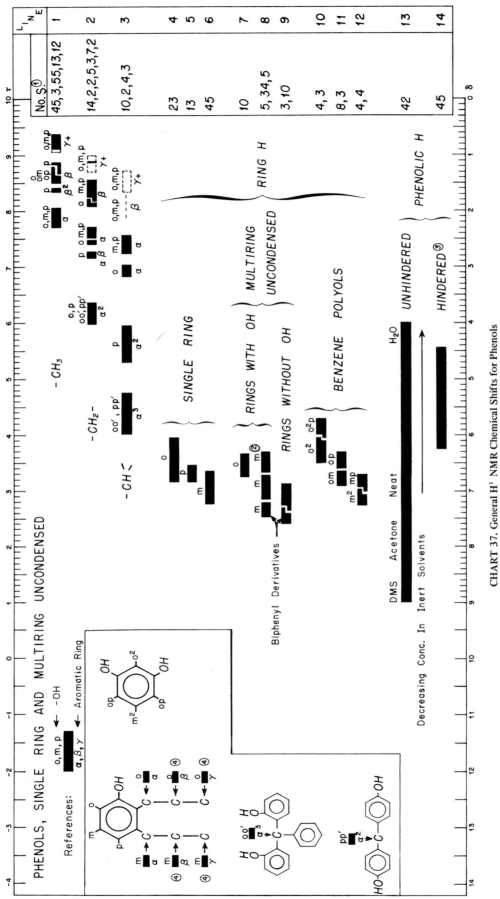

CHART 37. General H[1] NMR Chemical Shifts for Phenols

1. Total number of compound–solvent combinations included in each solid bar.
2. Shielded by ring currents in certain sterically hindered configurations.
3. Bulky substituents flanking OH on both sides. Shift is independent of solvent and concentration.
4. The unusual designation of these groups as *o*, *m*, or *p* is necessary to show the full extent of the effect of the OH.

CHART 37

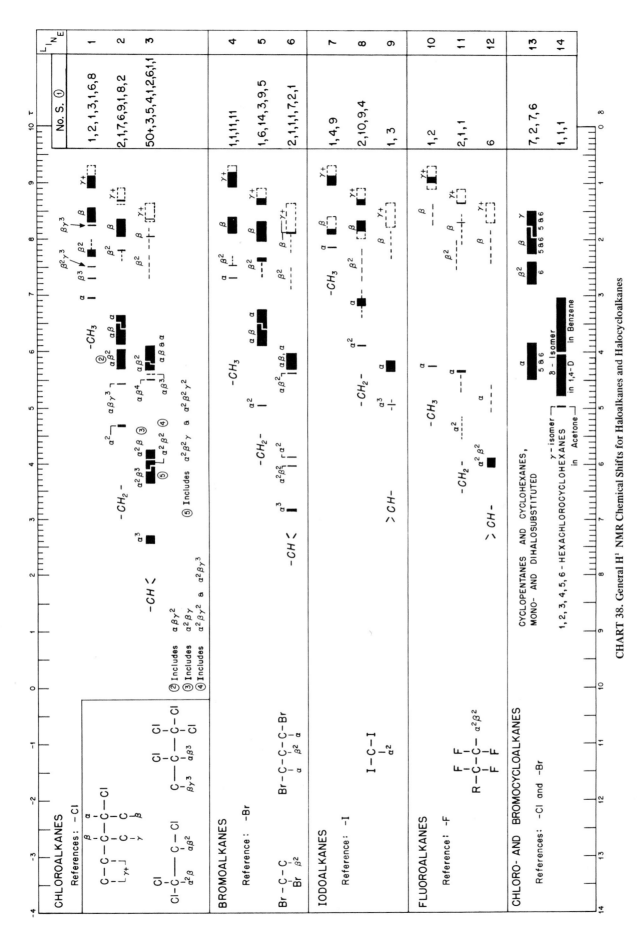

CHART 38. General H¹ NMR Chemical Shifts for Haloalkanes and Halocycloalkanes

1. Total number of compound-solvent combinations included in each solid bar.

CHART 38

236

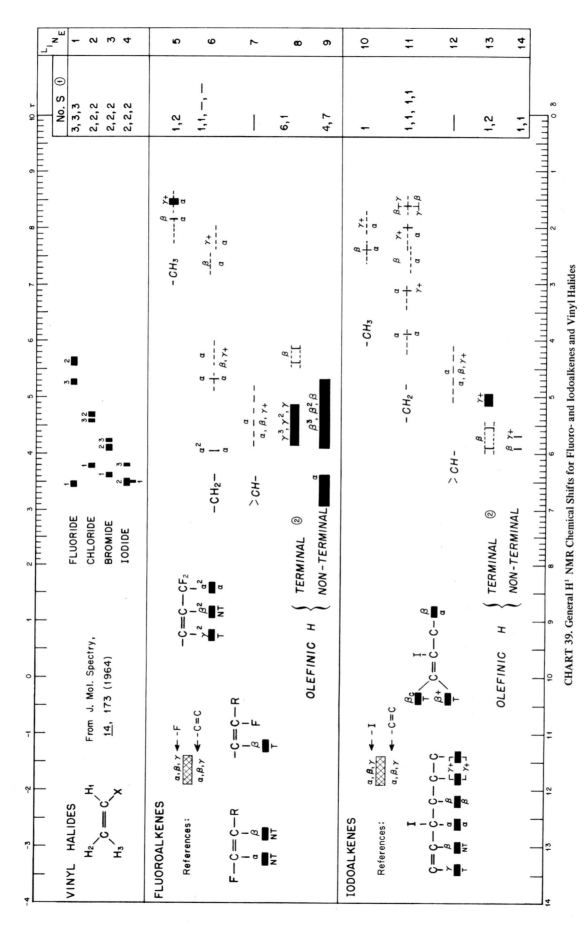

CHART 39. General H¹ NMR Chemical Shifts for Fluoro- and Iodoalkenes and Vinyl Halides

1. Total number of compound–solvent combinations included in each solid bar.
2. Terminal olefinic hydrogens normally have different chemical shifts, producing two spectral bands.

CHART 39

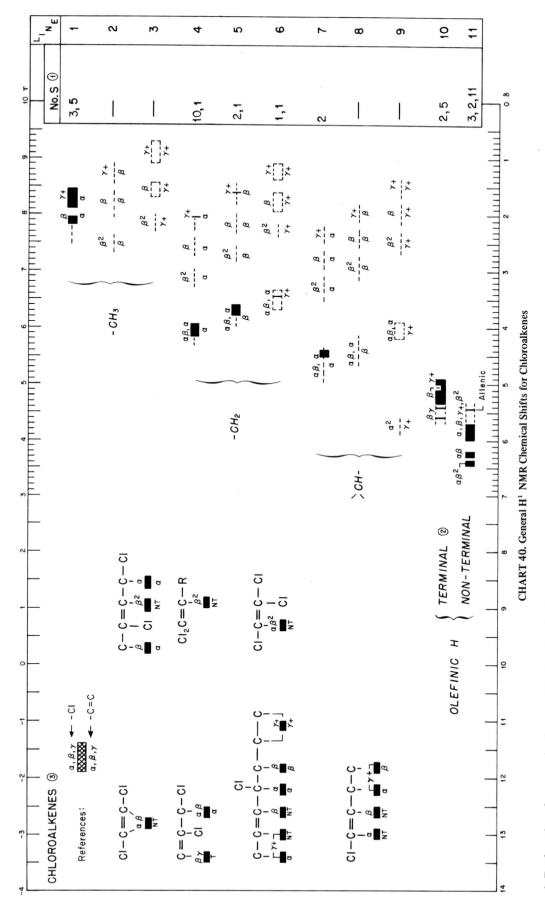

CHART 40. General H¹ NMR Chemical Shifts for Chloroalkenes

1. Total number of compound–solvent combinations included in each solid bar.
2. Terminal olefinic hydrogens normally have different chemical shifts, producing two spectral bands.
3. See Chart 39 for vinyl halides.

CHART 40

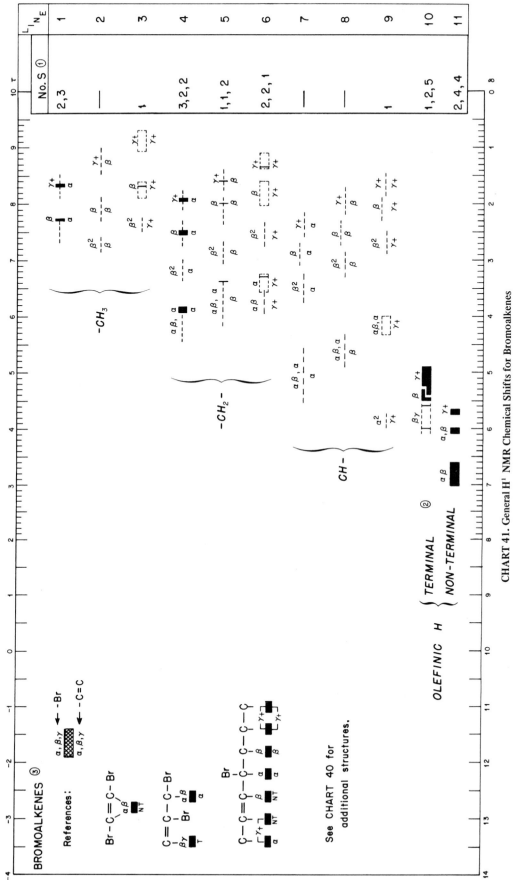

CHART 41. General H¹ NMR Chemical Shifts for Bromoalkenes

1. Total number of compound–solvent combinations included in each solid bar.
2. Terminal olefinic hydrogens normally have different chemical shifts, producing two spectral bands.
3. See Chart 39 for vinyl halides.

CHART 41

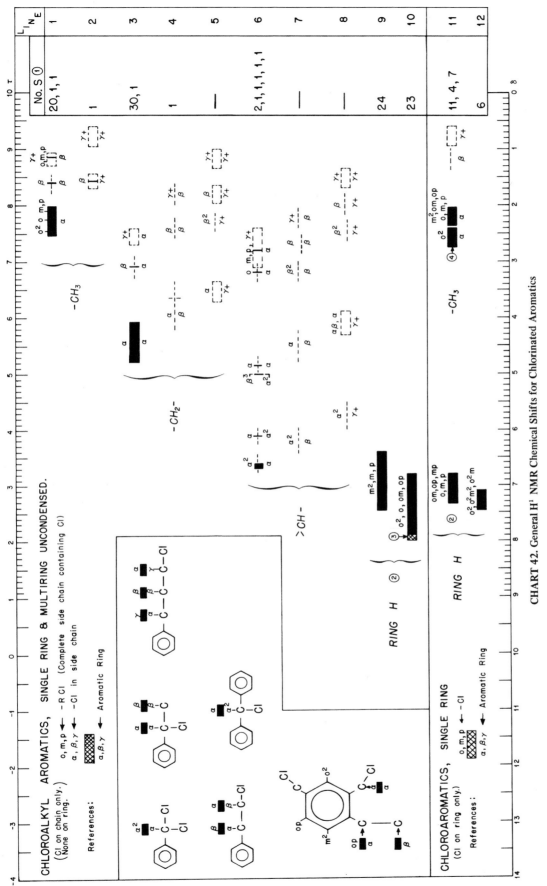

CHART 42. General H¹ NMR Chemical Shifts for Chlorinated Aromatics

1. Total number of compound–solvent combinations included in each solid bar.
2. Sometimes produces only a single narrow band. When more than one band is produced, hydrogens ortho to −RCl or to −Cl resonate at lower field than those meta or para.
3. Ortho to trichloromethyl group.
4. Includes o^2p, o^2m^2, and o^2m^2p.
NOTE: Chemical shifts for compounds having Cl on both ring and side chain are calculated by combining the proper shifts taken from this chart. See also Chart 44.

CHART 42

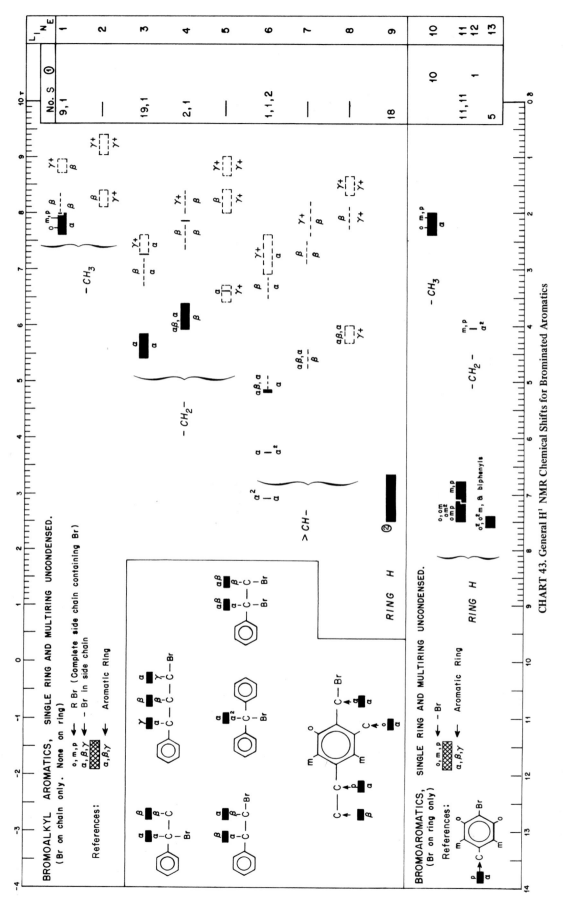

CHART 43. General H¹ NMR Chemical Shifts for Brominated Aromatics

1. Total number of compound–solvent combinations included in each solid bar.

2. Frequently produces a single narrow band. When more than one band is produced, hydrogens ortho to —RBr resonate at lower field than those meta or para.

NOTE: Chemical shifts for compounds having Br on both ring and side chain are calculated by combining the proper shifts taken from this chart. See also Chart 44.

CHART 43

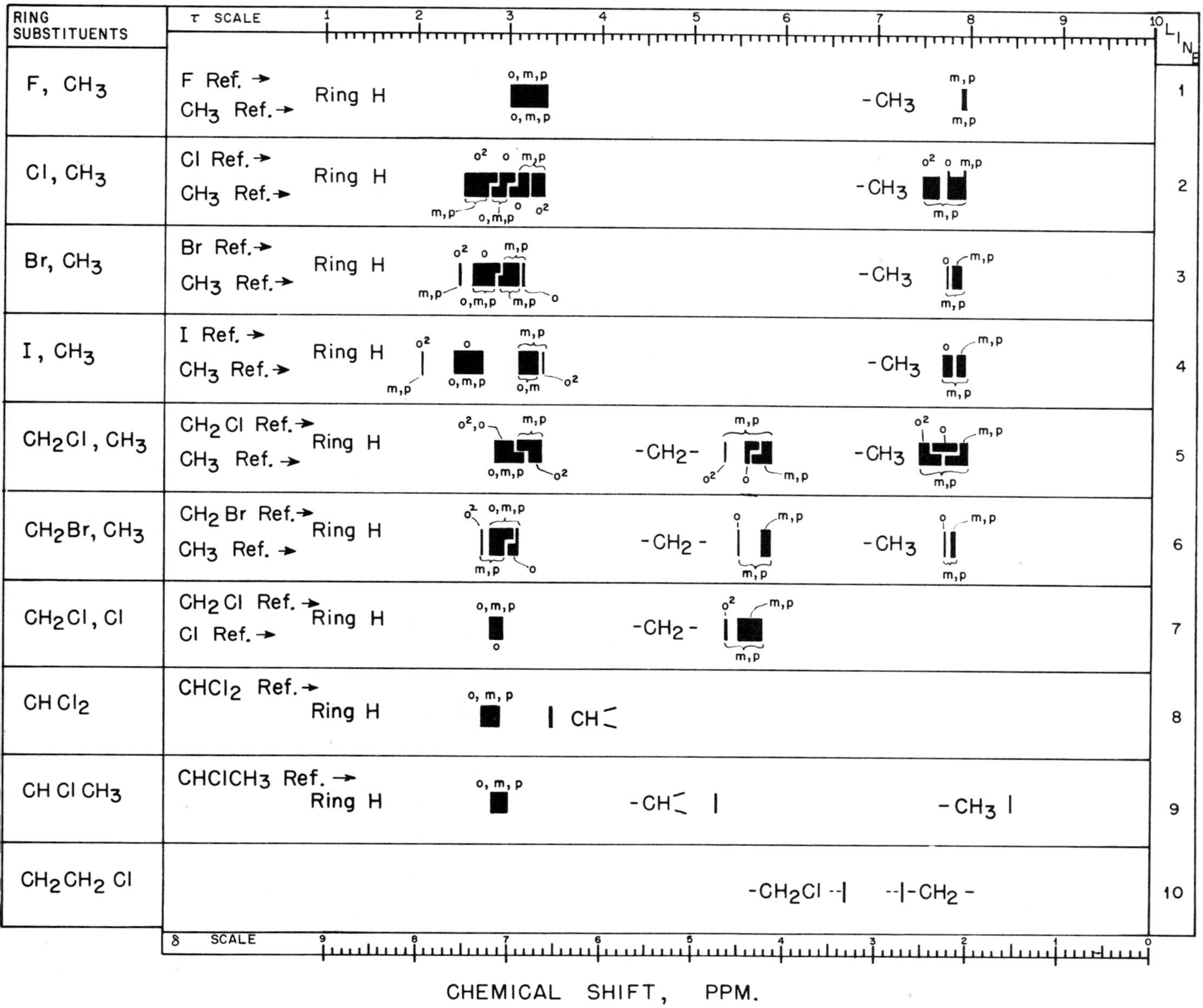

RING SUBSTITUENTS	τ SCALE			LINE
F, CH₃	F Ref. → Ring H o,m,p o,m,p	-CH₃ m,p / m,p		1
Cl, CH₃	Cl Ref. → Ring H o² o m,p / m,p o,m,p o o²	-CH₃ o² o m,p / m,p		2
Br, CH₃	Br Ref. → Ring H o² o m,p / m,p o,m,p m,p o	-CH₃ o m,p / m,p		3
I, CH₃	I Ref. → Ring H o² o m,p / m,p o,m,p o,m o²	-CH₃ o m,p / m,p		4
CH₂Cl, CH₃	CH₂Cl Ref. → Ring H o²,o m,p / o,m,p o²	-CH₂- m,p / o² o m,p -CH₃ o² m,p / m,p		5
CH₂Br, CH₃	CH₂Br Ref. → Ring H o² o,m,p / m,p o	-CH₂- o m,p / m,p -CH₃ o m,p / m,p		6
CH₂Cl, Cl	CH₂Cl Ref. → Ring H o,m,p o	-CH₂- o² m,p / m,p		7
CHCl₂	CHCl₂ Ref. → Ring H o,m,p	CH		8
CHCl CH₃	CHClCH₃ Ref. → Ring H o,m,p	-CH	-CH₃		9
CH₂CH₂Cl		-CH₂Cl -- -- CH₂-		10

CHEMICAL SHIFT, PPM.

Nomenclature:	o = ortho, m = meta, p = para.

Examples: In 1-chloro-3-methylbenzene the 2-hydrogen is ortho with reference to Cl and ortho with reference to CH₃; the 4-hydrogen is para to Cl and ortho to CH₃; the CH₃ group is meta to Cl. In 1,3-dichlorobenzene the 2-hydrogen is designated as O² to Cl.

ORTHO SUBSTITUENT EFFECTS

Substituent	Shift of 2-H, p.p.m.*
H	0.0
F	+0.2
CH₃	+0.15
CH₂Cl, CH₂Br	-0.10
Cl	-0.15
CHCl₂	-0.20
Br	-0.30
I	-0.45

* + means shift to higher field.
  - means shift to lower field.

CHART 44. Precise ① H¹ NMR Chemical Shifts for Halobenzenes and Haloalkyl Benzenes [A7]
1. All samples run as 50 vol. % solutions in CCl₄, using tetramethylsilane as internal reference.

CHART 44

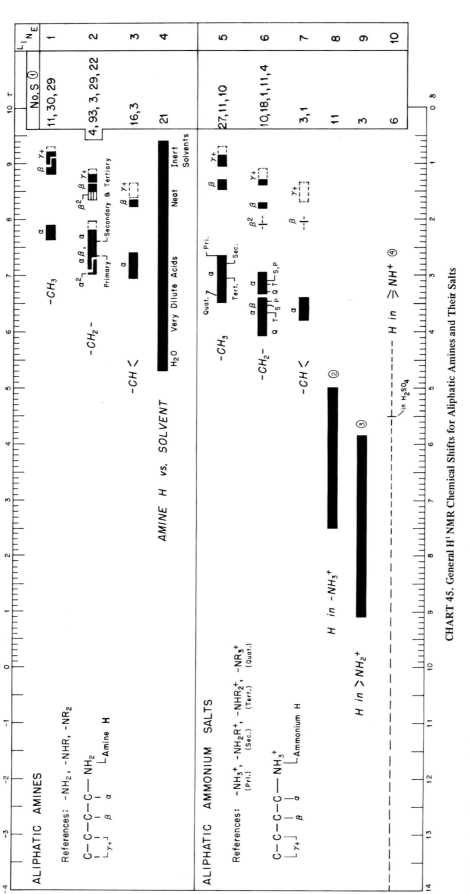

CHART 45. General H¹ NMR Chemical Shifts for Aliphatic Amines and Their Salts

1. Total number of compound–solvent combinations included in each solid or crosshatched bar.
2. This peak is distinct from those for water or acidic hydrogens sometimes, but sometimes they are coalesced. NH_3^+ hydrogens are spin-coupled to nitrogen at room temperature when in H_2O, HCl, or H_2SO_4 solvents, but not when in CF_3COOH solvent.
3. This peak is distinct from that of acidic H (note 2). NH_2^+ hydrogens are not spin-coupled to nitrogen in $CDCl_3$ or H_2SO_4.
4. H in NH^+ is frequently unobservable. It may be coalesced with the acidic H.

CHART 45

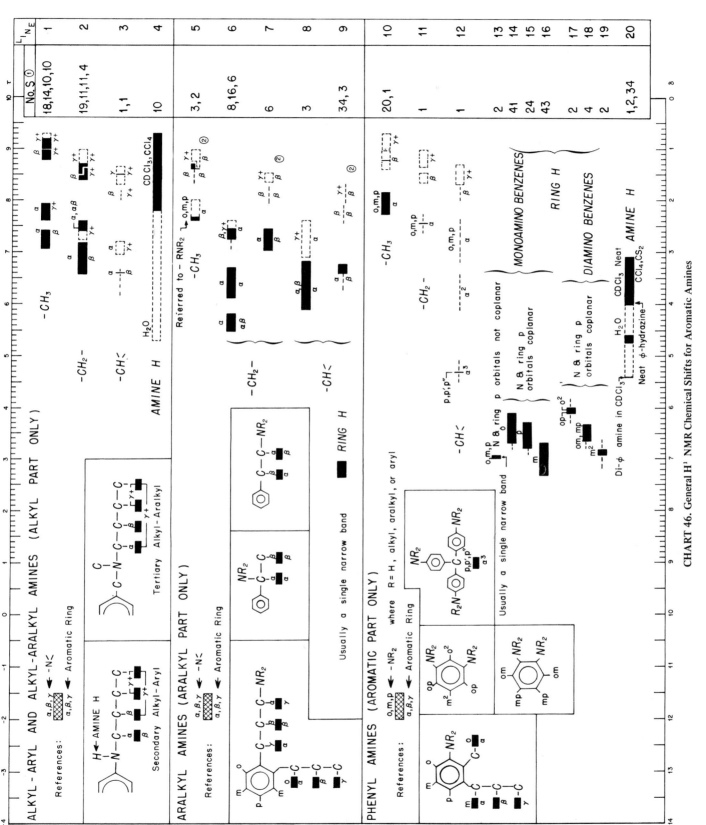

CHART 46. General H¹ NMR Chemical Shifts for Aromatic Amines

1. Total number of compound–solvent combinations included in each solid bar.
2. Chemical shifts for groups γ+ to the ring are shown in the top box of this chart (Alkyl Part).

CHART 46

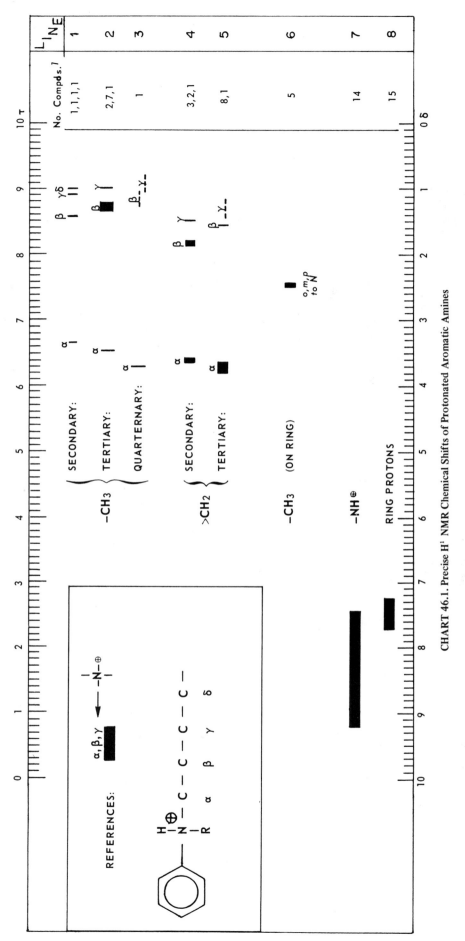

CHART 46.1. Precise H¹ NMR Chemical Shifts of Protonated Aromatic Amines

1. Refers to actual number of different compounds used for each block. All samples were run in glacial trifluoroacetic acid at 5% concentration, with TMS added as internal reference.

(Reproduced by permission of J. J. R. Reed.)

CHART 46.1

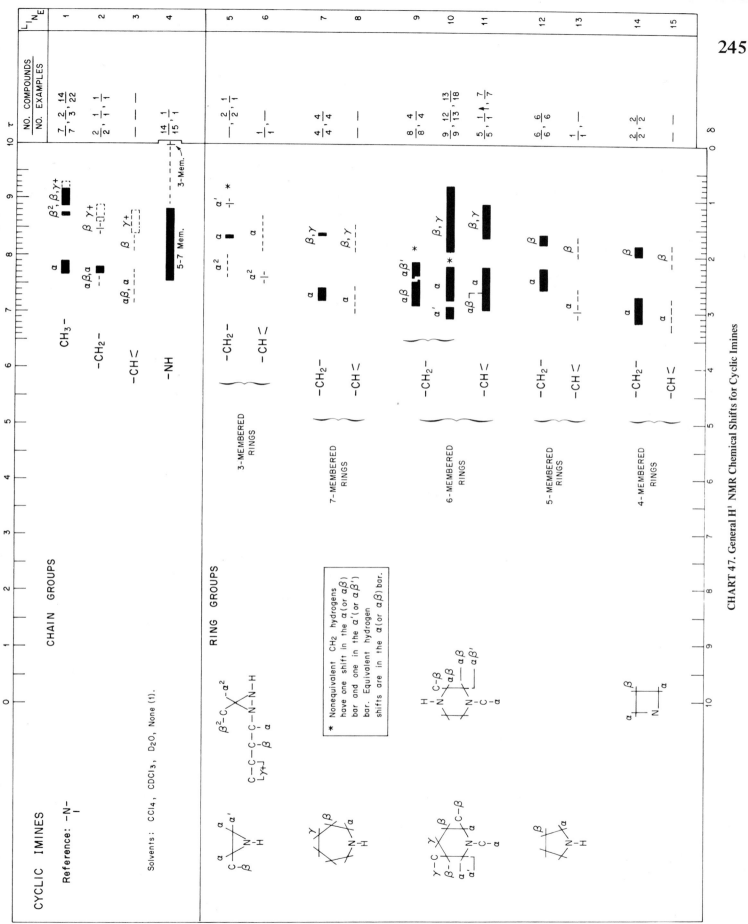

CHART 47. General H¹ NMR Chemical Shifts for Cyclic Imines

245

CHART 47

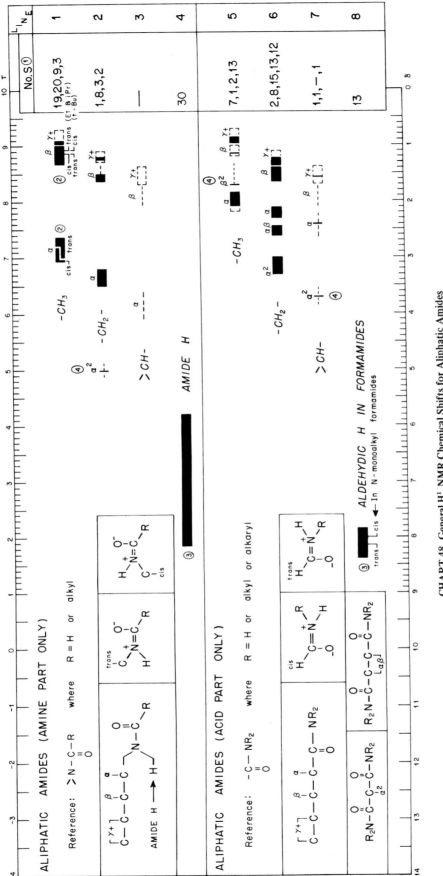

CHART 48. General H¹ NMR Chemical Shifts for Aliphatic Amides

1. Total number of compounds included in each solid bar. Solvents used: CCl_4, $CDCl_3$, D_2O, TFA (as noted).
2. In benzene solution the resonances are shifted 0.04—0.40 ppm upfield. The *cis* resonance is shifted farther than the *trans* so that, except for *t*-butyl substituents, the *cis-trans* relative shifts are reversed. See also Chart 49.
3. Exchanges very slowly, if at all, in neutral or weakly acidic solutions.
4. In TFA.

CHART 48

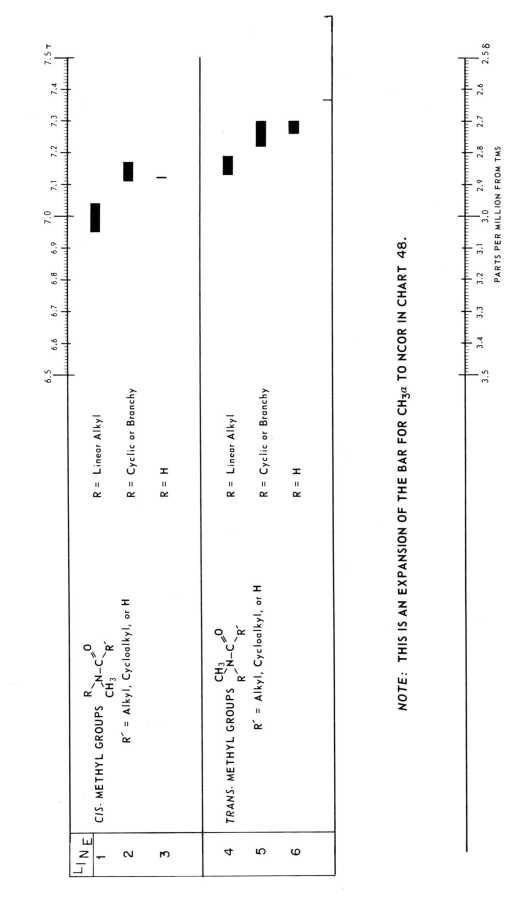

NOTE: THIS IS AN EXPANSION OF THE BAR FOR CH₃α TO NCOR IN CHART 48.

CHART 49. General H¹ NMR Chemical Shifts for Methyl Groups Alpha to Amide Nitrogen

(Reproduced by permission of J. J. R. Reed.)

CHART 49

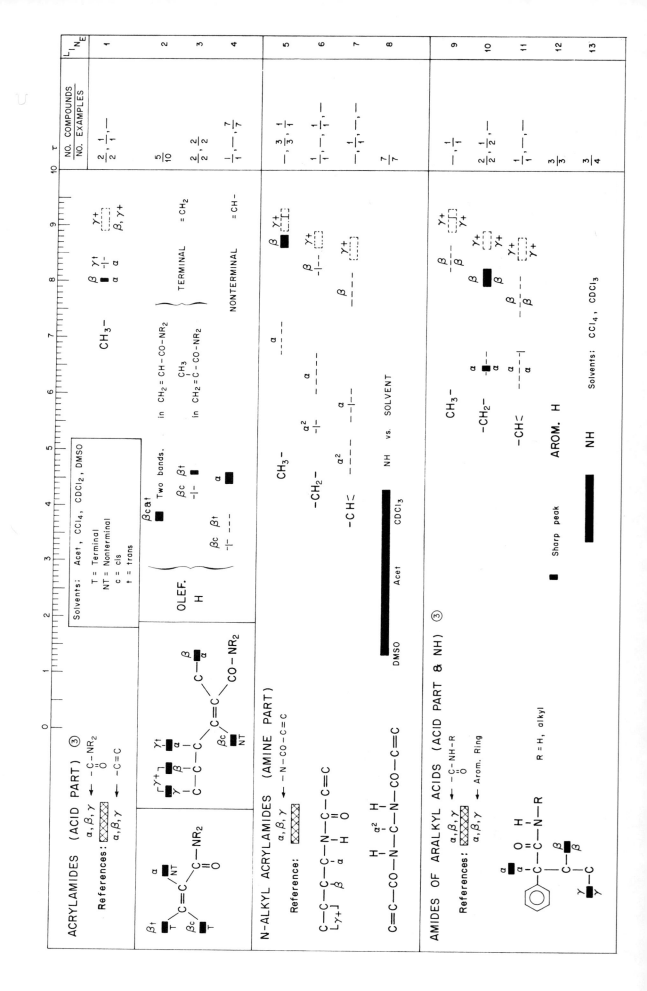

CHART 50

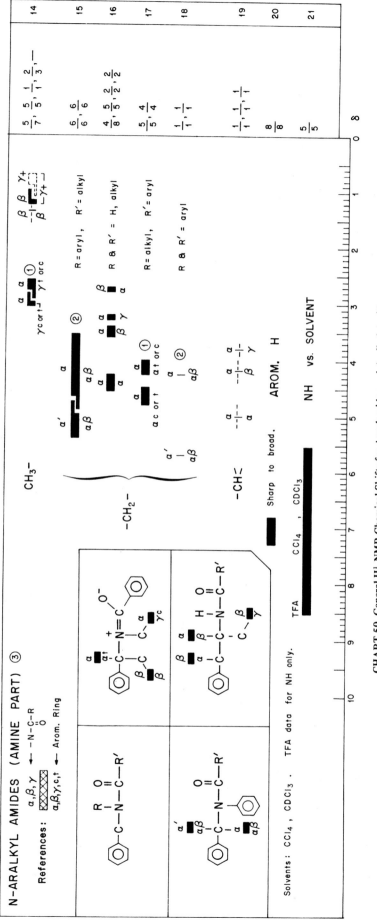

CHART 50. General H¹ NMR Chemical Shifts for Acrylamides and Aralkyl Amides

1. These bars represent shift for groups *cis* or *trans* to the R' group, but it is not clear which is which.
2. The benzyl methylene hydrogens usually have different chemical shifts, one in the α bar and one in the α' bar. This is due to nonequivalence rather than to cis-trans isomerism.
3. These correlations are incomplete because of lack of data.

CHART 50

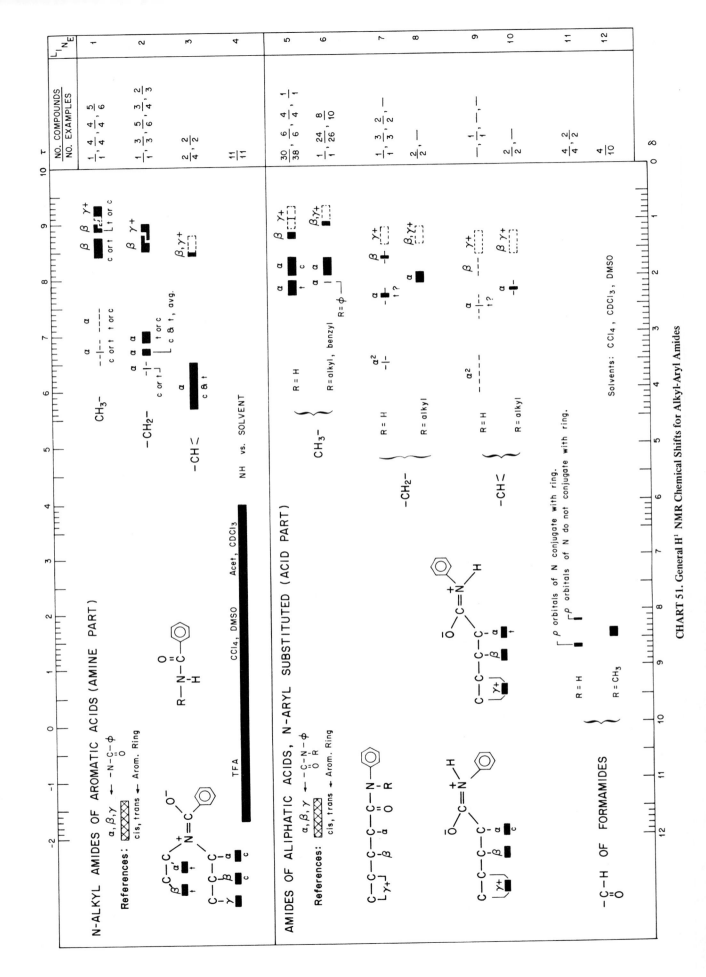

CHART 51. General H¹ NMR Chemical Shifts for Alkyl-Aryl Amides

CHART 51

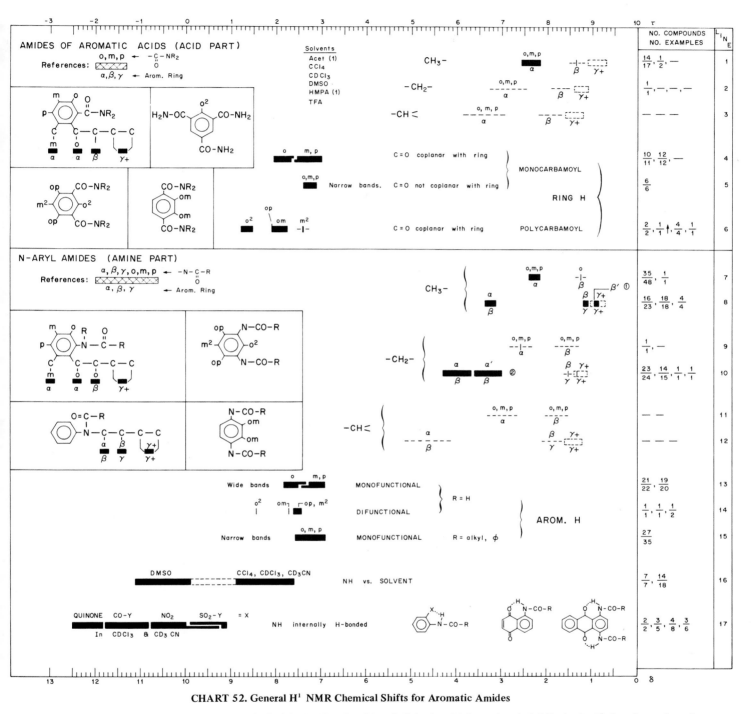

CHART 52. General H¹ NMR Chemical Shifts for Aromatic Amides

1. Methyls β to N (and therefore γ to the aromatic ring) in the N-aryl amides of aliphatic acids have chemical shifts in the β/γ bar. Isomerism of the *cis-trans* type is not evident in these shifts.

   In the corresponding amides of aromatic acids, however, *cis-trans* isomerism places one methyl shift in the β/γ bar and the other in the β'/γ bar. It is not clear which shift corresponds to the methyl *cis* to the acid ring and which corresponds to the methyl *trans*.

2. Hydrogens in methylenes α to N and β to aromatic frequently have different chemical shifts, on in the α/β bar and one in the α'/β bar. When these hydrogens have the same chemical shift it falls in the α/β bar.

CHART 52

252

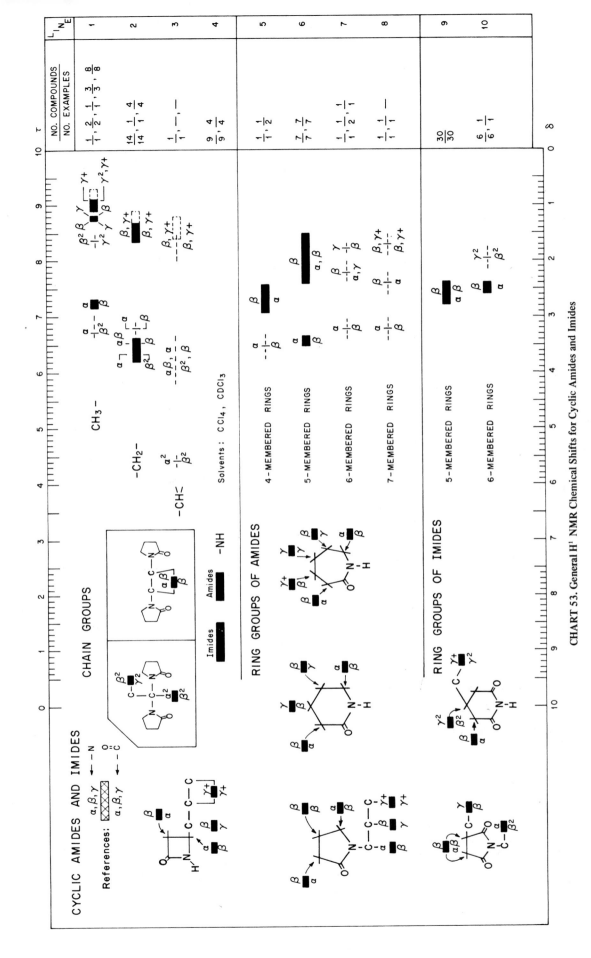

CHART 53. General H¹ NMR Chemical Shifts for Cyclic Amides and Imides

CHART 53

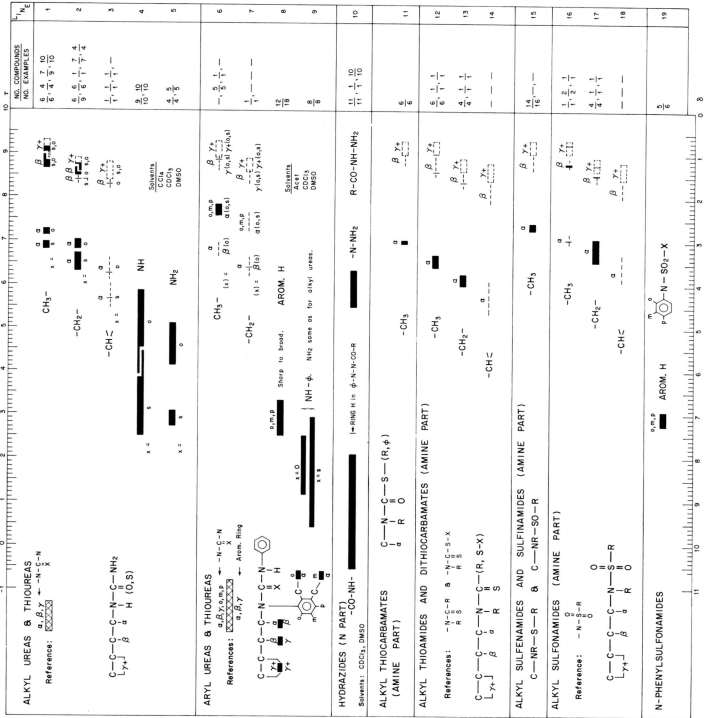

CHART 54. General H^1 NMR Chemical Shifts for Other Amide Types

CHART 54

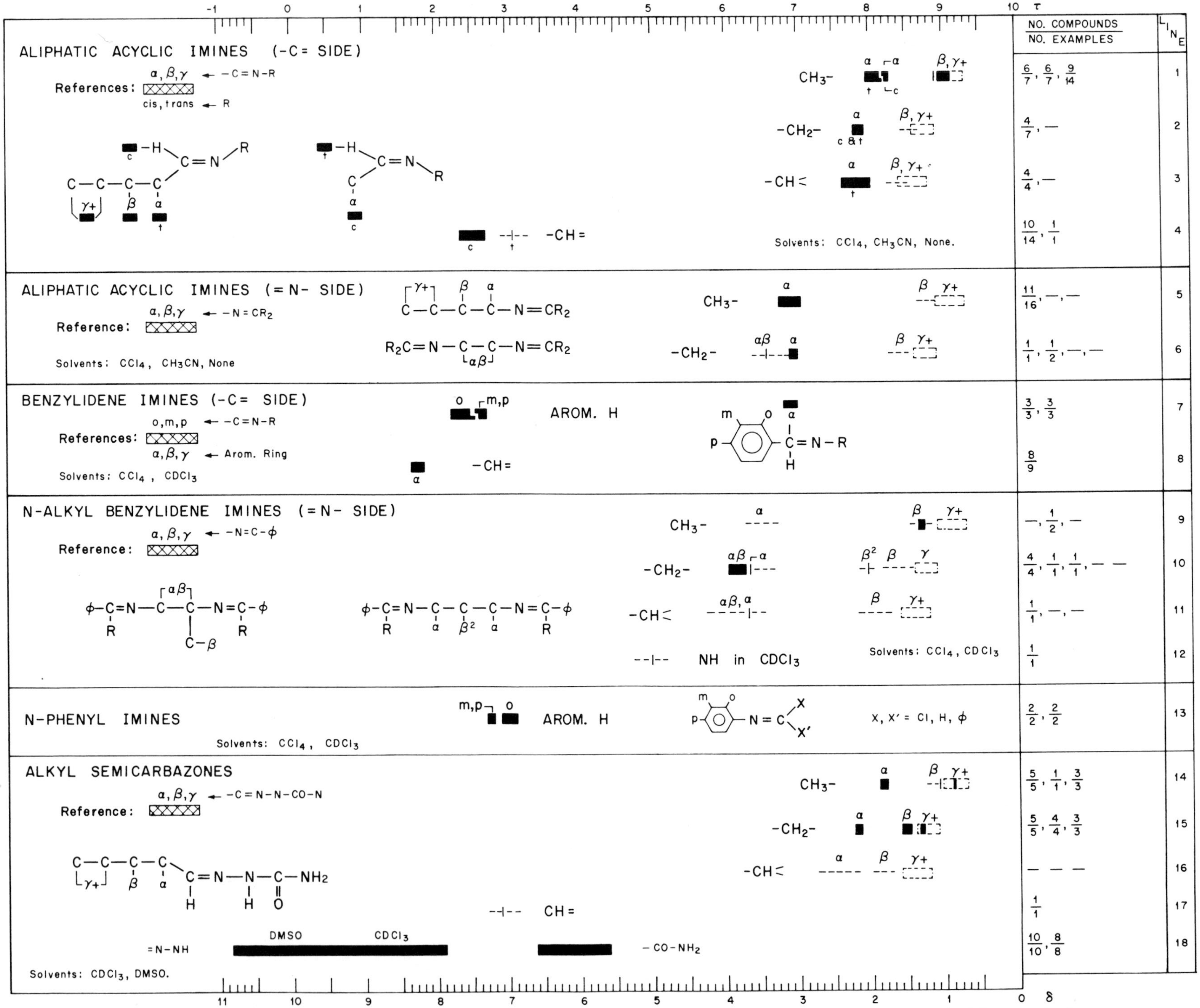

CHART 55. General H¹ NMR Chemical Shifts for Acyclic Imines and Semicarbazones

CHART 55

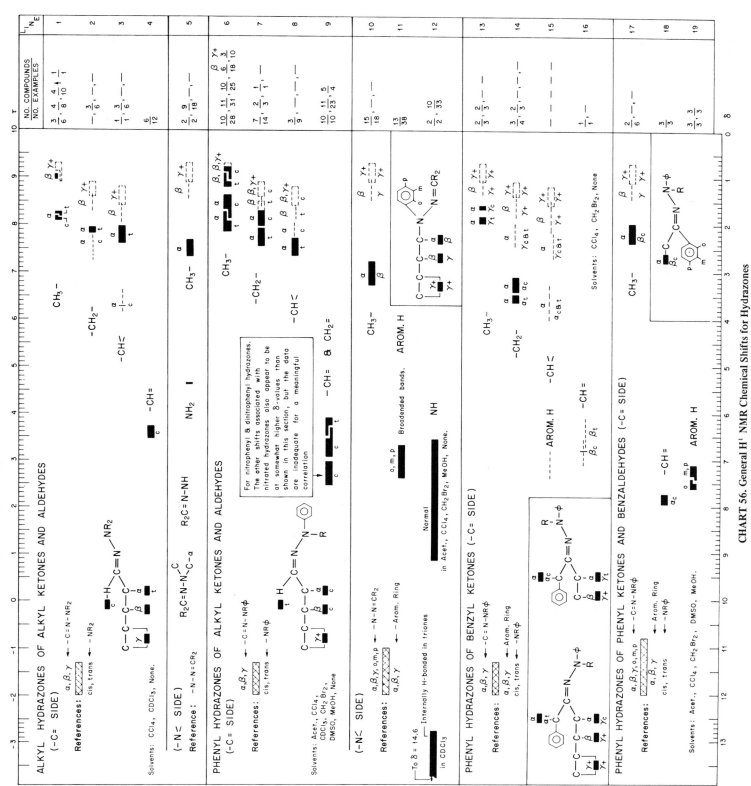

CHART 56. General H¹ NMR Chemical Shifts for Hydrazones

CHART 56

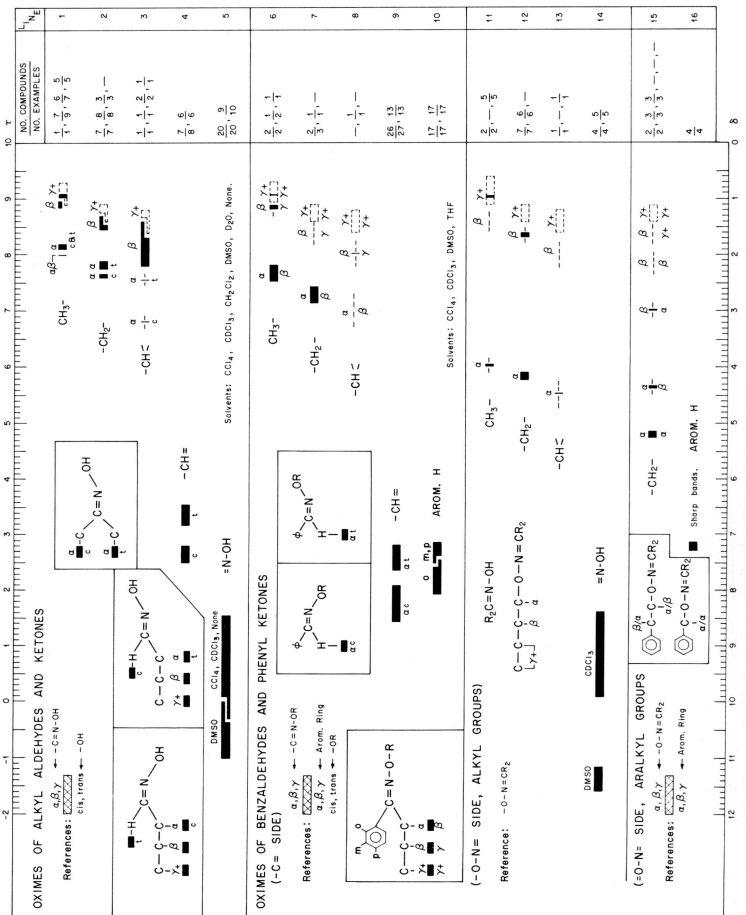

CHART 57. General H¹ NMR Chemical Shifts for Oximes

CHART 57

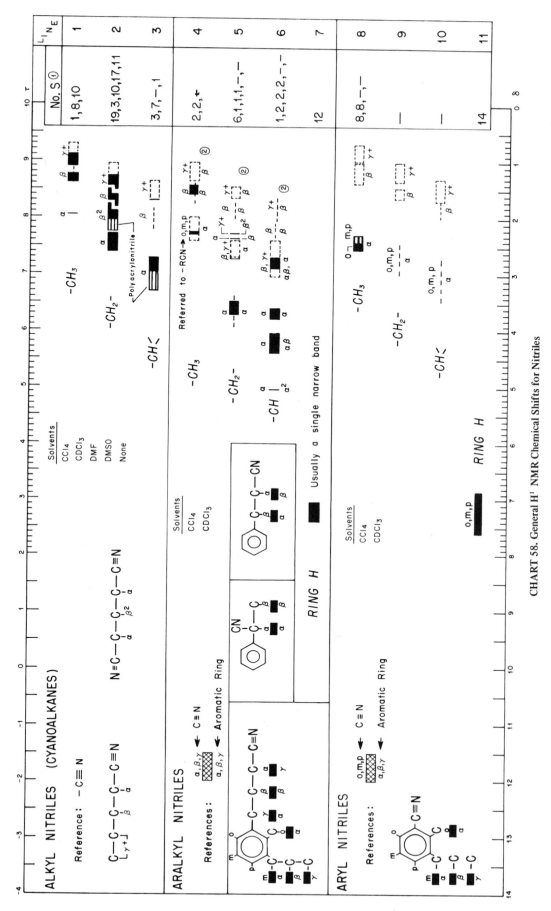

CHART 58. General H¹ NMR Chemical Shifts for Nitriles

1. Total number of compound–solvent combinations included in each solid bar.
2. Shifts of groups γ+/γ+ are shown by the γ+ bars for alkyl nitriles, above.

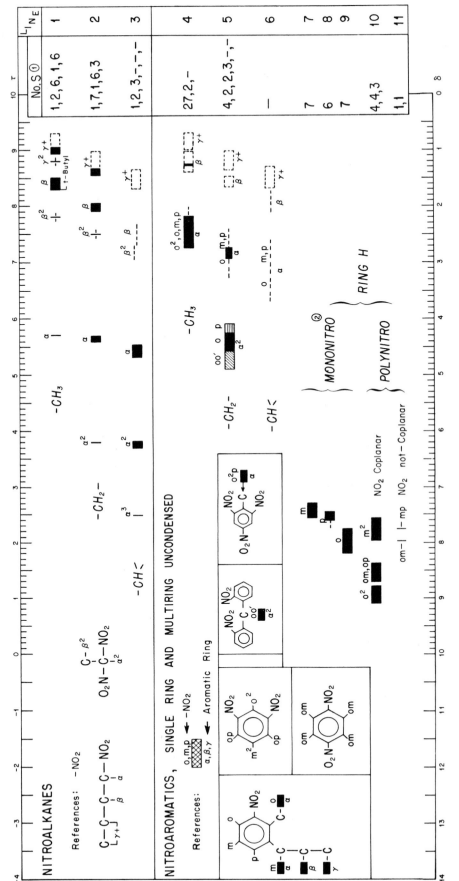

CHART 59. General H¹ NMR Chemical Shifts for Nitro Compounds

1. Total number of compound–solvent combinations included in each solid or crosshatched bar.
2. These data are for compounds in which the NO_2 group is coplanar with the ring. When the NO_2 is flanked by groups (CH_3, etc.) which force it out of the plane of the ring, most of its chemical shifting effect on ring hydrogens disappears.

CHART 59

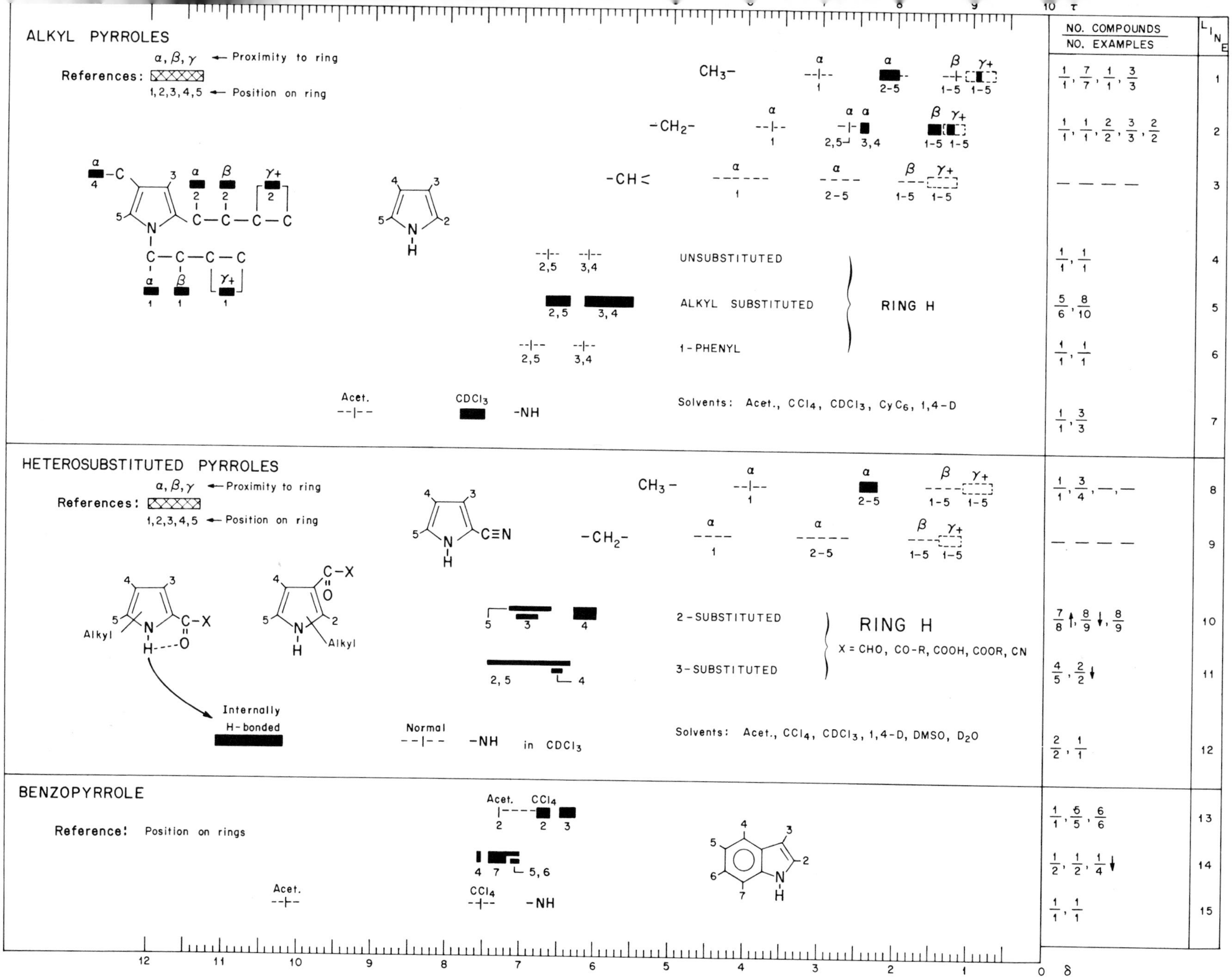

ALKYL PYRROLES

References:
α, β, γ — Proximity to ring
1,2,3,4,5 — Position on ring

CH₃—

—CH₂—

—CH≼

UNSUBSTITUTED

ALKYL SUBSTITUTED

1 - PHENYL

RING H

Solvents: Acet., CCl₄, CDCl₃, Cy C₆, 1,4 - D

Acet.

CDCl₃

—NH

HETEROSUBSTITUTED PYRROLES

References:
α, β, γ — Proximity to ring
1,2,3,4,5 — Position on ring

CH₃—

—CH₂—

2 - SUBSTITUTED

3 - SUBSTITUTED

RING H

X = CHO, CO-R, COOH, COOR, CN

Solvents: Acet., CCl₄, CDCl₃, 1,4 - D, DMSO, D₂O

Normal

—NH in CDCl₃

Internally
H — bonded

BENZOPYRROLE

Reference: Position on rings

Acet. CCl₄

CCl₄

Acet.

—NH

CHART 60. General H¹ NMR Chemical Shifts for Pyrroles

CHART 60

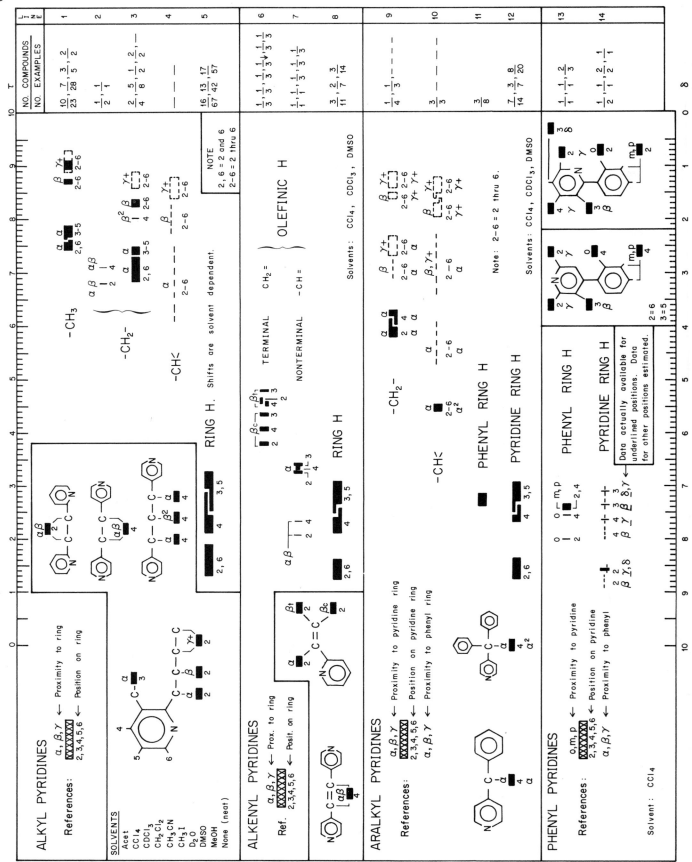

CHART 61. General H¹ NMR Chemical Shifts for Hydrocarbon Substituted Pyridines

CHART 61

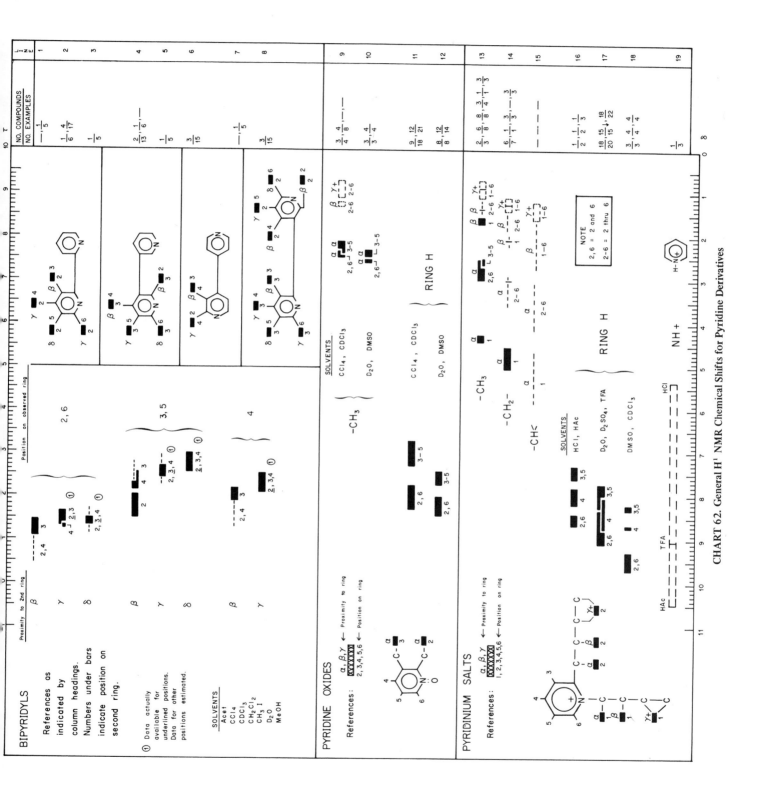

CHART 62. General H¹ NMR Chemical Shifts for Pyridine Derivatives

CHART 62

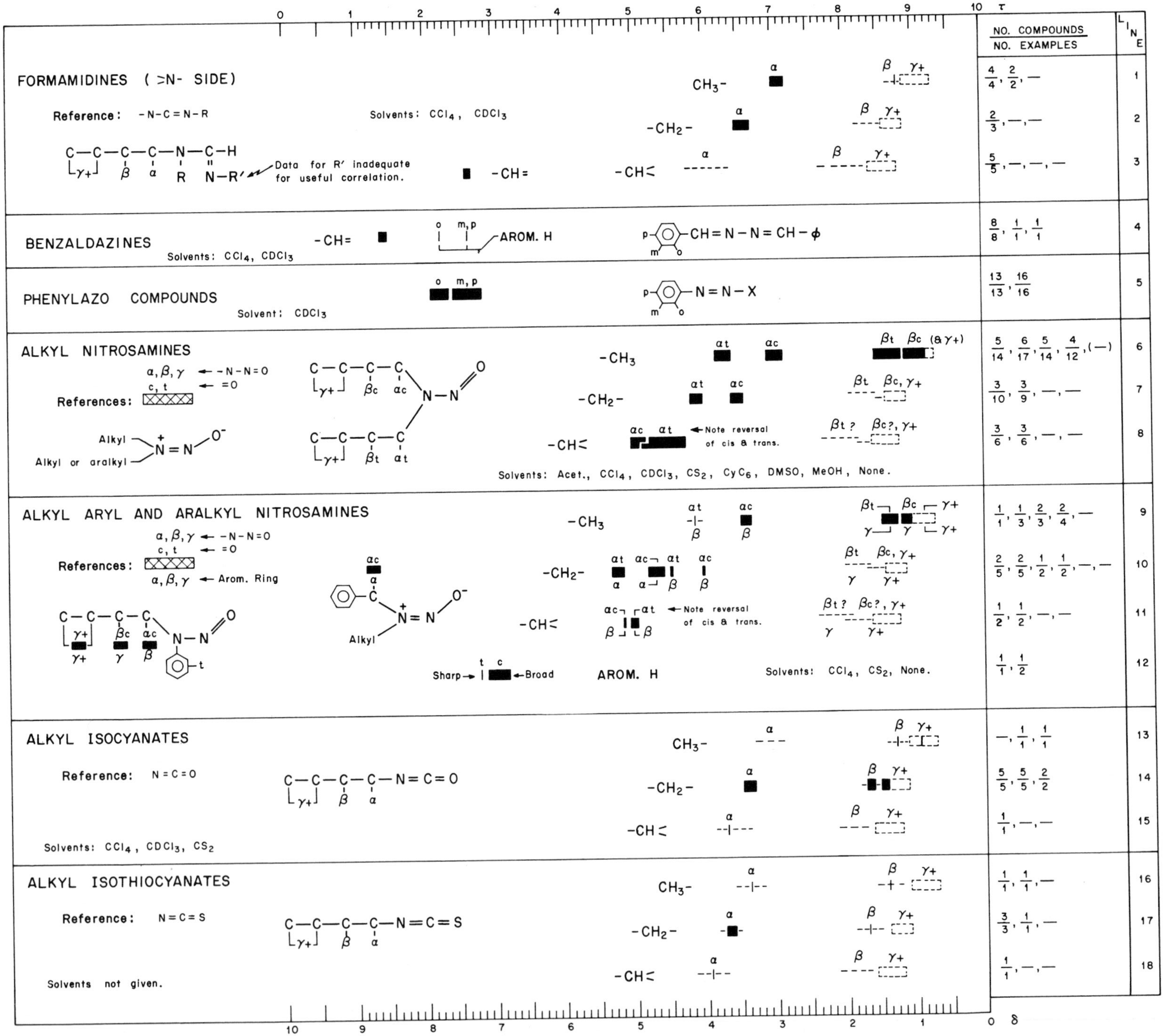

FORMAMIDINES (≥N- SIDE)

Reference: -N-C=N-R

Solvents: CCl₄, CDCl₃

$$C-C-C-C-N=C-N-C-H$$
$$\underset{\gamma+}{\bigsqcup}\ \beta \quad \alpha \quad R \quad N-R'$$
Data for R' inadequate for useful correlation.

BENZALDAZINES

Solvents: CCl₄, CDCl₃

-CH=

PHENYLAZO COMPOUNDS

Solvent: CDCl₃

ALKYL NITROSAMINES

α,β,γ ⟶ -N-N=O
c,t ⟶ =O

References:

$$C-C-C-C-\underset{\substack{\alpha c \\ \beta c}}{\overset{O}{N-N}}$$
$$\underset{\gamma+}{\bigsqcup}$$

Alkyl $\overset{+}{N}=\overset{-}{N}=\overset{-}{N}$

Alkyl or aralkyl

ALKYL ARYL AND ARALKYL NITROSAMINES

α,β,γ ⟶ -N-N=O
c,t ⟶ =O
α,β,γ ⟶ Arom. Ring

References:

$$C-C-C-C-\underset{\substack{\alpha c \\ \beta c \\ \gamma}}{\overset{O}{N-N}}$$
$$\underset{\gamma+}{\bigsqcup} \quad t$$

ALKYL ISOCYANATES

Reference: N=C=O

Solvents: CCl₄, CDCl₃, CS₂

$$C-C-C-C-N=C=O$$
$$\underset{\gamma+}{\bigsqcup}\ \beta \quad \alpha$$

ALKYL ISOTHIOCYANATES

Reference: N=C=S

Solvents not given.

$$C-C-C-C-N=C=S$$
$$\underset{\gamma+}{\bigsqcup}\ \beta \quad \alpha$$

CH₃−
−CH₂−
−CH≤
−CH=

o, m, p
AROM. H

−CH= ⌐AROM. H⌐

CH=N-N=CH-φ
−N=N-X
o
m,p

−CH₃
−CH₂−
−CH≤

Solvents: Acet., CCl₄, CDCl₃, CS₂, CyC₆, DMSO, MeOH, None.

Note reversal of cis & trans.

−CH₃
−CH₂−
−CH≤
AROM. H

Sharp → ← Broad

Solvents: CCl₄, CS₂, None.

Note reversal of cis & trans.

CH₃−
−CH₂−
−CH≤

CH₃−
−CH₂−
−CH≤

	NO. COMPOUNDS / NO. EXAMPLES	LINE
	4/4 , 2/2 , —	1
	2/3 , — , —	2
	5/5 , — , —	3
	8/8 , 1/1 , 1/1	4
	13/13 , 16/16	5
	5/14 , 6/17 , 5/14 , 4/12 , (—)	6
	3/10 , 3/9 , —	7
	3/6 , 3/6 , —	8
	1/1 , 2/3 , 3/4 , —	9
	2/5 , 2/2 , 1/2 , —	10
	1/2 , — , —	11
	1/1 , 1/2	12
	— , 1/1	13
	5/5 , 5/5 , 2/2	14
	1/1 , 1/1	15
	1/1 , 1/1	16
	3/3 , 1/1	17
	1/1 , 1/1	18

CHART 63. General H¹ NMR Chemical Shifts for Other Nitrogen Compounds

CHART 63

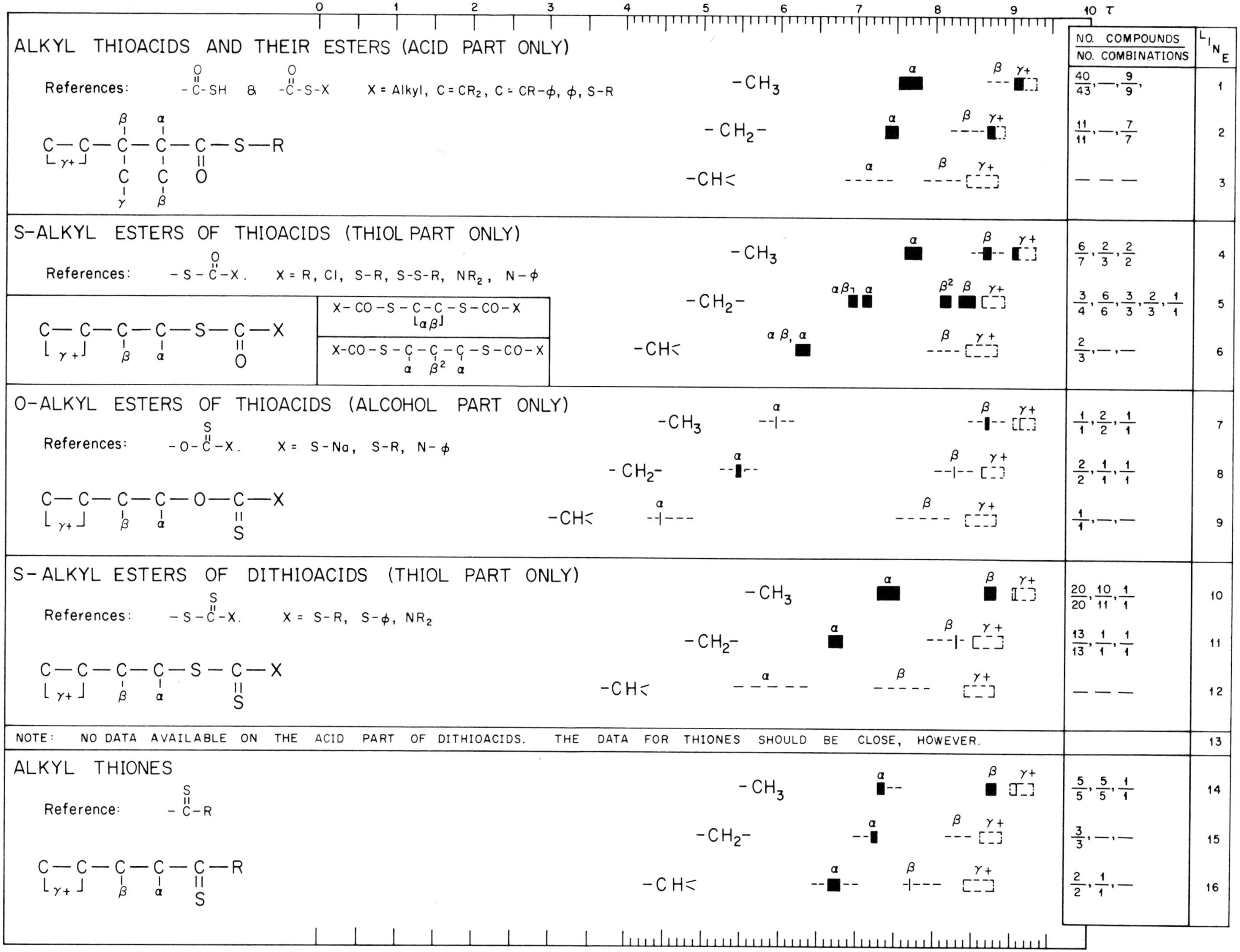

ALKYL THIOACIDS AND THEIR ESTERS (ACID PART ONLY)

References: $-\overset{O}{\overset{\|}{C}}-SH$ & $-\overset{O}{\overset{\|}{C}}-S-X$ X = Alkyl, C=CR$_2$, C=CR-ϕ, ϕ, S-R

$$\underset{\gamma+}{\underset{\lfloor}{C}}-\underset{\beta}{C}-\underset{\alpha}{C}-\underset{\gamma}{\overset{\|}{C}}-S-R$$

$-CH_3$

$-CH_2-$

$-CH<$

S-ALKYL ESTERS OF THIOACIDS (THIOL PART ONLY)

References: $-S-\overset{O}{\overset{\|}{C}}-X$.   X = R, Cl, S-R, S-S-R, NR$_2$, N-ϕ

$$X-CO-S-\underset{\alpha}{C}-\underset{\beta}{C}-S-CO-X$$

$$X-CO-S-\underset{\alpha}{C}-\underset{\beta^2}{C}-\underset{\beta}{C}-S-CO-X$$

$-CH_3$

$-CH_2-$

$-CH<$

O-ALKYL ESTERS OF THIOACIDS (ALCOHOL PART ONLY)

References: $-O-\overset{S}{\overset{\|}{C}}-X$.   X = S-Na, S-R, N-$\phi$

$$\underset{\gamma+}{\underset{\lfloor}{C}}-\underset{\beta}{C}-\underset{\alpha}{C}-O-\overset{S}{\overset{\|}{C}}-X$$

$-CH_3$

$-CH_2-$

$-CH<$

S-ALKYL ESTERS OF DITHIOACIDS (THIOL PART ONLY)

References: $-S-\overset{S}{\overset{\|}{C}}-X$.   X = S-R, S-$\phi$, NR$_2$

$$\underset{\gamma+}{\underset{\lfloor}{C}}-\underset{\beta}{C}-\underset{\alpha}{C}-S-\overset{S}{\overset{\|}{C}}-X$$

$-CH_3$

$-CH_2-$

$-CH<$

NOTE: NO DATA AVAILABLE ON THE ACID PART OF DITHIOACIDS. THE DATA FOR THIONES SHOULD BE CLOSE, HOWEVER.

ALKYL THIONES

Reference: $-\overset{S}{\overset{\|}{C}}-R$

$$\underset{\gamma+}{\underset{\lfloor}{C}}-\underset{\beta}{C}-\underset{\alpha}{C}-\overset{S}{\overset{\|}{C}}-R$$

$-CH_3$

$-CH_2-$

$-CH<$

LINE	NO. COMPOUNDS / NO. COMBINATIONS
1	40/43, 9/9
2	11/11, 7/7
3	—/—, —/—
4	6/7, 2/3, 2/2
5	3/4, 6/6, 3/3, 2/3, 1/1
6	2/3, —/—
7	1/1, 2/2, 1/1
8	2/2, 1/1, 1/1
9	1/1, —/—, —/—
10	20/20, 10/11, 1/1
11	13/13, 1/1, 1/1
12	—/—, —/—, —/—
13	
14	5/5, 5/5, 1/1
15	3/3, —/—, —/—
16	2/2, 1/1, —/—

CHART 64. General H¹ NMR Chemical Shifts for Alkyl Thioacids, Thioesters and Thiones

CHART 64

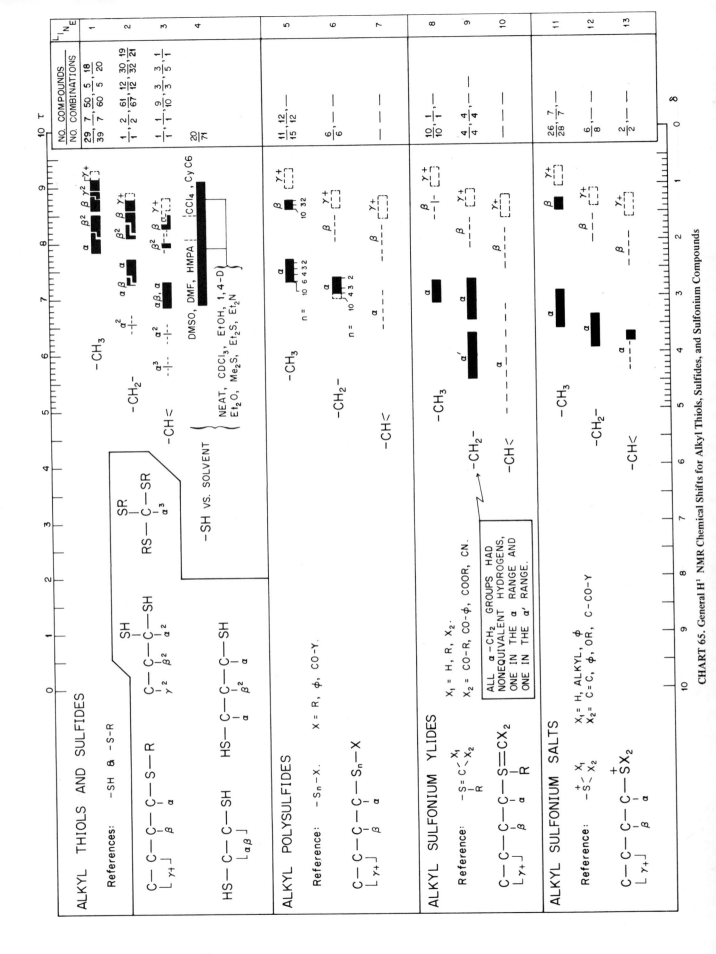

CHART 65. General H¹ NMR Chemical Shifts for Alkyl Thiols, Sulfides, and Sulfonium Compounds

CHART 65

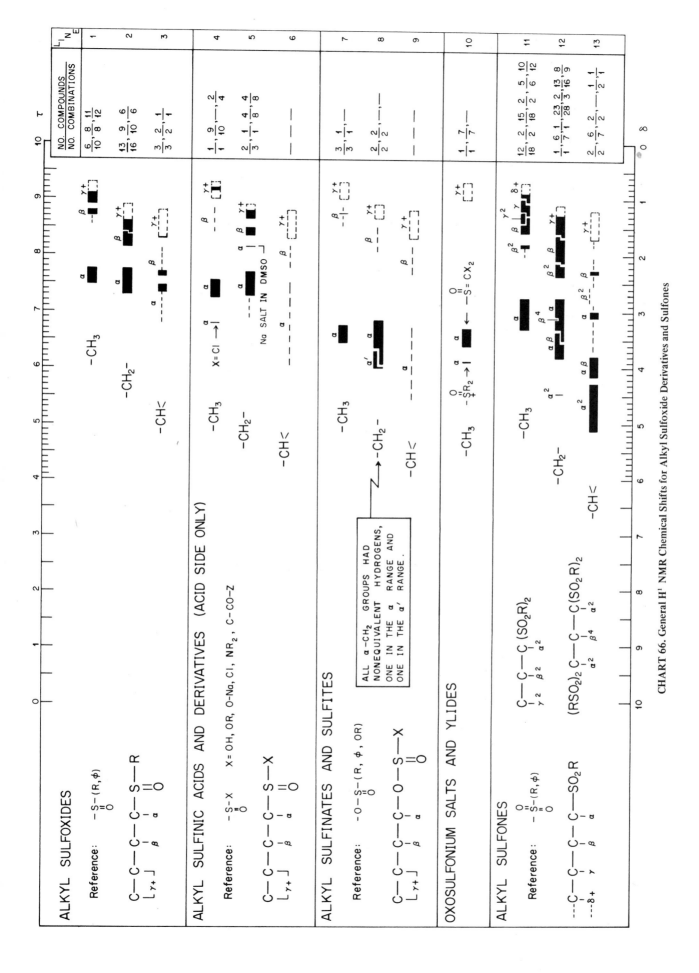

CHART 66. General H¹ NMR Chemical Shifts for Alkyl Sulfoxide Derivatives and Sulfones

CHART 66

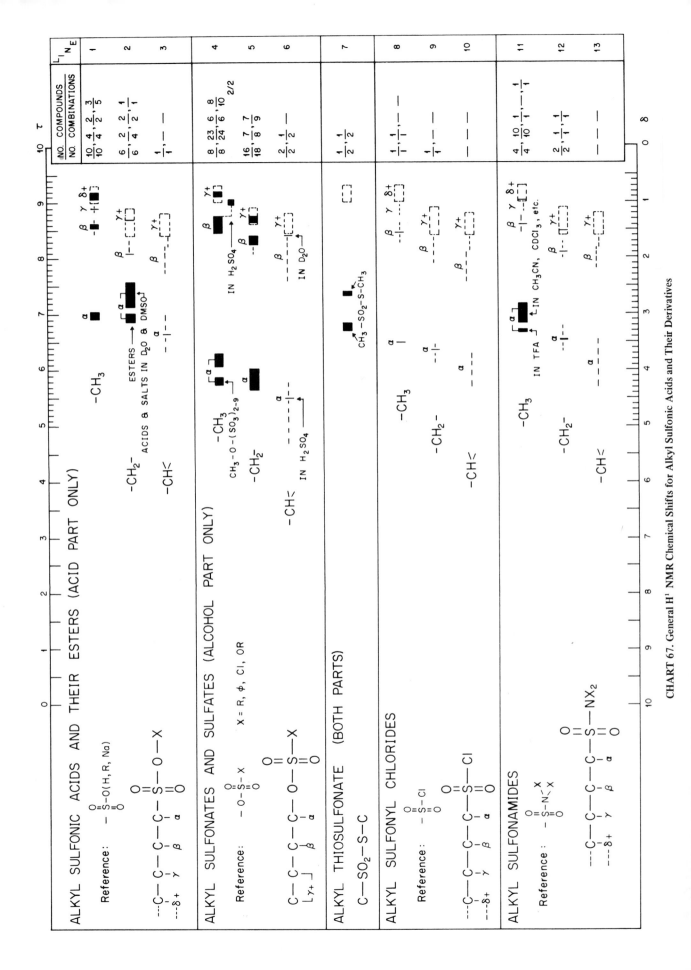

CHART 67. General H¹ NMR Chemical Shifts for Alkyl Sulfonic Acids and Their Derivatives

CHART 67

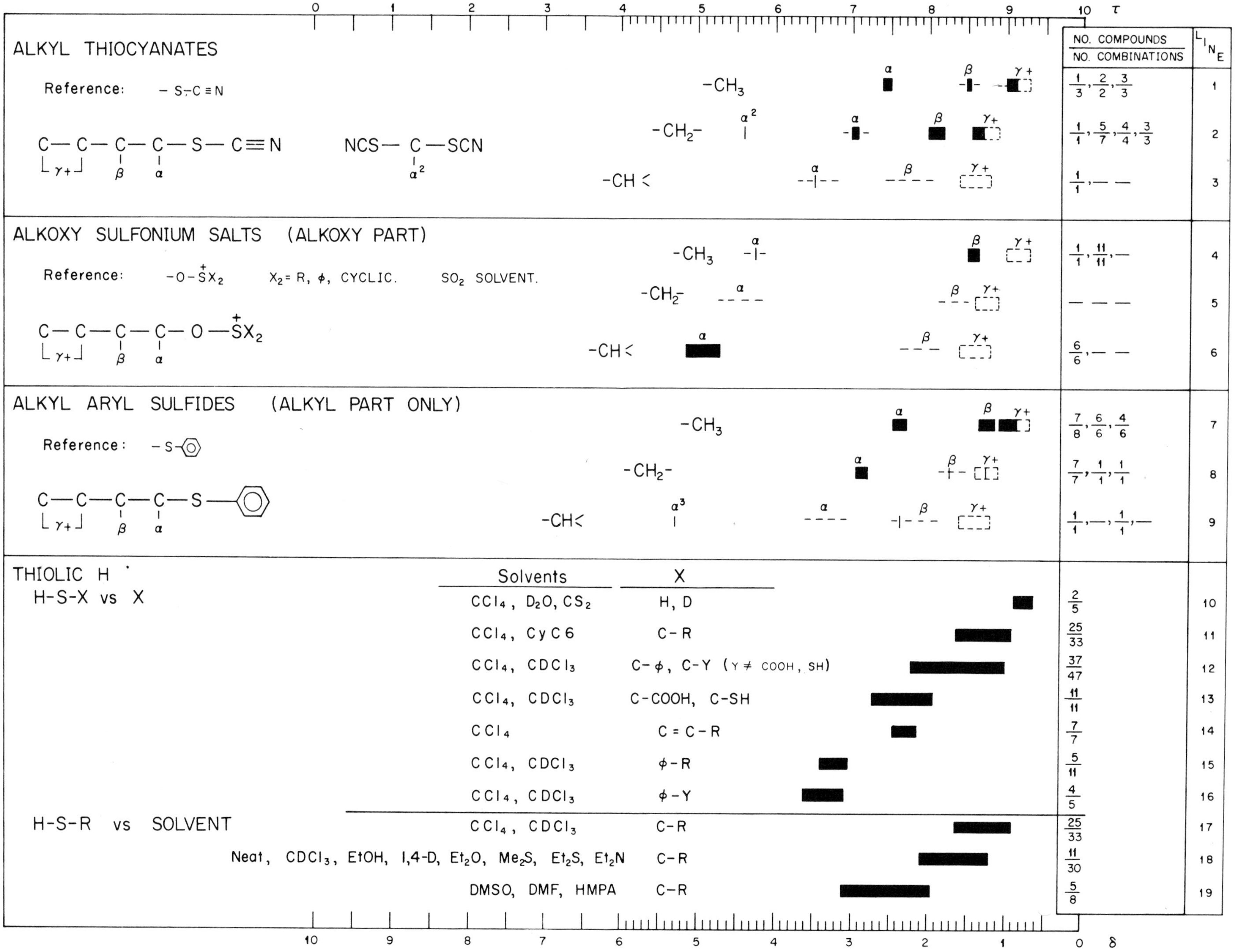

ALKYL THIOCYANATES

Reference: $-S-C\equiv N$

$\underset{\gamma+}{C}-\underset{\beta}{C}-\underset{\alpha}{C}-S-C\equiv N$

$NCS-\underset{\alpha^2}{C}-SCN$

$-CH_3$

$-CH_2-$

$-CH<$

ALKOXY SULFONIUM SALTS (ALKOXY PART)

Reference: $-O-\overset{+}{S}X_2$ $X_2 = R, \phi$, CYCLIC. SO_2 SOLVENT.

$\underset{\gamma+}{C}-\underset{\beta}{C}-\underset{\alpha}{C}-O-\overset{+}{S}X_2$

$-CH_3$

$-CH_2-$

$-CH<$

ALKYL ARYL SULFIDES (ALKYL PART ONLY)

Reference: $-S-\bigcirc$

$\underset{\gamma+}{C}-\underset{\beta}{C}-\underset{\alpha}{C}-S-\bigcirc$

$-CH_3$

$-CH_2-$

$-CH<$

THIOLIC H

H—S—X vs X

Solvents	X
CCl_4, D_2O, CS_2	H, D
CCl_4, $Cy\,C6$	C—R
CCl_4, $CDCl_3$	$C-\phi$, C—Y (Y $\neq$ COOH, SH)
CCl_4, $CDCl_3$	C—COOH, C—SH
CCl_4	C = C—R
CCl_4, $CDCl_3$	ϕ—R
CCl_4, $CDCl_3$	ϕ—Y

H—S—R vs SOLVENT

CCl_4, $CDCl_3$	C—R
Neat, $CDCl_3$, EtOH, 1,4-D, Et_2O, Me_2S, Et_2S, Et_2N	C—R
DMSO, DMF, HMPA	C—R

LINE	NO. COMPOUNDS / NO. COMBINATIONS
1	$\frac{1}{3}, \frac{2}{2}, \frac{3}{3}$
2	$\frac{1}{1}, \frac{5}{7}, \frac{4}{4}, \frac{3}{3}$
3	$\frac{1}{1}, \frac{-}{-}$
4	$\frac{1}{1}, \frac{11}{11}$
5	$\frac{-}{-}, \frac{-}{-}$
6	$\frac{6}{6}, \frac{-}{-}$
7	$\frac{7}{8}, \frac{6}{6}, \frac{4}{6}$
8	$\frac{7}{7}, \frac{1}{1}, \frac{1}{1}$
9	$\frac{1}{1}, \frac{-}{-}, \frac{-}{-}$
10	$\frac{2}{5}$
11	$\frac{25}{33}$
12	$\frac{37}{47}$
13	$\frac{11}{11}$
14	$\frac{7}{7}$
15	$\frac{5}{11}$
16	$\frac{4}{5}$
17	$\frac{25}{33}$
18	$\frac{11}{30}$
19	$\frac{5}{8}$

CHART 68. General H¹ NMR Chemical Shifts for Other Alkyl Sulfur Compounds and Thiolic H

CHART 68

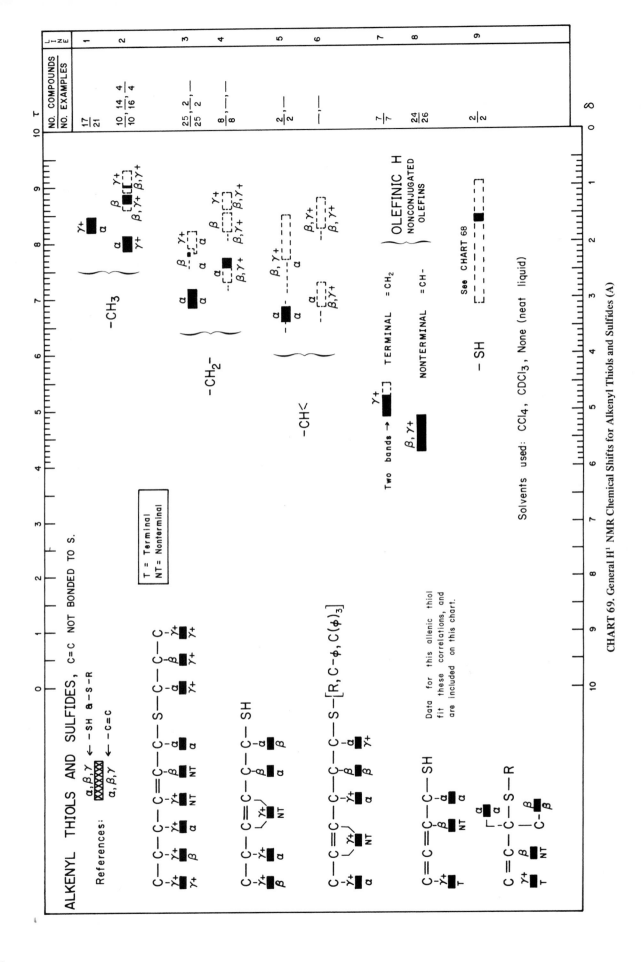

CHART 69. General H¹ NMR Chemical Shifts for Alkenyl Thiols and Sulfides (A)

CHART 69

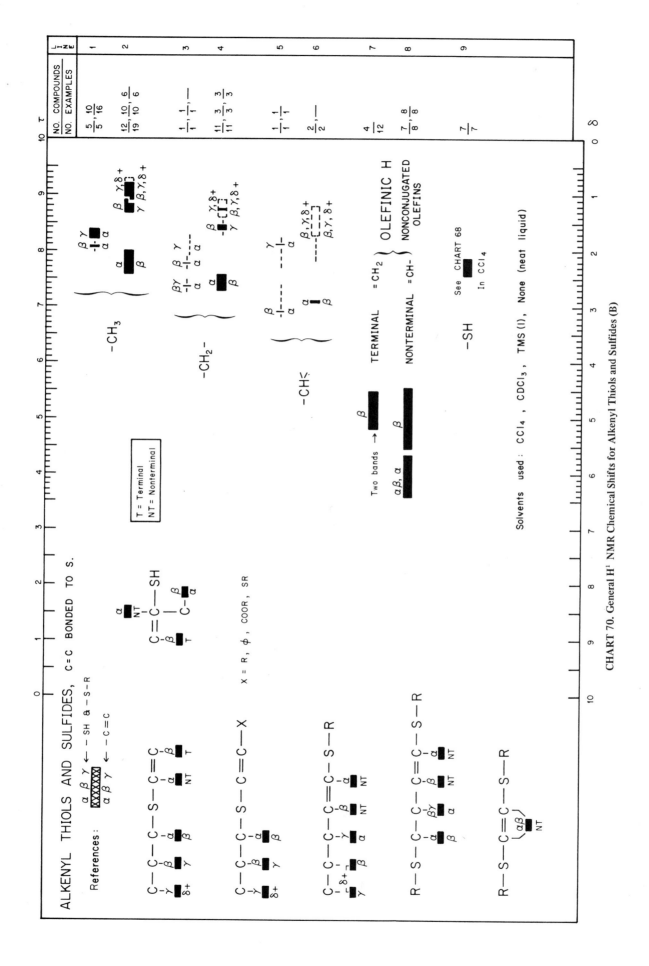

CHART 70. General H¹ NMR Chemical Shifts for Alkenyl Thiols and Sulfides (B)

CHART 70

270

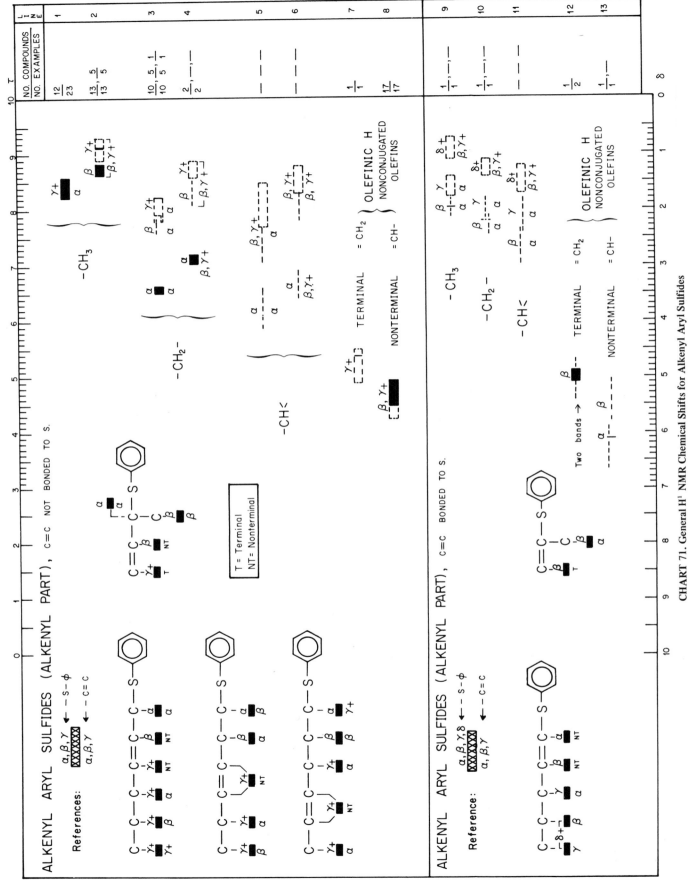

CHART 71. General H¹ NMR Chemical Shifts for Alkenyl Aryl Sulfides

Solvents used: CCl₄, None (neat liquid).

CHART 71

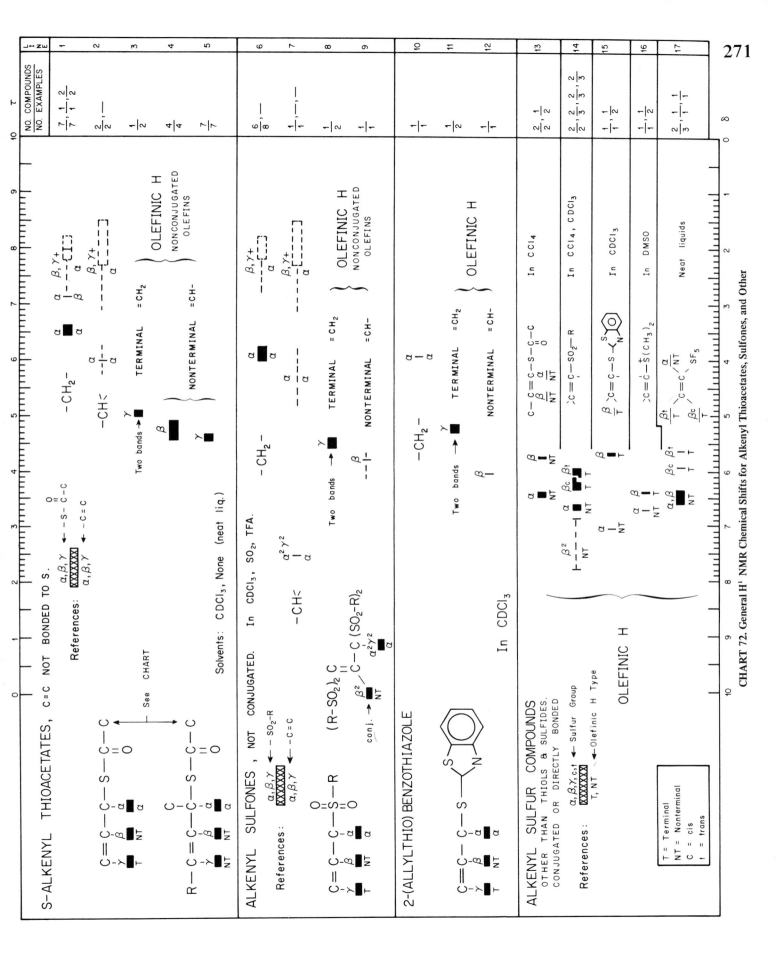

CHART 72. General H¹ NMR Chemical Shifts for Alkenyl Thioacetates, Sulfones, and Other

271

CHART 72

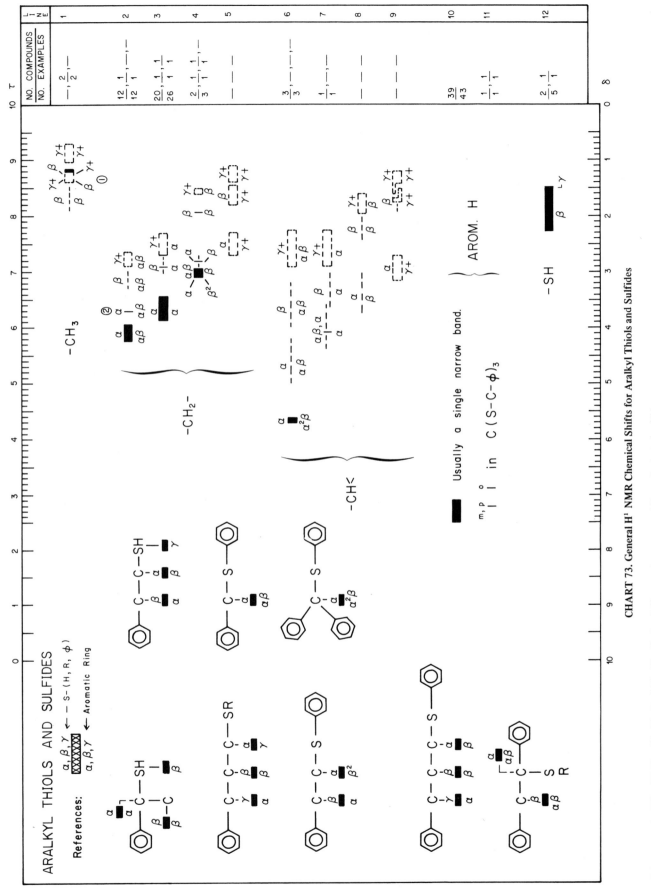

CHART 73. General H¹ NMR Chemical Shifts for Aralkyl Thiols and Sulfides

Solvents used: CCl₄, CDCl₃, CS₂ (1), D₂O(1). The D₂O soluble sample was an amidine substituted sulfide.
1. Methyls γ+/β have the same shift range as those β/γ+.
2. In C—⌬—S—C—⌬—NH—C. Apparently shifted as result of steric conformation rather than by NH—C substituent.

CHART 73

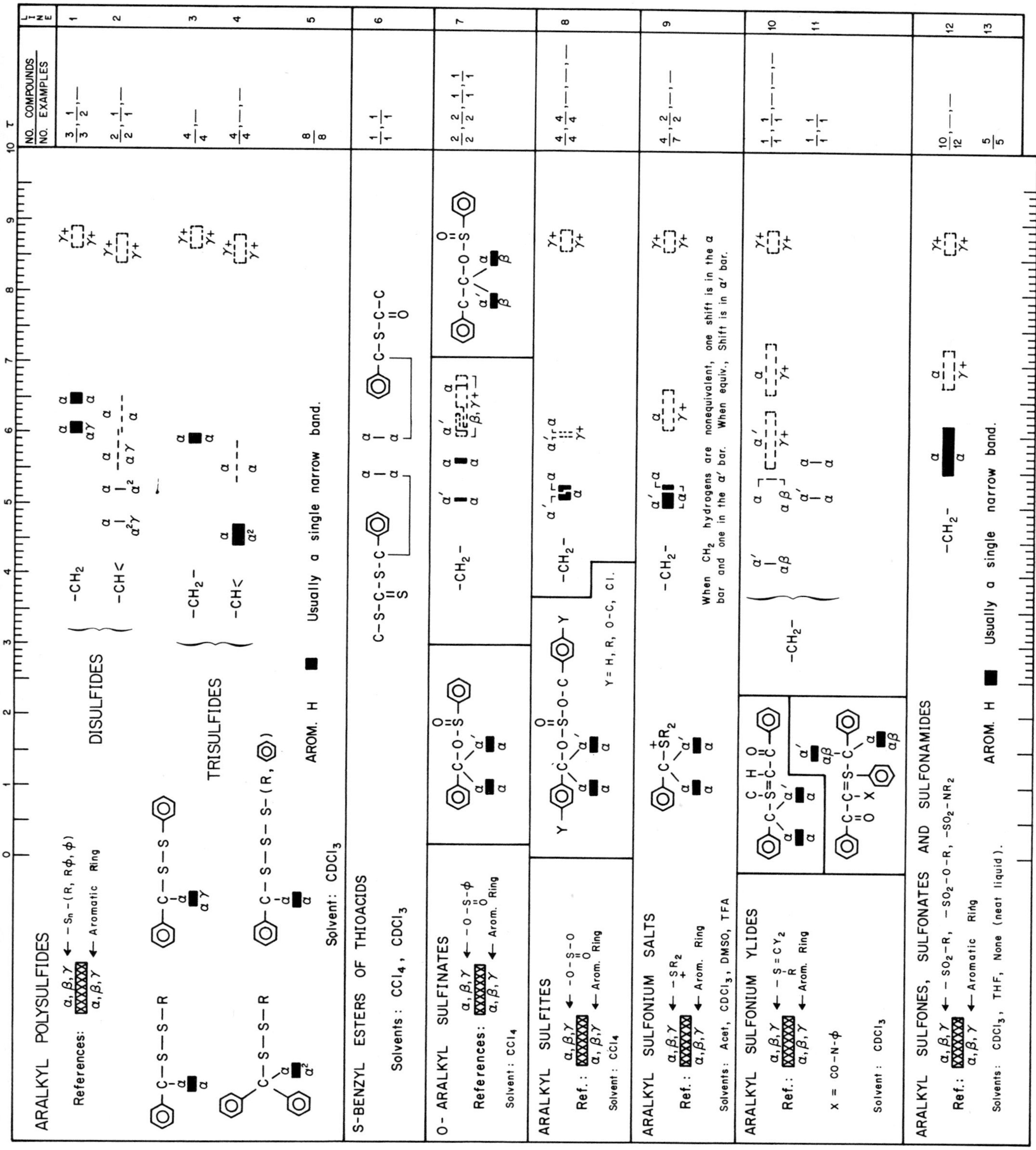

ARALKYL POLYSULFIDES

References: α,β,γ ◼ ← -Sₙ-(R, Rφ, φ) ; α,β,γ --- ← Aromatic Ring

DISULFIDES

TRISULFIDES

Solvent: CDCl₃

S-BENZYL ESTERS OF THIOACIDS

Solvents: CCl₄, CDCl₃

O-ARALKYL SULFINATES

References: α,β,γ ⬚ ← -O-S-φ (‖O) ; α,β,γ --- ← Arom. Ring

Solvent: CCl₄

Y = H, R, O-C, Cl.

ARALKYL SULFITES

Ref.: α,β,γ ⬚ ← -O-S=O ; α,β,γ --- ← Arom. Ring

Solvent: CCl₄

ARALKYL SULFONIUM SALTS

Ref.: α,β,γ ⬚ ← -SR₂⁺ ; α,β,γ --- ← Arom. Ring

Solvents: Acet, CDCl₃, DMSO, TFA

When CH₂ hydrogens are nonequivalent, one shift is in the α bar and one in the α′ bar. When equiv., shift is in the α′ bar.

ARALKYL SULFONIUM YLIDES

Ref.: α,β,γ ⬚ ← -S=CY₂ ; α,β,γ --- ← Aromatic Ring

x = CO-N-φ

Solvent: CDCl₃

ARALKYL SULFONES, SULFONATES AND SULFONAMIDES

Ref.: α,β,γ ⬚ ← -SO₂-R, -SO₂-O-R, -SO₂-NR₂ ; α,β,γ --- ← Aromatic Ring

Solvents: CDCl₃, THF, None (neat liquid).

AROM. H ◼ Usually a single narrow band.

CHART 74. General H¹ NMR Chemical Shifts for Aralkyl Sulfur Compounds Except Thiols and Sulfides

CHART 74

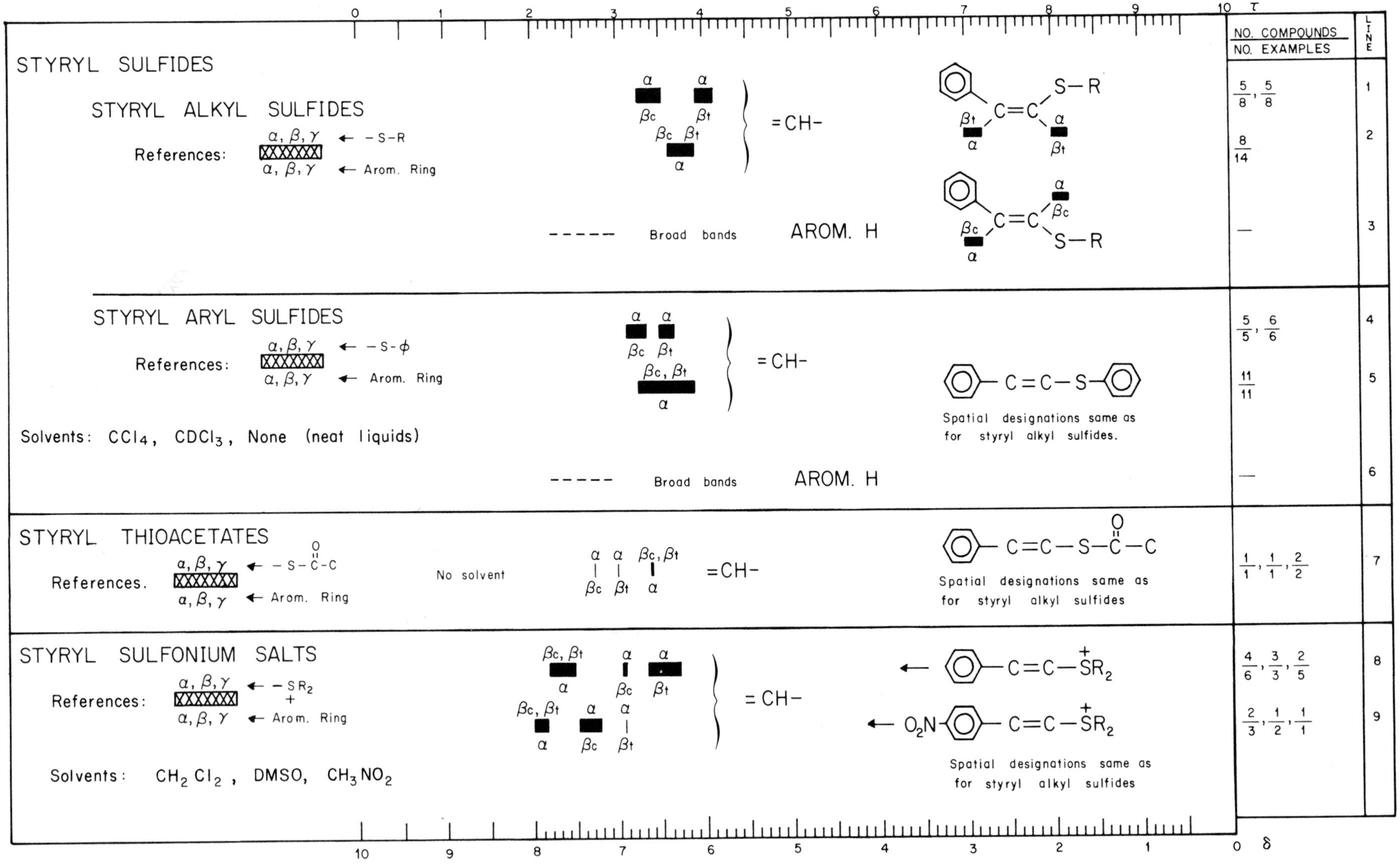

STYRYL SULFIDES

STYRYL ALKYL SULFIDES

References:
α,β,γ ← –S–R
α,β,γ ← Arom. Ring

Solvents: CCl₄, CDCl₃, None (neat liquids)

STYRYL ARYL SULFIDES

References:
α,β,γ ← –S–φ
α,β,γ ← Arom. Ring

STYRYL THIOACETATES

References:
α,β,γ ← –S–C–C (O)
α,β,γ ← Arom. Ring

No solvent

STYRYL SULFONIUM SALTS

References:
α,β,γ ← ⁺SR₂
α,β,γ ← Arom. Ring

Solvents: CH₂Cl₂, DMSO, CH₃NO₂

Spatial designations same as for styryl alkyl sulfides.

Spatial designations same as for styryl alkyl sulfides

Spatial designations same as for styryl alkyl sulfides

LINE	NO. COMPOUNDS / NO. EXAMPLES
1	5/8 , 5/8
2	8/14
3	—
4	5/5 , 6/6
5	11/11
6	—
7	1/1 , 1/1 , 2/2
8	4/6 , 3/3 , 5/5
9	2/3 , 1/2 , 1/1

=CH–

AROM. H

Broad bands

=CH–

AROM. H

Broad bands

=CH–

=CH–

CHART 75. General H¹ NMR Chemical Shifts for Styryl Sulfides, Thioacetates, and Sulfonium Salts

CHART 75

275

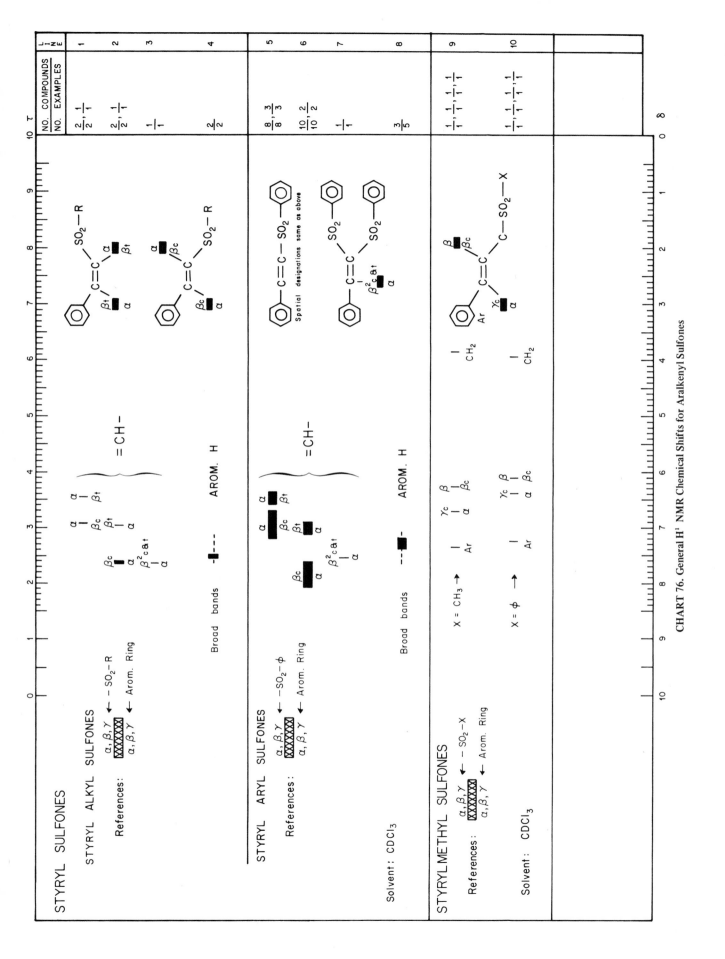

CHART 76. General H¹ NMR Chemical Shifts for Aralkenyl Sulfones

CHART 76

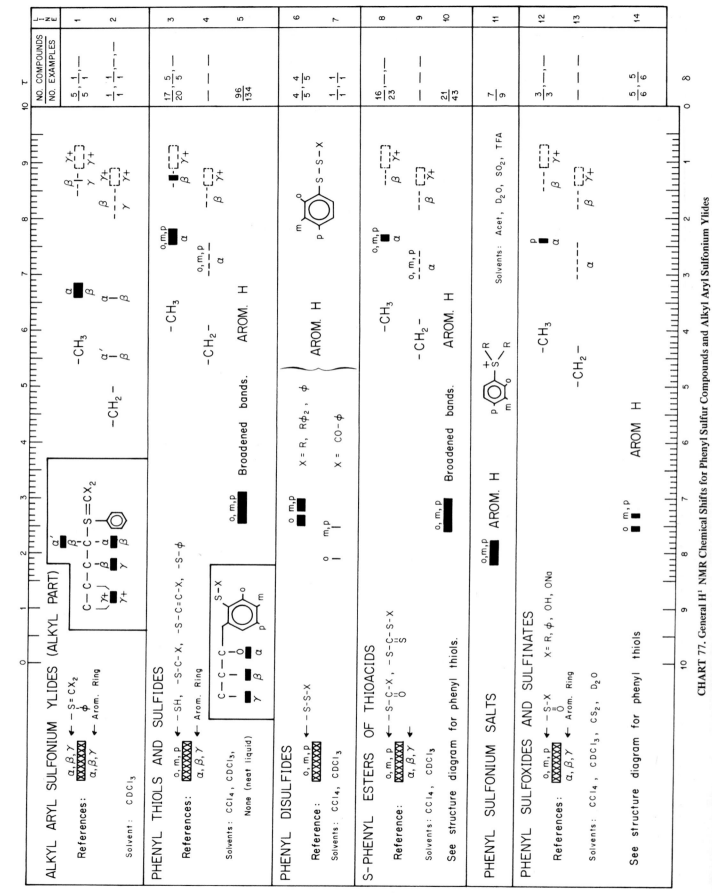

CHART 77. General H^1 NMR Chemical Shifts for Phenyl Sulfur Compounds and Alkyl Aryl Sulfonium Ylides

CHART 77

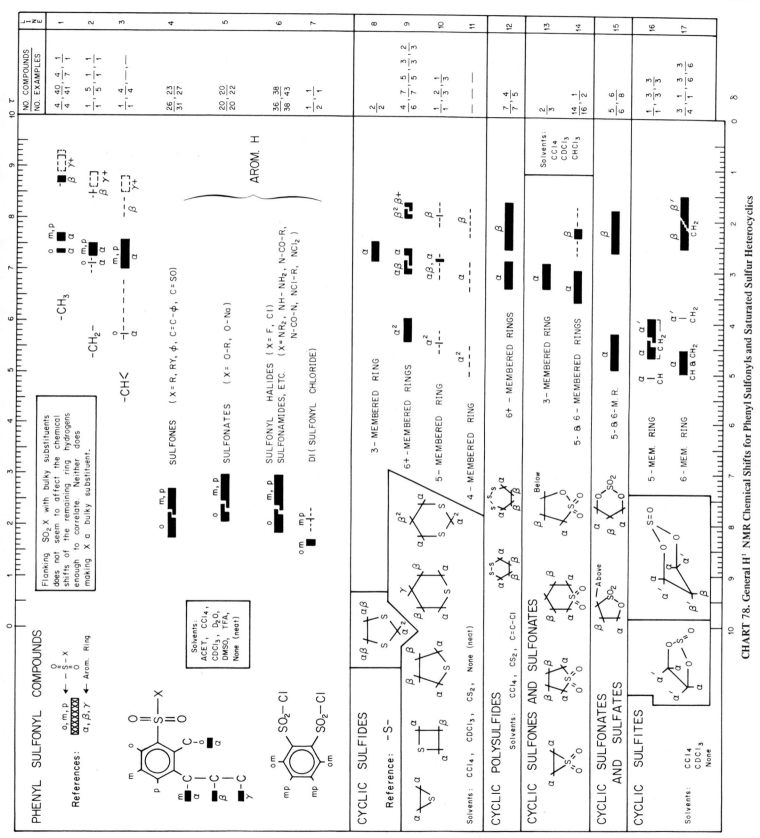

CHART 78. General H¹ NMR Chemical Shifts for Phenyl Sulfonyls and Saturated Sulfur Heterocyclics

CHART 78

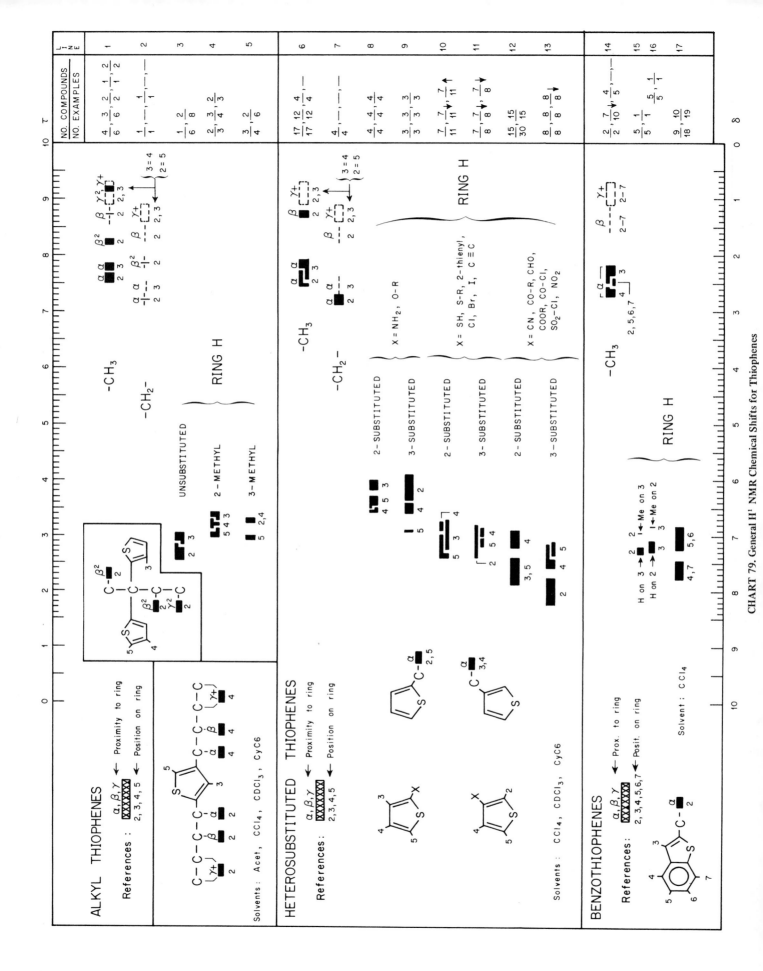

CHART 79. General H¹ NMR Chemical Shifts for Thiophenes

CHART 79

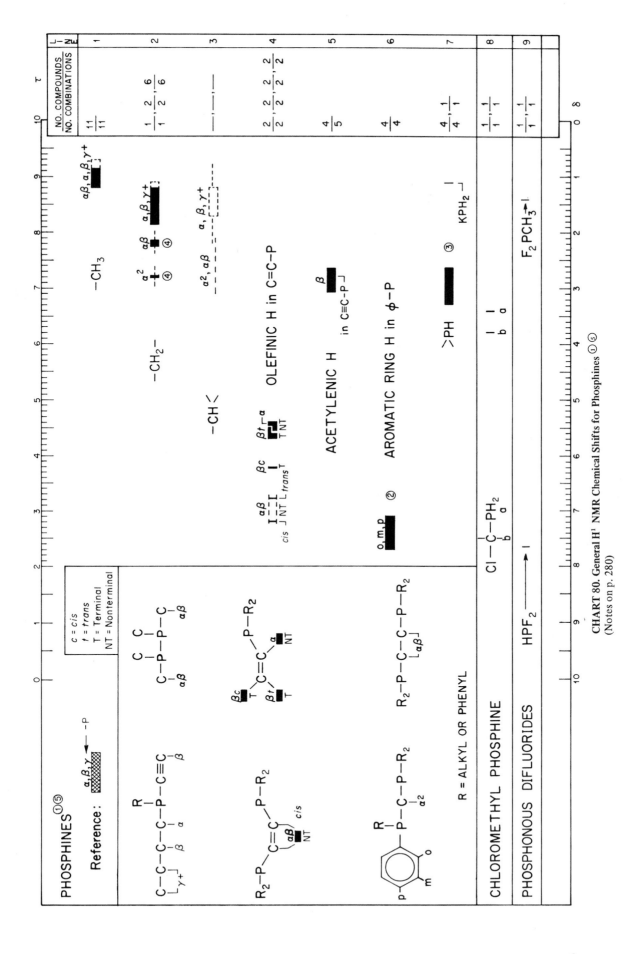

CHART 80. General H¹ NMR Chemical Shifts for Phosphines ①⑤
(Notes on p. 280)

CHART 80

Notes for Chart 80–Phosphines

1. This chart simply presents the few data available for phosphines. A more detailed correlation is not possible at this time.
2. Relative shifts of ring hydrogens are small. They could be the result of anistropy of adjacent structures rather than the effect of the phosphorus atom.
3. Centers of **PH** doublets. Because J_{PH} is 137 to 192 Hz the individual peaks of each doublet are far removed from this center.
4. The $\alpha\beta$ bar includes data for both $H_2P-C-C-PH_2$ and $\Phi_2P-C-C-P\Phi_2$. In this case the effect of the phenyls on the chemical shift is negligible. The α^2 bar is only for $\Phi_2P-C-P\Phi_2$. It is not known whether the effect of the phenyls is negligible in this case or not.
5. The chemical shifts associated with many alkyl phosphines are so small their spectra are not readily distinguishable from those of paraffins at 40 MHz and 60 MHz. The appearance of a sizable shoulder or peak on the low field side of the methylene band may indicate a phosphine.

Notes for Chart 81–Phosphine Oxides

1. Because of the scarcity of data, shifts for all types of phosphine oxides have been included on this chart.
2. When aromatic rings contribute significantly to the shifts, symbols below the data boxes show the distance of the specified group from the ring. When no lower symbol is used, aromatic rings are either absent or too far removed to affect the chemical shift.
3. Data are for the *cis* and *trans* isomers of this compound in acetic acid solvent.
4. Data were complete for only one compound, triphenylphosphine oxide. These data, which are plotted as short spikes above the main bar, indicate that ring hydrogens ortho to P=O resonate at slightly lower field than those meta and para to P=O. J_{PH} for the ortho hydrogens appears to be about 15 cps, which is in line with literature values.
5. The chemical shifts of the ring hydrogens of $\Phi_3P=N-R$ and $\Phi_3P=N-N=CH_2$ also fall within the block for aromatic ring hydrogen on this chart.
6. This shift, for Me_3PO, is questionable.

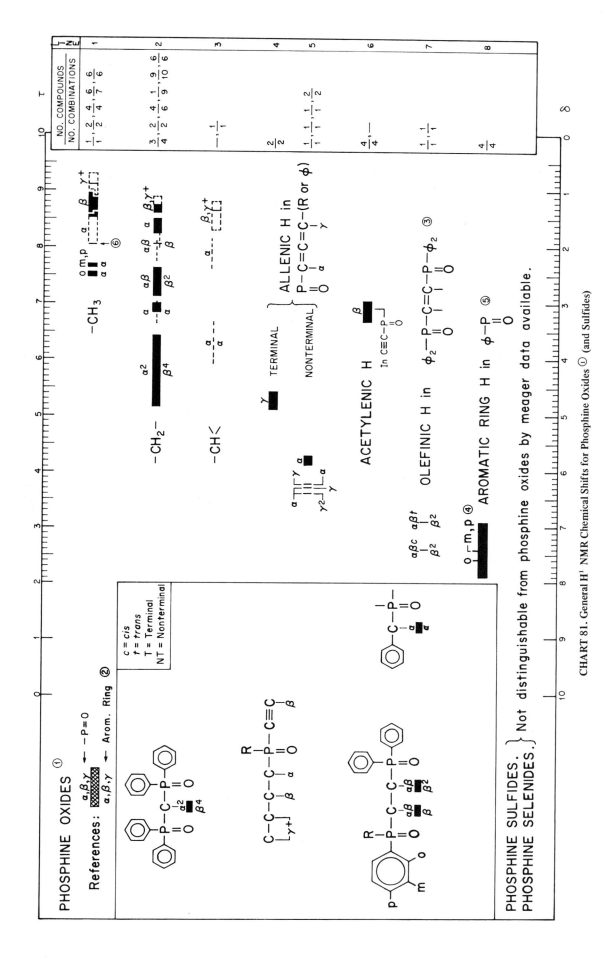

CHART 81. General H¹ NMR Chemical Shifts for Phosphine Oxides ① (and Sulfides)

281

CHART 81

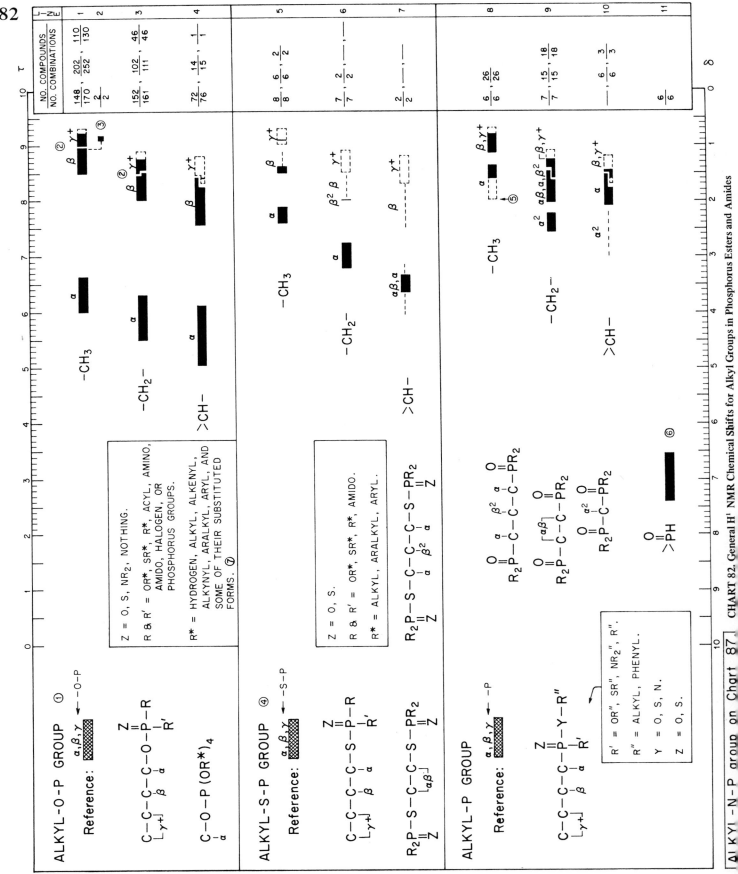

CHART 82. General H¹ NMR Chemical Shifts for Alkyl Groups in Phosphorus Esters and Amides

△ I KYI −N−P group on Chart 87

CHART 82

Notes for Chart 82—Alkyl Groups in Phosphorus Esters and Amides

1. Includes data for the alkoxy portions of alkyl, alkyl aralkyl, and alkyl aryl esters of all the acids of phosphorus and the oxyphosphoranes. Also includes the alkoxy portion of alkoxy cyclic esters. Scatter in the available data does not permit distinction among the esters of the various acids from chemical shift alone.

2. In aryl dialkyl phosphates, phosphonates, and phosphinates the two alkyl groups are generally observably nonequivalent at the β and farther positions. The methyls of individual isopropyl groups are also generally observably nonequivalent in such compounds. The difference in chemical shift between the nonequivalent groups is 0.1 to 0.2 ppm. The alpha groups are not observably nonequivalent, but are spin-coupled to the phosphorus. Also true for some aralkyl dialkyl phosphonates in which alkyl groups are bulky.

3. The resonance of one pair of the isopropyl methyls in diisopropyl diphenylmethylphosphonate and in diisopropyl phthalidylphosphonate appeared to be at this position. This is the highest field shift observed for methyls beta to $-$OP. It is not included in the data block for β methyls because it is a unique case. The observation is also open to some question.

4. Includes data for the alkylthio portions portions of $(RO)_2-P(O)-S-$alkyl, $(RO)_2-P(S)-S-$alkyl, $R-P(S)-(S-$alkyl)$_2$, and $R_2-P(S)-S-$alkyl. Chemical shifts of alkylthio groups in esters of other phosphorus thioacids should also fall in or close to these data blocks.

5. In $CH_3-P-[O-\bigcirc-SO_2-NH-NH_2]_2$

6. Centers of **PH** doublets. Because J_{PH} is 528 to 745 Hz, observation of both peaks of one of these doublets in a single scan including TMS requires scan widths of 700 to 800 Hz at 60 MHz and 1000 to 1100 Hz at 100 MHz.

7. Includes cyclic groups as well as acyclic.

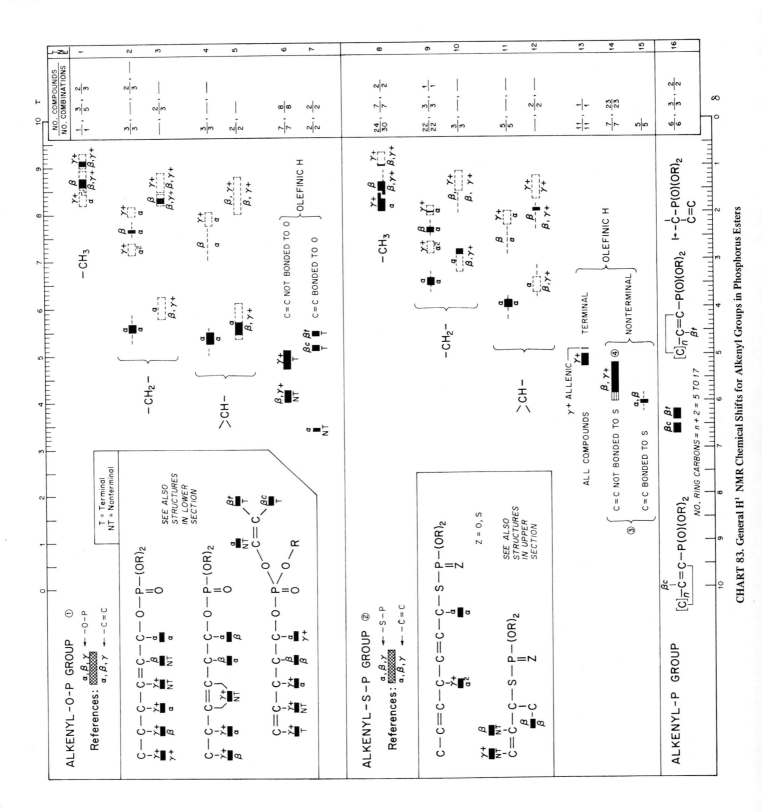

CHART 83. General H¹ NMR Chemical Shifts for Alkenyl Groups in Phosphorus Esters

CHART 83

Notes for Chart 83–Alkenyl Groups in Phosphorus Esters

1. Although data were available for only the phosphates, this chart should apply reasonably well to all alkenyl–O–P groups.
2. This chart should apply reasonably well to all alkenyl–S–P groups.
3. Bonding C=C directly to S appears to have little effect on the related chemical shifts.
4. The crosshatched bar is the data for unsubstituted alkyl groups. The solid bar shows the upfield shift which accompanies substitution of the double-bonded carbons with methyl or ethyl groups.

286

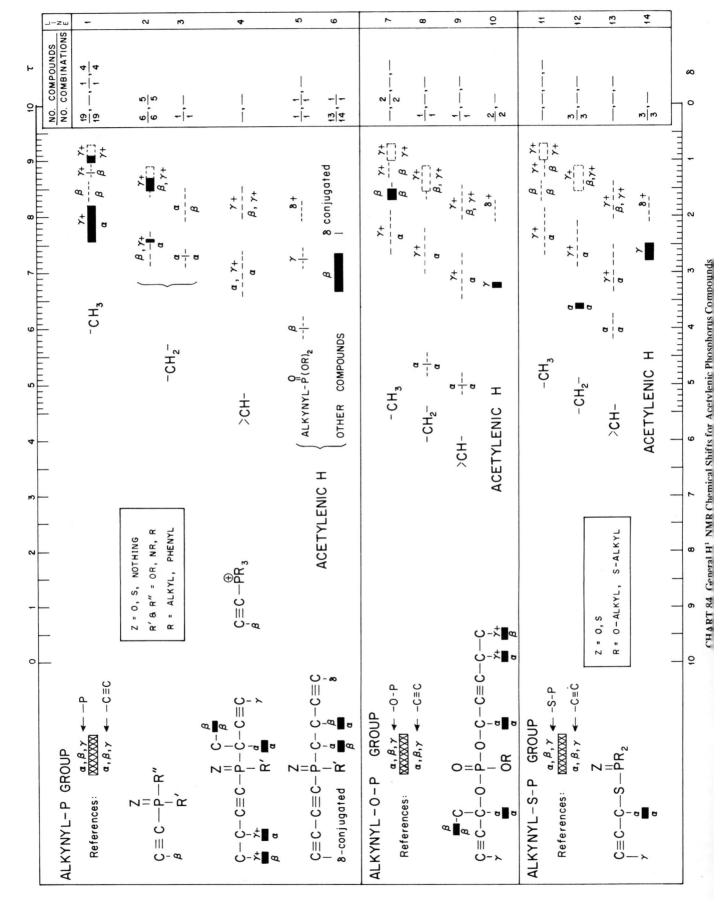

CHART 84 General H¹ NMR Chemical Shifts for Acetylenic Phosphorus Compounds

CHART 84

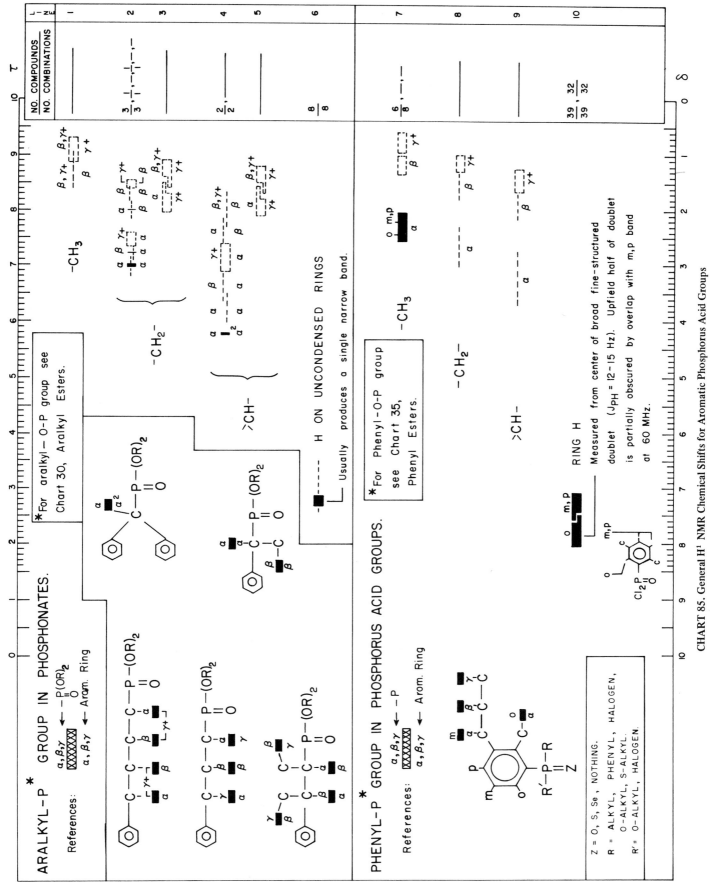

CHART 85. General H¹ NMR Chemical Shifts for Aromatic Phosphorus Acid Groups

CHART 85

288

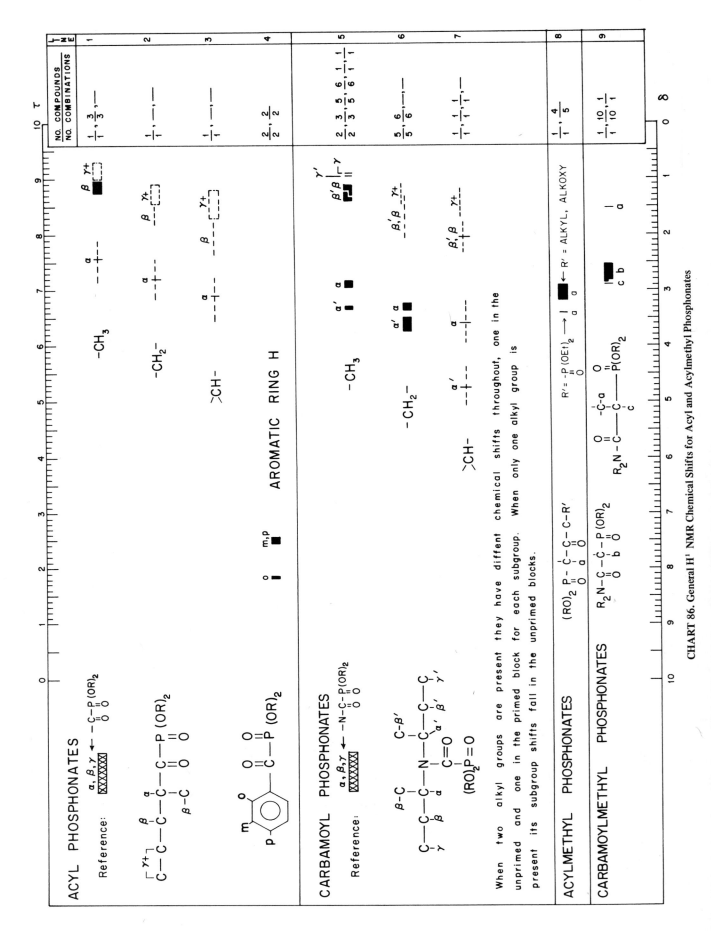

CHART 86. General H¹ NMR Chemical Shifts for Acyl and Acylmethyl Phosphonates

CHART 86

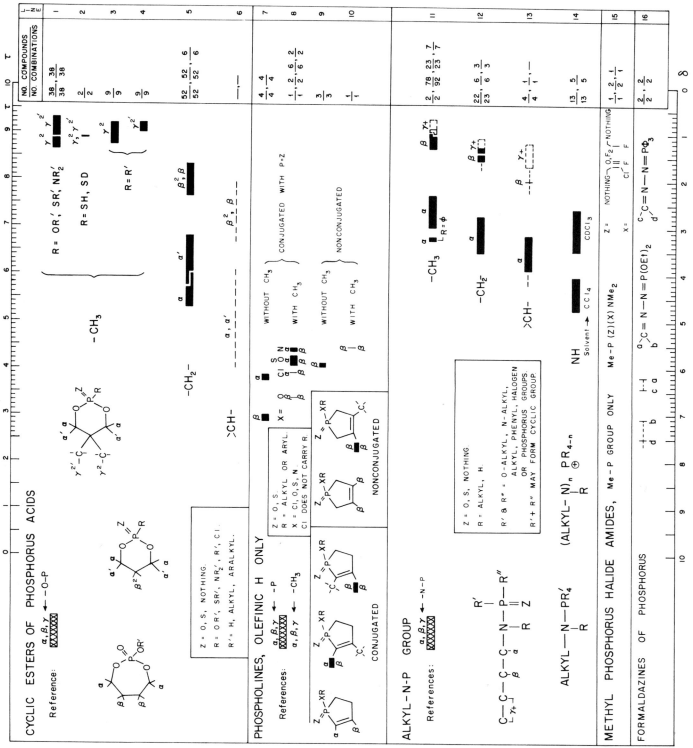

CHART 87. General H¹ NMR Chemical Shifts for Cyclic Esters and Alkyl Amides of Phosphorus

CHART 87

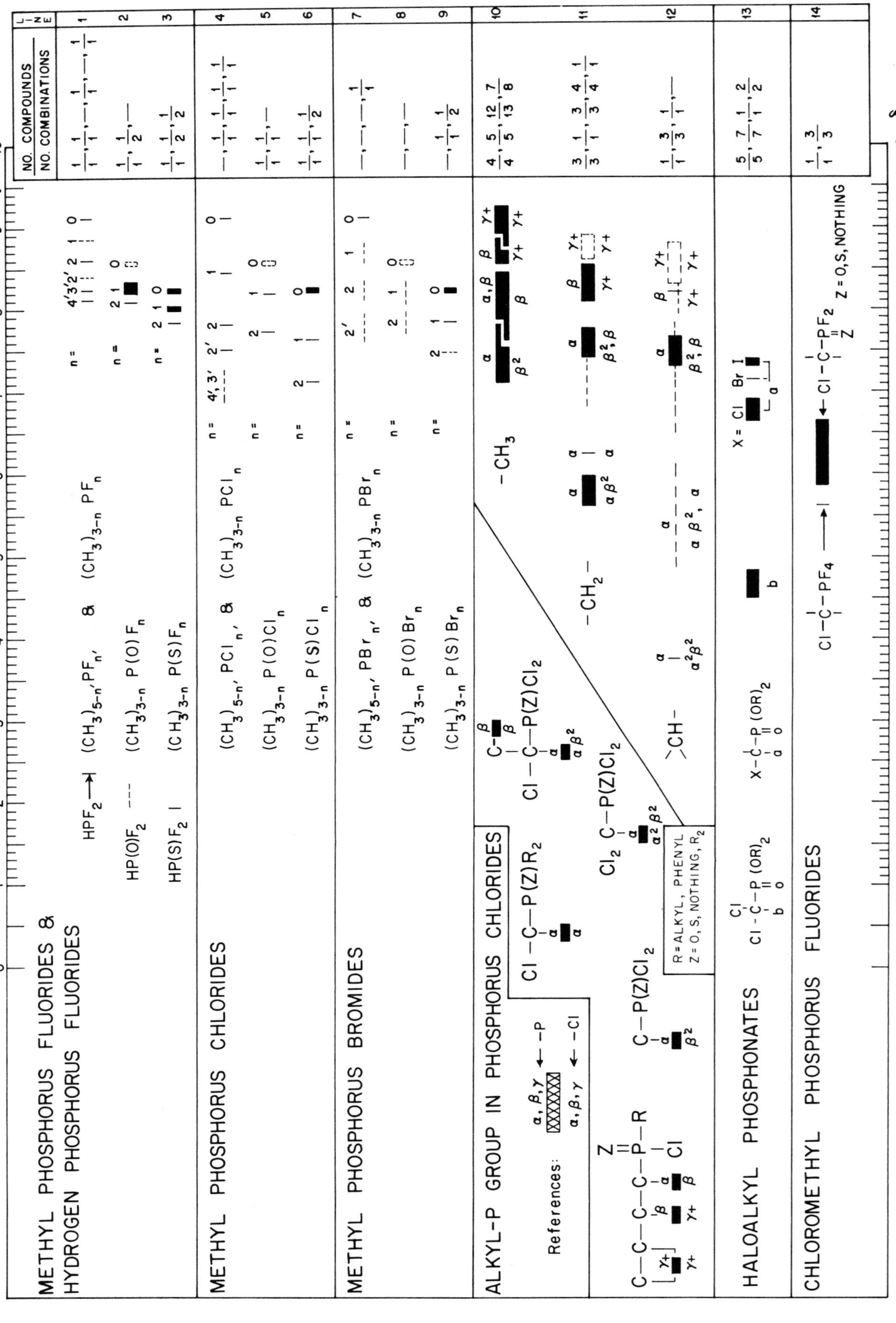

	NO. COMPOUNDS		
	NO. COMBINATIONS		

METHYL PHOSPHORUS FLUORIDES & HYDROGEN PHOSPHORUS FLUORIDES

$HPF_2 \longrightarrow$ $(CH_3)_{5-n}PF_n$, & $(CH_3)_{3-n}PF_n$

$HP(O)F_2$

$HP(S)F_2$

$(CH_3)_{3-n}P(O)F_n$

$(CH_3)_{3-n}P(S)F_n$

METHYL PHOSPHORUS CHLORIDES

$(CH_3)_{5-n}, PCl_n,$ & $(CH_3)_{3-n}PCl_n$

$(CH_3)_{3-n}P(O)Cl_n$

$(CH_3)_{3-n}P(S)Cl_n$

METHYL PHOSPHORUS BROMIDES

$(CH_3)_{5-n}, PBr_n,$ & $(CH_3)_{3-n}PBr_n$

$(CH_3)_{3-n}P(O)Br_n$

$(CH_3)_{3-n}P(S)Br_n$

ALKYL-P GROUP IN PHOSPHORUS CHLORIDES

$Cl-C-P(Z)R_2$

$Cl-C-C-P(Z)Cl_2$

$Cl_2C-P(Z)Cl_2$

$C-P(Z)Cl_2$

$-CH_3$

$-CH_2-$

$>CH-$

R = ALKYL, PHENYL
Z = O, S, NOTHING, R₂

References:

$C-C-C-C-P-R$

$Z=O$ ... $-P$ ... $-Cl$

HALOALKYL PHOSPHONATES

$X-\overset{Cl}{\underset{O}{C}}-\overset{}{\underset{}{P}}(OR)_2$

$Cl-\overset{Cl}{\underset{O}{C}}-P(OR)_2$

X = Cl Br I

CHLOROMETHYL PHOSPHORUS FLUORIDES

$Cl-C-PF_2$ Z = O, S, NOTHING

$Cl-C-PF_4 \longrightarrow$

CHART 88. General H¹ NMR Chemical Shifts for Alkyl Phosphorus Halides

CHART 88

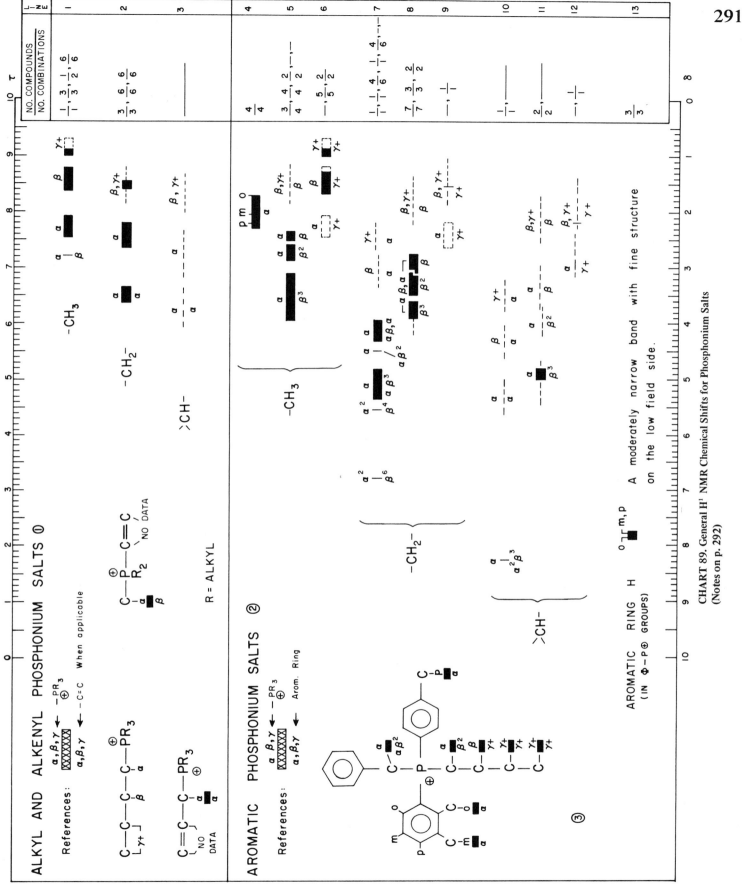

CHART 89. General H¹ NMR Chemical Shifts for Phosphonium Salts
(Notes on p. 292)

CHART 89

Notes for Chart 89—Phosphonium Salts

1. Scarcity of data does not permit a more detailed correlation.
2. Includes available data for all phosphonium salts which include an aromatic ring.
3. Additional illustrative structures for aromatic phosphonium salts:

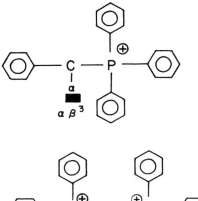

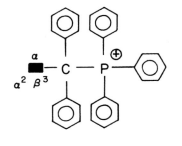

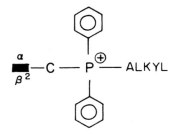

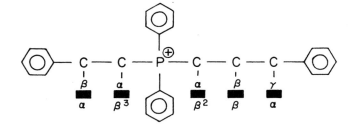

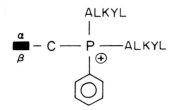

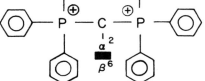

Coupling Constant Correlation Charts

CHART I 2 (top)

HYDROGEN–HYDROGEN COUPLINGS

Aliphatic

1. H–C–H
2. H–Ċ–H
3. H–C–C–H, H–C–Ċ–H, & H–Ċ–Ċ–H
4. H–C=C–H, cis
5. H–C=C–H, trans
6. H–C≡C–H
7. 4- to 9-Bond

Aromatic

8. Ortho
9. Meta
10. Para
11. Other 4- to 6-Bond
12. Furans, Thiophenes, Pyrroles

Across Heteroatoms

13. H–X–H, H–C–X–H, H–Ċ–X–H, H–C=N–H   X = N, O, S
14. 4- and 5-Bond

HYDROGEN–CARBON-13 COUPLINGS

Aliphatic

15. H–C–
16. H–C=
17. H–C≡
18. H–C–(C–, C=, C≡)
19. H–C=C
20. H–Ċ–(C–, C≡)
21. H–C≡C
22. 3-Bond

Aromatic

23. H–C in benzenes
24. H–C in heterocyclics
25. 2- to 4-Bond

HYDROGEN–FLUORINE COUPLINGS

Aliphatic

26. H–F
27. H–Ċ–F
28. H–C–F
29. H–C–C–F, H–Ċ–C–F,
30. H–C=C–F, cis
31. H–C=C–F, trans
32. 4- and 5-Bond

LINE	SUMMARY CHART (LINE)	NO. CPDS.
1	J1 (1-12), J2 (1-3), J3	563
2	J2 (9-17)	577
3	J4, J5, J6, J9 (1-3)	1814
4	J7, J8	463
5	J7, J8 (1-7)	351
6	J9 (4)	2
7	J9 (5-17), J10	621
8	J11(1-6), J12(1-3, 7-9, J13(8-10, 15,16)	1025
9	J11(7,8),J12(4,10),J13(11-13,17)	789
10	J11(9), J12(5,11), J13(14,18)	346
11	J12(6,12,13), J14(1-10)	211
12	J13(1-7)	384
13	J15(1-4),J16(1-8), J18(1-5)	175
14	J15(5-12),J16(9-11),J17, J18(6-9)	148
15	J19, J20	312
16	J21 (1-12)	59
17	J21 (13-15)	12
18	J22 (1-3,8)	24
19	J22 (5-7)	8
20	J22 (4,12)	4
21	J22 (9-11)	11
22	J22 (13-16)	47
23	J23(1-8)	133
24	J24 (1-10)	113
25	J23 (9-11), J24 (11-14)	13
26	J25 (1)	1
27	J25(2-9)	62
28	J25 (10-12)	14
29	J26, J27 (1-4,7)	122
30	J27 (5,6)	9
31	J27 (5,6)	16
32	J27 (8-15)	40

Hz   700   600   ⟵ Scale compressed & broken   0   10   20   40   60   80   100   200   Hz

1 — 615 Hz

⟵ Scale compressed & broken   600   700   Hz

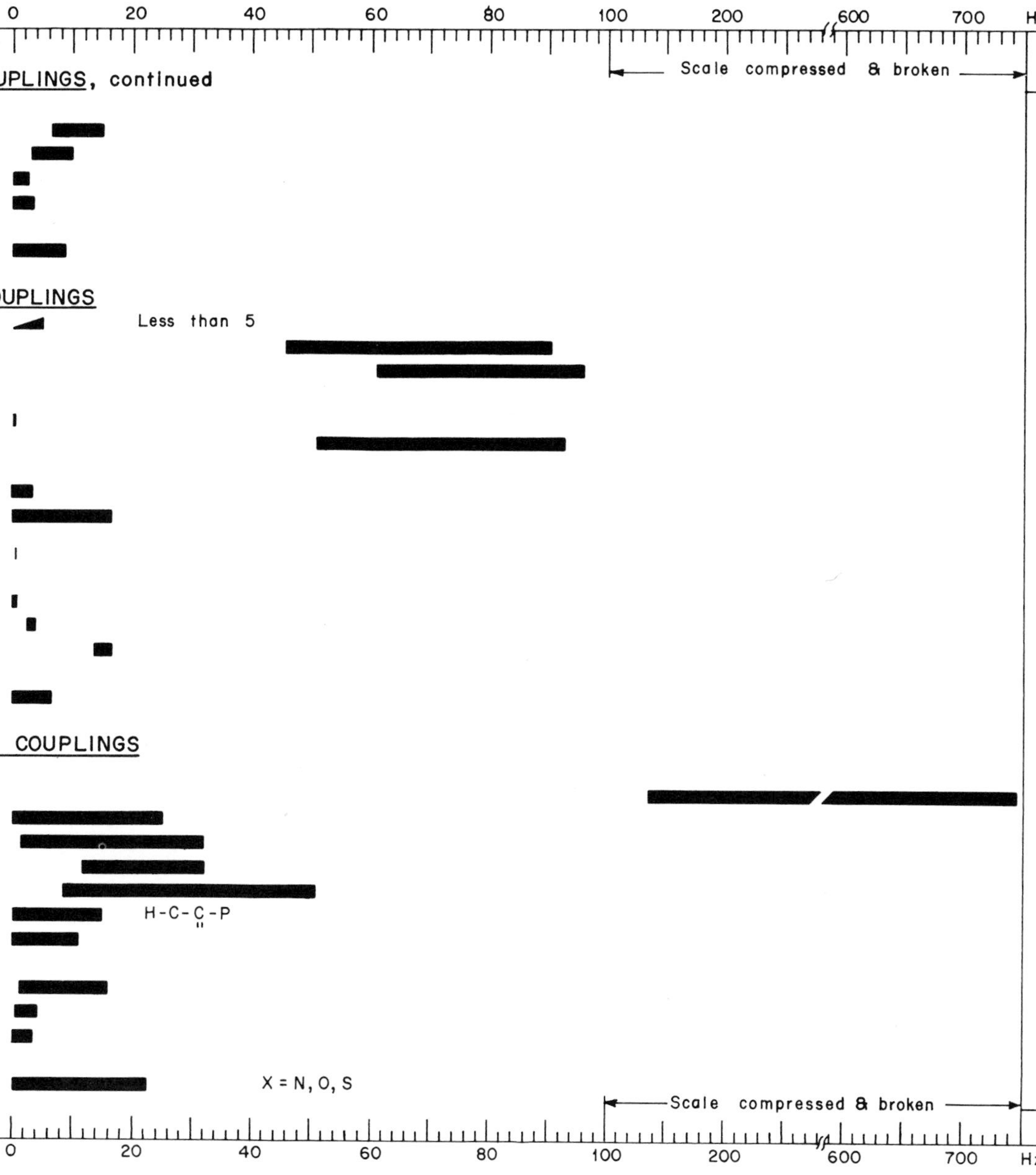

HYDROGEN-FLUORINE COUPLINGS, continued

		J ref.	Hz
Aromatic			
Ortho	33	J 28 (1,2,7,9,11,12)	57
Meta	34	J 28 (3-5,8,13)	49
Para	35	J 28 (6)	12
Other 4- to 6-Bond	36	J 28 (10,14-16)	19
Across Heteroatoms			
3- to 5-Bond	37	J 29	33

HYDROGEN-NITROGEN COUPLINGS

		J ref.	Hz
H-N- N-14 Amides	38	J 30 (6)	>10
N-14 Other	39	J 30 (1-5)	9
N-15 All	40	J 30 (7-11)	36
H-N= N-14	41	J 30 (12)	1
N-15	42	J 30 (13,14)	2
H-C-N- N-14 & N-15	43	J 31 (1-8)	28
H-C=N N-14 & N-15	44	J 31 (9-11)	16
H-C-N= N-15	45	J 31 (12)	2
H-C=N N-14	46	J 31 (13)	26
N-15 syn	47	J 31 (14)	15
N-15 anti	48	J 31 (15)	9
3- & 4-Bond N-14 & N-15	49	J 32	13

Less than 5

HYDROGEN-PHOSPHORUS COUPLINGS

		J ref.	Hz
Aliphatic			
H-P	50	J 33 (1,6-9)	63
H-C-P	51	J 34 (1-9)	219
H-C-P (=O)	52	J 34 (10-12)	34
H-C-C-P	53	J 35 (1-4)	47
H-C=C-P	54	J 35 (5-8)	34
H-C≡C-P, H-C-C-P,	55	J 35 (9-12)	19
4- and 5-Bond	56	J 38 (6-21)	94
Aromatic			
Ortho	57	J 38 (1-3)	57
Meta	58	J 38 (1-3)	23
Other 4- & 5-Bond	59	J 38 (4,5)	9
Across Heteroatoms			
H-C-X-P	60	J 36, J 37	475

H-C-C-P (with =O)

X = N, O, S

Scale compressed & broken

Hz — 0 20 40 60 80 100 200 600 700

CHART 12. Index to and Condensed Summary of Coupling Constants

CHART 12

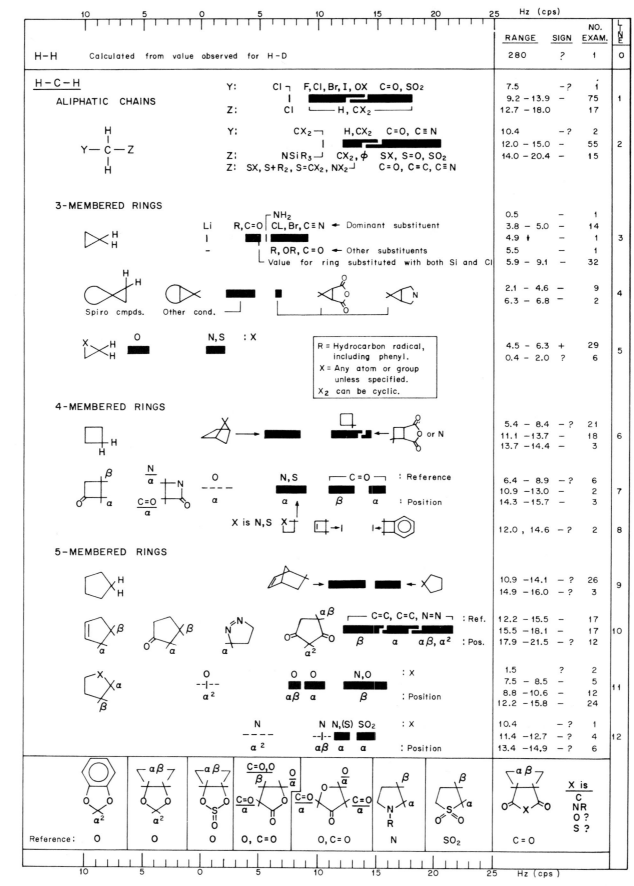

		RANGE	SIGN	NO. EXAM.	LINE
H–H	Calculated from value observed for H–D	280	?	1	0

H–C–H

ALIPHATIC CHAINS

Y: Cl ⌐ F,Cl,Br,I, OX C=O, SO₂
Z: Cl ⌐ H, CX₂

	7.5	–?	1	
	9.2 – 13.9	–	75	1
	12.7 – 18.0	–	17	

Y: CX₂ ⌐ H,CX₂ C=O, C≡N
Z: NSiR₃ ⌐ CX₂, φ SX, S=O, SO₂
Z: SX, S+R₂, S=CX₂, NX₂ ⌐ C=O, C=C, C≡N

	10.4	–?	2	
	12.0 – 15.0	–	55	2
	14.0 – 20.4	–	15	

3-MEMBERED RINGS

Li R,C=O ⌐NH₂ CL,Br,C≡N ← Dominant substituent
I ⌐ R, OR, C=O ← Other substituents
Value for ring substituted with both Si and Cl

	0.5	–	1	
	3.8 – 5.0	–	14	
	4.9	–	1	3
	5.5	–	1	
	5.9 – 9.1	–	32	

Spiro cmpds. Other cond.

| | 2.1 – 4.6 | – | 9 | 4 |
| | 6.3 – 6.8 | – | 2 | |

X: O N,S

R = Hydrocarbon radical, including phenyl.
X = Any atom or group unless specified.
X₂ can be cyclic.

| | 4.5 – 6.3 | + | 29 | 5 |
| | 0.4 – 2.0 | ? | 6 | |

4-MEMBERED RINGS

or N

	5.4 – 8.4	–?	21	
	11.1 – 13.7	–	18	6
	13.7 – 14.4	–	3	

N/α C=O/α O/α N,S ⌐C=O⌐ : Reference
β α α β α : Position

	6.4 – 8.9	–?	6	
	10.9 – 13.0	–	2	7
	14.3 – 15.7	–	3	

X is N,S

| | 12.0 , 14.6 | –? | 2 | 8 |

5-MEMBERED RINGS

| | 10.9 – 14.1 | –? | 26 | 9 |
| | 14.9 – 16.0 | –? | 3 | |

αβ ⌐C=C, C=C, N=N⌐ : Ref.
β α αβ, α² : Pos.

	12.2 – 15.5	–	17	
	15.5 – 18.1	–	17	10
	17.9 – 21.5	–?	12	

X O O N,O : X
α² αβ α β : Position

	1.5	?	2	
	7.5 – 8.5	–	5	11
	8.8 – 10.6	–	12	
	12.2 – 15.8	–	24	

N N N,(S) SO₂ : X
α² αβ α α : Position

	10.4	–?	1	
	11.4 – 12.7	–?	4	12
	13.4 – 14.9	–?	6	

X is
C
NR
O ?
S ?

Reference: O   O   O   O, C=O   O, C=O   N   SO₂   C=O

CHART J1. Hydrogen–Hydrogen Coupling Constants, H–C–H Series (A)

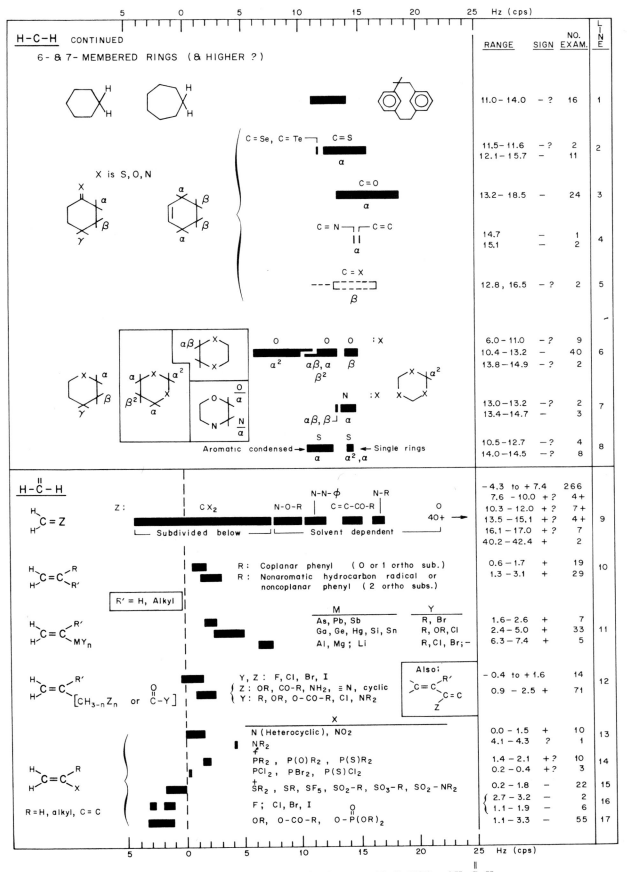

CHART J2. Hydrogen–Hydrogen Coupling Constants, H–C–H(B) and H–C–H

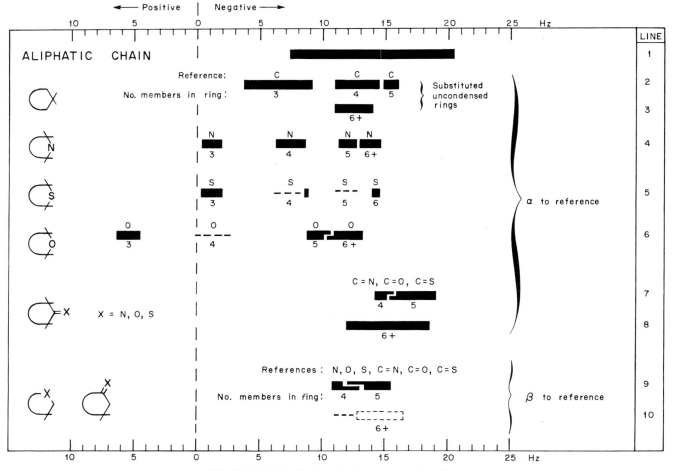

CHART J3. H–C–H Coupling Constant vs. Ring Size

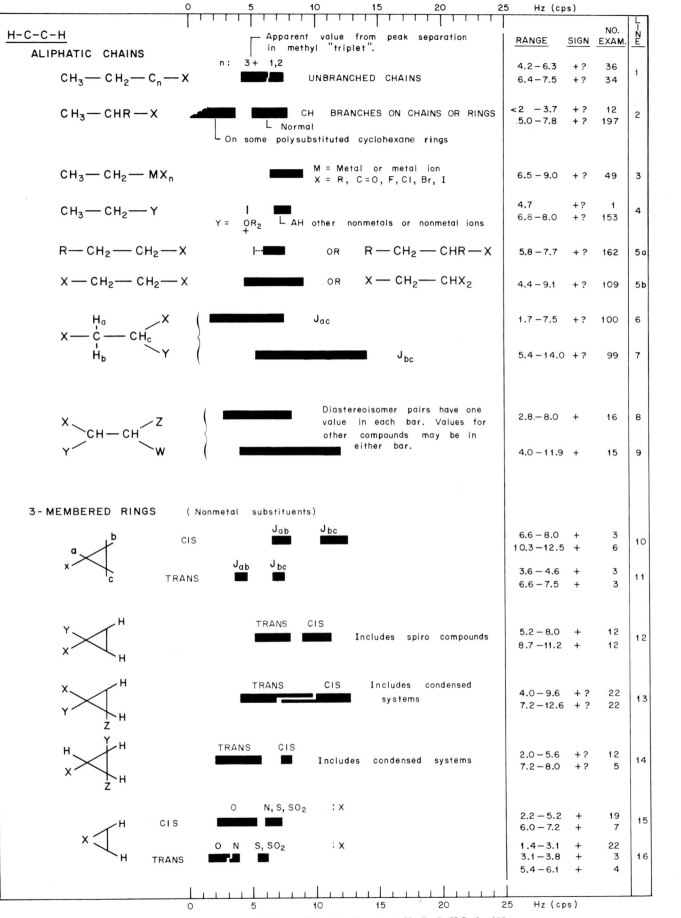

CHART J4. Hydrogen–Hydrogen Coupling Constants, H−C−H Series (A)

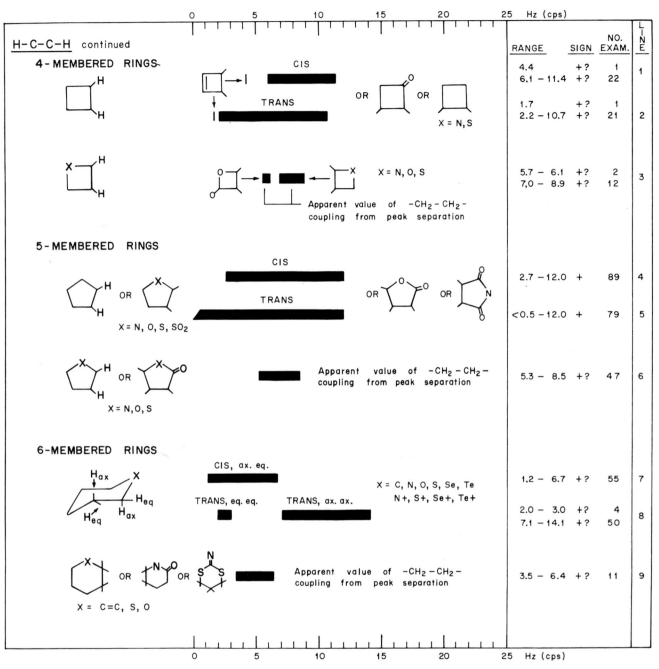

CHART J5. Hydrogen–Hydrogen Coupling Constants, H–C–C–H Series (B)

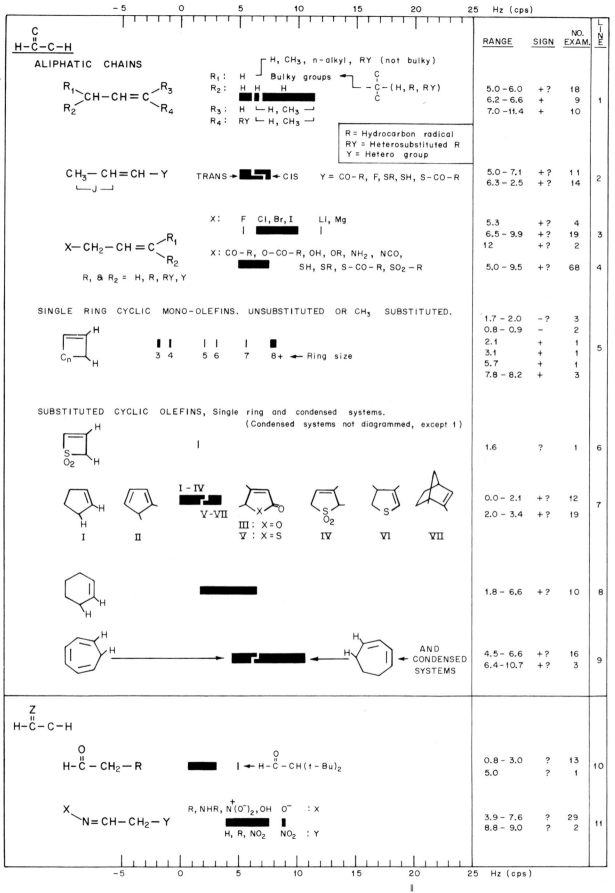

CHART J6. Hydrogen–Hydrogen Coupling Constants, H–C–C–H Series

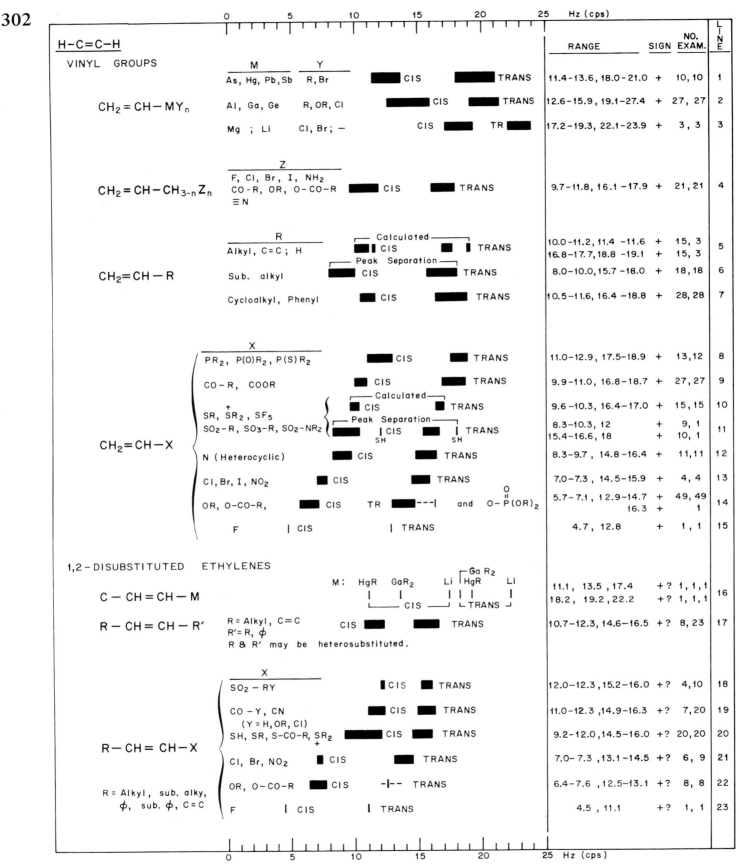

CHART J7. Hydrogen–Hydrogen Coupling Constants, H−C=C−H Series (A)

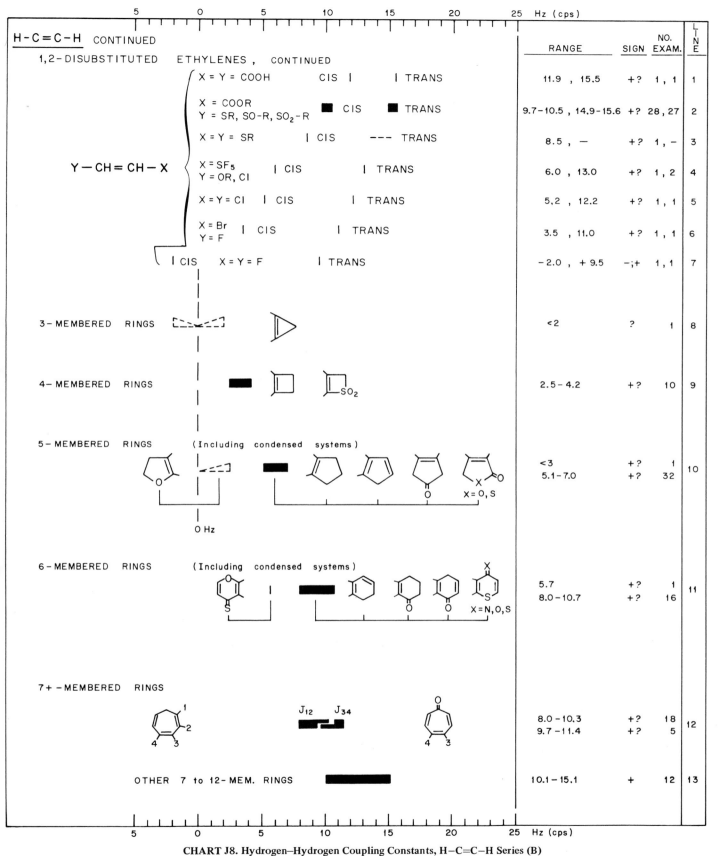

CHART J8. Hydrogen–Hydrogen Coupling Constants, H–C=C–H Series (B)

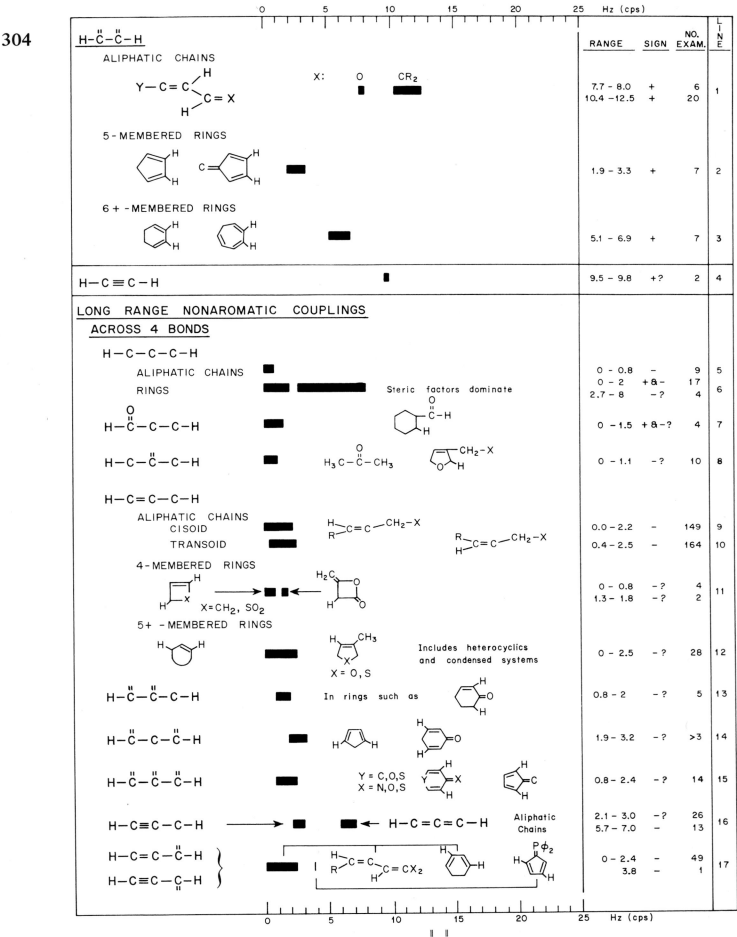

CHART J9. H–H Coupling Constants, H–C̈–C̈–H, H–C≡C–H and 4-Bond

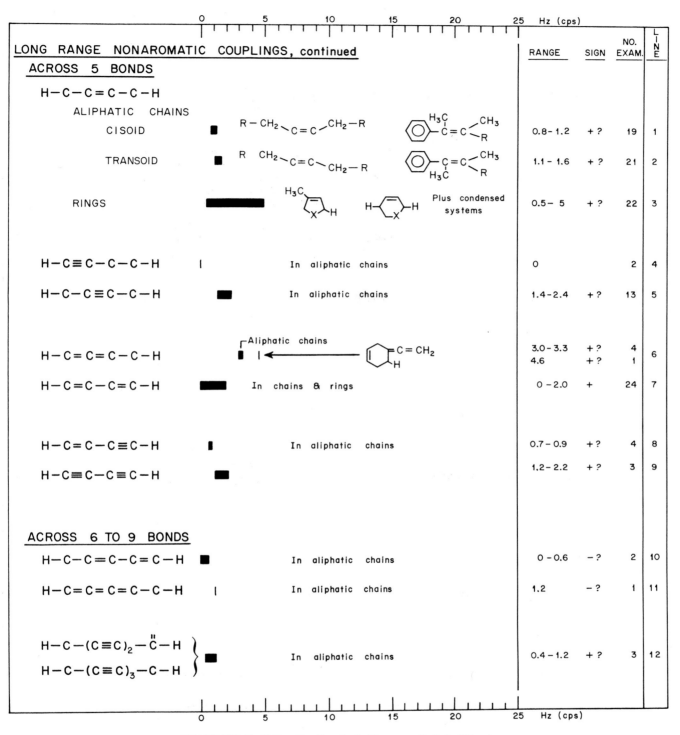

CHART J10. H−H Coupling Constants, Nonaromatic 5- to 9-Bond

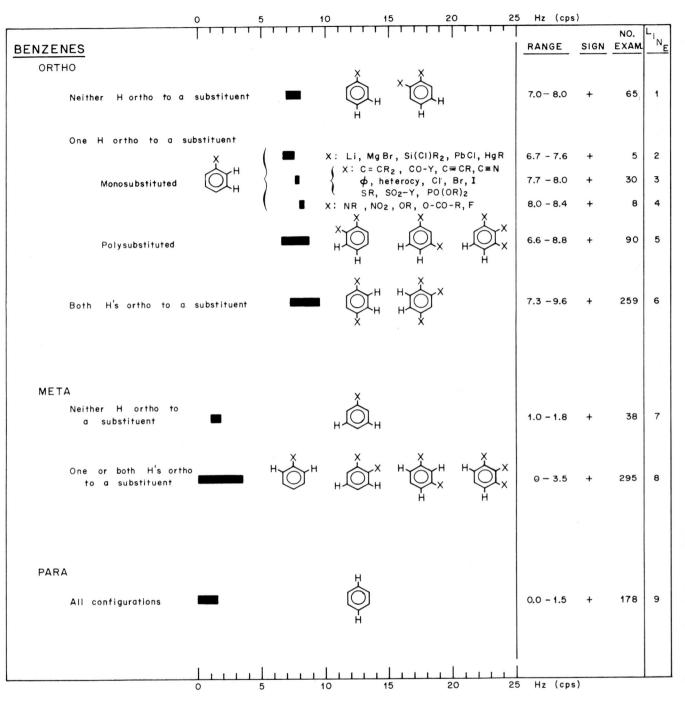

CHART J11. Hydrogen–Hydrogen Coupling Constants for Benzenes

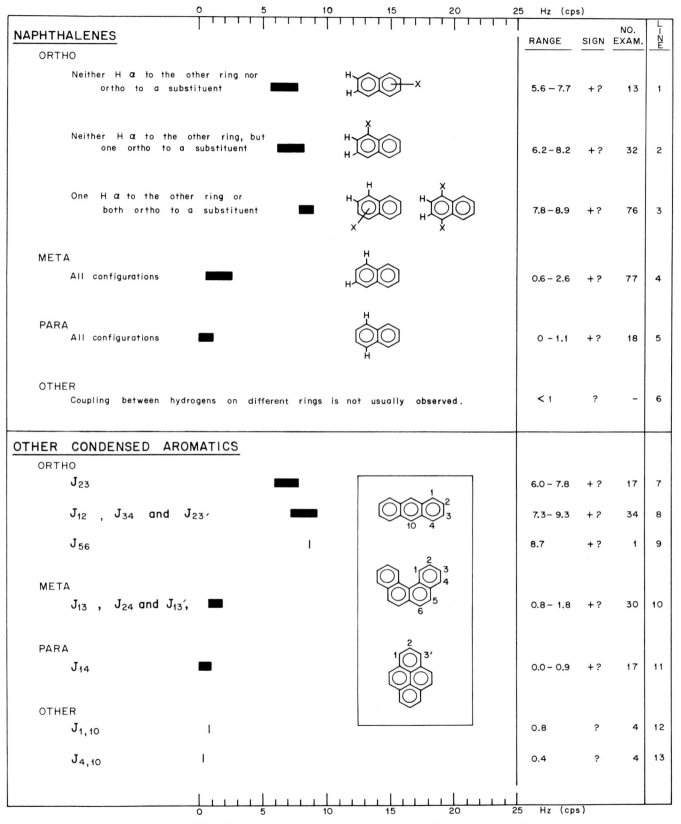

	RANGE	SIGN	NO. EXAM.	LINE

NAPHTHALENES

ORTHO

Neither H α to the other ring nor ortho to a substituent — 5.6 – 7.7 | + ? | 13 | 1

Neither H α to the other ring, but one ortho to a substituent — 6.2 – 8.2 | + ? | 32 | 2

One H α to the other ring or both ortho to a substituent — 7.8 – 8.9 | + ? | 76 | 3

META

All configurations — 0.6 – 2.6 | + ? | 77 | 4

PARA

All configurations — 0 – 1.1 | + ? | 18 | 5

OTHER

Coupling between hydrogens on different rings is not usually observed. — < 1 | ? | – | 6

OTHER CONDENSED AROMATICS

ORTHO

J_{23} — 6.0 – 7.8 | + ? | 17 | 7

J_{12} , J_{34} and $J_{23'}$ — 7.3 – 9.3 | + ? | 34 | 8

J_{56} — 8.7 | + ? | 1 | 9

META

J_{13} , J_{24} and $J_{13'}$ — 0.8 – 1.8 | + ? | 30 | 10

PARA

J_{14} — 0.0 – 0.9 | + ? | 17 | 11

OTHER

$J_{1,10}$ — 0.8 | ? | 4 | 12

$J_{4,10}$ — 0.4 | ? | 4 | 13

CHART J12. Hydrogen–Hydrogen Coupling Constants for Condensed Aromatics

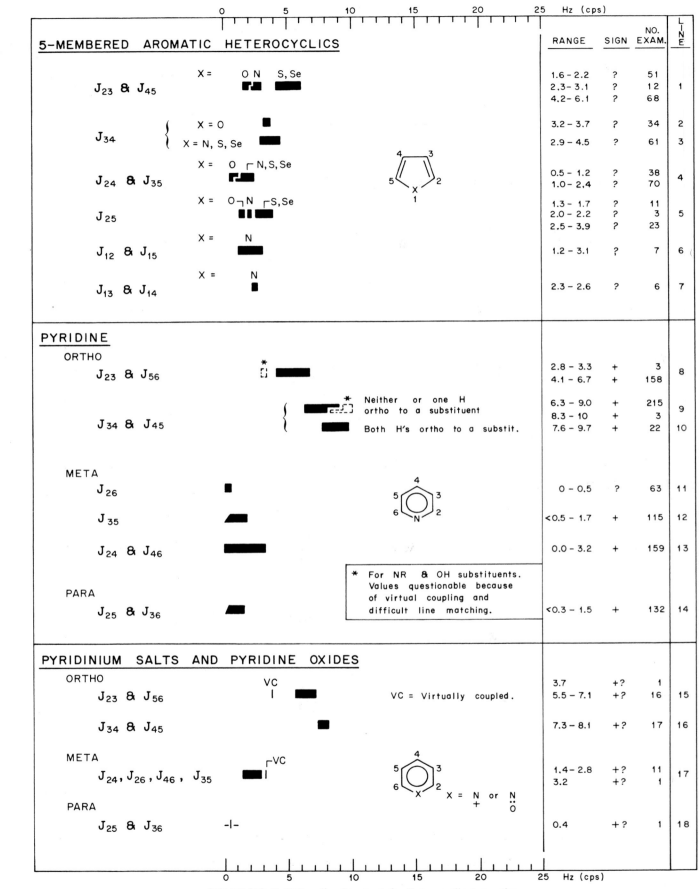

CHART J13. H–H Coupling Constants for Heterocyclic Aromatics

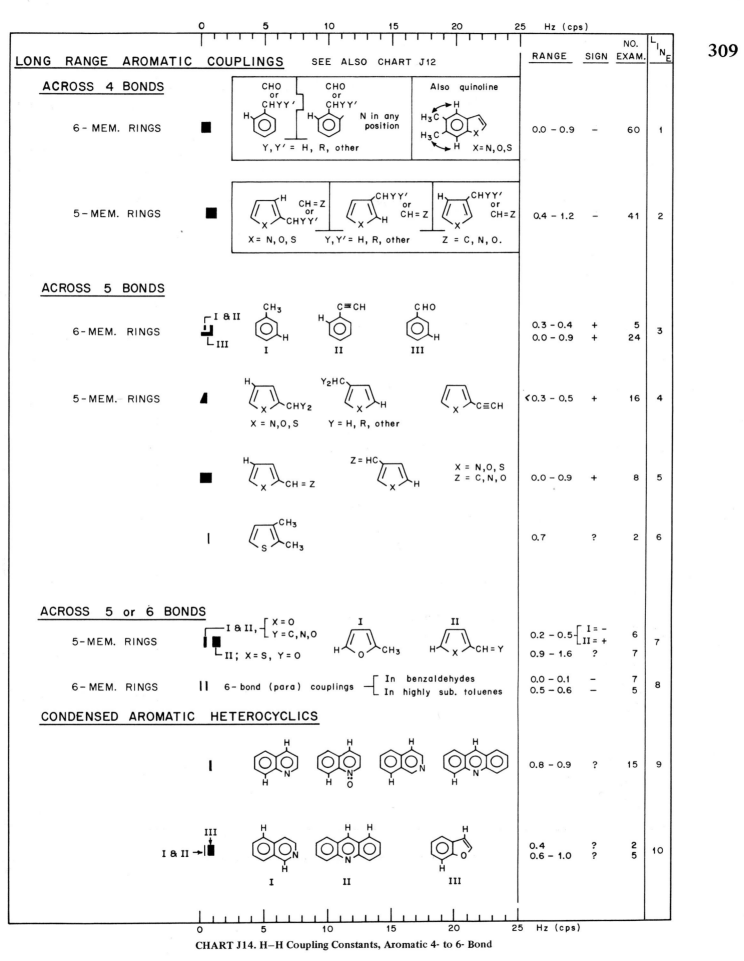

CHART J14. H–H Coupling Constants, Aromatic 4- to 6- Bond

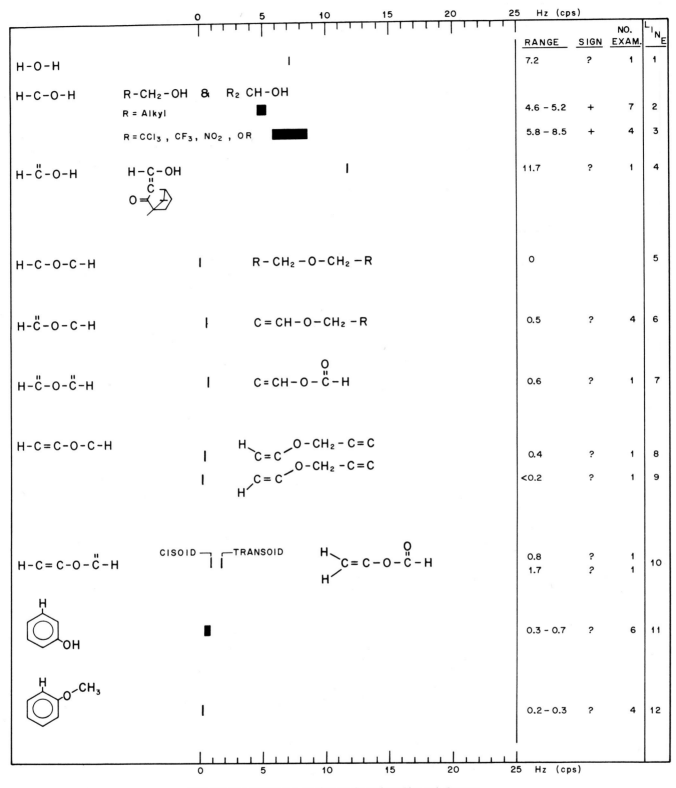

CHART J15. Hydrogen–Hydrogen Couplings Through Oxygen

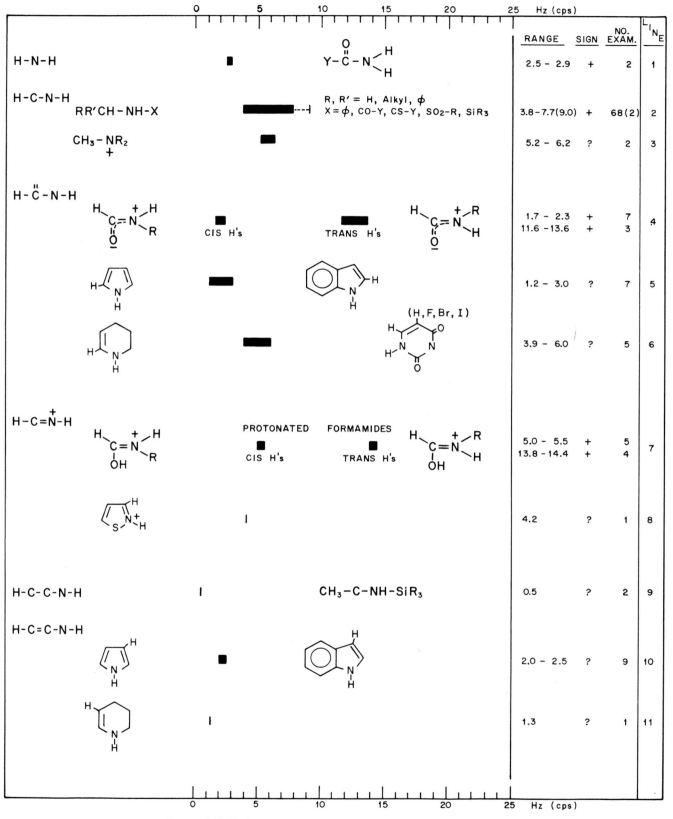

CHART J16. Hydrogen–Hydrogen Couplings Through Nitrogen (A)

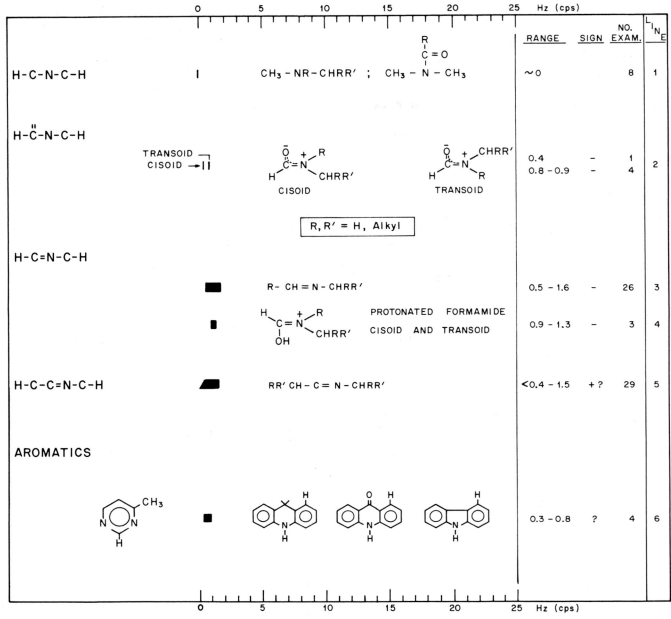

CHART J17. Hydrogen–Hydrogen Couplings Through Nitrogen (B)

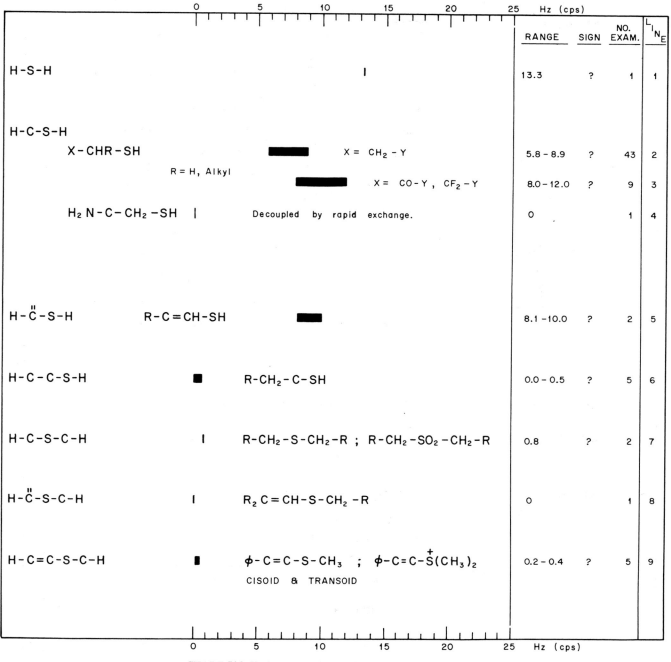

CHART J18. Hydrogen–Hydrogen Couplings Through Sulfur

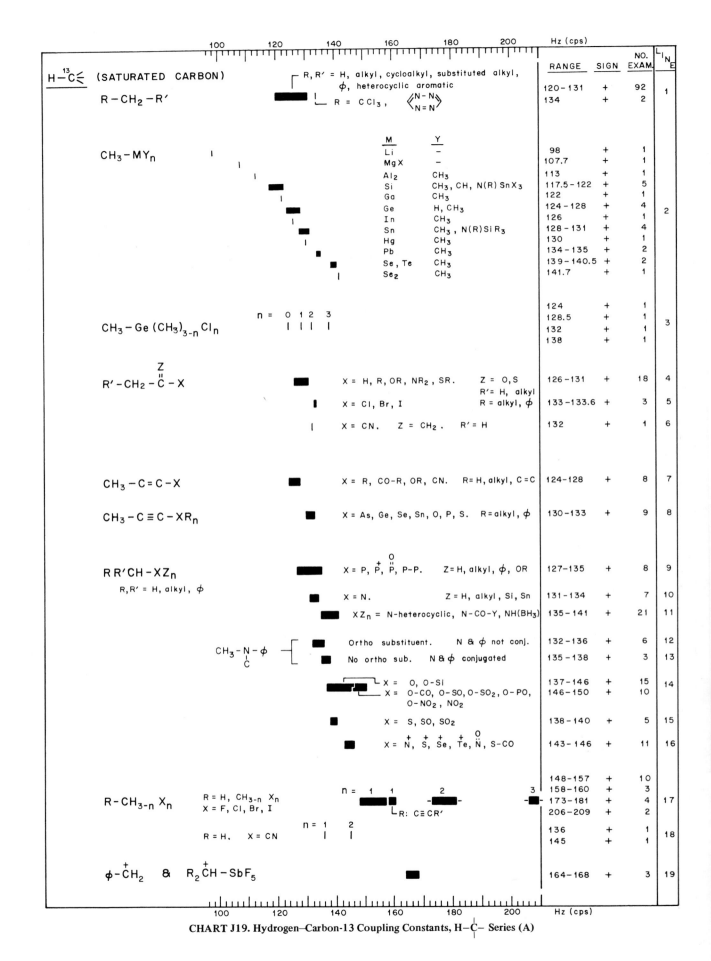

CHART J19. Hydrogen–Carbon-13 Coupling Constants, H–C– Series (A)

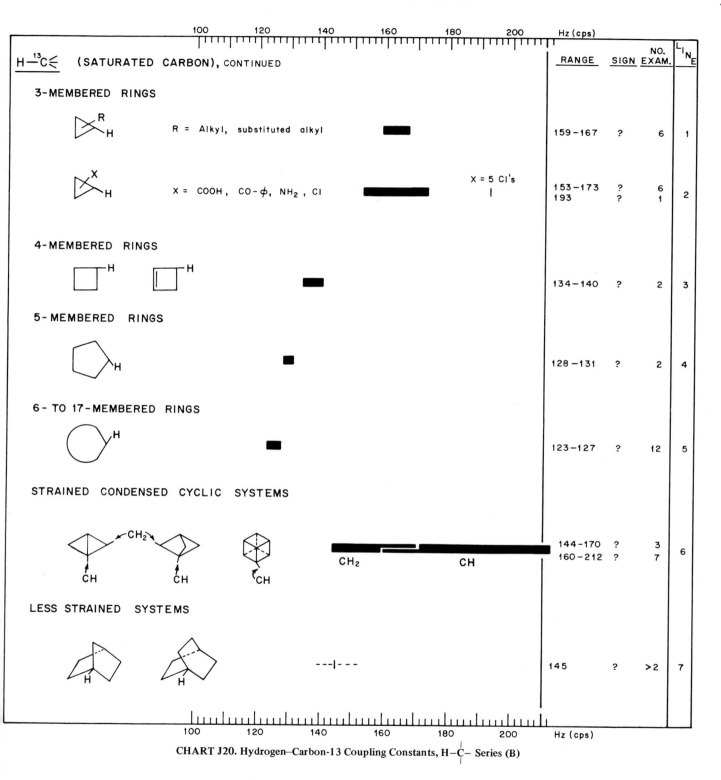

CHART J20. Hydrogen–Carbon-13 Coupling Constants, H–C– Series (B)

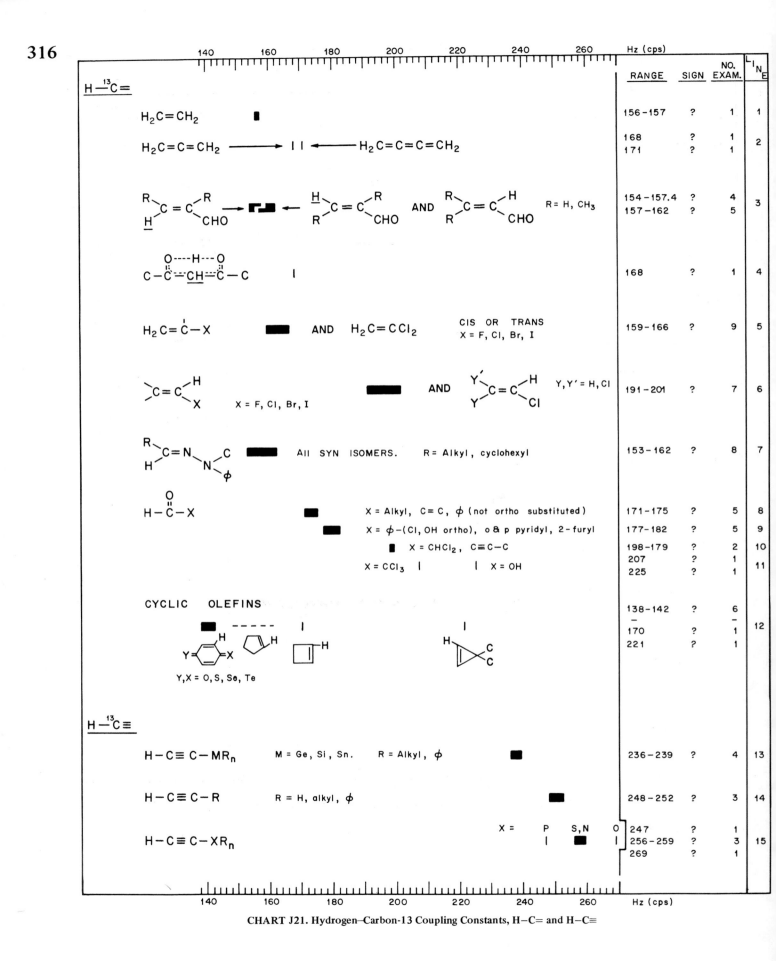

CHART J21. Hydrogen–Carbon-13 Coupling Constants, H–C= and H–C≡

Hz (cps)

H–C–^{13}C

	RANGE	SIGN	NO. EXAM.	LINE
Y–CH$_2$–^{13}C–X Y = H, Cl. X = Cl, Br, I, OH	2.0–6.0	–	7	1
X$_2$CH–^{13}CX$_2$ X = Cl, Br	1.1–1.2	+	2	2

H–C–^{13}C=

CH$_3$–^{13}C–X X = H, alkyl, ϕ, OH, Cl, Br, I (with =O)	5.8–7.5	?	10	3

H–$\overset{\text{O}}{\overset{\|}{\text{C}}}$–^{13}C

H–$\overset{\text{O}}{\overset{\|}{\text{C}}}$–^{13}CH$_{3-n}Cl_n$ n = 0 1 2 3	26.6	?	1	
	–	?	–	4
	35.8	?	1	
	46.8	?	1	

H–C=^{13}C

H$_2$C=^{13}C–R R = H, CH$_2$–Br. CIS & TRANS	1.0–2.9	–?	3	5
structures &	0.8	+?	1	6
	16.0–17.5	+?	2	
H(2)–C=^{13}C(3) H(3)–C=^{13}C(2)	6.2	?	1	7
	9.0	?	1	

H–C–^{13}C≡

CH$_3$–^{13}C≡X X = N, C–Y	10.0–11.4	?	5	8

H–C≡^{13}C

H–C≡^{13}C–MR$_n$ M = Ge, Si, Sn	41–42.5	?	4	9
H–C≡^{13}C–R R = H, ϕ	49.3–49.9	?	2	10
H–C≡^{13}C–Xϕ_n X = P S, N	45.8	?	1	11
	51.6–55.5	?	3	
	61	?	1	

H–$\overset{\text{O}}{\overset{\|}{\text{C}}}$–^{13}C≡

H–$\overset{\text{O}}{\overset{\|}{\text{C}}}$–^{13}C≡C–C	32.8	?	1	12

H–C–C–^{13}C

CH$_3$–$\overset{\text{C}}{\underset{\text{C}}{\text{C}}}$–^{13}CX X = H, RY, CO–Y, COOR, =CR OH, O–Si, O–CO–R, O–SO–OR, O–SO–Cl, O–SO$_2$–R N(+)H$_3$, =N–NR$_2$, Cl, Br, I	3.6–6.4	+?	42	13

H–C–$\overset{\text{O}}{\overset{\|}{\text{C}}}$–^{13}C

C–$\overset{\text{O}}{\overset{\|}{\text{C}}}$–CH$_2$–$\overset{\text{O}}{\overset{\|}{\text{C}}}$–^{13}C	<0.5	?	1	14

H–$\overset{\|}{\text{C}}$–$\overset{\|}{\text{C}}$–^{13}C & H–C–$\overset{\|}{\text{C}}$–^{13}C

^{13}CH$_3$–C=CH=C–C O--H--O	2–3	?	2	15

H–C–C≡^{13}C

CH$_3$–C≡^{13}C–X	4.8–5.8	?	2	16

CHART J22. Hydrogen–Carbon-13 Coupling Constants, 2- and 3-Bond

AROMATICS, H — ^{13}C

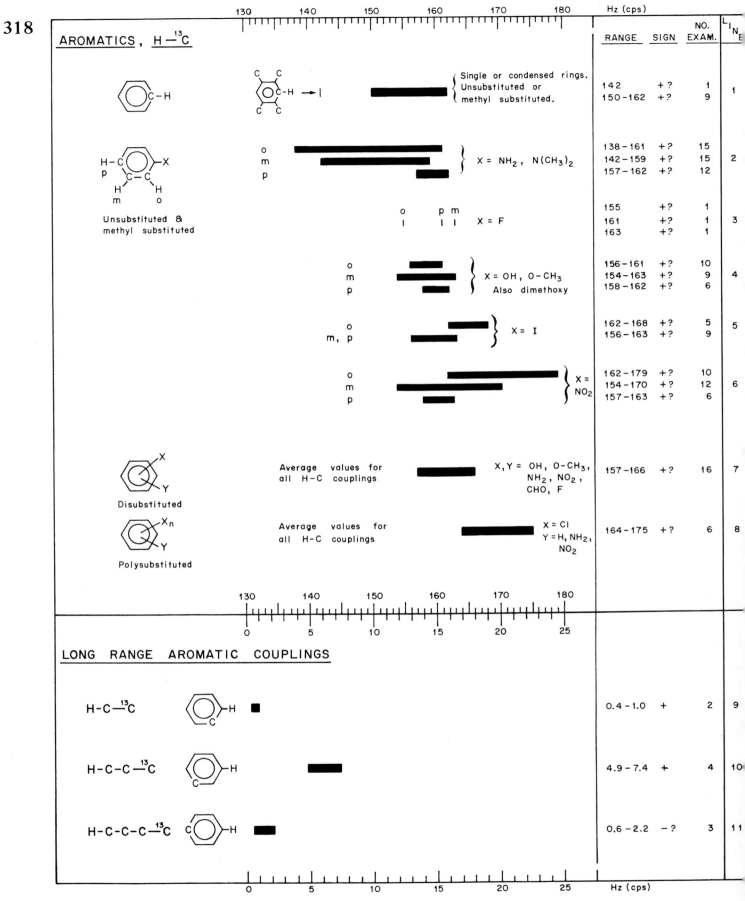

	RANGE	SIGN	NO. EXAM.	LINE
Single or condensed rings. Unsubstituted or methyl substituted.	142	+?	1	1
	150-162	+?	9	
o	138-161	+?	15	2
m	142-159	+?	15	
p	157-162	+?	12	
X = F o	155	+?	1	3
p	161	+?	1	
m	163	+?	1	
X = OH, O-CH$_3$ Also dimethoxy o	156-161	+?	10	4
m	154-163	+?	9	
p	158-162	+?	6	
X = I o	162-168	+?	5	5
m, p	156-163	+?	9	
X = NO$_2$ o	162-179	+?	10	6
m	154-170	+?	12	
p	157-163	+?	6	
X, Y = OH, O-CH$_3$, NH$_2$, NO$_2$, CHO, F	157-166	+?	16	7
X = Cl Y = H, NH$_2$, NO$_2$	164-175	+?	6	8

C-H → I   Single or condensed rings. Unsubstituted or methyl substituted.

H-C   X   X = NH$_2$, N(CH$_3$)$_2$
p C-C
   H   H
   m   o
Unsubstituted & methyl substituted

o   p m   X = F

Disubstituted — Average values for all H-C couplings

Polysubstituted — Average values for all H-C couplings

LONG RANGE AROMATIC COUPLINGS

	RANGE	SIGN	NO. EXAM.	LINE
H-C—^{13}C	0.4-1.0	+	2	9
H-C-C—^{13}C	4.9-7.4	+	4	10
H-C-C-C—^{13}C	0.6-2.2	-?	3	11

CHART J23. Hydrogen–Carbon-13 Coupling Constants for Aromatics

AROMATIC HETEROCYCLICS, H–^{13}C

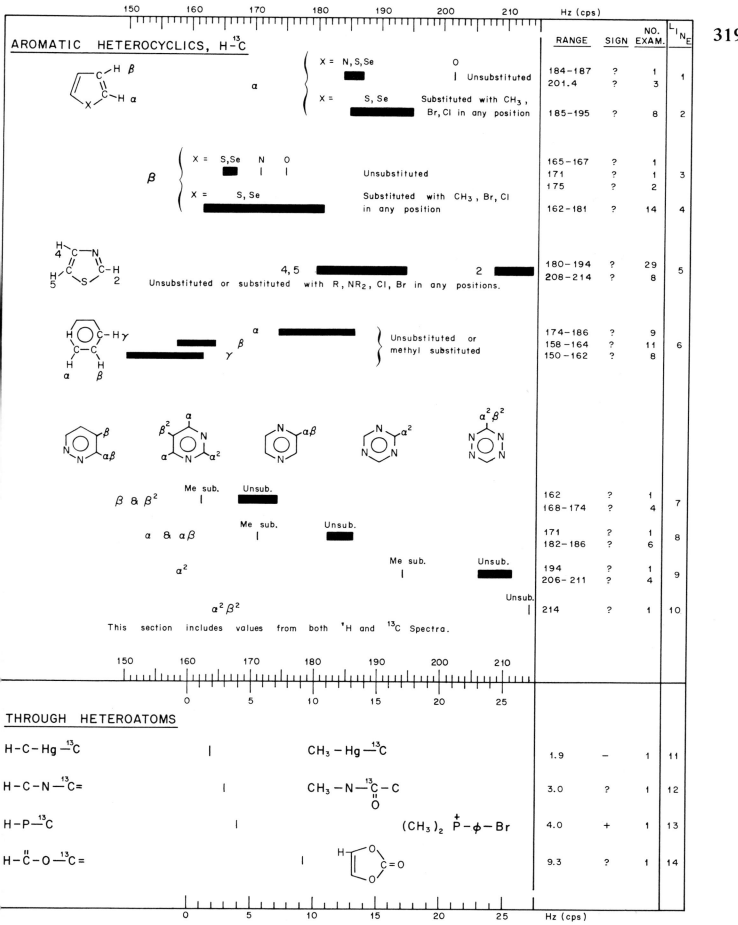

	RANGE	SIGN	NO. EXAM.	LINE
	184–187	?	1	1
	201.4	?	3	
	185–195	?	8	2
	165–167	?	1	3
	171	?	1	
	175	?	2	
	162–181	?	14	4
	180–194	?	29	5
	208–214	?	8	
	174–186	?	9	6
	158–164	?	11	
	150–162	?	8	
	162	?	1	7
	168–174	?	4	
	171	?	1	8
	182–186	?	6	
	194	?	1	9
	206–211	?	4	
	214	?	1	10

This section includes values from both ^{1}H and ^{13}C Spectra.

THROUGH HETEROATOMS

		RANGE	SIGN	NO. EXAM.	LINE
H–C–Hg—^{13}C	CH_3–Hg—^{13}C	1.9	–	1	11
H–C–N—^{13}C=	CH_3–N—^{13}C–C (with O double bond)	3.0	?	1	12
H–P—^{13}C	$(CH_3)_2\overset{+}{P}$–ϕ–Br	4.0	+	1	13
H–$\overset{O}{C}$–O—^{13}C=		9.3	?	1	14

CHART J24. Hydrogen–Carbon-13 Coupling Constants for Heterocompounds

CHART J24

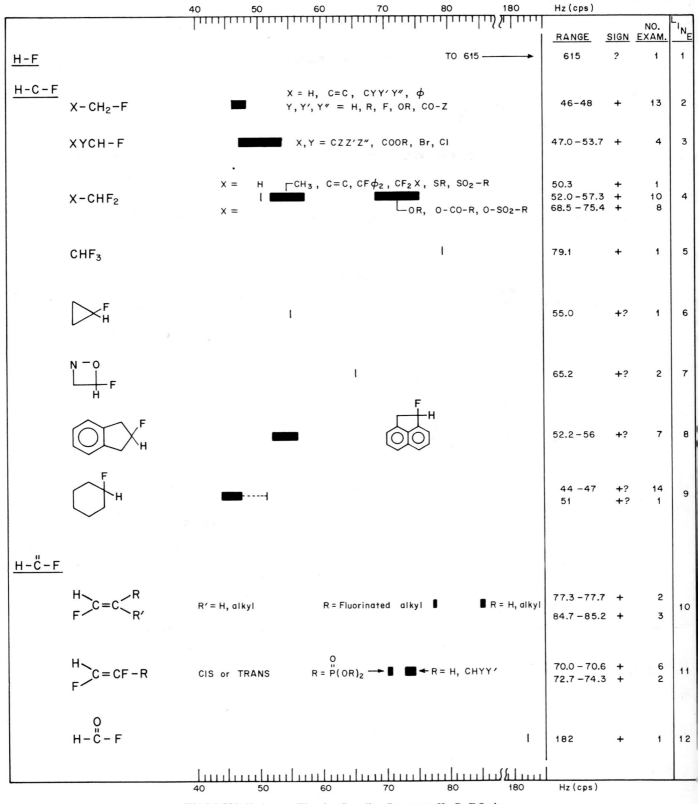

CHART J25. Hydrogen–Fluorine Coupling Constants, H–C–F Series

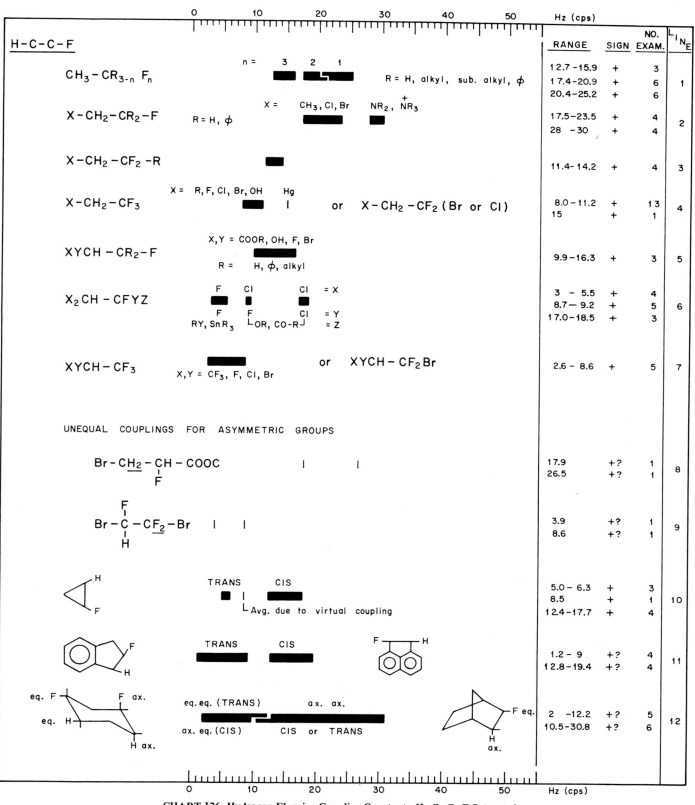

CHART J26. Hydrogen-Fluorine Coupling Constants, H–C–F Saturated

322

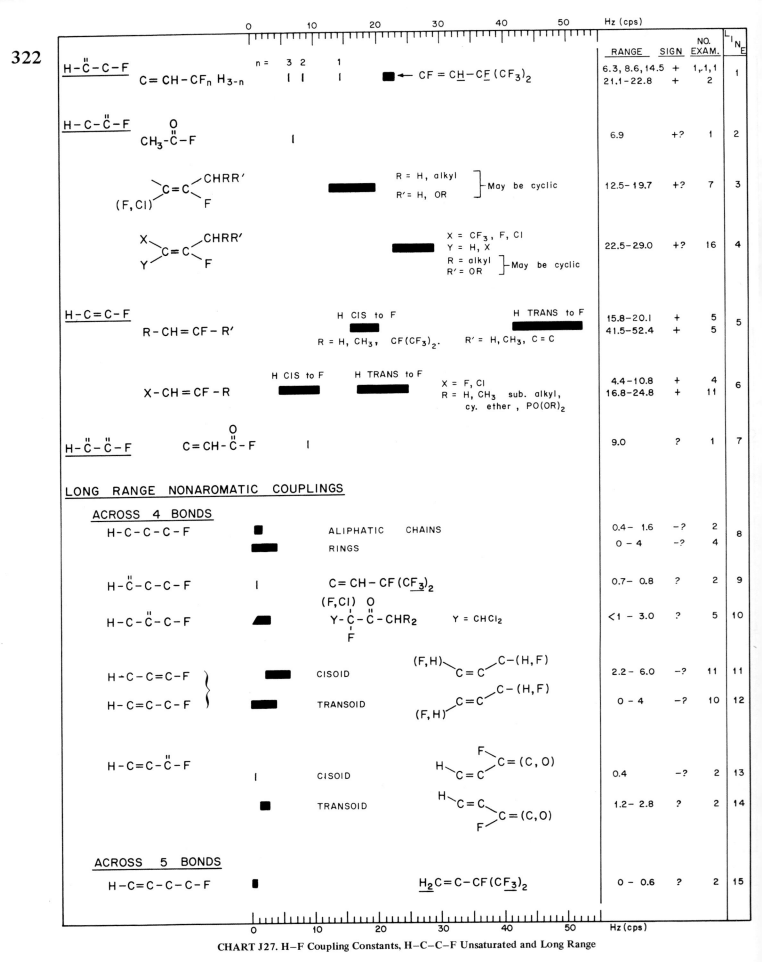

CHART J27. H—F Coupling Constants, H—C—C—F Unsaturated and Long Range

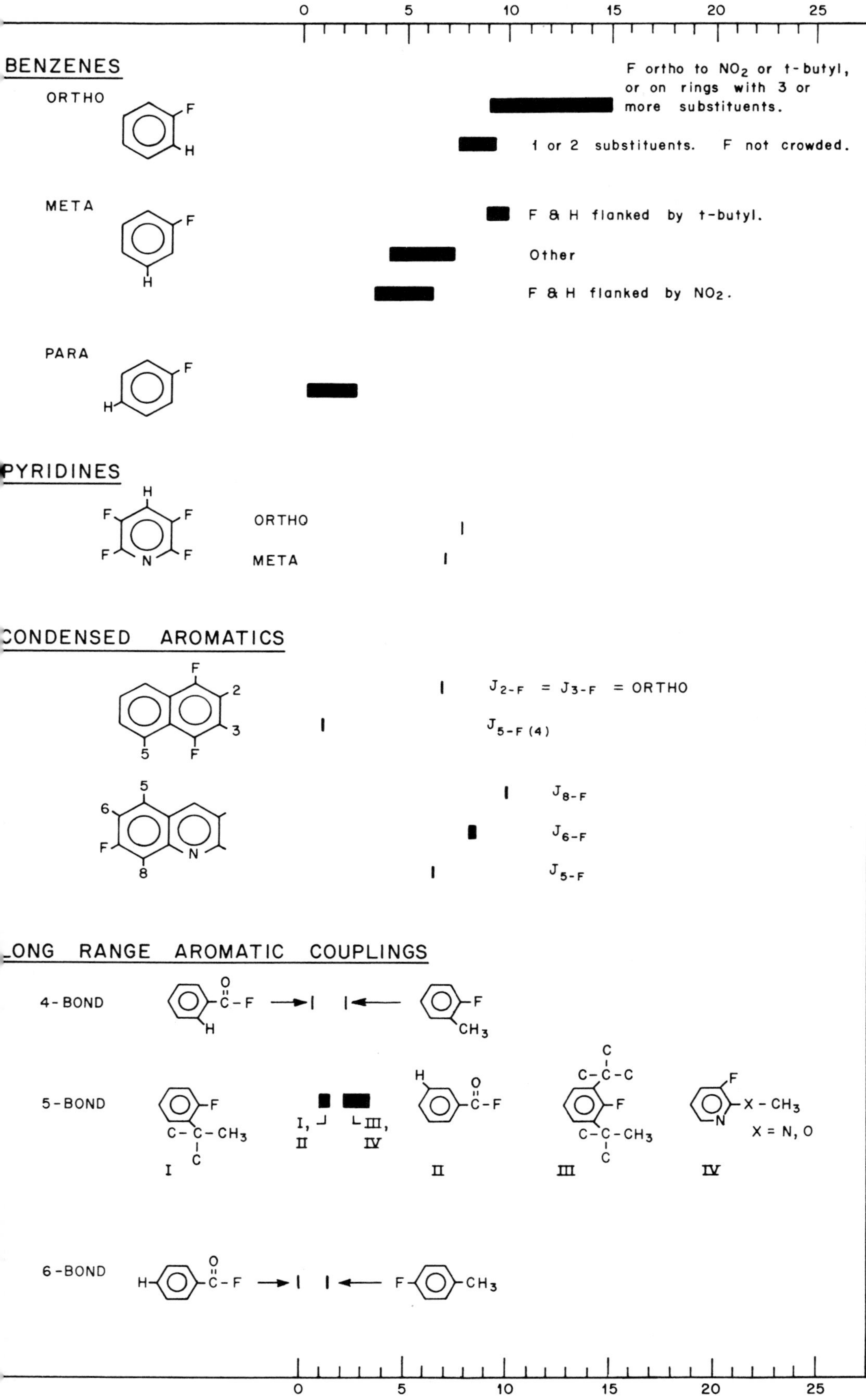

Hz (cps)

	RANGE	SIGN	NO. EXAM.	LINE

BENZENES

ORTHO

F ortho to NO₂ or t-butyl, or on rings with 3 or more substituents.

	RANGE	SIGN	NO. EXAM.	LINE
F ortho to NO₂ or t-butyl, or on rings with 3 or more substituents.	9.2 – 15	+	8	1
1 or 2 substituents. F not crowded.	7.6 – 9.4	+	43	2

META

F & H flanked by t-butyl.	9 – 10	+	4	3
Other	4.3 – 7.4	+	40	4
F & H flanked by NO₂.	3.6 – 6.4	+	2	5

PARA

	0.2 – 2.7	+	12	6

PYRIDINES

ORTHO	7.8	?	1	7
META	7.0	?	1	8

CONDENSED AROMATICS

$J_{2-F} = J_{3-F}$ = ORTHO	6.9	?	1	9
$J_{5-F(4)}$	1.1	?	1	10
J_{8-F}	10	?	2	11
J_{6-F}	8.2 – 8.4	?	2	12
J_{5-F}	6.5	?	2	13

LONG RANGE AROMATIC COUPLINGS

4-BOND

	0.6	–	1	
	2.3	?	1	14

5-BOND

I, II, III, IV; X = N, O

	1.0 – 1.5	+?	6	
	2.2 – 3.4	?	8	15

6-BOND

	0.02	–	1	
	1.4	?	1	16

Hz (cps)

CHART J28. Hydrogen–Fluorine Coupling Constants for Aromatics

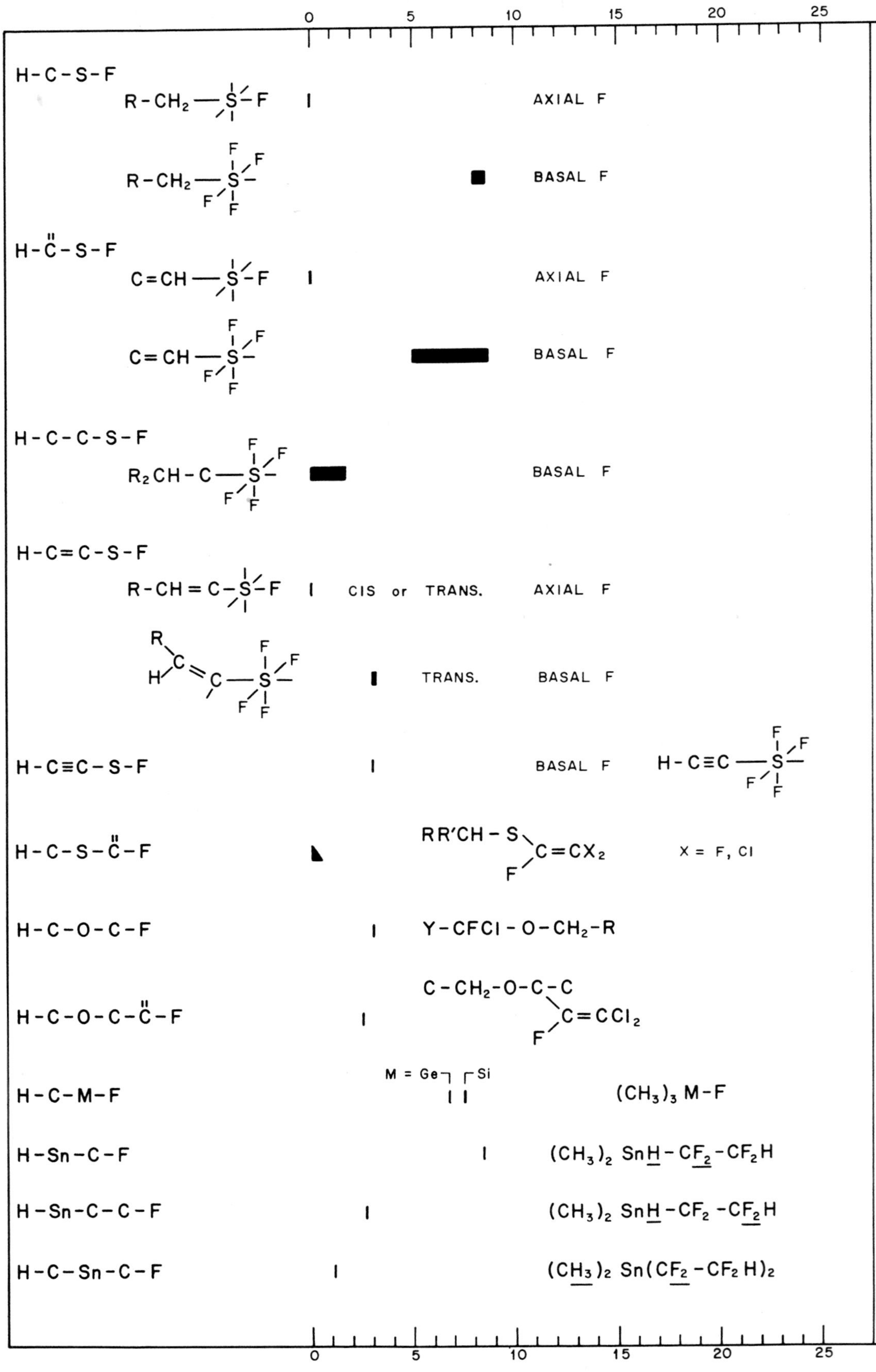

Hz (cps)

Structure		Notes	RANGE	SIGN	NO. EXAM.	LINE
H-C-S-F	R-CH₂—S'—F	AXIAL F	0.0		1	1
	R-CH₂—SF₄ (BASAL)	BASAL F	8.0 – 8.6	?	4	2
H-C''-S-F	C=CH—S'—F	AXIAL F	0.0		1	3
	C=CH—SF₄	BASAL F	5 – 8.7	?	4	4
H-C-C-S-F	R₂CH-C—SF₄	BASAL F	0 – 1.7	?	2	5
H-C=C-S-F	R-CH=C—S'—F	CIS or TRANS. AXIAL F	0.0		5	6
	(vinyl)—SF₄	TRANS. BASAL F	3.0 – 3.2	?	2	7
H-C≡C-S-F	H-C≡C—SF₄	BASAL F	3	?	1	8
H-C-S-C''-F	RR'CH-S-C(F)=CX₂ X = F, Cl		0 – <1	?	6	9
H-C-O-C-F	Y-CFCl-O-CH₂-R		3.0	?	1	10
H-C-O-C-C''-F	C-CH₂-O-C-C C(F)=CCl₂		2.5	?	1	11
H-C-M-F	M = Ge ⌐ ⌐ Si (CH₃)₃ M-F		6.8 / 7.5	? / ?	1 / 1	12
H-Sn-C-F	(CH₃)₂ SnH-CF₂-CF₂H		8.4	?	1	13
H-Sn-C-C-F	(CH₃)₂ SnH-CF₂-CF₂H		2.7	?	1	14
H-C-Sn-C-F	(CH₃)₂ Sn(CF₂-CF₂H)₂		1.1	?	1	15

Hz (cps)

CHART J29. Hydrogen–Fluorine Couplings Through Heteroatoms

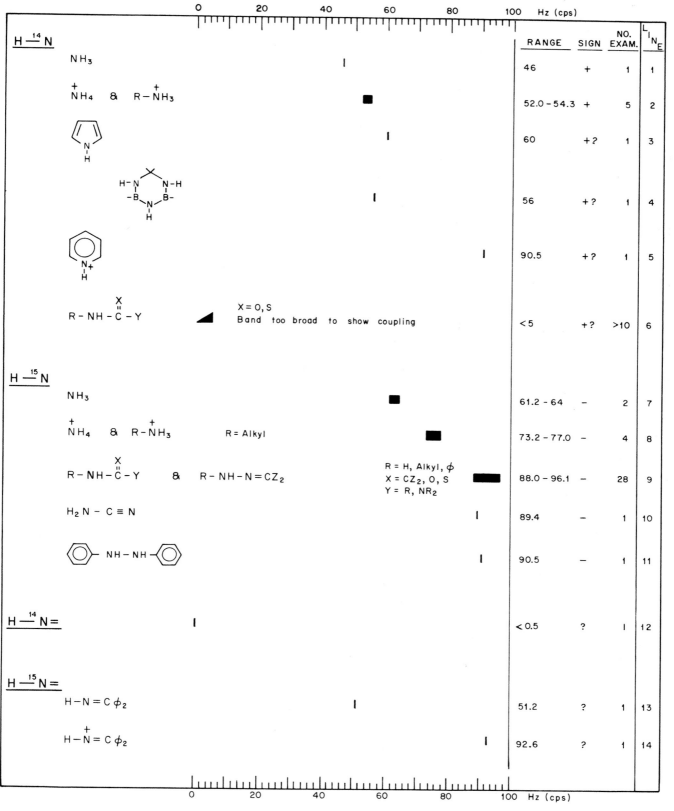

CHART J30. Hydrogen–Nitrogen Coupling Constants, H–N Series

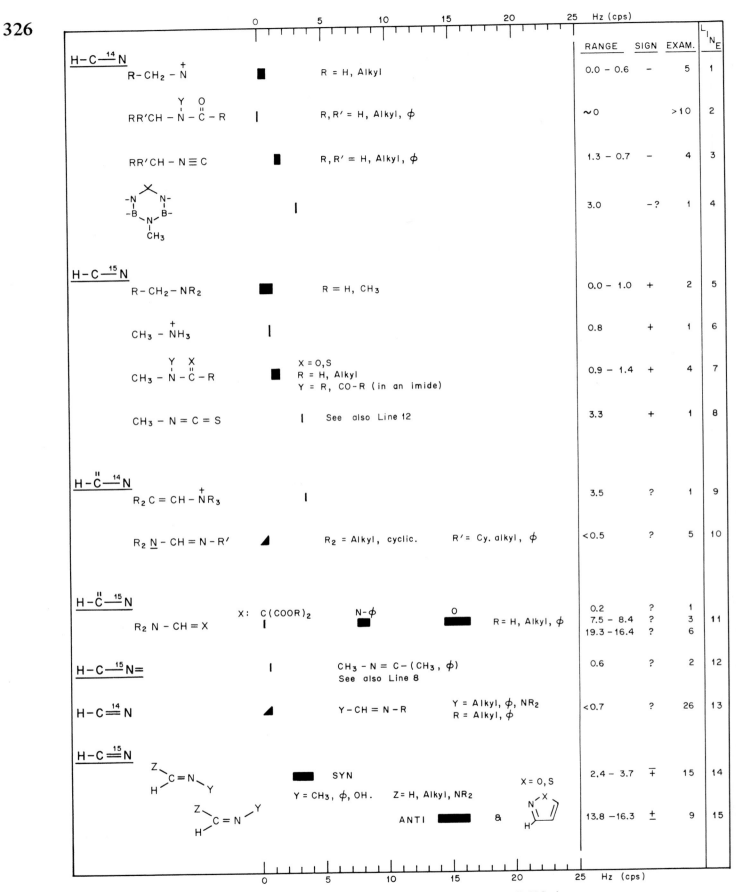

CHART J31. Hydrogen–Nitrogen Coupling Constants, H–C–N Series

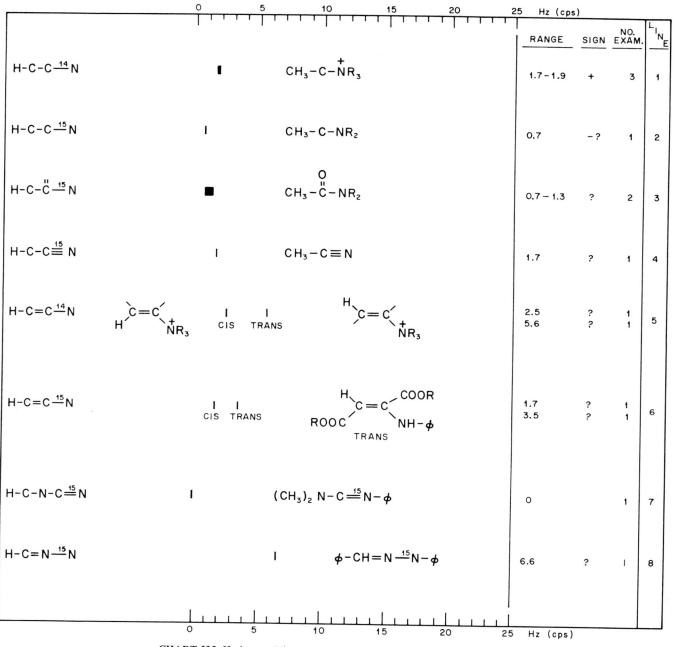

CHART J32. Hydrogen–Nitrogen Coupling Constants, H–C–C–N and Other

SUMMARY

LINE	NO. CMPDS.	SIGN	RANGE HZ (cps)
1	63	+	137–745
2	255, 1	+ & –	0.1–32, 94
3	631	H-C-C-P=+ OTHERS=?	0–51
4	117	?	0–11.2
5	10 WITH NONZERO VALUES	?	0–6.5
6	3	+	137–140
	8	+	177–225
7	6	+	427–490
	17	+	492–576
	7	+	600–657
	13	+	642–745
8	4	+	500–550
9	1	+	94
	1	+	444
	2	+	620–685
	1	+	725

Hz ... (scale 0 100 200 300 400 500 600 700 800) ... HZ (cps)

H–P

H–P

H–X–P X= C, Si, Ge, P

H–C–X–P X= C,O,S,N,P,As

H–C–C–X–P X= C,O,S,N

5 OR MORE BONDS NONZERO VALUES FOR H-C-C=C-P, H-C-C-C=C=C-P, H-C≡C-C≡C-P, ETC.

R = HYDROCARBON RADICAL; SAT., UNSAT., OR AROM.

H–P

$\underset{Y}{\overset{X}{H-P}}$ X= METAL OR ⊖, H, R, RCl, F   Y= H, R, RCl, F, PR$_2$

$Z=\underset{Y}{\overset{X}{H-P}}$ X= R, OR, OR, OR   Y= R, R, OR, OR   Z= O,S, O,S, S, O

$\overset{⊕}{\underset{Y}{H-P}}\overset{X}{}$ X= H, R   Y= H, R

MISCELLANEOUS

$\left[\underset{O\ O}{\overset{O\ O}{H-P-P-O}}\right]^{-3}$ J_{H-P-P}

$\left[\underset{O\ O}{\overset{O\ O}{X-P-O-P-H}}\right]$ W = –2, –3 X= H, O J_{H-P}

$\underset{S}{\overset{\|}{H-P}}F_2$

CHART J33. Hydrogen–Phosphorus Coupling Constants. *Summary* and H–P Series

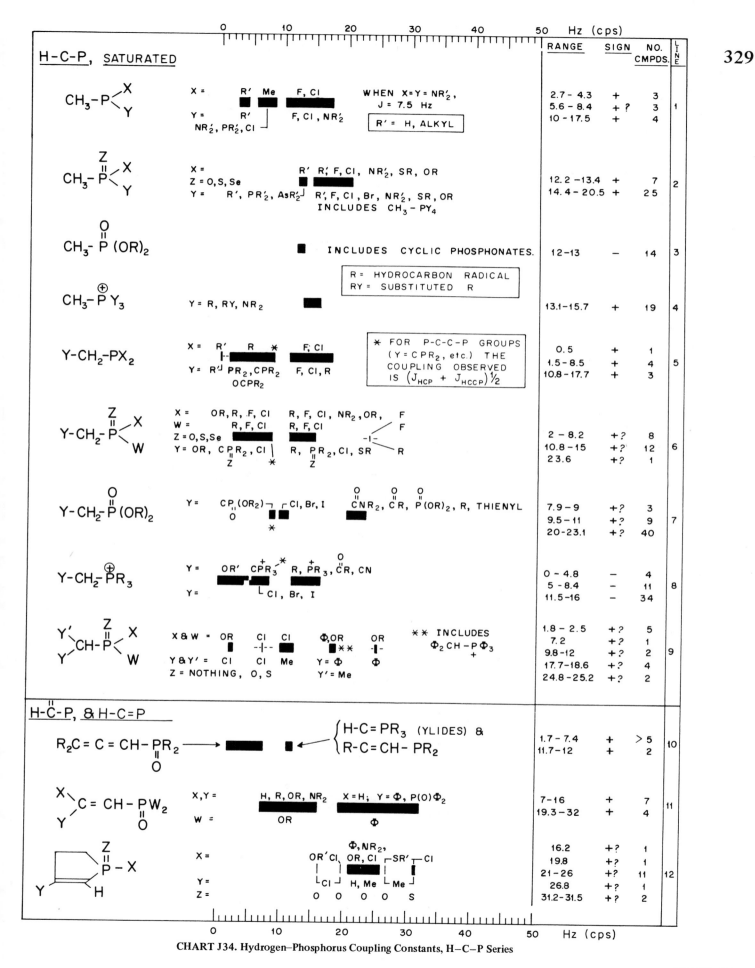

CHART J34. Hydrogen–Phosphorus Coupling Constants, H–C–P Series

330

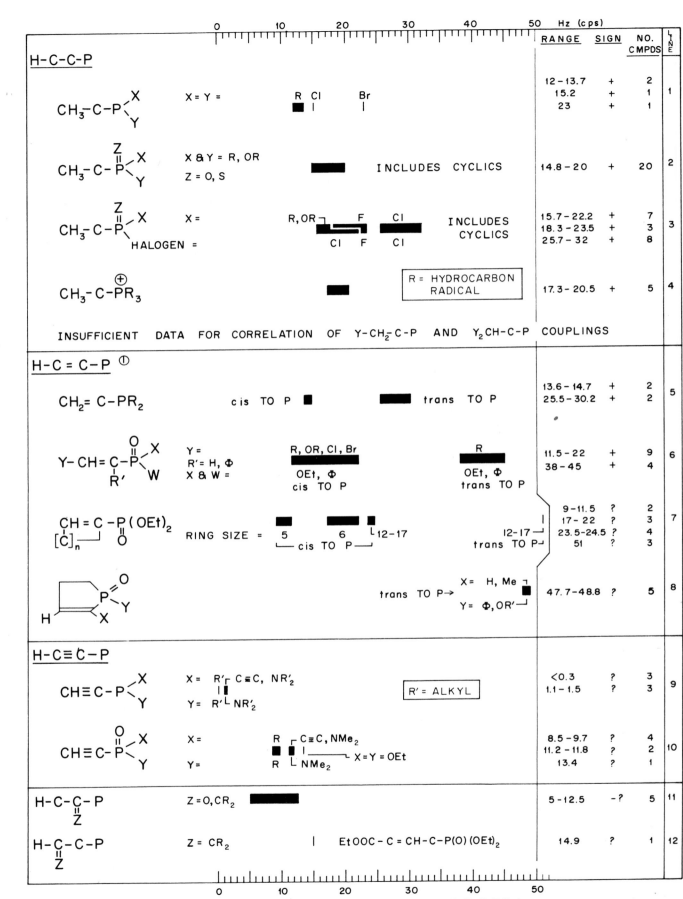

CHART J35. Hydrogen-Phosphorus Coupling Constants, H–C–C–P Series

1. Coupling constants associated with aromatic rings are in Chart J38.

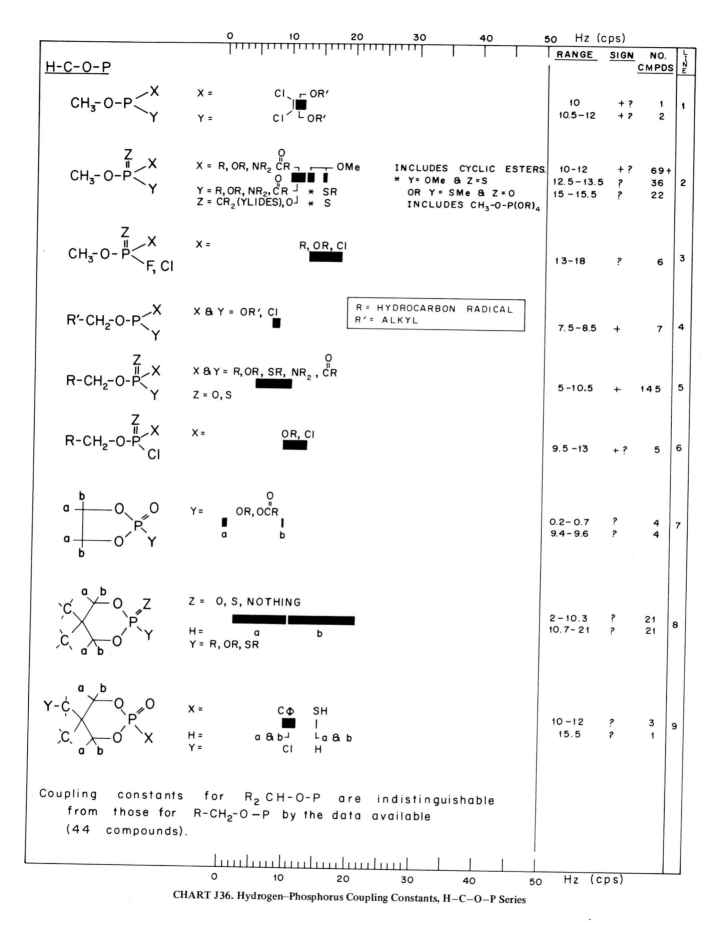

CHART J36. Hydrogen–Phosphorus Coupling Constants, H–C–O–P Series

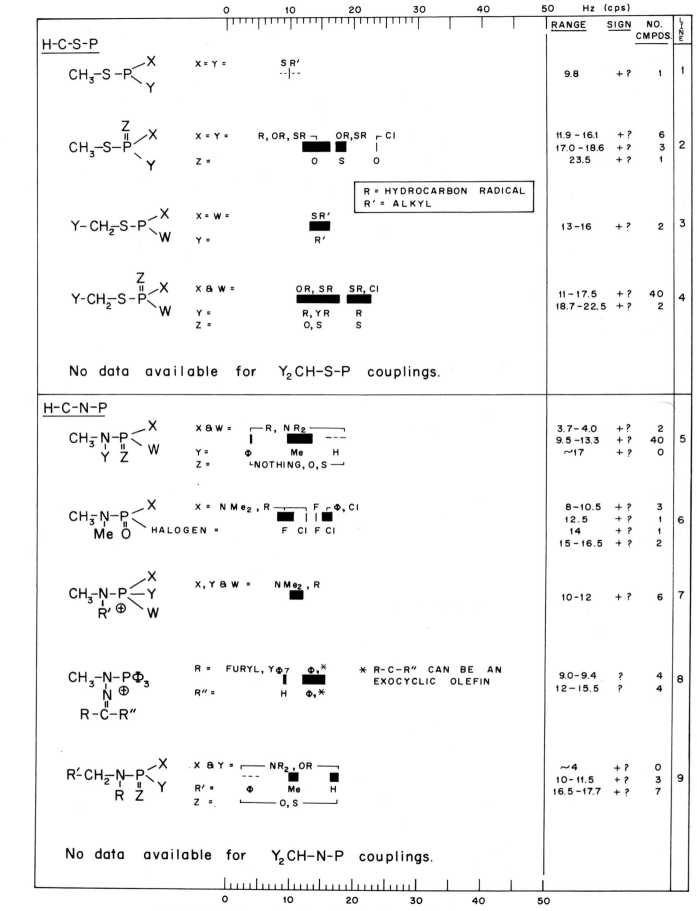

CHART J37. Hydrogen–Phosphorus Coupling Constants, H–C–S–P and H–C–N–P

0 10 20 30 40 50 Hz (cps)

	RANGE	SIGN	NO. CMPDS	LINE

H—⬡—P

Line 1:
Y = H, C-O-C
X = WΦ *, H, WΦ
W = WΦ, H
H = m, o

* IN TRIPHENYL PHOSPHINE
$J_o \sim J_m \sim J_p = 3.5$ Hz.

** H_m-P & H_o-P HAVE THE SAME SIGN.

	RANGE	SIGN	NO.	LINE
	1.1 – 3.5	?	2	1
	6.6 – 7.9	?	3	
		**		

Line 2:
X = YΦ OR Cl, YΦ OR, SR, R
W = YΦ OR Cl, YΦ OR, Cl
H = m m o, o o
Y = H, R, OH, OR, COOR, NR_2, Br
Z = O, S, Se

	RANGE	SIGN	NO.	LINE
	2.1 – 3.4	?	13	2
	4.2	?	1	
	6.7	?	1	
	10.5 – 11.5	?	14	
	12 – 16	?	35	

Line 3:
X & Y = OR, NR_2, Cl
H = m o

	RANGE	SIGN	NO.	LINE
	0.9 – 1.6	?	6	3
	1.4 – 2.8	?	5	

LONG RANGE AROMATIC COUPLINGS

Line 4:
X = R, OR, Cl
Y = OR, Cl
Me = m o o

R = HYDROCARBON RADICAL

	RANGE	SIGN	NO.	LINE
	0		—	4
	1 – 2	?	3	
	3.7	?	1	

Line 5:
X & Y = R, OR
Z = O C
H = Ortho

	RANGE	SIGN	NO.	LINE
	0.8 – 1.1	?	3	5
	2.2 – 2.8	?	2	

LONG RANGE ALIPHATIC COUPLINGS

Structure	RANGE	SIGN	NO.	LINE
H-C-C-C-P	0		—	6
H-C-C=C-P(O)(YR)$_2$ Y = C, O	1.9 – 3.2	?	2	7
Y-CH=C-C-P(O)(OR)$_2$ Y = Me, H, Br, COOR	5.0 – 7.4	?	4	8
R-CH=C=C-P(O)Φ$_2$	10.9 – 11.2	?	3	9
CH$_2$=C=C-PΦ$_2$	1.6	?	1	10
Y$_2$CH-C≡C-P(Z)X$_2$ Y = H,R,OR,Cl. X = R,OR,NR$_2$. Z = O,S,NOTHING	0.9 – 4.6	?	23	11
CH$_3$-(C≡C)$_2$-P(Z)X$_2$ X = R,OR. Z = O,S,NOTHING.	1.3 – 2.2	?	4	12
H-C-C=C-C-P INCLUDES H-C-C≡C-C-P & H-C-C=C=C-P	5.6 – 6.5	?	5	13
H-C-C-O-P * HCCOP$_{III}$ = +. HCCOP(O) = –. HCCOP(S) = +	0 – 1	*	28	14
H$_2$C=C-O-P(O)(OEt)$_2$ trans/cis	1, 2.7	?	2, 2	15
H-C≡C-O-P(O)(OEt)$_2$	0.35	?	1	16
H-C-C-S-P(O)(OR)$_2$ <?	1.5, 6.7?	?	3,1	17
H-C-C(=CY$_2$)-S-P(O)(OR)$_2$	1.5 – 3	?	2	18
H-C=C-S-P(Z)(OR)$_2$ Z = O, S. cis ≤ trans	2 – 7.3	?	3	19
H-C-C=C-S-P cis/trans CH$_3$-C=C-S-P(S)(OEt)$_2$	2.2, 3.8	?	1	20
H-C-C-N-P	<1	?	9	21

0 10 20 30 40 50 Hz (cps)

CHART J38. Hydrogen–Phosphorus Coupling Constants, Aromatic and Long Range

Typical Spectra

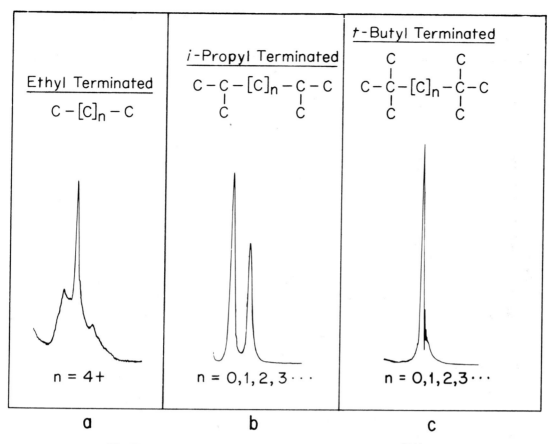

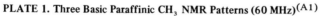

PLATE 1. Three Basic Paraffinic CH_3 NMR Patterns (60 MHz)[A1]

PLATE 1

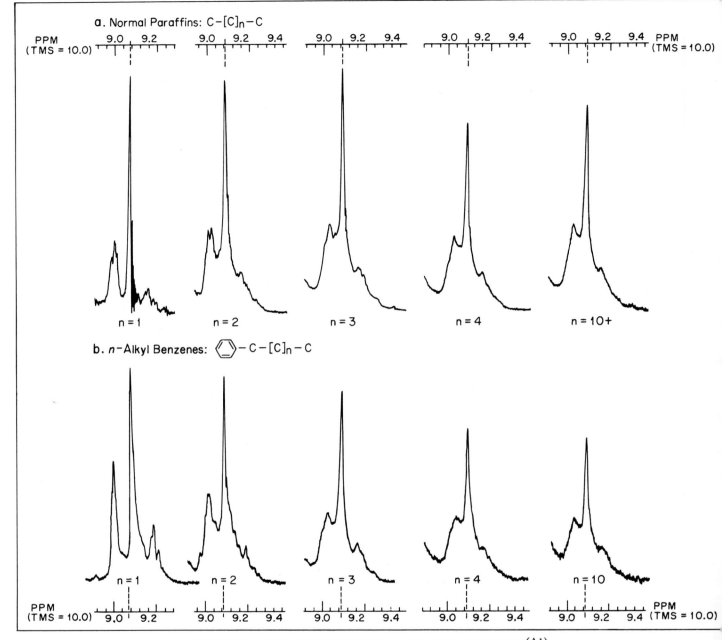

PLATE 2. Alkyl CH₃ NMR Triplet Patterns vs. Chain Length (60 MHz)[A1]

PLATE 2

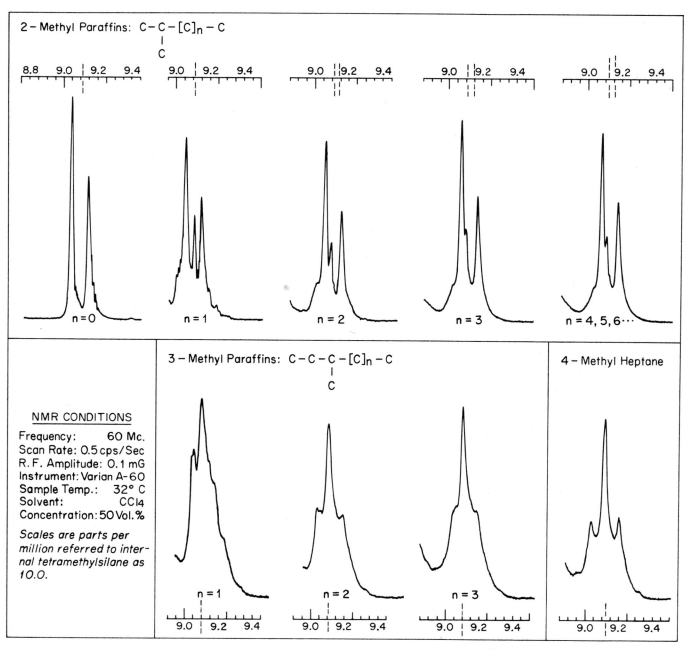

PLATE 3. CH$_3$ NMR Patterns for Monomethyl Paraffins (60 MHz)[A1]

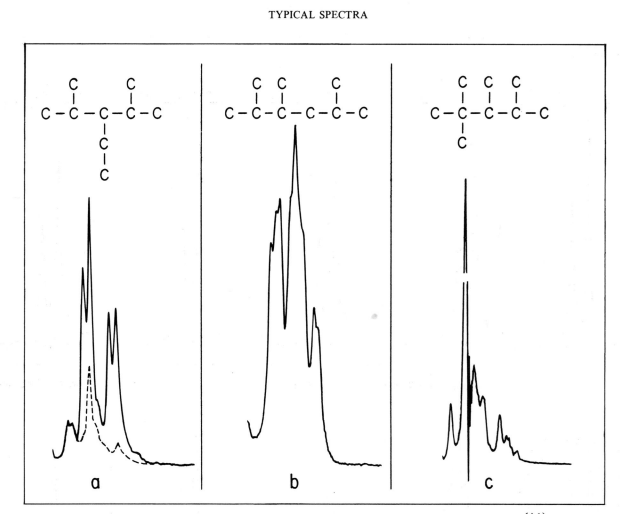

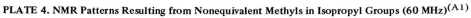

PLATE 4. NMR Patterns Resulting from Nonequivalent Methyls in Isopropyl Groups (60 MHz)[A1]

PLATE 4

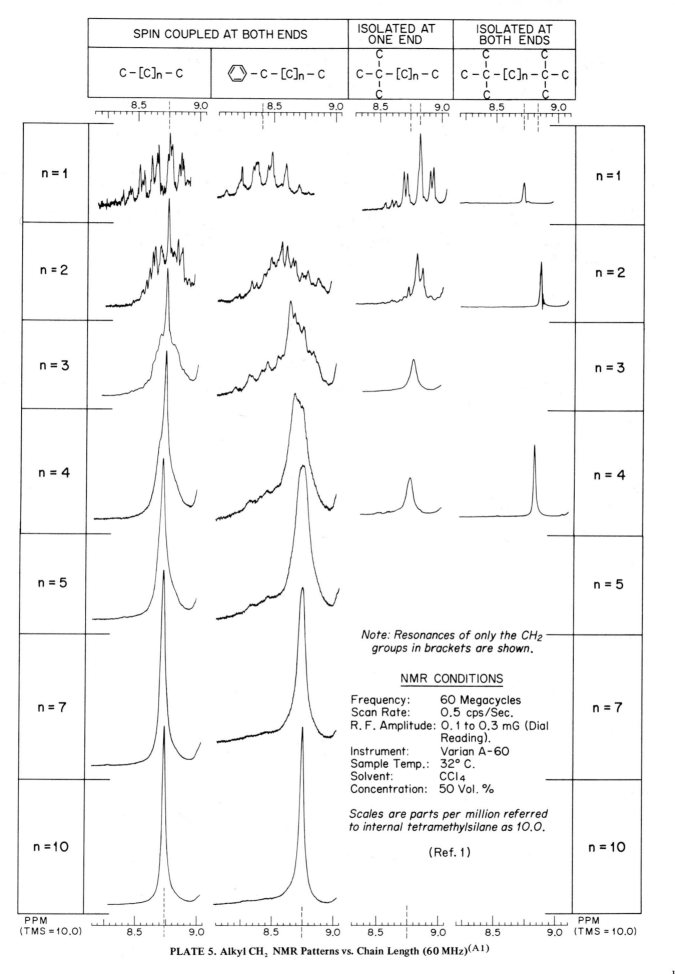

PLATE 5. Alkyl CH₂ NMR Patterns vs. Chain Length (60 MHz)[A1]

PLATE 5

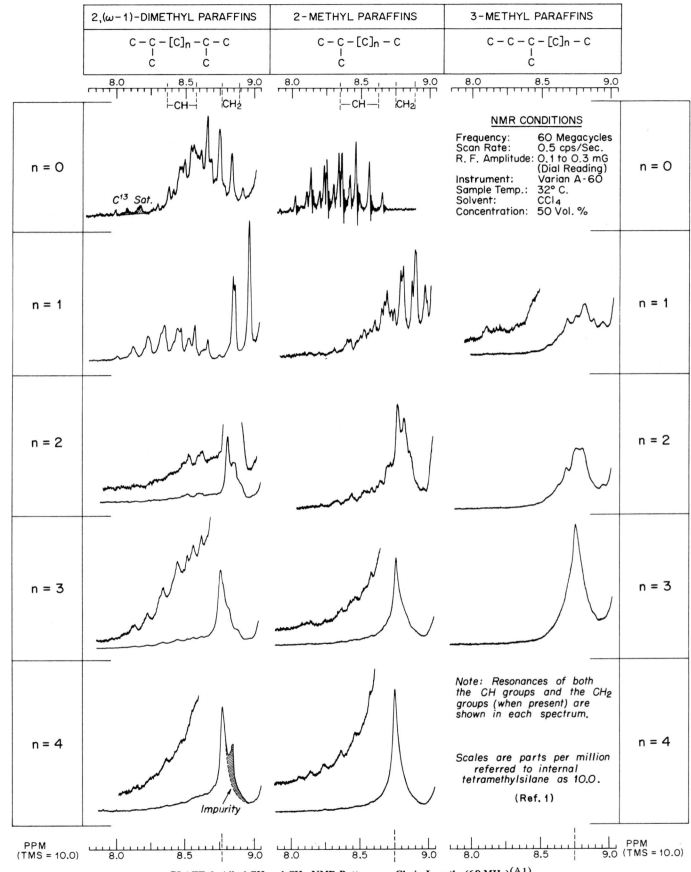

2,(ω−1)-DIMETHYL PARAFFINS	2-METHYL PARAFFINS	3-METHYL PARAFFINS
C − C − [C]$_n$ − C − C | | | C C	C − C − [C]$_n$ − C | C	C − C − C − [C]$_n$ − C | C

NMR CONDITIONS

Frequency: 60 Megacycles
Scan Rate: 0.5 cps/Sec.
R. F. Amplitude: 0.1 to 0.3 mG
(Dial Reading)
Instrument: Varian A-60
Sample Temp.: 32° C.
Solvent: CCl$_4$
Concentration: 50 Vol. %

Note: Resonances of both the CH groups and the CH$_2$ groups (when present) are shown in each spectrum.

Scales are parts per million referred to internal tetramethylsilane as 10.0.

(Ref. 1)

C^{13} Sat.

Impurity

PPM
(TMS = 10.0)

PLATE 6. Alkyl CH and CH$_2$ NMR Patterns vs. Chain Length. (60 MHz)[A1]

PLATE 6

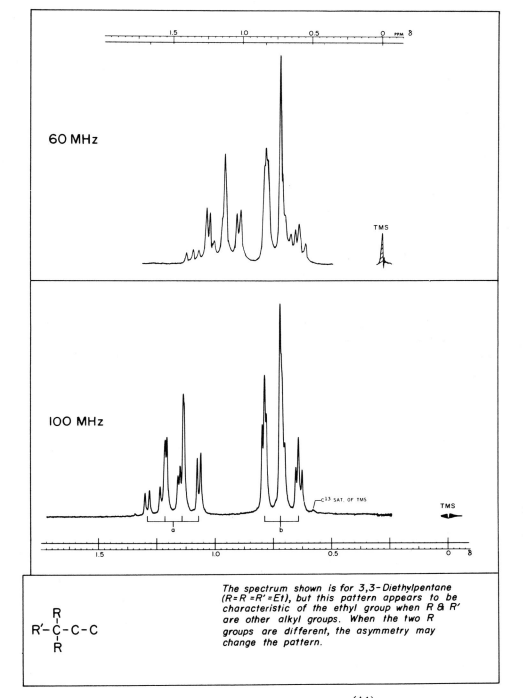

The spectrum shown is for 3,3-Diethylpentane
(R=R =R' =Et), but this pattern appears to be
characteristic of the ethyl group when R & R'
are other alkyl groups. When the two R
groups are different, the asymmetry may
change the pattern.

PLATE 7. The Isolated Paraffinic Ethyl Group[A1]

PLATE 7

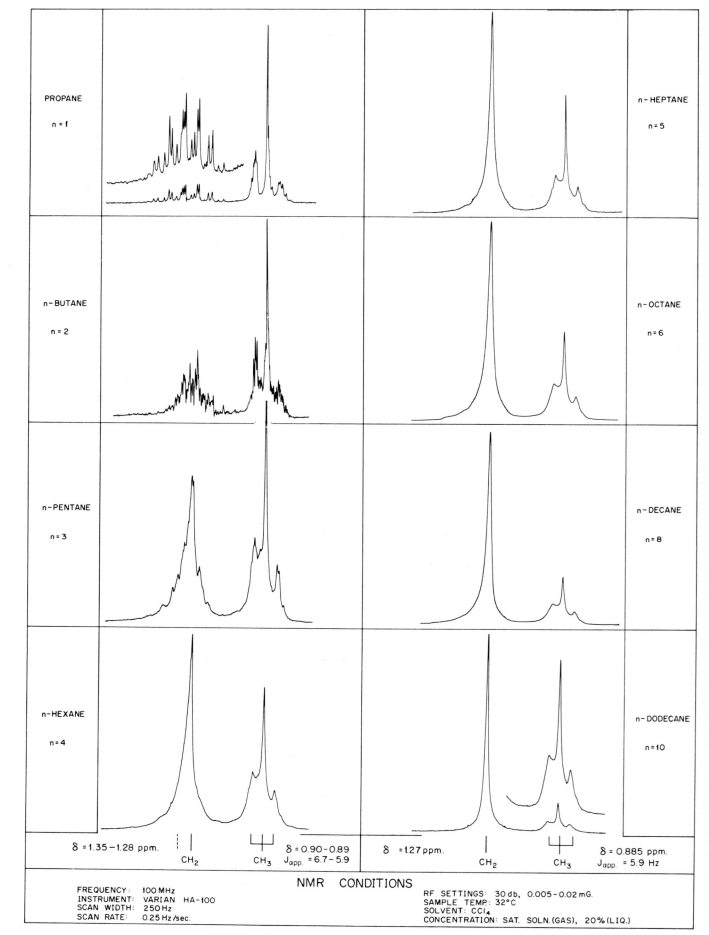

PROPANE

n = 1

n-BUTANE

n = 2

n-PENTANE

n = 3

n-HEXANE

n = 4

n-HEPTANE

n = 5

n-OCTANE

n = 6

n-DECANE

n = 8

n-DODECANE

n = 10

$\delta = 1.35-1.28$ ppm. $\delta = 0.90-0.89$ $J_{app.} = 6.7-5.9$

CH_2 CH_3

$\delta = 1.27$ ppm. $\delta = 0.885$ ppm. $J_{app.} = 5.9$ Hz

CH_2 CH_3

NMR CONDITIONS

FREQUENCY: 100 MHz
INSTRUMENT: VARIAN HA-100
SCAN WIDTH: 250 Hz
SCAN RATE: 0.25 Hz/sec.

RF SETTINGS: 30 db, 0.005-0.02 mG.
SAMPLE TEMP: 32°C
SOLVENT: CCl₄
CONCENTRATION: SAT. SOLN. (GAS), 20% (LIQ.)

PLATE 8. NMR Spectra of n-Paraffins (100 MHz) C–[C] n–C

PLATE 8

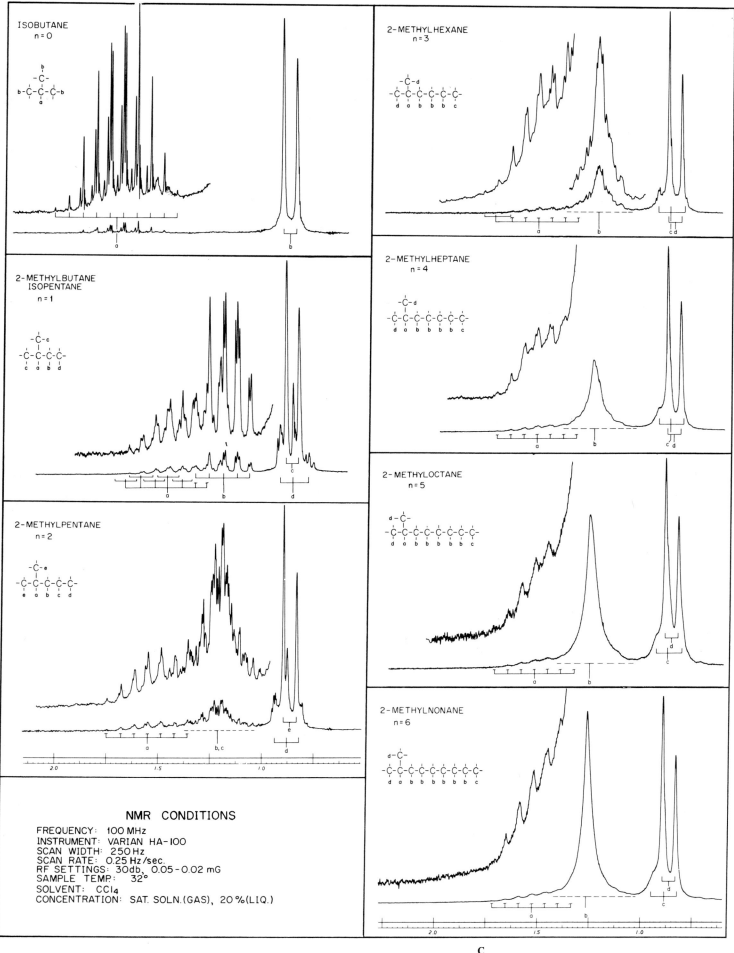

PLATE 9. NMR Spectra of 2-Methyl Paraffins (100 MHz) C–C–[C]n–C

ISOBUTANE
n = 0

2-METHYLBUTANE
ISOPENTANE
n = 1

2-METHYLPENTANE
n = 2

2-METHYLHEXANE
n = 3

2-METHYLHEPTANE
n = 4

2-METHYLOCTANE
n = 5

2-METHYLNONANE
n = 6

NMR CONDITIONS

FREQUENCY: 100 MHz
INSTRUMENT: VARIAN HA-100
SCAN WIDTH: 250 Hz
SCAN RATE: 0.25 Hz/sec.
RF SETTINGS: 30db, 0.05-0.02 mG
SAMPLE TEMP: 32°
SOLVENT: CCl₄
CONCENTRATION: SAT. SOLN.(GAS), 20%(LIQ.)

PLATE 9

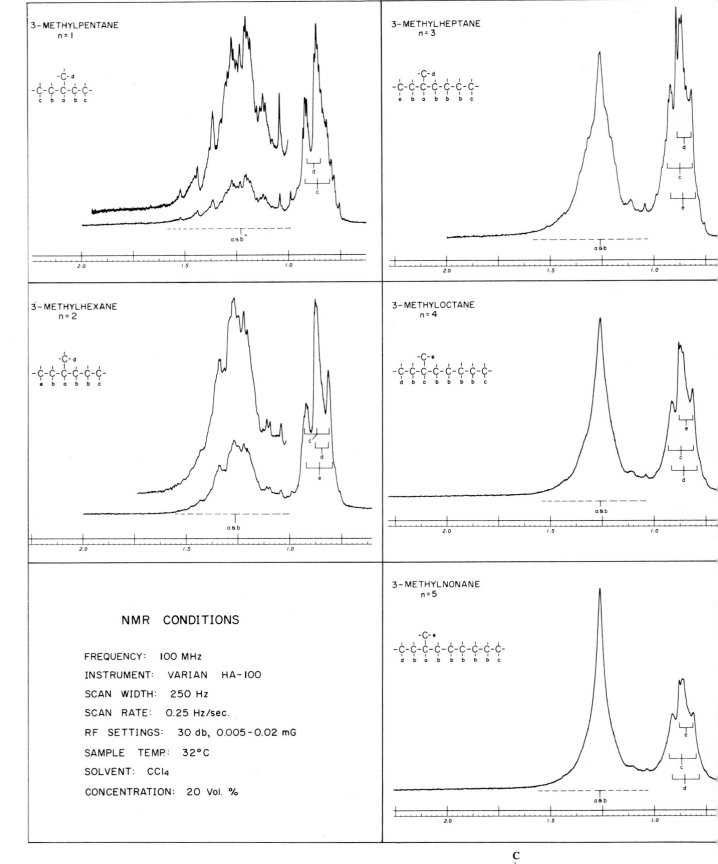

PLATE 10. NMR Spectra of 3-Methyl Paraffins (100 MHz) C–C–C–[C]$_n$–C

PLATE 10

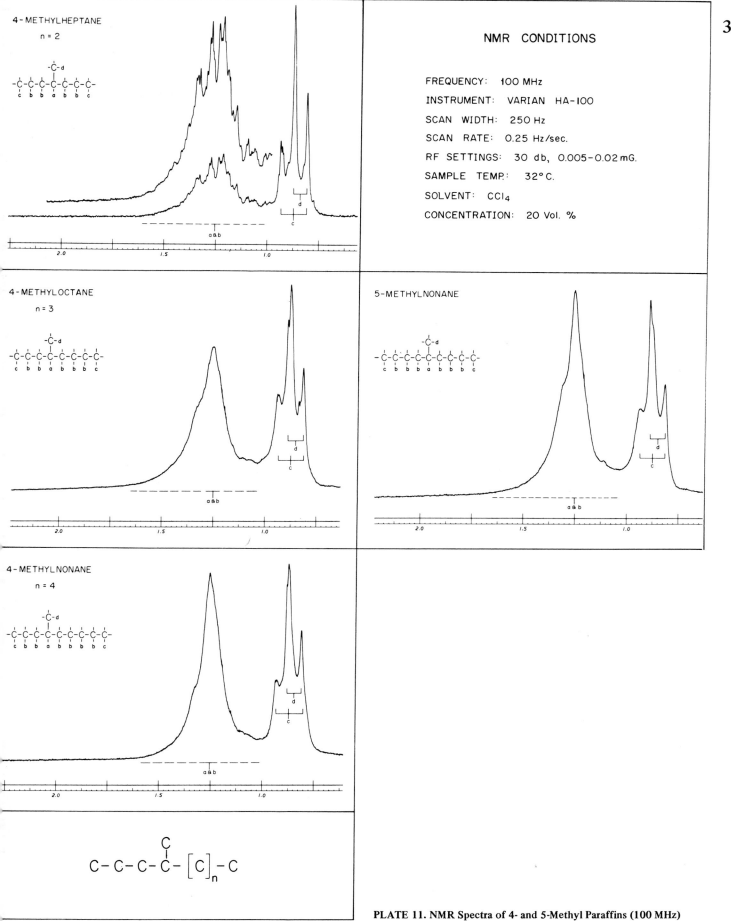

PLATE 11. NMR Spectra of 4- and 5-Methyl Paraffins (100 MHz)

PLATE 11

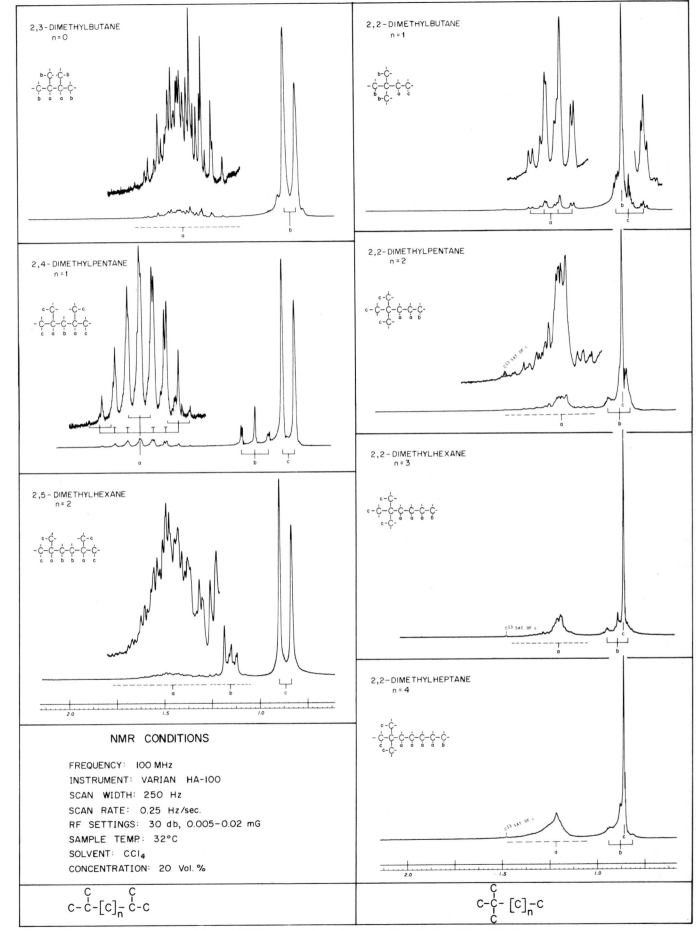

348

NMR CONDITIONS

FREQUENCY: 100 MHz
INSTRUMENT: VARIAN HA-100
SCAN WIDTH: 250 Hz
SCAN RATE: 0.25 Hz/sec.
RF SETTINGS: 30 db, 0.005-0.02 mG
SAMPLE TEMP: 32°C
SOLVENT: CCl₄
CONCENTRATION: 20 Vol. %

PLATE 12. NMR Spectra of Some Important Dimethyl Paraffins (100 MHz)

PLATE 12

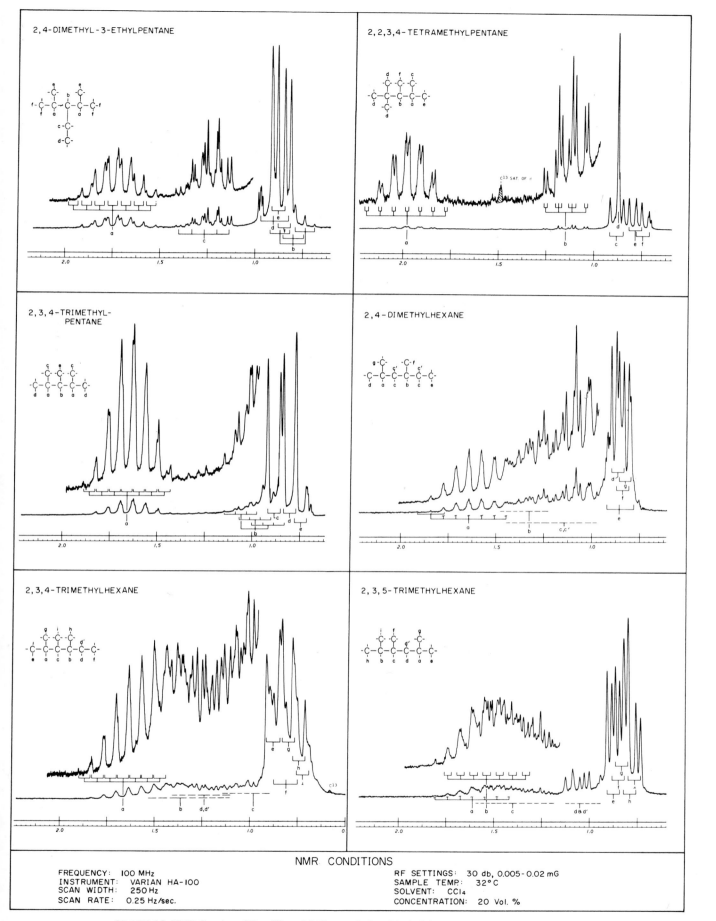

PLATE 13. NMR Spectra of Paraffins with Nonequivalent Methyls in Isopropyl Groups (100 MHz)

PLATE 13

350

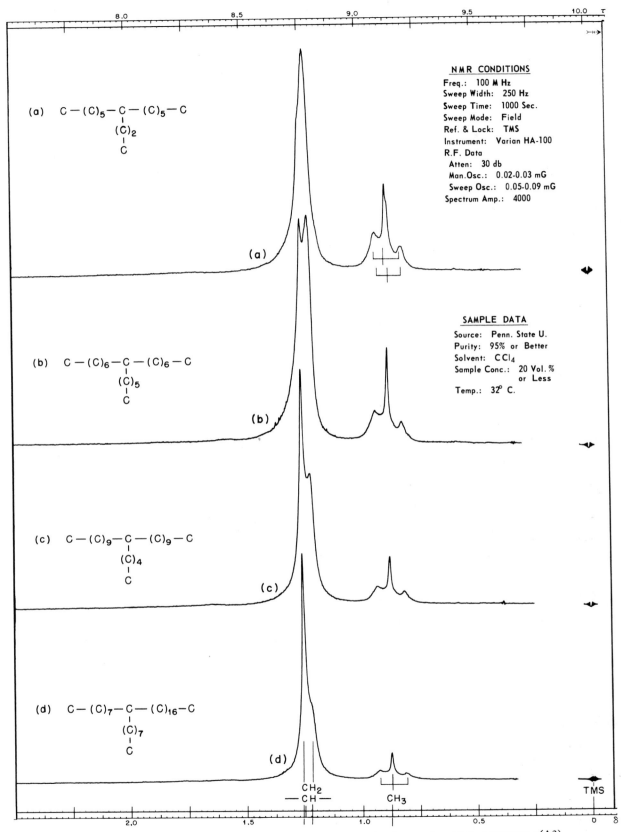

PLATE 14. Typical NMR Spectra of Paraffins with Single Branches Longer than Methyl (100 MHz)[A3]

PLATE 14

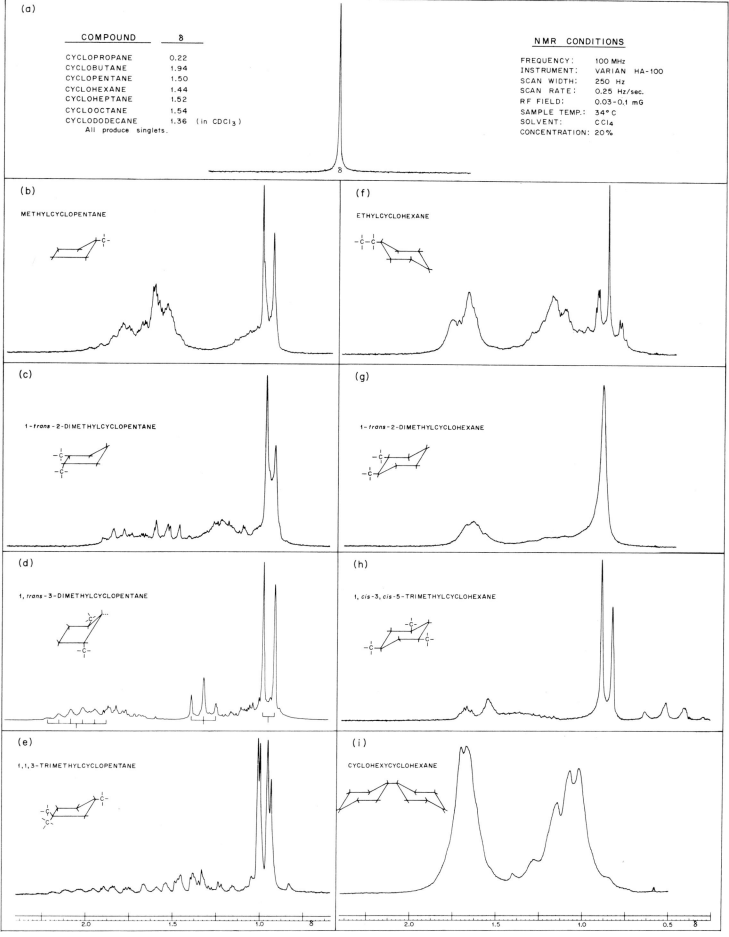

PLATE 15. Representative NMR Spectra of Cycloparaffins (100 MHz)

PLATE 15

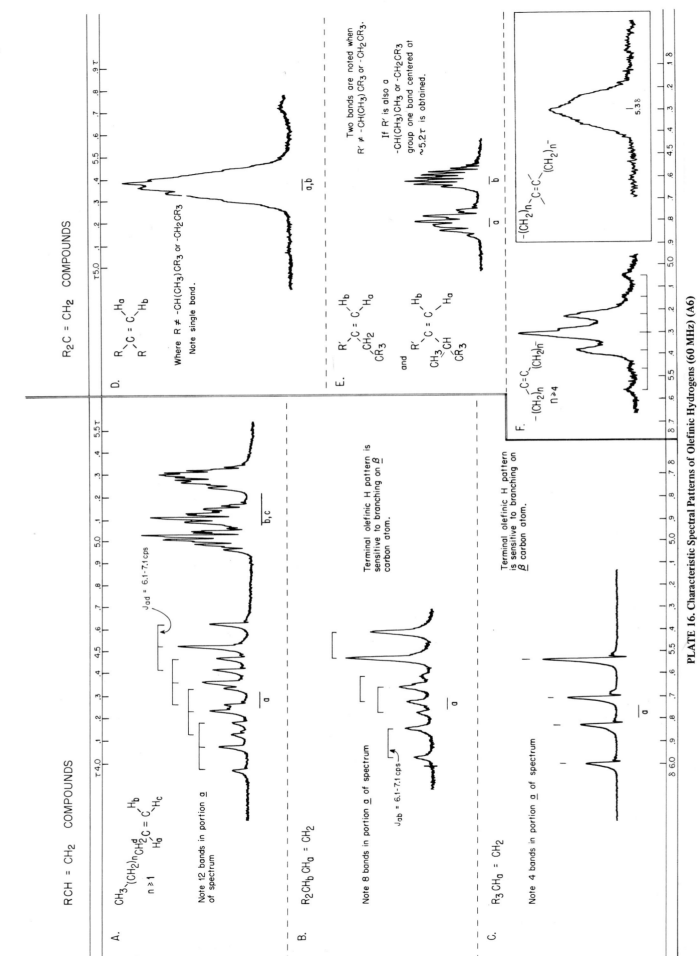

PLATE 16. Characteristic Spectral Patterns of Olefinic Hydrogens (60 MHz) (A6)

PLATE 16

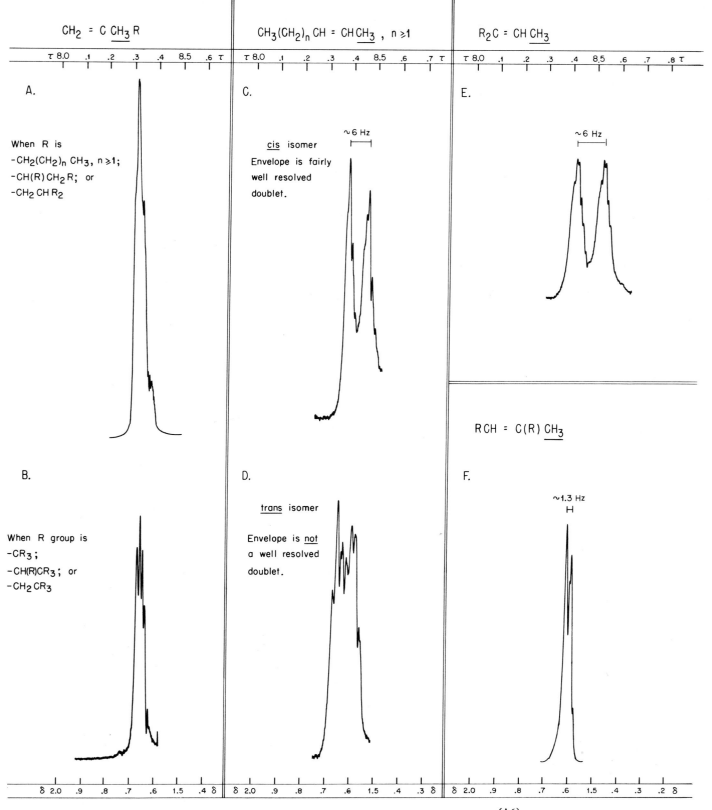

PLATE 17. Characteristic Spectral Patterns for α−CH₃ Groups (60 MHz)[A6]

PLATE 17

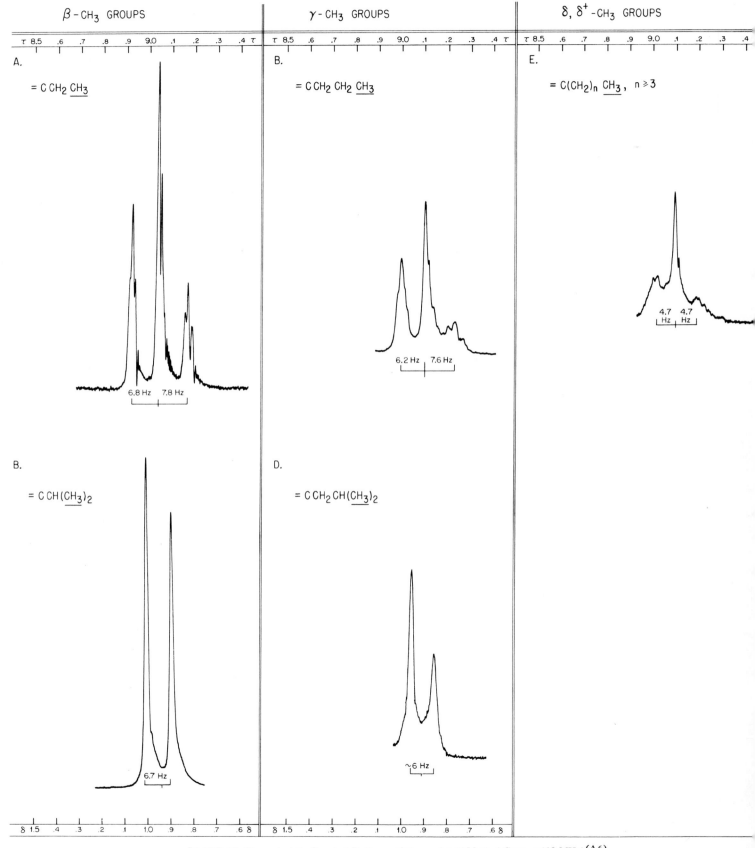

PLATE 18. Characteristic Spectral Patterns of β,γ, and δ,δ⁺ Methyl Groups (60 MHz)[A6]

PLATE 18

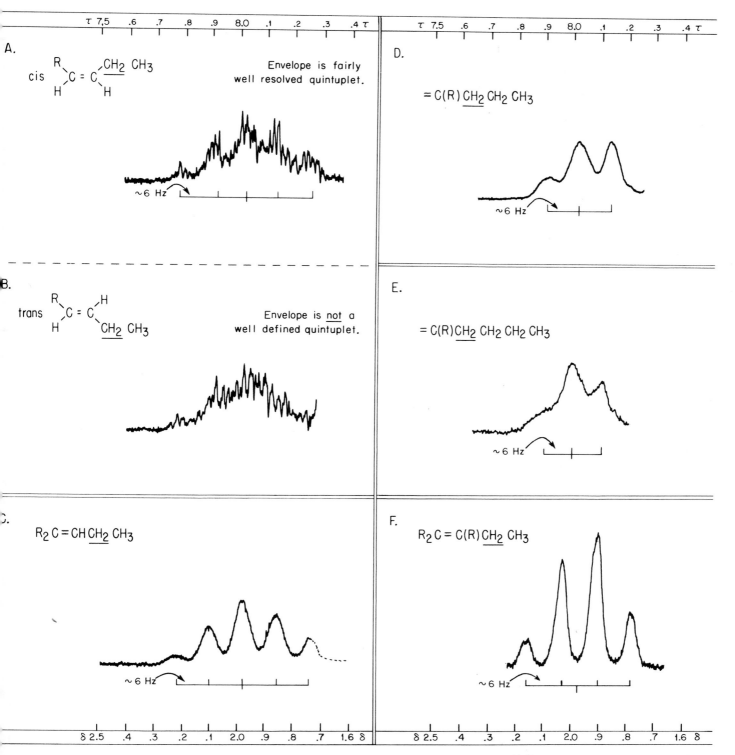

PLATE 19. Characteristic Spectral Patterns of α-Methylene Groups (60 MHz)[A6]

PLATE 19

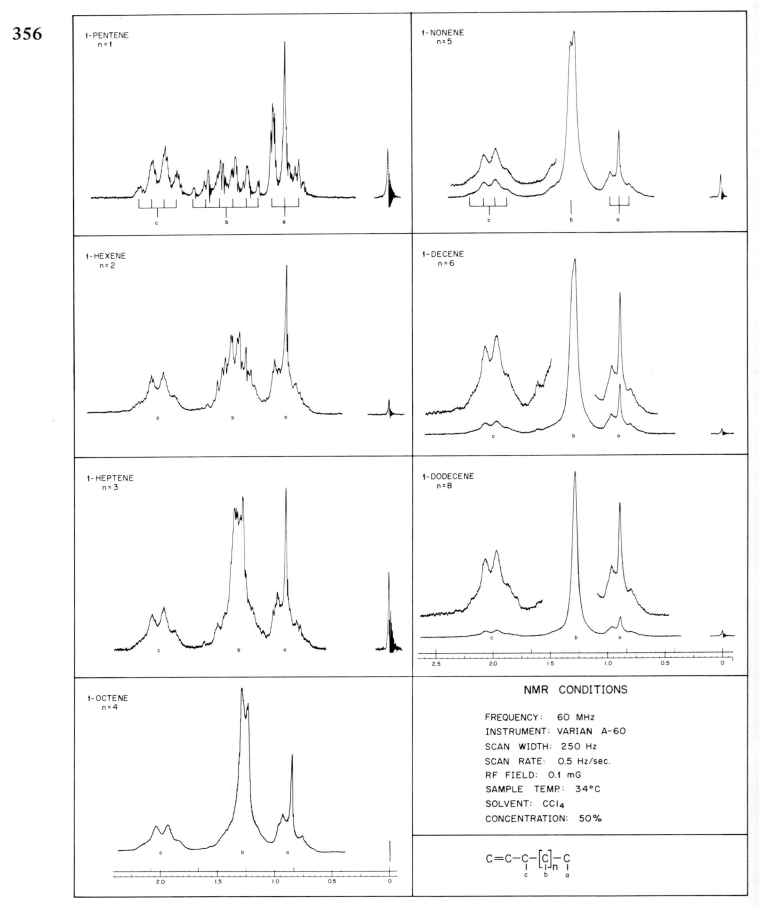

PLATE 20. NMR Spectra of the Saturated Part of Linear α-Olefins (60 MHz)

PLATE 20

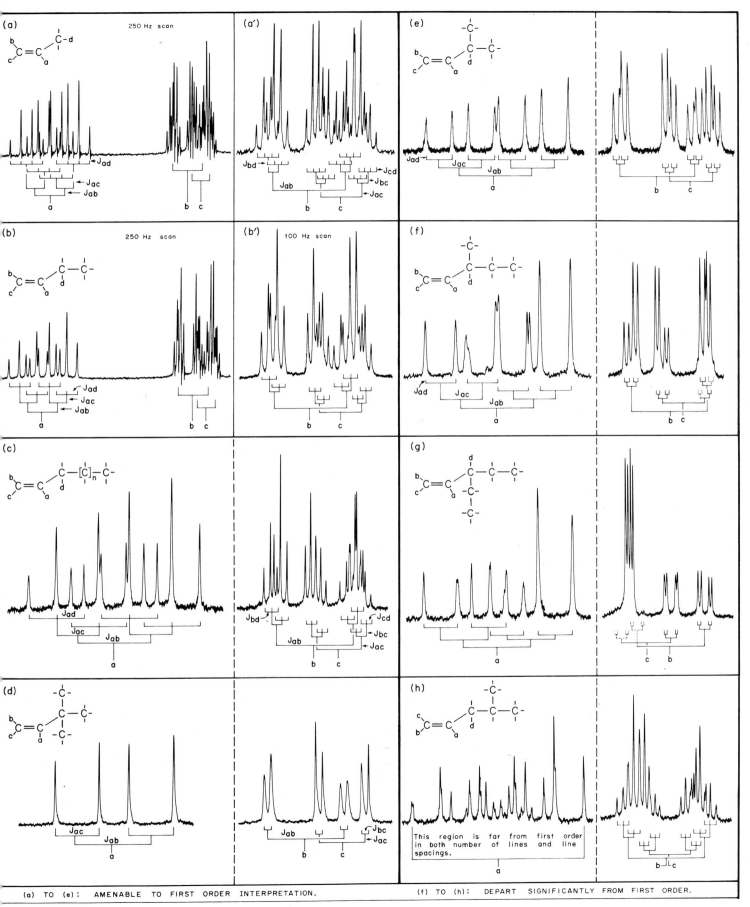

PLATE 21. NMR Spectra of the Vinyl Group in Aliphatic Olefins (in CCl₄ at 100 MHZ, 100 Hz scan width.)

PLATE 21

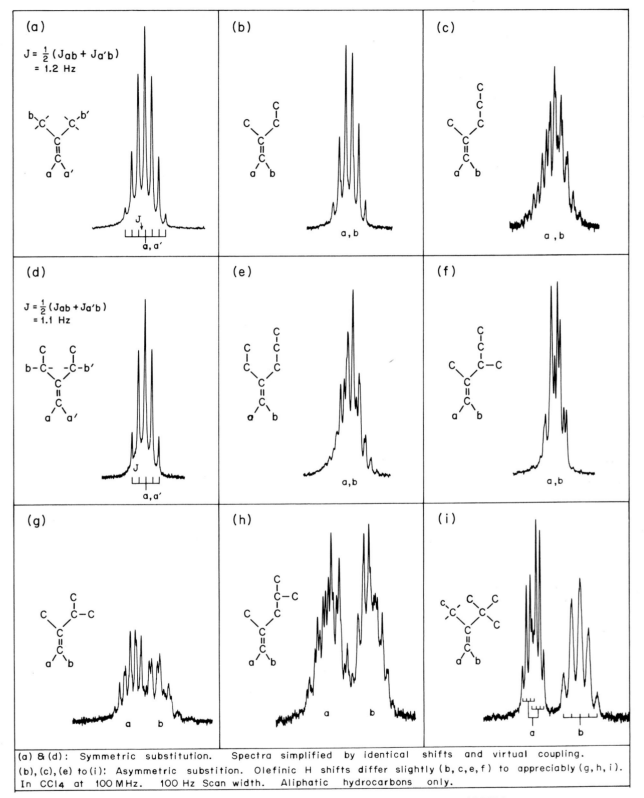

PLATE 22. Spectra of Olefinic H in 1,1-Disubstituted Ethylenes (100 MHz)

PLATE 22

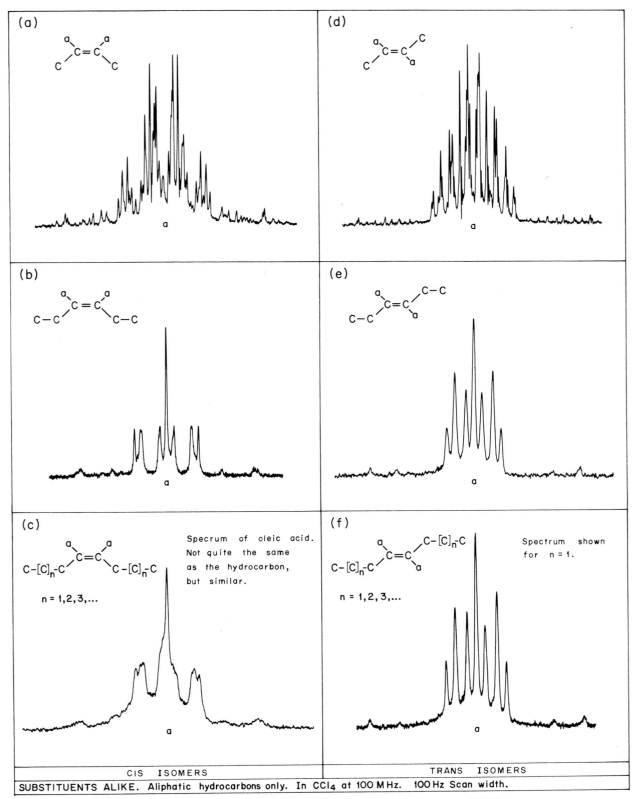

PLATE 23. Spectra of Olefinic H in 1,2-Disubstituted Ethylenes (A) (100 MHz)

PLATE 23

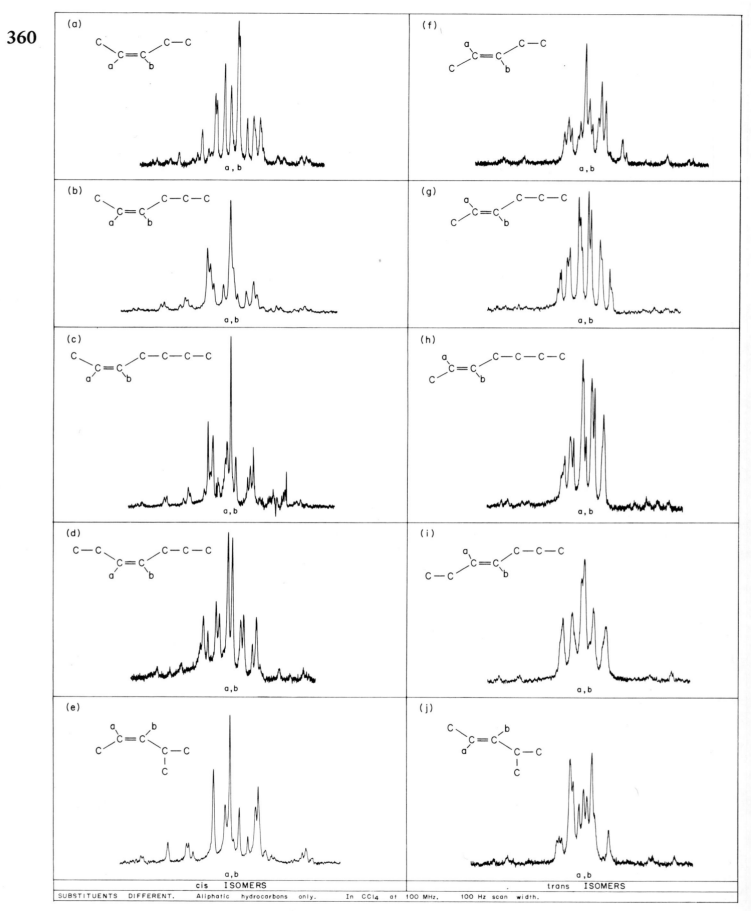

PLATE 24. Spectra of Olefinic H in, 1,2-Disubstituted Ethylenes (B) (100 MHz)

PLATE 24

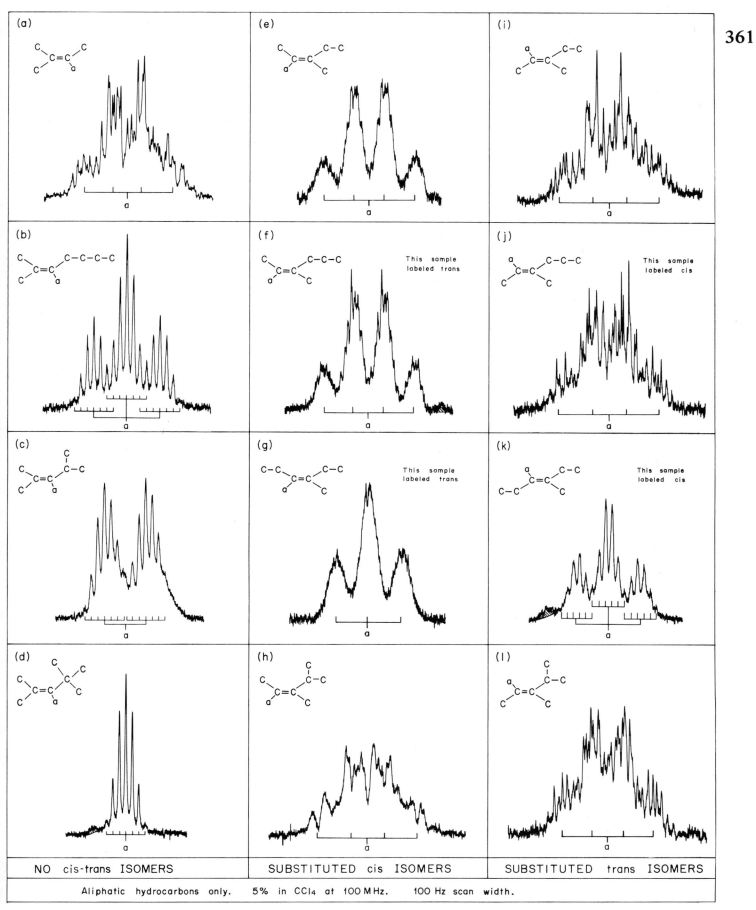

PLATE 25. Spectra of Olefinic H in Trisubstituted Ethylenes (100 MHz)

PLATE 25

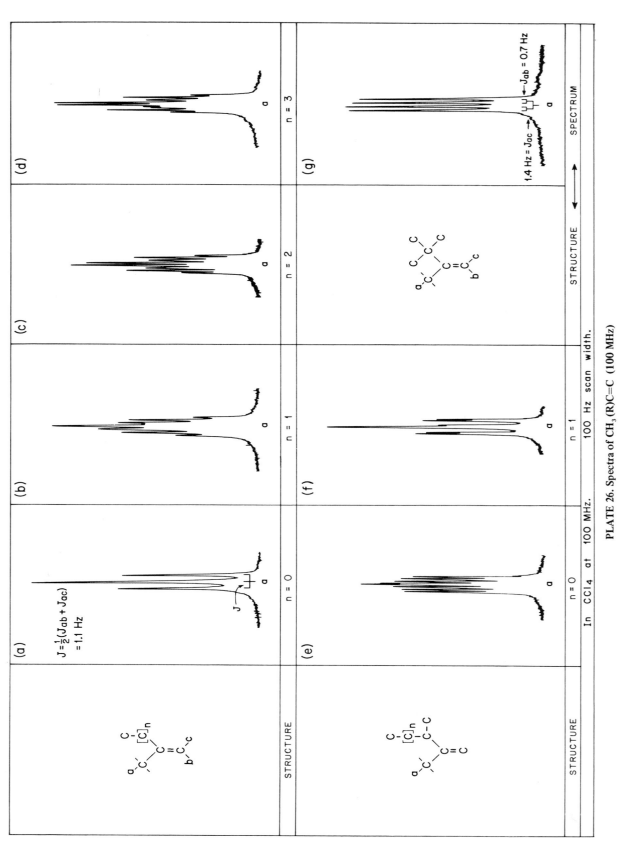

PLATE 26. Spectra of CH₃ (R)C=C (100 MHz)

PLATE 26

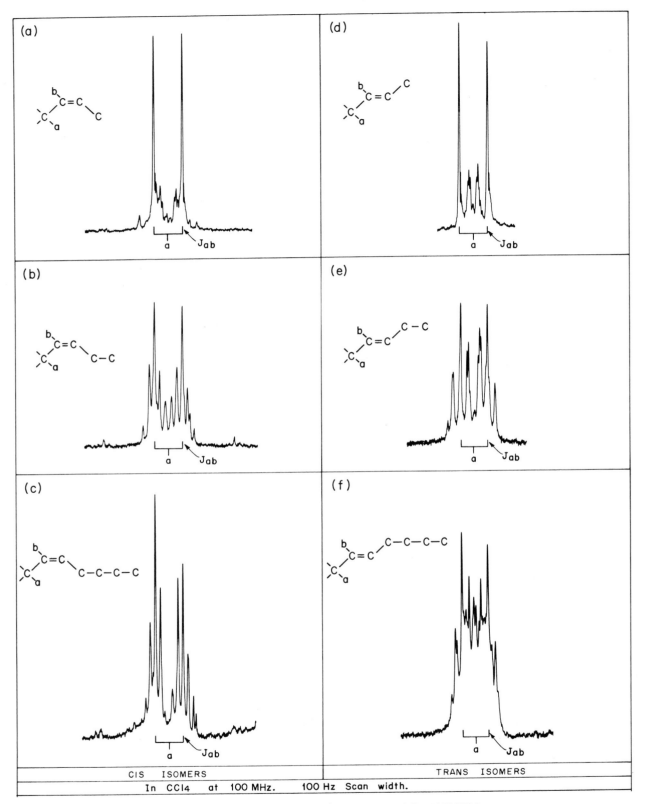

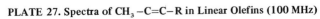

PLATE 27. Spectra of CH₃ −C=C−R in Linear Olefins (100 MHz)

PLATE 27

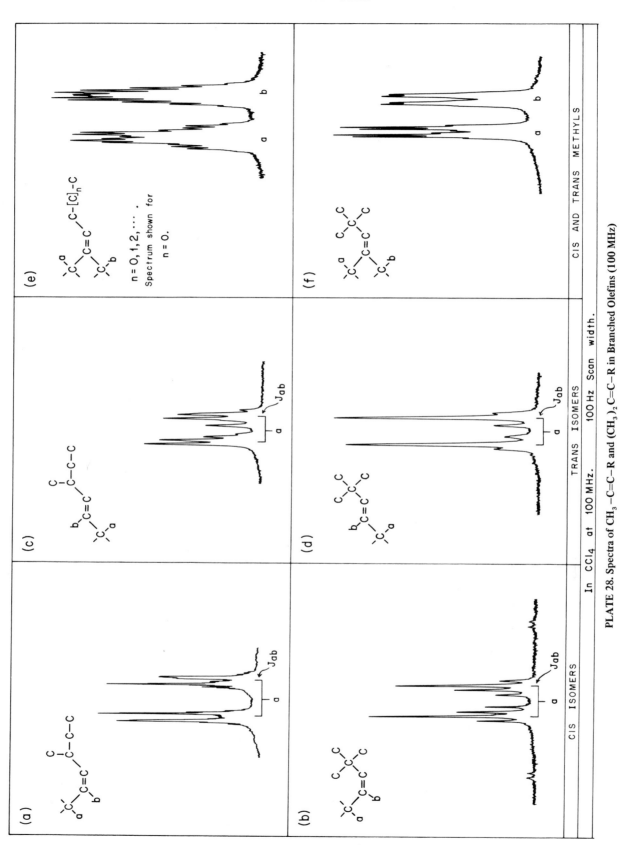

PLATE 28. Spectra of $CH_3-C=C-R$ and $(CH_3)_2C=C-R$ in Branched Olefins (100 MHz)

In CCl_4 at 100 MHz.   100 Hz Scan width.

PLATE 28

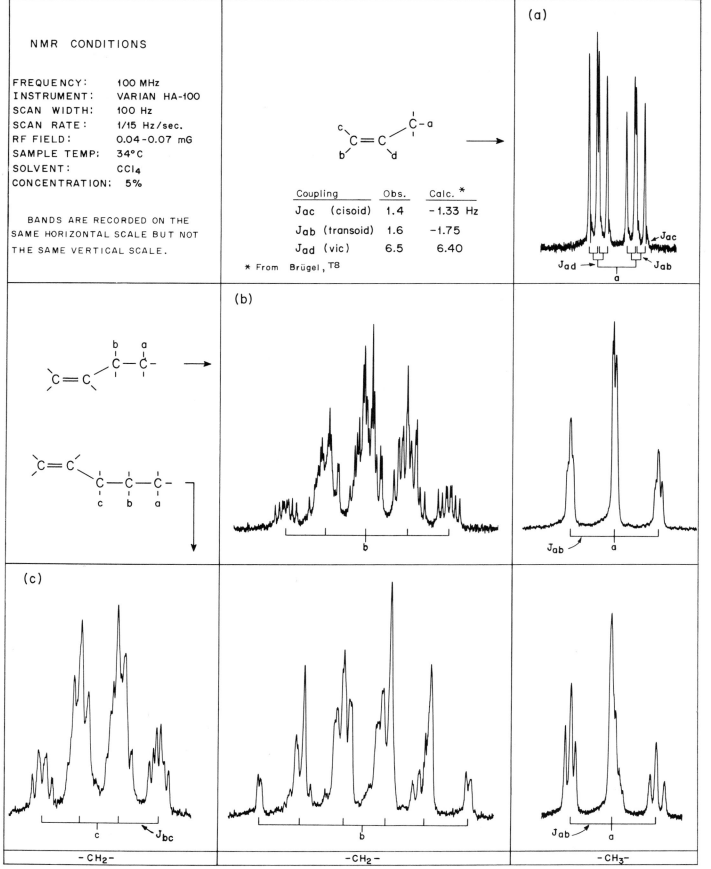

PLATE 29. NMR Spectra of Saturated Part of $C_3 - C_5$ Linear α-Olefins (100 MHz)

PLATE 29

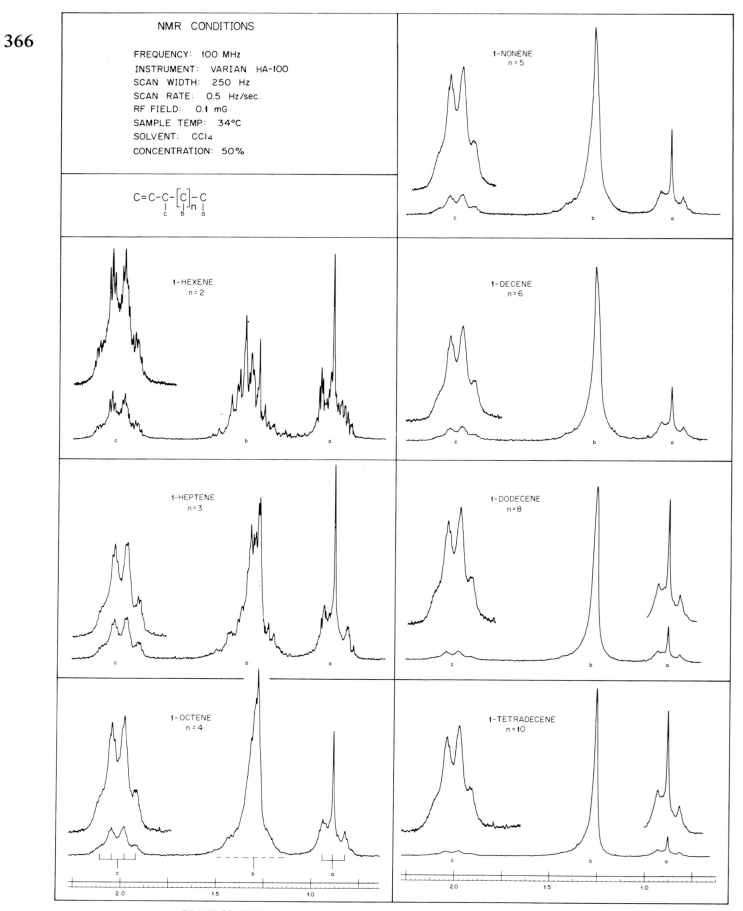

PLATE 30. NMR Spectra of the Saturated Part of $C_6 - C_{14}$ Linear α-Olefins (100 MHz)

PLATE 30

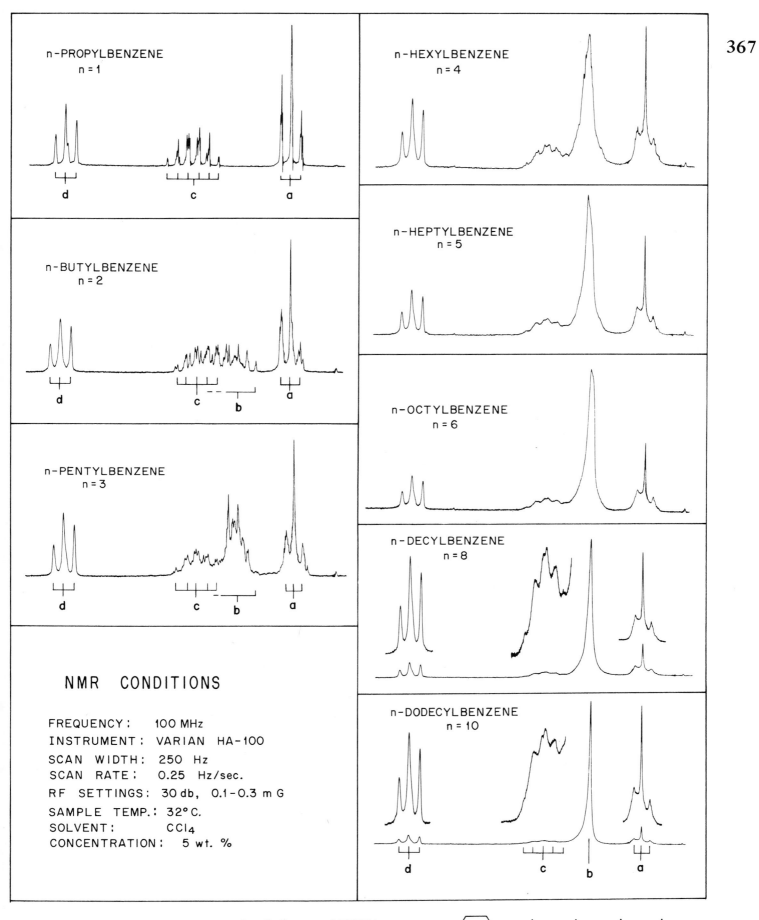

PLATE 31. NMR Spectra of the Alkyl Part of *n*-Alkylbenzenes (100 MHz)

PLATE 31

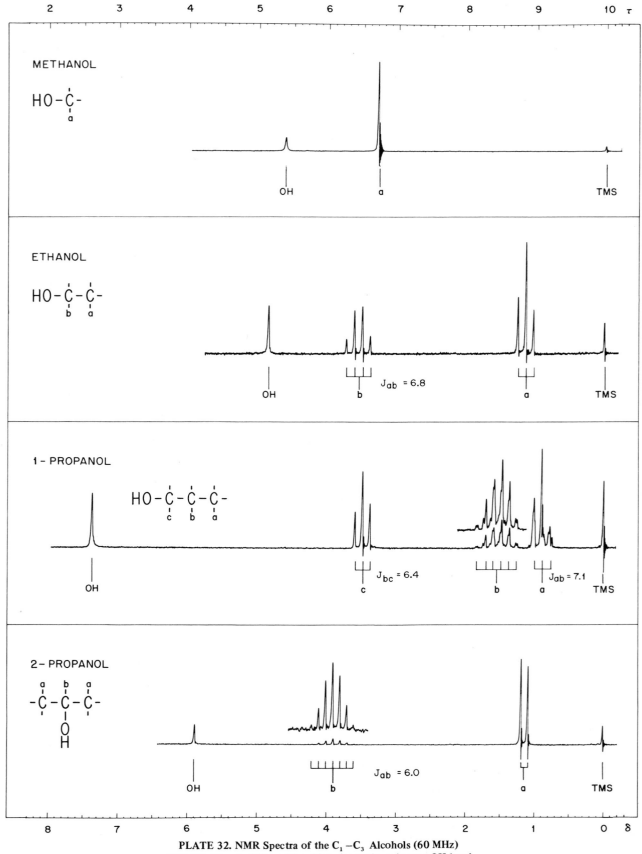

PLATE 32. NMR Spectra of the $C_1 - C_3$ Alcohols (60 MHz)
10% Solutions in CCl_4, with acid to decouple and move OH band

PLATE 32

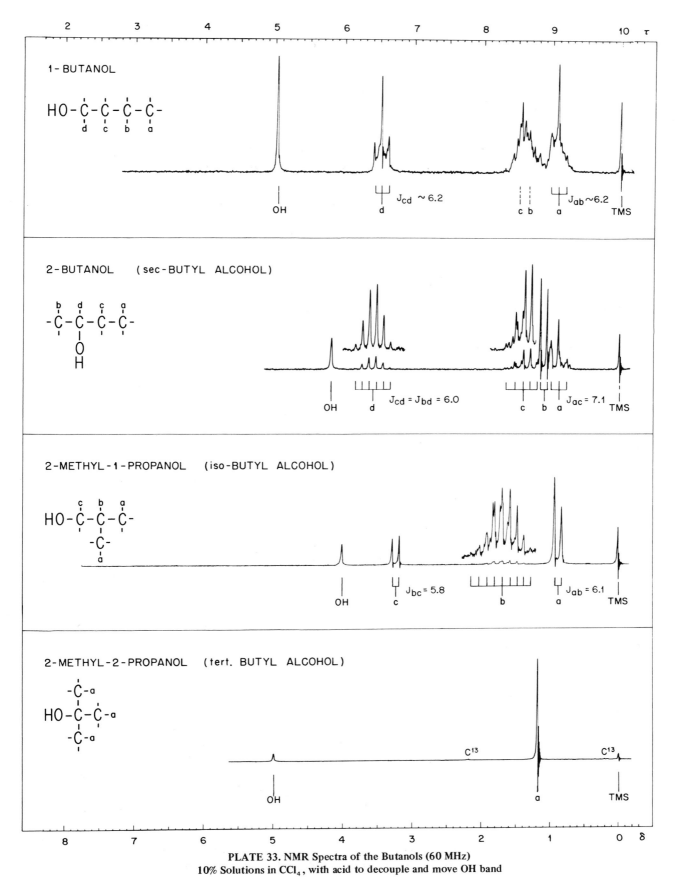

PLATE 33. NMR Spectra of the Butanols (60 MHz)
10% Solutions in CCl$_4$, with acid to decouple and move OH band

PLATE 33

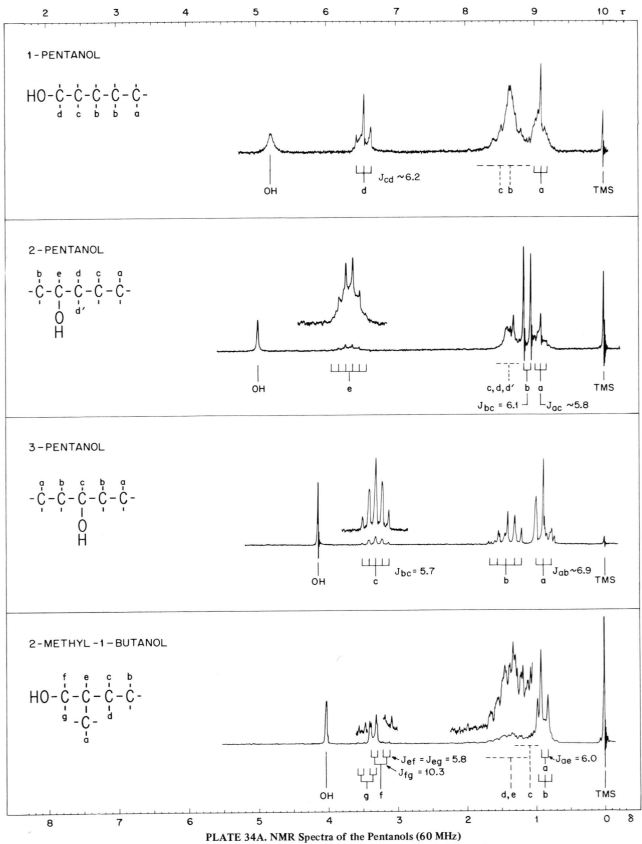

PLATE 34A. NMR Spectra of the Pentanols (60 MHz)
10% Solutions in CCl$_4$, with acid to decouple and move OH band

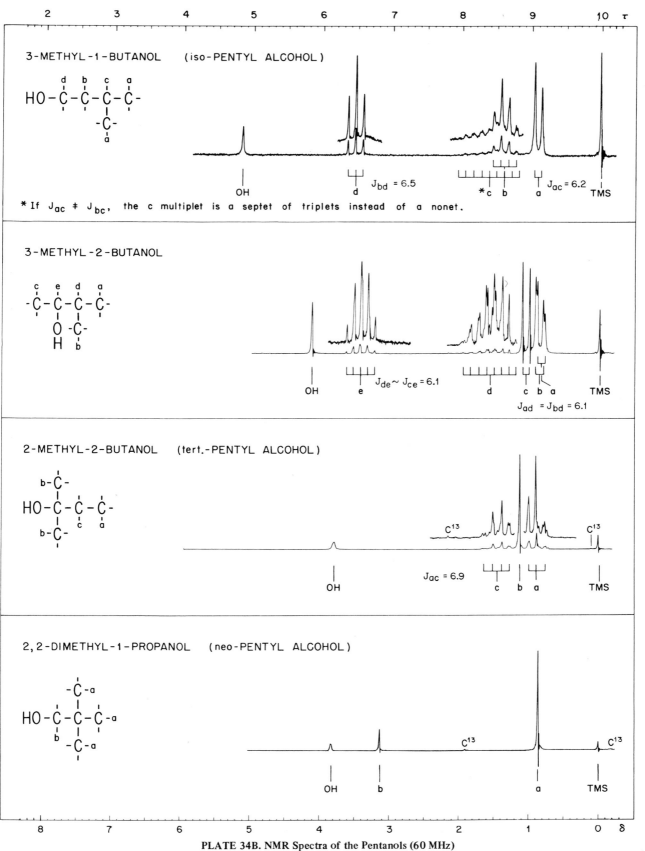

3-METHYL-1-BUTANOL (iso-PENTYL ALCOHOL)

$J_{bd} = 6.5$

OH   d   *c b   a $J_{ac} = 6.2$ TMS

*If $J_{ac} \neq J_{bc}$, the c multiplet is a septet of triplets instead of a nonet.

3-METHYL-2-BUTANOL

OH   e $J_{de} \sim J_{ce} = 6.1$ d   c b a   TMS

$J_{ad} = J_{bd} = 6.1$

2-METHYL-2-BUTANOL (tert.-PENTYL ALCOHOL)

C^{13} C^{13}

OH $J_{ac} = 6.9$ c   b   a   TMS

2,2-DIMETHYL-1-PROPANOL (neo-PENTYL ALCOHOL)

C^{13} C^{13}

OH   b   a   TMS

PLATE 34B. NMR Spectra of the Pentanols (60 MHz)
10% Solutions in CCl_4, with acid to decouple and move OH band

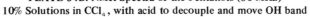

PLATE 34B

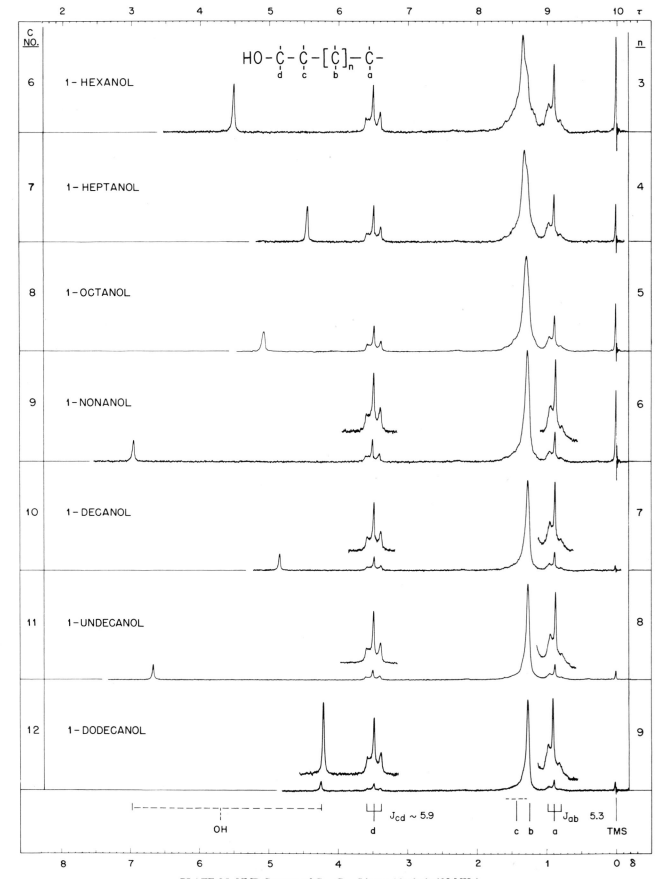

PLATE 35. NMR Spectra of C_6–C_{12} Linear Alcohols (60 MHz)
10% Solutions in CCl_4, with acid to decouple and move OH band

PLATE 35

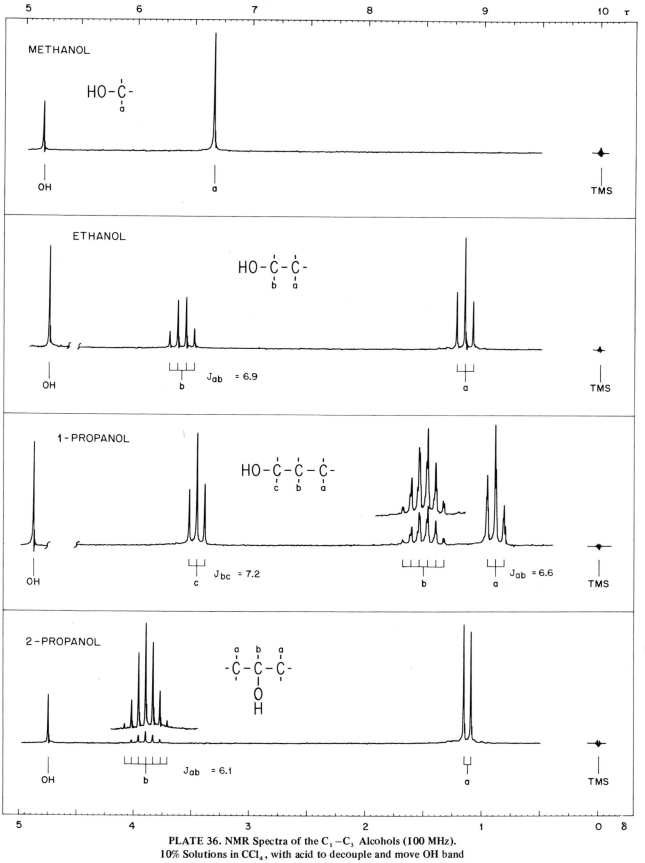

PLATE 36. NMR Spectra of the C_1-C_3 Alcohols (100 MHz).
10% Solutions in CCl_4, with acid to decouple and move OH band

PLATE 36

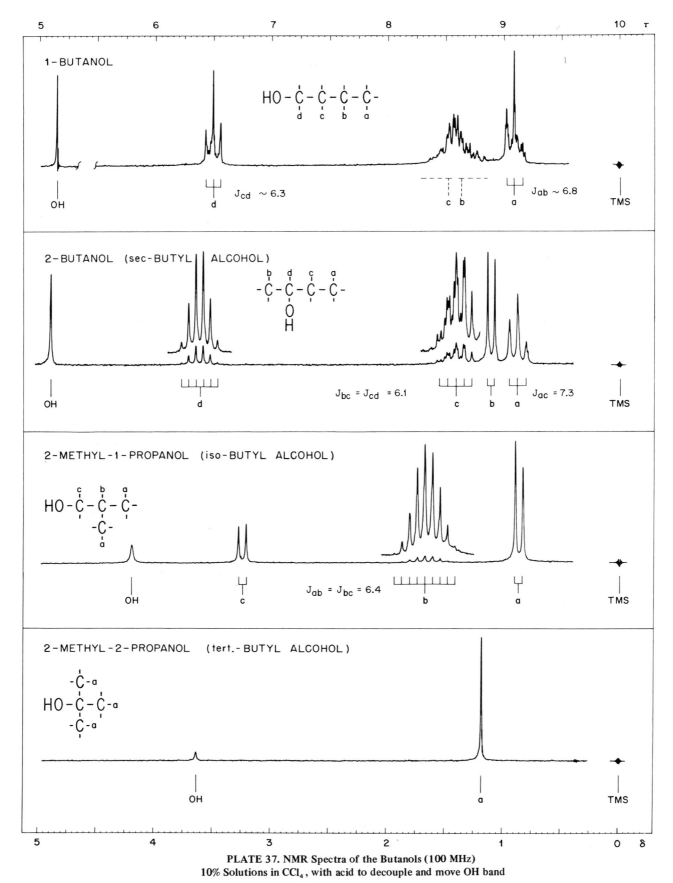

PLATE 37. NMR Spectra of the Butanols (100 MHz)
10% Solutions in CCl₄ , with acid to decouple and move OH band

PLATE 37

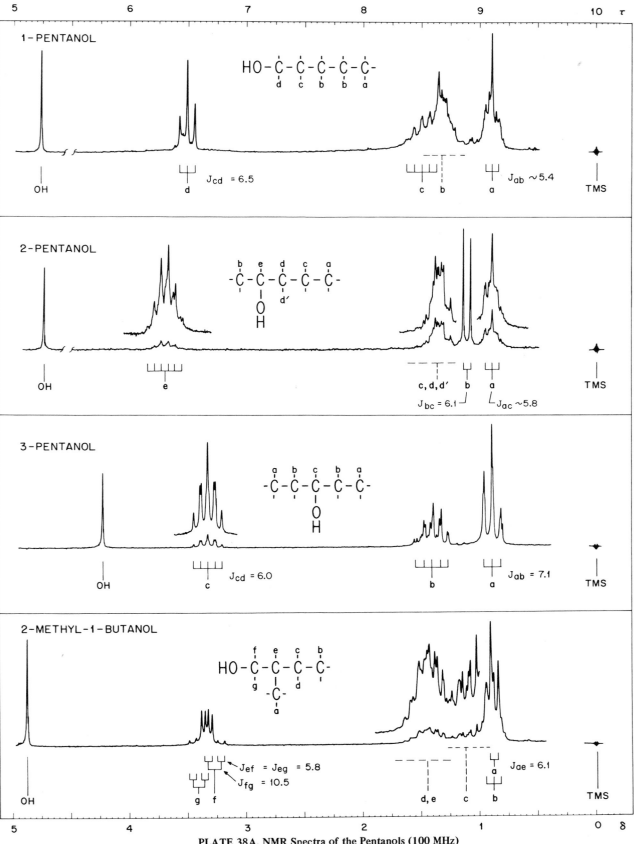

PLATE 38A. NMR Spectra of the Pentanols (100 MHz)
10% Solutions in CCl$_4$, with acid to decouple and move OH band

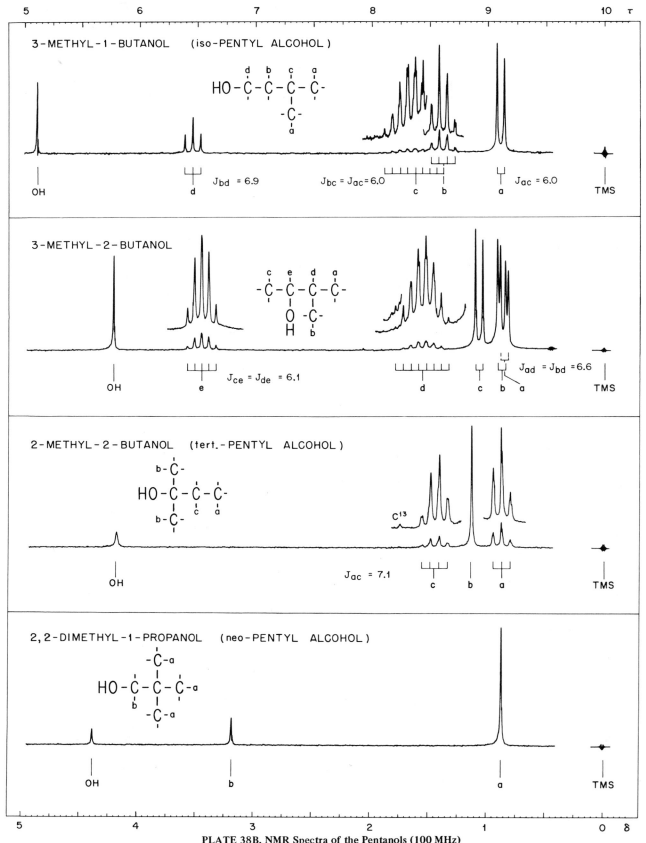

PLATE 38B. NMR Spectra of the Pentanols (100 MHz)
10% Solutions in CCl$_4$, with acid to decouple and move OH band

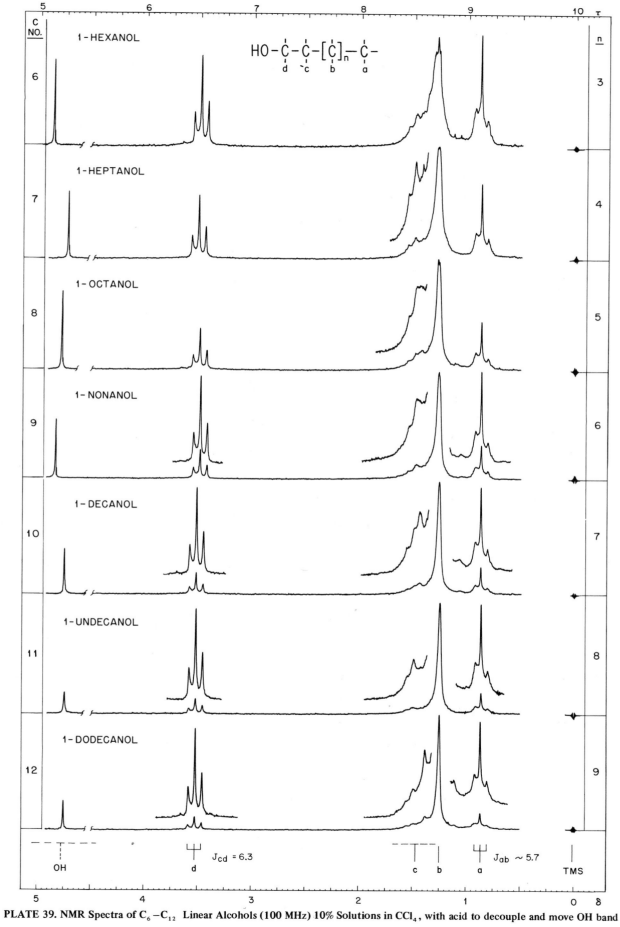

PLATE 39. NMR Spectra of $C_6 - C_{12}$ Linear Alcohols (100 MHz) 10% Solutions in CCl_4, with acid to decouple and move OH band

PLATE 39

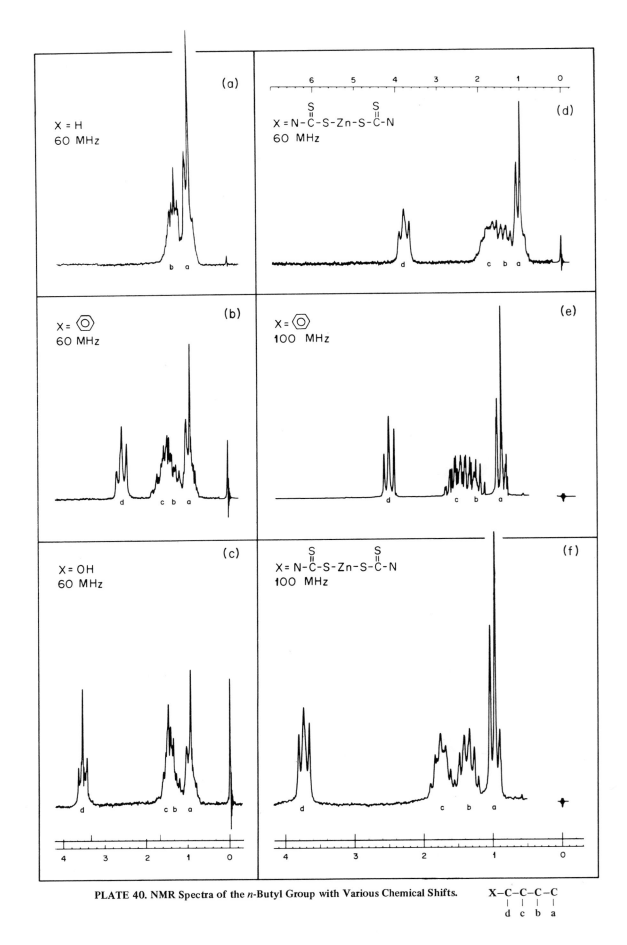

PLATE 40. NMR Spectra of the *n*-Butyl Group with Various Chemical Shifts.

X–C–C–C–C
    | | | |
    d   c   b   a

PLATE 40

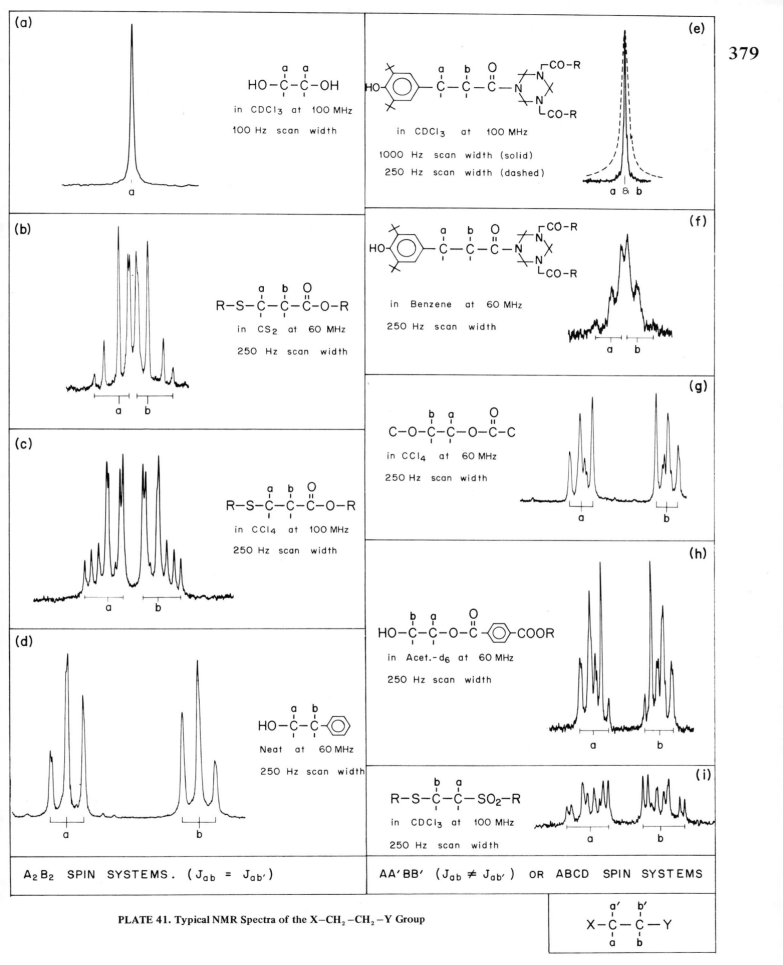

(a)

HO–C–C–OH

in CDCl₃ at 100 MHz

100 Hz scan width

(b)

R–S–C–C–C–O–R

in CS₂ at 60 MHz

250 Hz scan width

(c)

R–S–C–C–C–O–R

in CCl₄ at 100 MHz

250 Hz scan width

(d)

HO–C–C–⟨phenyl⟩

Neat at 60 MHz

250 Hz scan width

(e)

HO–⟨ring⟩–C–C–C–N⟨CO–R / CO–R⟩

in CDCl₃ at 100 MHz

1000 Hz scan width (solid)

250 Hz scan width (dashed)

(f)

HO–⟨ring⟩–C–C–C–N⟨CO–R / CO–R⟩

in Benzene at 60 MHz

250 Hz scan width

(g)

C–O–C–C–O–C–C

in CCl₄ at 60 MHz

250 Hz scan width

(h)

HO–C–C–O–C–⟨ring⟩–COOR

in Acet.-d₆ at 60 MHz

250 Hz scan width

(i)

R–S–C–C–SO₂–R

in CDCl₃ at 100 MHz

250 Hz scan width

A₂B₂ SPIN SYSTEMS. (J_{ab} = J_{ab'})

AA'BB' (J_{ab} ≠ J_{ab'}) OR ABCD SPIN SYSTEMS

X–C–C–Y

PLATE 41. Typical NMR Spectra of the X–CH₂–CH₂–Y Group

PLATE 41

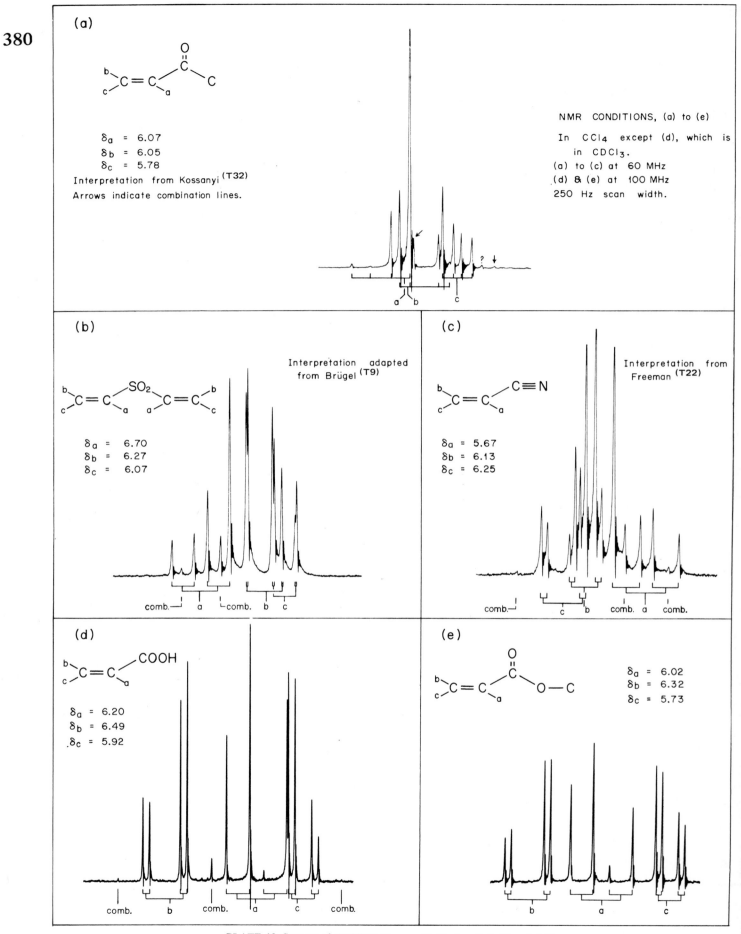

(a)

$\delta_a = 6.07$
$\delta_b = 6.05$
$\delta_c = 5.78$

Interpretation from Kossanyi [T32]
Arrows indicate combination lines.

NMR CONDITIONS, (a) to (e)

In CCl_4 except (d), which is in $CDCl_3$.
(a) to (c) at 60 MHz
(d) & (e) at 100 MHz
250 Hz scan width.

(b)

Interpretation adapted from Brügel [T9]

$\delta_a = 6.70$
$\delta_b = 6.27$
$\delta_c = 6.07$

comb. a comb. b c

(c)

Interpretation from Freeman [T22]

$\delta_a = 5.67$
$\delta_b = 6.13$
$\delta_c = 6.25$

comb. c b comb. a comb.

(d)

$\delta_a = 6.20$
$\delta_b = 6.49$
$\delta_c = 5.92$

comb. b comb. a c comb.

(e)

$\delta_a = 6.02$
$\delta_b = 6.32$
$\delta_c = 5.73$

b a c

PLATE 42. Spectra of the Vinyl Group with Various Substituents (A)

PLATE 42

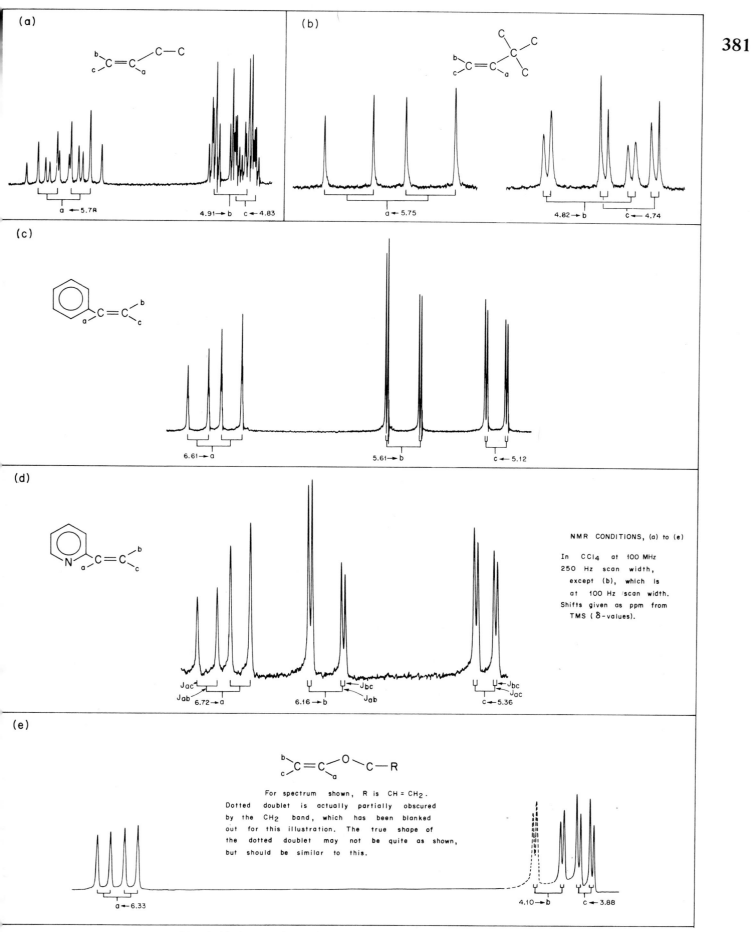

PLATE 43. Spectra of the Vinyl Group with Various Substituents (B)

PLATE 43

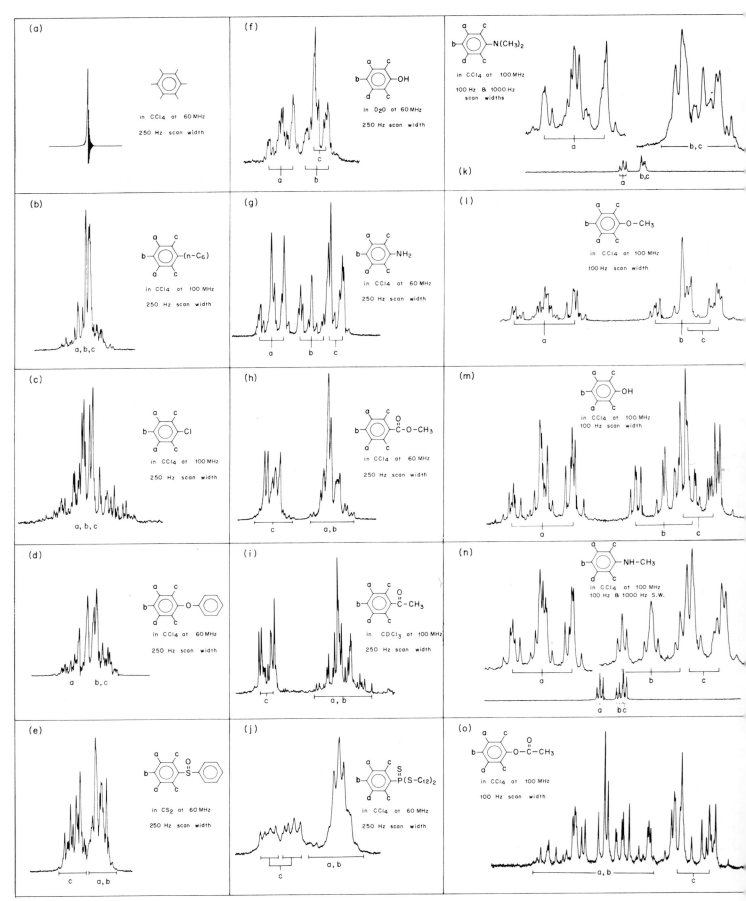

PLATE 44. Typical NMR Spectra of Monosubstituted Benzene Rings AA′BCC′ Systems Interpreted as A₂BC₂

PLATE 44

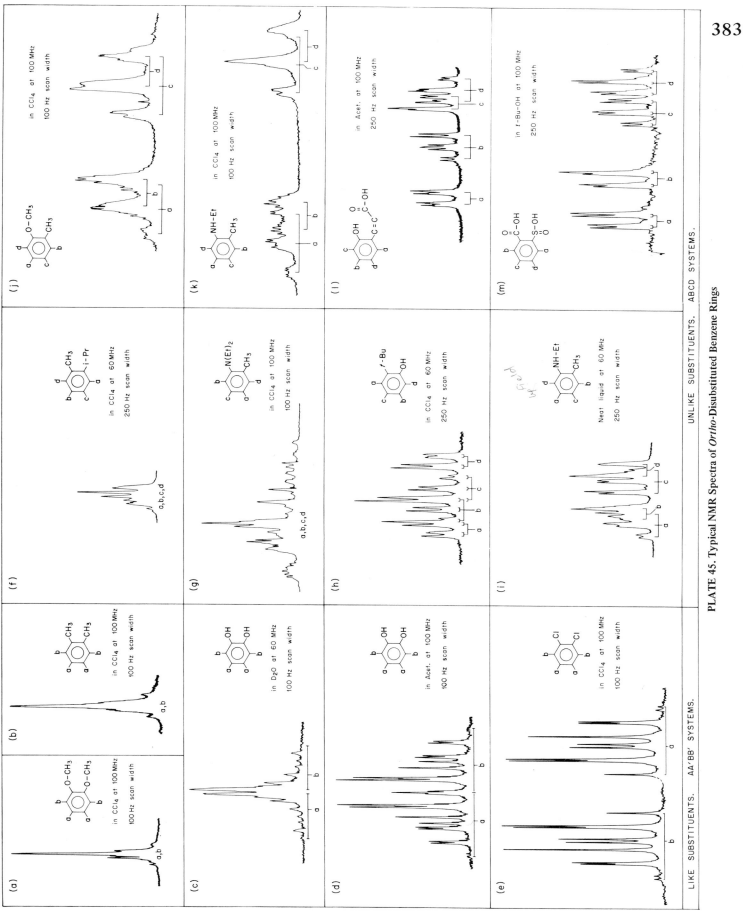

PLATE 45. Typical NMR Spectra of *Ortho*-Disubstituted Benzene Rings

PLATE 45

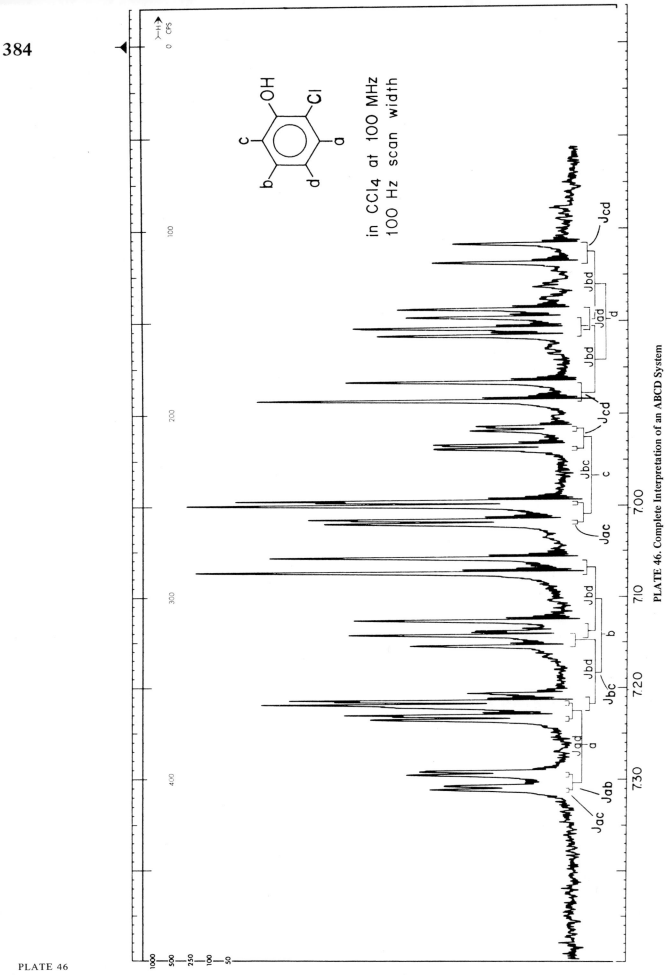

in CCl₄ at 100 MHz
100 Hz scan width

PLATE 46. Complete Interpretation of an ABCD System

PLATE 46

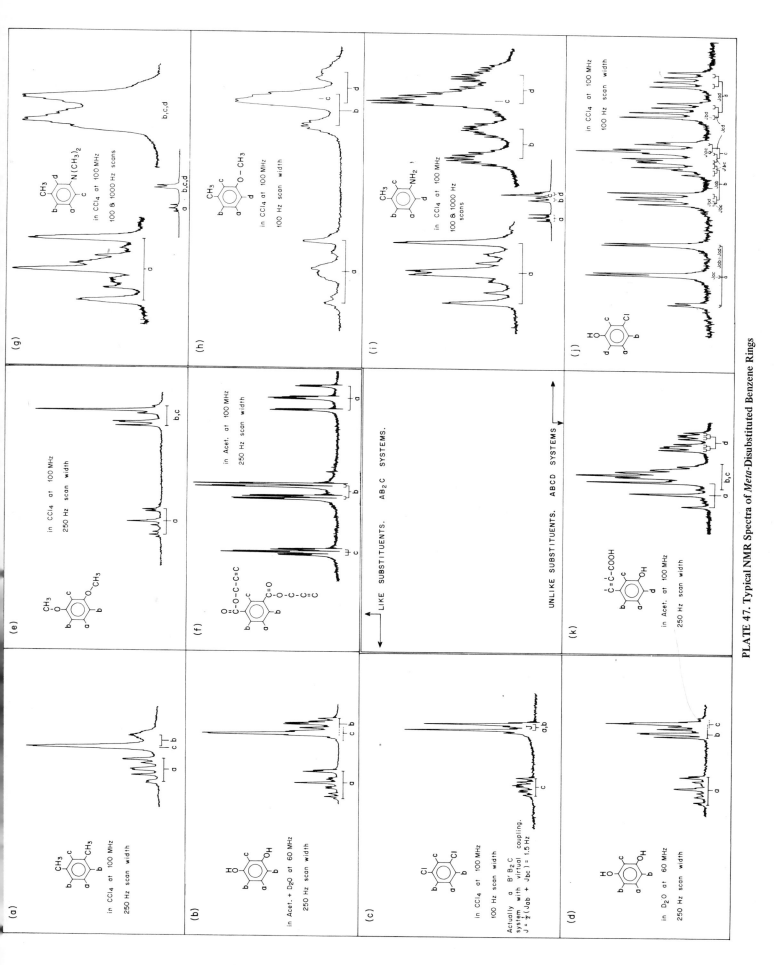

PLATE 47. Typical NMR Spectra of *Meta*-Disubstituted Benzene Rings

PLATE 47

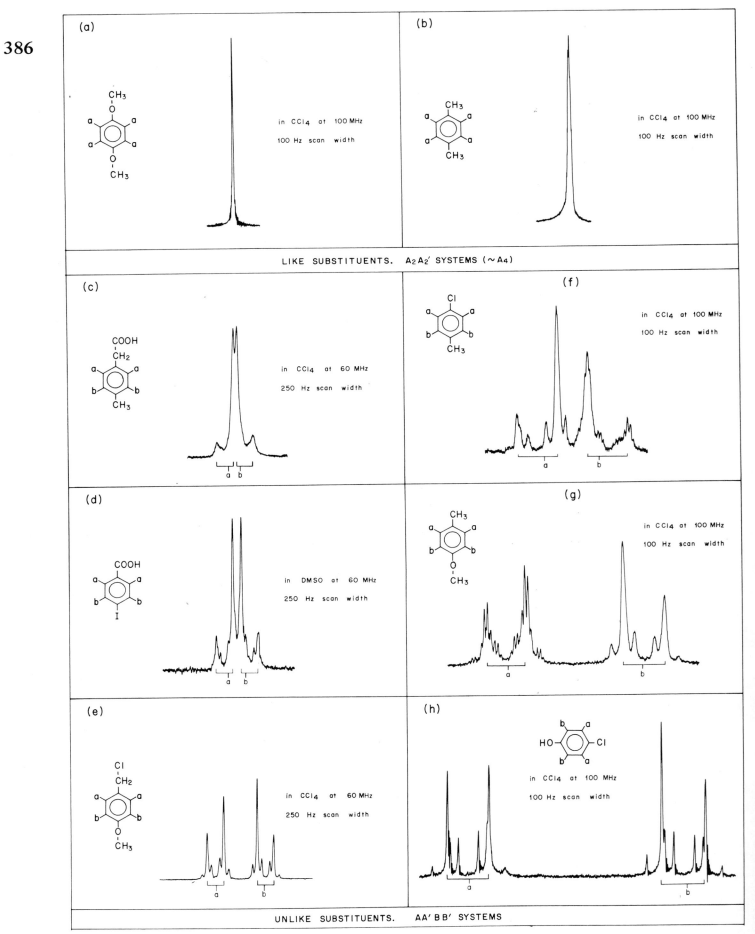

(a) in CCl₄ at 100 MHz 100 Hz scan width

(b) in CCl₄ at 100 MHz 100 Hz scan width

LIKE SUBSTITUENTS. A₂A₂′ SYSTEMS (∼A₄)

(c) in CCl₄ at 60 MHz 250 Hz scan width

(f) in CCl₄ at 100 MHz 100 Hz scan width

(d) in DMSO at 60 MHz 250 Hz scan width

(g) in CCl₄ at 100 MHz 100 Hz scan width

(e) in CCl₄ at 60 MHz 250 Hz scan width

(h) in CCl₄ at 100 MHz 100 Hz scan width

UNLIKE SUBSTITUENTS. AA′BB′ SYSTEMS

PLATE 48. Typical NMR Spectra of *Para*-Disubstituted Benzene Rings

PLATE 48

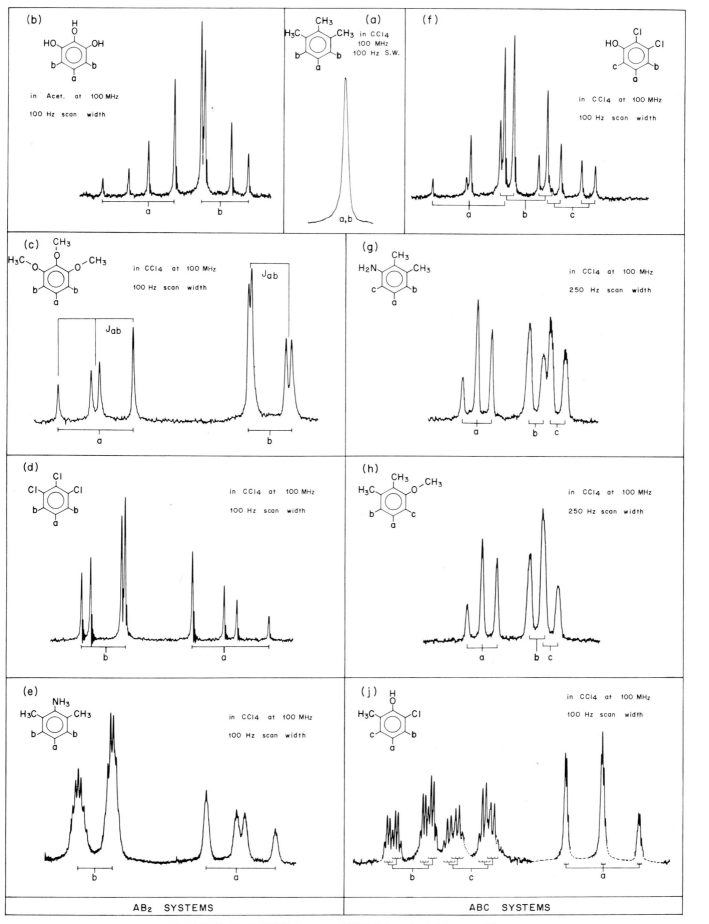

PLATE 49. Typical NMR Spectra of 1,2,3-Trisubstituted Benzene Rings

PLATE 49

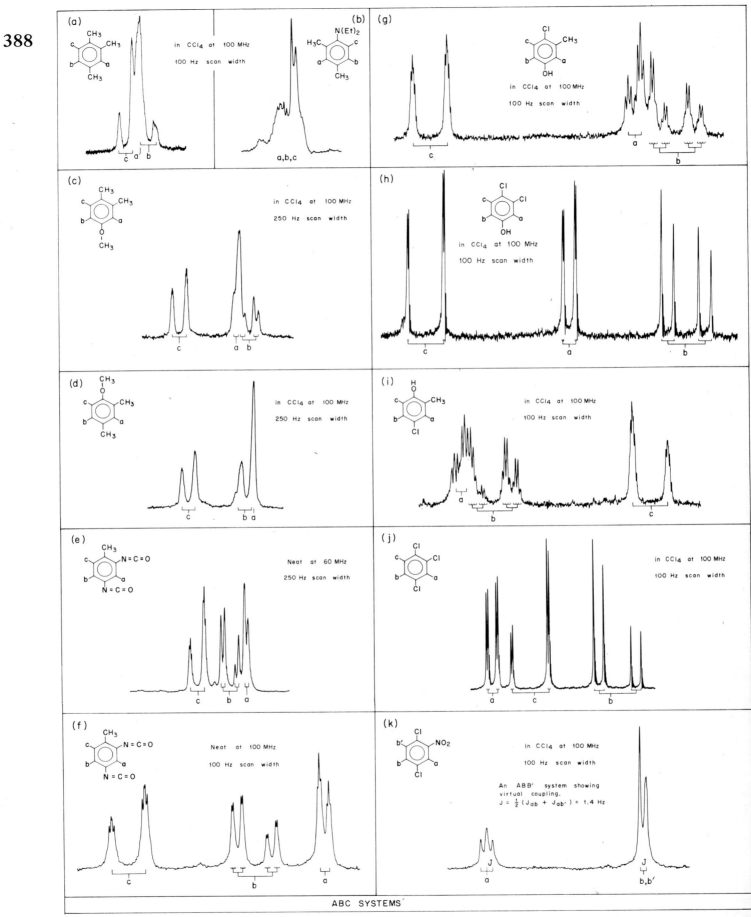

ABC SYSTEMS

PLATE 50. Typical NMR Spectra of 1,2,4-Trisubstituted Benzene Rings

PLATE 50

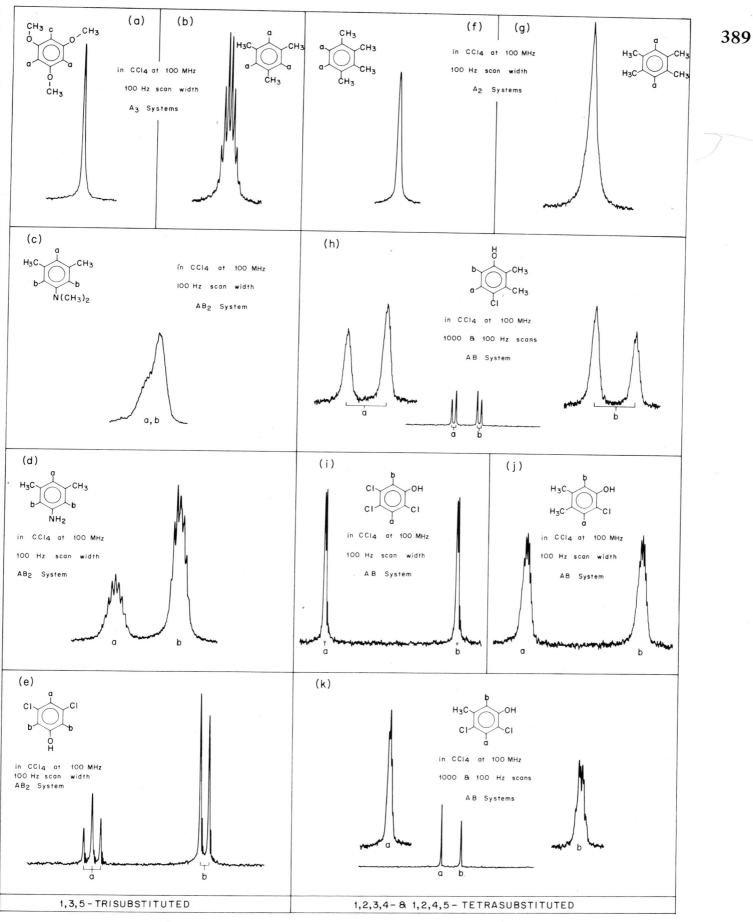

PLATE 51. NMR Spectra of Some Tri- and Tetrasubstituted Benzene Rings

PLATE 51

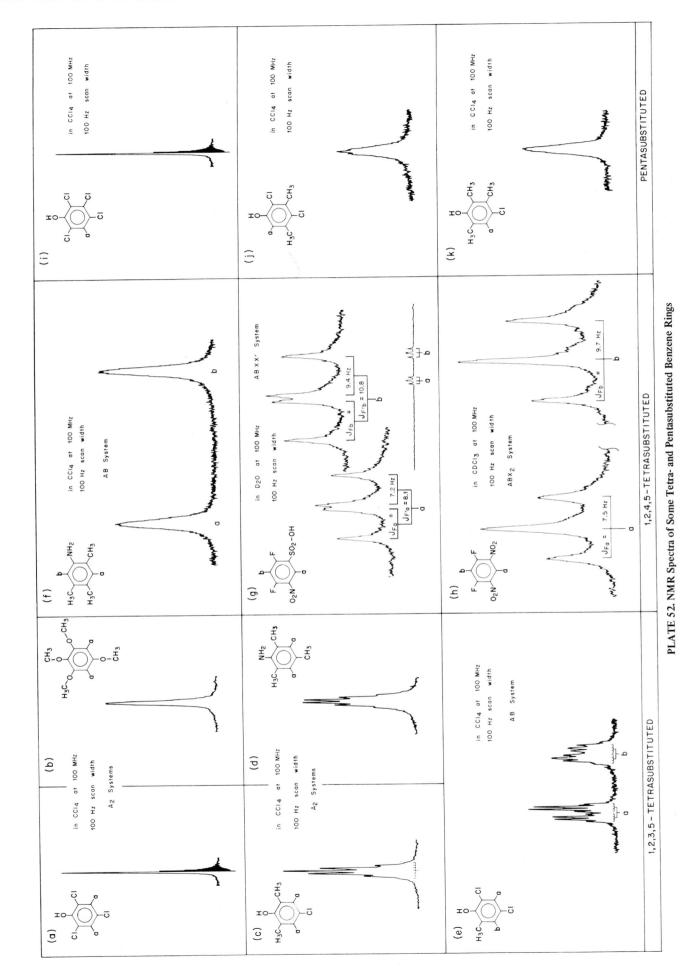

PLATE 52. NMR Spectra of Some Tetra- and Pentasubstituted Benzene Rings

PLATE 52

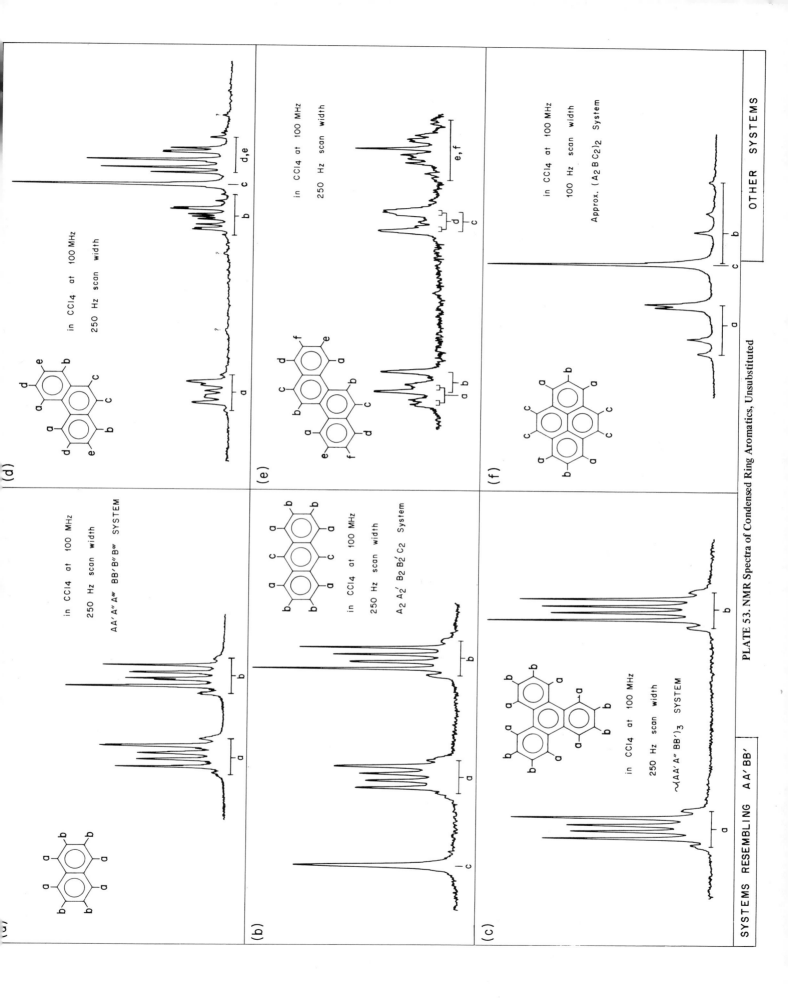

PLATE 53. NMR Spectra of Condensed Ring Aromatics, Unsubstituted

SYSTEMS RESEMBLING AA′BB′   OTHER SYSTEMS

(a)

in CCl₄ at 100 MHz

250 Hz scan width

AA′A″A‴ BB′B″B‴ SYSTEM

(b)

in CCl₄ at 100 MHz

250 Hz scan width

A₂ A₂′ B₂ B₂′ C₂ System

(c)

in CCl₄ at 100 MHz

250 Hz scan width

∿(AA′A″ BB′)₃ SYSTEM

(d)

in CCl₄ at 100 MHz

250 Hz scan width

(e)

in CCl₄ at 100 MHz

250 Hz scan width

(f)

in CCl₄ at 100 MHz

100 Hz scan width

Approx. (A₂ B C₂)₂ System

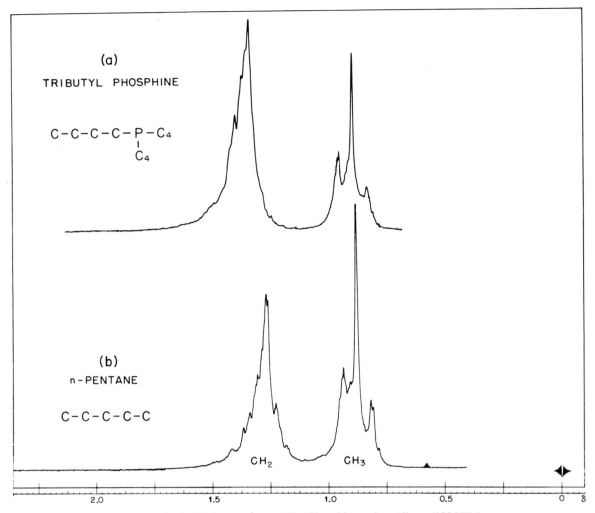

(a)

TRIBUTYL PHOSPHINE

$C-C-C-C-P-C_4$
$\quad\quad\quad\quad\quad |$
$\quad\quad\quad\quad\ C_4$

(b)

n-PENTANE

$C-C-C-C-C$

CH_2

CH_3

2.0   1.5   1.0   0.5   0 δ

PLATE 54. Similar Spectra for an Alkyl Phosphine and an Alkane (100 MHz)

PLATE 54

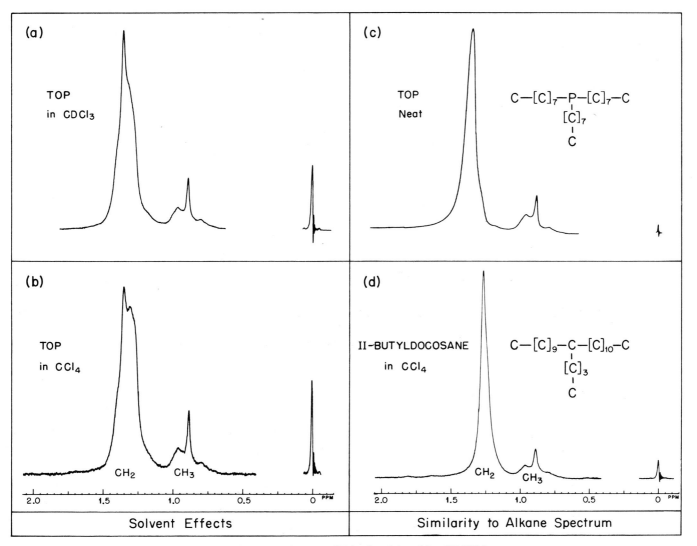

PLATE 55. NMR Spectra of Tri-*n*-Octylphosphine (60 MHz)

PLATE 55

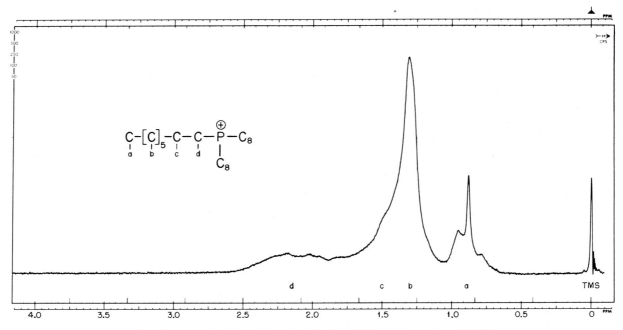

PLATE 56. Tri-*n*-Octylphosphine in CDCl₃ + Trifluoroacetic Acid (60 MHz)

PLATE 56

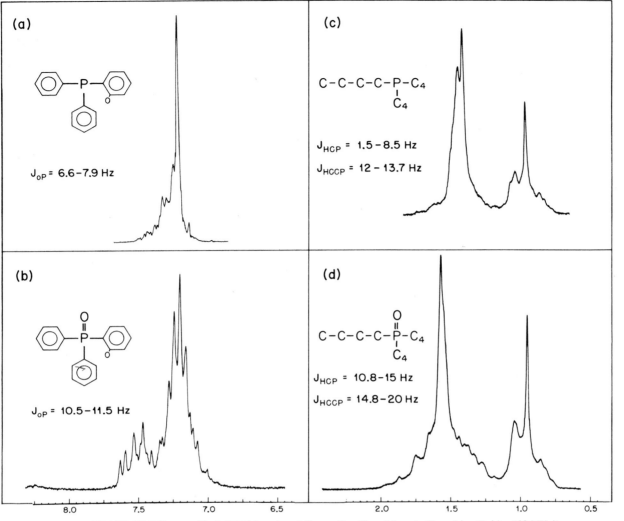

PLATE 57

PLATE 57. Effect on Their NMR Spectra of Converting Phosphines to Phosphine Oxides (60 MHz)

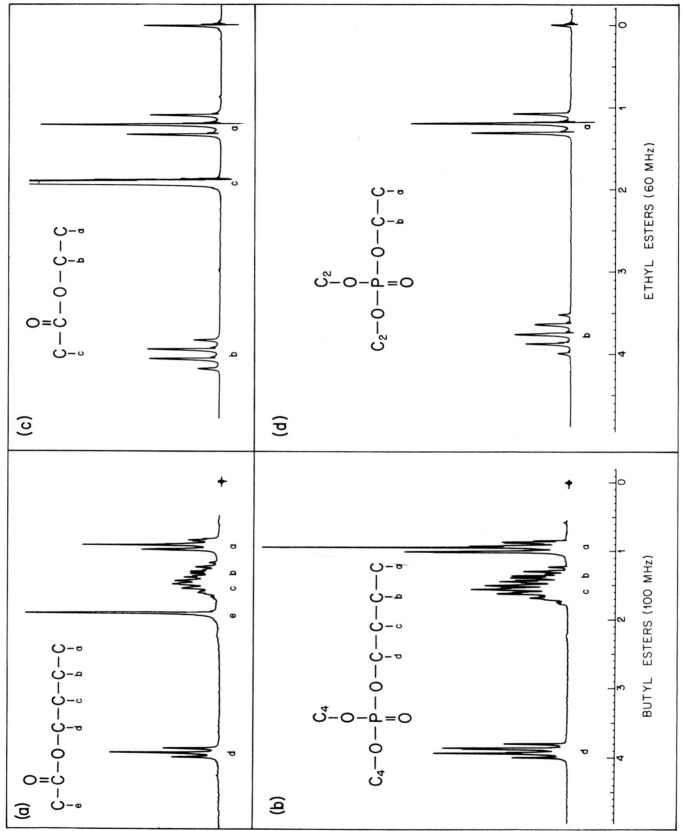

PLATE 58. Phosphorus–Hydrogen Spin Coupling Helps Identify Phosphorus Esters

PLATE 58

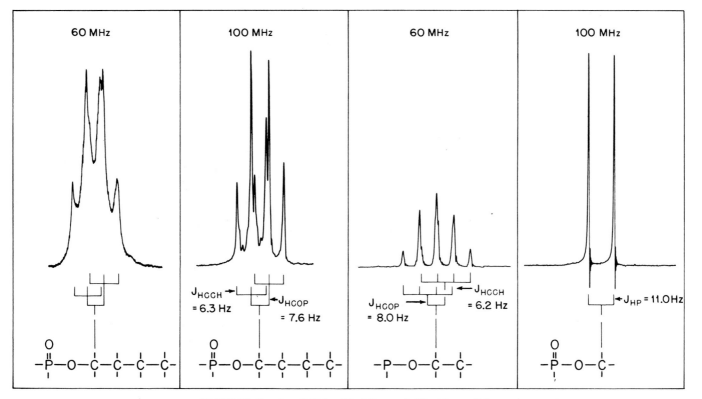

PLATE 59. Spectra of Alpha Alkyl Groups in Phosphorus Esters

PLATE 59

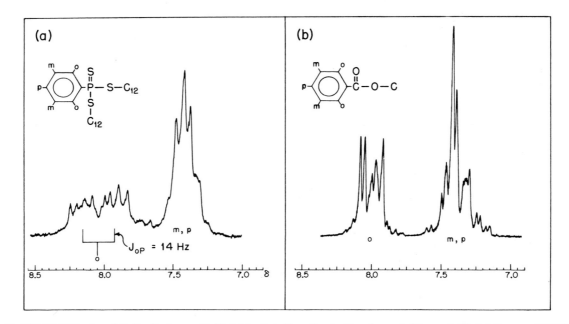

PLATE 60. Widening of Ortho Hydrogen Multiplet by Spin Coupling to Phosphorus in Monosubstituted Benzene (60 MHz)

PLATE 60

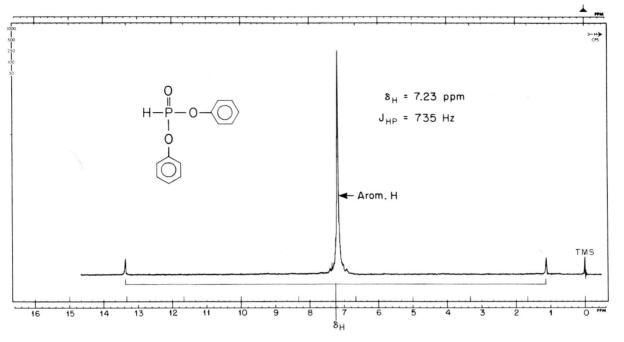

PLATE 61. The Very Wide H–P Doublet

PLATE 61

References, Appendix, and Indexes

References

I. INTRODUCTION

A. ORGANIZATION

This chapter contains references to all the literature cited in or used in compiling this work. For the convenience of the reader the list has been divided into Text References, Acknowledgments, and Data References.

Text References are the usual citations made to support statements in the text or illustrations or to call the reader's attention to additional information on the subject under discussion. They are identified by a "T" before the number.

Acknowledgments indicate the sources of illustrations which have been copied, with permission, from other publications. They are identified by an "A" before the number.

Data References show the sources of the data used to compile the correlations of NMR parameters with molecular structure. They will permit the reader to review the original data actually used, but they do not necessarily include an exhaustive survey of the literature on each subject. Thorough literature surveys were conducted only when necessary to obtain enough data for an adequate correlation, as discussed in the next section.

Text References and Acknowledgments are listed alphabetically by senior author, without further subdivision. Data References are divided into major sources, such as textbooks and data compilations, and scientific papers. Major sources are divided according to type, and then listed in order of publication date. Scientific papers are divided according to the class of compounds of interest in this work, and then listed alphabetically by senior author.

B. SELECTION AND PROCESSING OF DATA

The data used in compiling the charts and figures in this book came from many sources, including textbooks, catalogs of NMR spectra and scientific papers. In general, data were used only if measured on hydrogen signal stabilized spectrometers, at room temperature, and referenced to internal tetramethylsilane. In a few cases data measured by the sideband technique on less stable spectrometers, and some data referenced to internal cyclohexane and converted to the equivalent TMS references values, were used. For published spectra the interpretations were checked against comparable spectra and incomplete interpretations were completed where possible.

The data so selected were then examined critically to reduce errors and inconsistencies. The preparation of the detailed correlation charts constituted one cross-check of the data. Gross deviations stand out clearly in such plots, forcing a reconsideration of the deviated points in terms of structural differences or errors in measurement, interpretation or transcribing. The correlations themselves were cross-checked during preparation of the summary charts, and additional errors and inconsistencies were eliminated. The resulting set of correlations are believed to be self-consistent and as accurate as is required for normal analytical work. This belief is supported by several years of successful use of the correlations.

Original spectra or major tabulations of data were preferred to smaller tabulations. When an adequate volume of data was available from the preferred sources no effort was made to conduct a thorough search of the literature. As a result, some good data sources have undoubtedly been omitted because the data would have been redundant in this work. For phosphorus and sulfur compounds and unsaturated oxygenated aliphatics, however, it was necessary to search the literature thoroughly to obtain enough data to support useful correlations.

All spectra reproduced in this book were run in the author's laboratory.

II. TEXT REFERENCES

T1. F.A.L. Anet, *Can. J. Chem. 39,* 2262 (1961).

T2. N. S. Angerman, S. S. Danyluk and T. A. Victor, Paper delivered before 13th Experimental NMR Conference, Pacific Grove, California, April 30–May 4, 1972.

T3. J. S. Babiec, Jr., J. R. Barrante and G. D. Vickers, *Anal. Chem. 40,* 610 (1968).

T4. K. W. Bartz and N. F. Chamberlain, *Anal. Chem. 36,* 2151 (1964).

T5. A. A. Bothner-By, Geminal and Vicinal Proton-Proton Coupling Constants in Organic Compounds, in: *Advances in Magnetic Resonance,* Vol. 1, J. A. Waugh, Ed., Academic Press, New York (1965), pp. 195-316.

T6. F. A. Bovey, *Nuclear Magnetic Resonance Spectroscopy,* Academic Press, New York (1969).

T7. F. A. Bovey, *High Resolution NMR of Macromolecules,* Academic Press, New York (1972).

T8. W. Brügel, *Nuclear Magnetic Resonance Spectra and Chemical Structure,* Academic Press, New York (1967), 235 pp.

T9. W. Brügel, T. Ankel, and F. Krückeberg, *Zeit. Electrochem. 64,* 1121 (1960).

T10. P. E. Butler and W. H. Mueller, *Anal. Chem. 38,* 1407 (1966).

T11. J. R. Campbell, *Aldrichimica Acta 4,* No. 4, 55 (1971).

T12. N. F. Chamberlain, Nuclear Magnetic Resonance and Electron Paramagnetic Resonance, in: *Treatise on Analytical Chemistry,* Chapter 39, I. M. Kolthoff and P. J. Elving, Eds., Part I, Volume 4, pp. 1885–1958, Wiley Interscience, New York (1963).

T13. O. L. Chapman, *J. Amer. Chem. Soc. 86,* 1256 (1964).

T14. D. R. Clutter, L. Petrakis, R. L. Stenger, and R. K. Jensen, *Anal. Chem. 44,* 1395 (1972).

T15. R. E. Davis and M. R. Willcott, Paper delivered before 13th Experimental NMR Conference, Pacific Grove, California, April 30–May 4, 1972.

T16. R. J. Day and C. N. Reilley, *Anal. Chem. 38,* 1323 (1966).

T17. M. W. Dietrich, J. S. Nash, and R. E. Keller, *Anal. Chem. 38,* 1479 (1966).

T18. A. W. Douglas, *J. Chem. Phys. 45,* 3465 (1966).

T19. J. W. Emsley, J. Feeney, and L. H. Sutcliffe, *High Resolution Nuclear Magnetic Resonance Spectroscopy,* Pergamon Press, Oxford (1965).

T20. R. R. Ernest, R. Freeman, Bo Gestblom, and T. R. Lusebrink, *Mol Phys. 13,* 283 (1967).

T21. H. M. Fales and T. Luukkainen, *Anal. Chem. 37,* 955 (1965).

T22. R. Freeman, *J. Chem. Phys. 40,* 3571 (1964).

T23. V. W. Goodlett, *Anal. Chem. 37,* 431 (1965).

T24. G. A. Haley, *Anal. Chem. 44,* 580 (1972).

T25. F. A. Hart, G. P. Moss, and M. L. Staniforth, *Tetrahedron Letters* No. 37, 3389 (1971).

T26. C. C. Hinckley, *J. Amer. Chem. Soc. 91,* 5160 (1969).

T27. E. Hirsch and K. H. Altgelt, *Anal. Chem. 42,* 1330 (1970).

T28. L. M. Jackman and S. Sternhell, *Applications of Nuclear Magnetic Resonance Spectroscopy in Organic Chemistry,* Pergamon Press, Oxford (1969).

T29. J. L. Jungnickel, *Anal. Chem. 35,* 1985 (1963).

T30. M. Karplus, *J. Chem. Phys. 30,* 11 (1959).

T31. S. A. Knight, *Chem. Ind., 1967,* 1920.

T32. J. Kossanyi, *Bull. Soc. Chim. France 1965,* 704.

T33. S. L. Manatt, *J. Amer. Chem. Soc. 89,* 4544 (1967).

T34. Merck Sharp and Dohme of Canada, Shift Reagents, Literature References, February 1972 (193 references).

T35. J. I. Musher, *Spectrochim. Acta 16,* 835 (1960).

T36. J. I. Musher and E. J. Corey, *Tetrahedron 18,* 791 (1962).

T37. T. F. Page and W. E. Bresler, *Anal. Chem. 36,* 1981 (1964).

T38. J. A. Pople, *Proc. Roy. Soc. (London) A239,* 541(1957).

T39. M-M. Rousselot and N. Bellavita, *Compt. Rend. C265,* 853 (1967).

T40. S. Sternhell, Long Range H^1-H^1 Spin-Spin Coupling Constants in Nuclear Magnetic Resonance Spectroscopy, *Rev. Pure and Appl. Chem. 14,* 15 (1964).

T41. J. L. Sudmeier and C. N. Reilley, *Anal. Chem. 36,* 1698 and 1707 (1964).

T42. F. Taddei and L. Pratt, *J. Chem. Soc. (London) 1964,* 1553.

T43. K. B. Wiberg and B. J. Nist, *The Interpretation of NMR Spectra,* W. A. Benjamin, New York (1962).

T44. R. B. Williams, *ASTM Special Technical Publication No. 224,* 168–194 (1958).

T45. R. B. Williams and N. F. Chamberlain, *Proc. 6th World Petroleum Congress,* Section V, Paper 17 (1963).

T46. B. J. Zwolinski, et al, *Catalog of Selected Nuclear Magnetic Resonance Spectral Data,* Thermodynamics Research Center, Department of Chemistry, Texas A&M University, College Station, Texas 77843. (continuing publication). American Petroleum Institute Research Project 44 Catalog.

III. ACKNOWLEDGMENTS

A1. K. W. Bartz and N. F. Chamberlain, *Anal. Chem. 36,* 2151 (1964). Figures reproduced by permission of the American Chemical Society.

A2. N. F. Chamberlain, *Proc. Amer. Petrol. Inst. 44* (III), 361 (1964). Reproduced by permission of the American Petroleum Institute.

A3. N. F. Chamberlain, *Proc. 7th World Petroleum Congress 9,* 13 (1967). Reproduced by permission of the World Petroleum Congress.

A4. N. F. Chamberlain, *Anal. Chem. 40,* 1317 (1968). Reproduced by permission of the American Chemical Society.

A5. W. R. Edwards and N. F. Chamberlain, *J. Polymer Sci. A1,* 2299 (1963). Reproduced by permission of Interscience Publishers.

A6. F. C. Stehling and K. W. Bartz, *Anal. Chem. 38,* 1467 (1966). Reproduced by permission of the American Chemical Society.

A7. F. C. Stehling, *Anal. Chem. 35,* 773 (1963). Reproduced by permission of the American Chemical Society.

A8. C. Heathcock, *Can. J. Chem. 40,* 1865 (1962). Reproduced by permission of the National Research Council of Canada.

IV. DATA REFERENCES—MAJOR RESOURCES

A. TEXTBOOKS

1. C. P. Slichter, *Principles of Magnetic Resonance,* Harper & Row, New York (1963).
2. R. H. Bible, *Interpretation of NMR Spectra,* Plenum Press, New York (1965).
3. J. W. Emsley, J. Feeney, and L. H. Sutcliffe, *High Resolution Nuclear Magnetic Resonance Spectroscopy,* Pergamon Press, Oxford (1965).
4. A. Carrington and A. D. McLachlan, *Introduction to Magnetic Resonance,* Harper and Row, London (1967).
5. D. W. Mathieson, Editor, *Nuclear Magnetic Resonance for Organic Chemists,* Academic Press, London (1967).
6. E. D. Becker, *High Resolution NMR,* Academic Press, New York (1969).
7. F. A. Bovey, *Nuclear Magnetic Resonance Spectroscopy,* Academic Press, New York (1969).
8. L. M. Jackman and S. Sternhell, *Applications of Nuclear Magnetic Resonance Spectroscopy in Organic Chemistry,* Pergamon Press, Oxford (1969).
9. T. C. Farrar and E. D. Becker, *Pulse and Fourier Transform NMR,* Academic Press, New York (1971).

B. CATALOGS OF NMR SPECTRA

10. B. J. Zwolinski, *et al., Catalog of Selected Nuclear Magnetic Resonance Spectral Data,* Thermodynamics Research Center, Department of Chemistry, Texas A & M University, College Station, Texas 77843 (continuing publication).
    a. American Petroleum Institute Research Project 44 Catalog.
    b. Thermodynamics Research Center Data Project Catalog.
11. N. S. Bhacca, L. F. Johnson, J. N. Schoolery, D. P. Hollis, and E. A. Pier, *High Resolution NMR Spectral Catalog,* Volumes I and II, Varian Associates, 611 Hansen Way, Palo Alto, California 94303. (1962 and 1963).
12. *Nuclear Magnetic Resonance Spectra,* Sadtler Research Laboratories, 3316 Spring Garden Street, Philadelphia, Pennsylvania 19104 (continuing publication).
13. *High Resolution NMR Spectra,* JEOL, Inc., 477 Riverside Avenue, Medford, Mass. 02155.
    a. Volumes 1 and 2 combined and distributed by Sadtler Research Laboratories (1967).
    b. Two Minimar volumes, distributed by JEOL (1970 and 1971).

C. TABULATIONS OF NMR DATA

14. G.V.D. Tiers, Characteristic Nuclear Magnetic Resonance Shielding Values for Hydrogen in Organic Structures, Published privately in 1958, and reproduced in Refs. 3 and 7.
15. S. Sternhell, Long Range H^1-H^1 Spin-Spin Coupling Constants in Nuclear Magnetic Resonance Spectroscopy, *Rev. Pure and Appl. Chem. 14,* 15 (1964).
16. A. A. Bothner-By, Geminal and Vicinal Proton-Proton Coupling Constants in Organic Compounds, in: *Advances in Magnetic Resonance,* Vol. 1, J. S. Waugh, Ed., Academic Press, New York (1965), pp. 195–316.
17. G. Mavel, Studies of Phosphorus Compounds Using the Magnetic Resonance Spectra of Nuclei Other Than Phosphorus-31, in: *Progress in Nuclear Magnetic Resonance Spectroscopy,* Vol. 1, Ed. by J. W. Emsley, J. Feeney, and L. H. Sutcliffe, Pergamon Press, Oxford (1966), pp. 251–373.
18. W. Brügel, *Nuclear Magnetic Resonance Spectra and Chemical Structure,* Academic Press, New York 1967), 235 pp.
19. F. A. Bovey, *NMR Data Tables for Organic Compounds,* Wiley-Interscience, New York (1967), 610 pp.
20. E. F. Mooney and P. H. Winson, "Fluorine-19 Nuclear Magnetic Resonance Spectroscopy" in: *Annual Review of NMR Spectroscopy,* Vol. 1, E. F. Mooney, Ed., Academic Press, London (1968), pp. 243–311.
21. H. A. Szymanski and R. Yelin, *NMR Band Handbook,* Plenum Press, New York (1968), 432 pp.
22. H. Booth, Applications of ^{1}H Nuclear Magnetic Resonance Spectroscopy to the Conformational Analysis of Cyclic Compounds, in: *Progress in Nuclear Magnetic Resonance Spectroscopy,* Vol. 5, Ed. by J. W. Emsley, J. Feeney, and L. H. Sutcliffe, Pergamon Press, Oxford (1969), pp. 149–381.
23. E. F. Mooney and P. H. Winson, Carbon-13 Chemical Shifts and Coupling Constants, in: *Annual Review of NMR Spectroscopy,* Vol. 2, E. F. Mooney, Ed., Academic Press, London (1969), pp. 153–218.
24. K. Jones and E. F. Mooney, "Fluorine-19 Nuclear Magnetic Resonance Spectroscopy, in: *Annual Review of NMR Spectroscopy,* Vol. 3, E. F. Mooney, Ed., Academic Press, London (1970), pp. 261–421. Now titled *Annual Reports on NMR Spectroscopy.*
25. N. F. Chamberlain and J. J. R. Reed, Nuclear Magnetic Resonance Data of Sulfur Compounds, in: *The Analytical Chemistry of Sulfur and Its Compounds,* Part III, J. H. Karcher, Ed., Wiley-Interscience, New York (1971), 308 pp.
26. Functional Group and Chemical Shift Indexes to Refs. 11 and 12.

D. CHEMICAL SHIFT CHARTS

27. L. H. Meyer, A. Saika, and H. S. Gutowsky, *J. Amer. Chem. Soc. 75,* 4567 (1953).
28. N. F. Chamberlain, *Anal. Chem. 31,* 56 (1959).
29. K. Nukada, O. Yamamoto, T. Suzuki, M. Takeuchi, and M. Ohnishi, *Anal. Chem. 35,* 1892 (1963).
30. M. W. Dietrich and R. E. Keller, *Anal. Chem. 36,* 258 (1964).
31. E. Mohacsi, *J. Chem. Ed. 41,* 38 (1964).
32. E. Mohacsi, *Analyst 91,* No. 1078, 57 (Jan. 1966).
33. G. Slomp and J. G. Lindberg, *Anal. Chem. 39,* 60 (1967).
34. O. Yamamoto, T. Suzuki, M. Yanagisawa, and K. Hayamizu, *Anal. Chem. 40,* 568 (1968).
35. N. F. Chamberlain, F. C. Stehling, K. W. Bartz, and J. J. R. Reed, *Nuclear Magnetic Resonance Data for Hydrogen-1,* Esso Research and Engineering Company, Box 4255, Baytown, Texas 77520 (1965, revised 1969). Includes tabulated coupling constant correlations and typical spectra.

E. THEORETICAL SPECTRA

36. K. B. Wiberg and B. J. Nist, *The Interpretation of NMR Spectra,* W. A. Benjamin, New York (1962).

37. P. L. Corio, *Structure of High-Resolution NMR Spectra,* Academic Press, New York (1966).
38. Textbooks, References 1–9.

F. SPECTRAL PATTERN GUIDE

39. W. W. Simons and M. Zanger, *The Sadtler Guide to NMR Spectra,* Sadtler Research Laboratories, 3316 Spring Garden St., Philadelphia, Pa., (1972).

V. DATA REFERENCES–SCIENTIFIC PAPERS

A. HYDROCARBONS

Cycloparaffins

40. F. A. L. Anet and M. St. Jacques, *J. Amer. Chem. Soc. 88,* 2585 (1966).
41. J. J. Burke and P. C. Lauterbur, *J. Amer. Chem. Soc. 86,* 1870 (1964).
42. D. T. Longone and A. H. Miller, *Chem. Comm. 1967,* 447.

Aliphatic Olefins and Acetylenes

43. N. F. Cywinski, *J. Org. Chem. 30,* 361 (1965).
44. S. A. Francis and E. D. Archer, *Anal. Chem. 35,* 1363 (1963).
45. E. J. Smutny and H. Chung, *Amer. Chem. Soc., Div. Petrol. Chem., Preprints 14*(2), B112 (1969).
46. Y. Vo-Quang, L. Vo-Quang, and G. Empotz, *Compt. Rend. 258,* 4586 (1964).

Cyclic Olefins and Annulenes

47. P. Arnaud, J-L. Pierre, and M. Vidal, *Bull. Soc. Chim. France 1967,* 3810.
48. J. C. Calder and F. Sondheimer, *Chem. Comm. 1966,* 904.
49. Y. Gaoni, A. Melera, F. Sondheimer, and R. Wolovsky, *Proc. Chem. Soc. (London) 1964,* 397.
50. G. W. Griffin and L. I. Peterson, *J. Amer. Chem. Soc. 85,* 2268 (1963).
51. E. A. Hill and J. D. Roberts, *J. Amer. Chem. Soc. 89,* 2047 (1967).
52. U. E. Matter, C. Pascual, E. Pretsch, A. Pross, W. Simon, and S. Sternhell, *Tetrahedron 25,* 2023 (1969).
53. J. A. Pople and K. G. Untch, *J. Amer. Chem. Soc. 88,* 4811 (1966).
54. G. Schroder, *Tetrahedron Letters, 4083* (1966).
55. E. Vogel, W. Schröck, and W. A. Böll, *Angew. Chem. 78,* 753 (1966).
56. M. R. Willcott, Doctoral Thesis, Yale University (1963).

Cycloalkyl- and Cyclanoaromatics

57. E. J. Eisenbraun, J. R. Mattox, R. C. Bansal, P. W. Flanagan, and A. B. Carel, *Amer. Chem. Soc., Div. Petrol. Chem., Preprints 10*(3), 157 (1965).
58. H. Friebolin and S. Kabuss, in: *Nuclear Magentic Resonance in Chemistry,* B. Pesce, Ed., Academic Press, New York (1965), pp. 125–132.
59. B. J. Mair and T. J. Mayer, *Anal. Chem. 36,* 351 (1964).
60. B. L. Shapiro, M. J. Gattuso, N. F. Hepfinger, R. L. Shone, and W. L. White, *Tetrahedron Letters 1971,* 219.

61. B. L. Shapiro, M. J. Gattuso, and G. R. Sullivan, *Tetrahedron Letters 1971,* 223.
62. T. F. Wood, W. M. Easter, Jr., M. S. Carpenter, and J. Angiolini, *J. Org. Chem. 28,* 2248 (1963).

Cyclophanes

63. N. L. Allinger, M. A. Da Rooge, and R. B. Hermann, *J. Amer. Chem. Soc. 83,* 1974 (1961).
64. D. J. Cram and R. C. Helgeson, *J. Amer. Chem. Soc. 88,* 3515 (1966).
65. J. C. Dignan and J. B. Miller, *J. Org. Chem. 32,* 490 (1967).
66. D. T. Longone and C. L. Warren, *J. Amer. Chem. Soc. 84,* 1507 (1962).
67. D. T. Longone and H. S. Chow, *J. Amer. Chem. Soc. 86,* 3898 (1964).
68. D. J. Wilson, V. Boekelheide, and R. W. Griffin, Jr., *J. Amer. Chem. Soc. 82,* 6302 (1960).

Aromatic Olefins

69. J. M. Bollinger, J. J. Burke, and E. M. Arnett, *J. Org. Chem. 31,* 1310 (1966).
70. Gurudata, J. B. Stothers, and J. D. Talman, *Can. J. Chem. 45,* 731 (1967).
71. G. P. Newsoroff and S. Sternhell, *Aust. J. Chem. 19,* 1667 (1966).
72. H. Prinzbach and U. Fischer, *Angew.Chem. 78,* 642 (1966).
73. H. Rottendorf, S. Sternhell, and J. R. Wilmshurst, *Aust. J. Chem. 18,* 1759 (1965).
74. R. van der Linde, O. Korver, P. K. Korver, P. J. van der Haak, J. U. Veenland, and Th. J. de Boer, *Spectrochim. Acta 21,* 1893 (1965).
75. R. H. Wiley, T. H. Crawford and N. F. Bray, *Polymer Letters 3,* 99 (1965).

Biphenyls and Condensed Aromatics

76. D. Cagniant, *Bull. Soc. Chim. France 1966,* 2325.
77. W. Carruthers, H. N. M. Stewart, P. G. Hansell, and K. M. Kelley, *J. Chem. Soc. 1967*(C), 2607.
78. T. B. Cobb and J. D. Memory, *J. Chem. Phys. 47,* 2020 (1967).
79. A. Cornu, H. Ulrich, and K. Persaud, *Chim. Anal. 47,* 357 (1965).
80. D. C. F. Garbutt, K. G. R. Pachler, and J. R. Parrish, *J. Chem. Soc. 1965,* 2324.
81. R. H. Martin, N. Defay, F. Geerts-Evrard, and S. Delavarenne, *Tetrahedron 20,* 1073 (1964).
82. R. H. Michel, *J. Polymer Sci. A2,* 2533 (1964).
83. I. Puskas and E. K. Fields, *Amer. Chem. Soc., Div. Petrol Chem., Preprints 12*(2), D32 (1957).
84. F. F. Yew, R. J. Kurland, and B. J. Mair, *Anal. Chem. 36,* 843 (1964).
85. F. F. Yew and B. J. Mair, *Amer. Chem. Soc., Div. Petrol. Chem., Preprints 10*(3), 99 (1965).

B. OXYGEN COMPOUNDS

Acids and Esters

86. A. Abrahams, S. E. Wiberley, and F. C. Nachod, *Applied Spectry. 18,* 13 (1964).

87. P. Bucci and R. Rossi, in: *Nuclear Magnetic Resonance in Chemistry,* B. Pesce, Ed., Academic Press, New York (1965), pp. 133–142.

88. P. J. Collins and S. Sternhell, *Aust. J. Chem. 19, 317* (1966).

89. D. Doskocilova and B. Schneider, *J. Polymer Sci. B3,* 213 (1965).

90. J. A. Elvidge and P. D. Ralph, *J. Chem. Soc.(B) 1966,* 243, and J. Chem. Soc. (C) *1966,* 387.

91. R. R. Frazer, *Can. J. Chem. 38,* 549 (1960).

92. R. R. Frazer and D. E. McGreer, *Can. J. Chem. 39,* 505 (1961).

93. L. M. Jackman and R. H. Wiley, *J. Chem. Soc. 1960,* 2882 and 2886.

94. A. A. Pavia, J. Wylde, R. Wylde, and E. Arnal, *Bull. Soc. Chim. France 1965,* 2709.

95. J. M. Purcell, S. G. Morris, and H. Susi, *Anal. Chem. 38,* 588 (1966).

96. D. Savastianoff and M. Pfau, *Bull. Soc. Chim. France 1967,* 4162.

97. G. Slomp, Research Department, The Upjohn Co., Kalamazoo, Mich., 49001, private communication, 1962.

98. M. van Gorkam and G. E. Hall, *Spectrochim. Acta 22,* 990 (1966).

Alcohols

99. A. M. de Roos, Doctoral Thesis, Univ. of Amsterdam, (1965).

100. K. B. Wiberg and D. E. Barth, *J. Amer. Chem. Soc. 91,* 5124 (1969).

Aldehydes

101. A. W. Douglas and J. H. Goldstein, *J. Mol. Spectry 16,* 1 (1965).

102. R. E. Klink and J. B. Stothers, *Can. J. Chem. 44,* 45 (1966).

103. J. Wiemann, O. Convert, H. Danechpejouh, and D. Lelandais, *Bull. Soc. Chim. France 1966,* 1760.

Ethers

104. F. A. Bovey, *Chem. and Eng. News 43,* 98 (1965).

105. J. Chuche, *Compt. Rend. C263,* 779 (1966).

106. Reference 58 above.

107. J. Feeney, A. Ledwith, and L. H. Sutcliffe, *J. Chem. Soc. 1962,* 2021.

108. J. L. Jungnickel and C. A. Reilly, in: *Nuclear Magnetic Resonance in Chemistry,* B. Pesce, Ed., Academic Press, New York (1965), pp. 83–109.

109. P. K. Korver, P. J. Vander Haak, H. Steinberg, and Th. J. de Boer, *Rec. Trav. Chim. 84,* 129 (1965).

110. M. Martin, G. Martin and P. Caubere, *Bull. Soc. Chim. France 1964,* 3066.

111. J. D. Prugh, A. C. Huitric, and W. C. McCarthy, *J. Org. Chem. 29,* 1991 (1964).

Ketones

112. I. B. Douglass, F. J. Ward, and R. V. Norton, *J. Org. Chem. 32,* 324 (1967).

113. E. J. Eisenbraun, J. M. Springer, C. W. Hinman, P. W. Flanagan, and M. C. Hamming, *Amer. Chem. Soc., Div. Petrol. Chem., Preprints 14*(3), A49 (1969).

114. E. V. Elam and H. E. Davis, *J. Org. Chem. 32,* 1562 (1967).

115. J. Kossanyi, *Bull. Soc. Chim. France 1965,* 704.

116. C. Lumbrose and P. Maitte, *Bull. Soc. Chim. France 1965,* 315.

117. E. Marcus, J. K. Chan, and C. B. Strow, *J. Org. Chem. 31,* 1369 (1966).

C. HALOGEN COMPOUNDS

118. N. Boden, J. W. Emsley, J. Feeney, and L. H. Sutcliffe, in: *Nuclear Magnetic Resonance in Chemistry,* B. Pesce, Ed., Academic Press, New York (1965), pp. 149–157.

119. A. A. Bothner-By, S. Castellano and H. Gunther, *J. Amer. Chem. Soc. 87,* 2439 (1965).

120. M. Y. De Wolf and J. D. Baldeschwieler, *J. Mol. Spectry. 13,* 344 (1964).

121. R. E. Mayo and J. H. Goldstein, *J. Mol Spectry. 14,* 173 (1964).

122. F. S. Mortimer, *J. Mol. Spectry. 3,* 335 (1959).

123. H. F. White, *Anal. Chem. 36,* 1291 (1964).

D. NITROGEN COMPOUNDS

Amines and Imines

124. M. C. Caserio, R. E. Pratt, and R. J. Holland, *J. Amer. Chem. Soc. 88,* 5747 (1966).

125. G. J. Karabatsos and S. S. Lande, *Tetrahedeon 24,* 3907 (1968).

126. A. Mathias, *Anal. Chem. 38,* 1931 (1966).

127. J. L. Sudmeier and C. N. Reilley, *Anal. Chem. 36,* 1698 and 1707 (1964).

128. M. Zanger, W. W. Simons, and A. R. Gennaro, *J. Org. Chem. 33,* 3673 (1968).

Amides

129. J. D. Bartels-Keith and R. F. W. Cieciuch, *Can. J. Chem. 46,* 2593 (1968).

130. A. J. R. Bourn, D. G. Gillies, and E. W. Randall, *Tetrahedron 22,* 1825 (1966).

131. R. F. C. Brown, L. Radom, S. Sternhell, and I. D. Rae, *Can. J. Chem. 46,* 2577 (1968).

132. R. E. Carter, *Acta Chem. Scand. 21,* 75 (1967).

133. C. E. Holloway and M. H. Gitlitz, *Can. J. Chem. 45,* 2659 (1967).

134. S. R. Johns, J. A. Lamberton, and A. A. Sioumis, *Chem. Comm. 1966,* 480.

135. H. Kessler and A. Rieker, *Z. Naturforsch. 22*b, 456 (1967).

136. L. A. LaPlanche and M. T. Rogers, *J. Amer. Chem. Soc. 85,* 3728 (1963).

137. L. A. LaPlanche and M. T. Rogers, *J. Amer. Chem. Soc. 86,* 337 (1964).

138. I. D. Rae, *Can. J. Chem. 44,* 1334 (1966).

139. T. H. Siddall, III *J. Mol. Spectry. 20,* 183 (1966).

140. T. H. Siddall, III and R. J. Garner, *Can. J. Chem. 44,* 2387 (1966).

141. T. H. Siddall, III and C. A. Prohaska, *J. Amer. Chem. Soc. 88,* 1172 (1966).

142. T. H. Siddall, III and M. L. Good, *J. Inorg. and Nucl. Chem.* *29,* 149 (1967).

143. H. A. Staab and D. Lauer, *Tetrahedron Letters 1966,* 4593.

144. B. Sunners, L. H. Piette, and W. G. Schneider, *Can. J. Chem. 38,* 681 (1960).

145. J. F. K. Wilshire, *Tetrahedron Letters 1958,* 475.

Hydrazones

146. G. J. Karabatsos and R. A. Taller, *J. Amer. Chem. Soc. 85,* 3624 (1963).

147. G. J. Karabatsos and K. L. Krumel, *Tetrahedron 23,* 1097 (1967).

148. G. J. Karabatsos and R. A. Taller, *Tetrahedron 24,* 3923 (1968).

149. E. W. Warnhoff, *Can. J. Chem. 42,* 1664 (1964).

Oximes

150. E. Buehler, *J. Org. Chem. 32,* 261 (1967).

151. J. P. Kintzinger and J. M. Lehn, *Chem. Comm. 1967,* 660.

152. G. G. Kleinspehn, J. A. Jung, and S. A. Studniarz, *J. Org. Chem. 32,* 460, (1967).

153. I. Pejkovic-Tadic, M. Hranisavljevic-Jakovljevic, S. Nesic, C. Pascual, and W. Simon, *Helv. Chim. Acta 48,* 1157 (1965).

154. H. Saito and K. Nukada, *J. Mol Spectry. 18,* 1 (1965).

Nitriles and Isonitriles

155. I. D. Kuntz, P. von R. Schleyer, and A. Allerhand, *J. Chem. Phys. 35,* 1533 (1961).

156. K. Matsuzaki, T. Uryu, and K. Ishigure, *J. Polymer Sci. 4B,* 93 (1966).

Nitro Compounds

157. D. G. Gehring and G. S. Reddy, (a) *Anal. Chem. 37,* 868 (1965); (b) *Anal. Chem. 40,* 792 (1968).

158. W. Hoffman, L. Stefaniak, T. Urbanski, and M. Witanowski, *J. Amer. Chem. Soc. 86,* 554 (1964).

Pyridines

159. R. A. Abramovitch and J. B. Davis, *J. Chem. Soc. (B) 1966,* 1137.

160. W. Brügel, *Zeit. Electrochem. 66,* 159 (1962).

161. S. Castellano, C. Sun, and R. Kostelnik, *J. Chem. Phys. 46,* 327 (1967).

162. V. M. S. Gil, *Mol. Phys. 9,* 97 (1965).

163. K. Griesbaum and G. W. Burton, *Amer. Chem. Soc., Div. Petrol. Chem., Preprints 13*(1), 216 (1968).

164. F. A. Karmer and R. West, *J. Phys. Chem. 69,* 673 (1965).

165. T. McL. Spotswood and C. I. Tanzer, *Aust. J. Chem. 20,* 1227 (1967).

Other

166. G. Casini and M. L. Salvi, in: *Nuclear Magnetic Resonance in Chemistry,* B. Pesce, Ed., Academic Press, New York (1965), pp. 255–262.

167. A. Mathias, *Tetrahedron 21,* 1073 (1965).

168. G. J. Karabatsos and R. A. Taller, *J. Amer. Chem. Soc. 86,* 4373 (1964).

E. SULFUR COMPOUNDS

General (Several compound types in each reference)

169. N. F. Chamberlain and J. J. R. Reed, Nuclear Magnetic Resonance Data of Sulfur Compounds, Part III of *The Analytical Chemistry of Sulfur and Its Compounds,* J. H. Karchmer, Ed., Wiley-Interscience, New York (1971). This book contains a tabulation of all the data taken from all the rest of these references.

170. J. F. Biellmann and H. Callot, *Chem. Commun.* 458 (1967).

171. Maud Brink, *Acta Chem. Scand. 20,* 2882 (1966).

172. E. D. Brown, S. M. Iqbal, and L. N. Owen, *J. Chem. Soc.* (c) *1966,* 415.

173. M. C. Caserio, R. E. Pratt, and R. J. Holland, *J. Amer. Chem. Soc. 88,* 5747 (1966).

174. E. V. Elam and H. E. Davis, *J. Org. Chem. 32,* 1562 (1967).

175. K. Griesbaum, A. A. Oswald, and B. E. Hudson, Jr., *J. Amer. Chem. Soc. 85,* 1969 (1963).

176. K. Hayamizu and O. Yamamoto, *J. Mol. Spectry. 25,* 422 (1968).

177. H. Hogeveen, G. Maccagnani, and F. Taddei, *Rec. Trav. Chim. 83,* 937 (1964).

178. J. J. Looker, *J. Org. Chem. 31,* 2973 (1966).

179. W. H. Mueller and P. E. Butler, *J. Org. Chem. 32,* 2925 (1967).

180. W. H. Mueller and P. E. Butler, *J. Org. Chem. 33,* 1533 (1968).

181. R. J. Niedzielski, R. S. Drago, and R. L. Middaugh, *J. Amer. Chem. Soc. 86,* 1694 (1964).

182. M. Oki and H. Iwamura, *Bull. Chem. Soc. Japan 35,* 1428 (1962).

183. T. Olijnsma, J. B. F. N. Engberts, and J. Strating, *Rec. Trav. Chim. 6,* 463 (1967).

184. A. A. Oswald, K. Griesbaum, W. A. Thaler, and B. E. Hudson, Jr., *J. Amer. Chem. Soc. 84,* 3897 (1962).

185. A. A. Oswald, K. Griesbaum, B. E. Hudson, Jr., and J. M. Bregman, *J. Amer. Chem. Soc. 86,* 2877 (1964).

186. G. R. Pettit, I. B. Douglass, and R. A. Hill, *Can. J. Chem. 42,* 2357 (1964).

187. J. M. Purcell and H. Susi, *Appl. Spectry. 19,* 105 (1965).

188. A. H. Raper and E. Rothstein, *J. Chem. Soc. 1963,* 1027.

189. G. A. Russell and G. J. Mikol, *J. Amer. Chem. Soc. 88,* 5498 (1966).

190. F. Taddei and C. Zauli, in: *Nuclear Magnetic Resonance in Chemistry,* B. Pesce, Ed., Academic Press, N.Y. (1965) pp. 180–181.

191. B. K. Tidd, *J. Chem. Soc. 1963,* 3909.

192. W. P. Trompen, C. Kruk, P. J. Van der Haak, and H. O. Huisman, *Rec. Trav. Chim. 85,* 185 (1966).

193. J. van der Veen, *Rec. Trav. Chim. 84,* 540 (1965).

194. J. R. Van Wazer and D. Grant, *J. Amer. Chem. Soc. 86,* 1450 (1964).

195. NMR at Work Series, No. 90, Varian Associates, Palo Alto, Calif.

196. J. W. Wilt and W. J. Wagner, *Chem. Ind. (London) 1964,* 1389.

Thiols

197. P. E. Butler and W. H. Mueller, *Anal. Chem. 38,* 1407 (1966).

198. M. Demuynck and J. Vialle, *Bull Soc. Chim. France 1967,* 1213.

199. M. Demuynck and J. Vialle, *Bull. Soc. Chim. France 1967,* 2748.

200. J. W. Forbes and J. L. Jungnickel, *Appl. Spectry. 19*, 124 (1965).
201. J. F. Harris, Jr., *J. Org. Chem. 30*, 2190 (1965).
202. S. H. Marcus and S. I. Miller, *J. Phys. Chem. 68*, 331 (1964).
203. S. H. Marcus and S. I. Miller, *J. Amer. Chem. Soc. 88*, 3719 (1966).
204. M-M. Rousselot, *Compt. Rend. C262*, 26 (1966).
205. M-M. Rousselot and M. Martin, *Compt. Rend. C262*, 1445 (1966).
206. H. Schmidbaur and W. Siebert, *Chem. Ber. 97*, 2090 (1964).
207. O. P. Strausz, T. Hikida, and H. E. Gunning, *Can. J. Chem. 43*, 717 (1965).
208. K. Takahashi and G. Hazato, *Bull. Chem. Res. Inst. Non-Aqu. Soln. Tohoku Univ. 16*, 1 (1966).

Sulfides

209. F. Bohlmann, C. Arndt, and J. Starnick, *Tetrahedron Letters 1963*, 1605.
210. M. G. Burdon and J. G. Moffatt, *J. Amer. Chem. Soc. 89*, 4725 (1967).
211. C. H. Bushweller, *J. Amer. Chem. Soc. 89*, 5978 (1967).
212. A. Deljac, Z. Stefanac, and K. Balenovic, *Tetrahedron*, Supplement 8 (Part I), 33, (1966).
213. Ben E. Edwards, Doctoral Dissertation, Univ. of Illinois, 1962.
214. H. Friebolin and S. Kabuss, in: *Nuclear Magnetic Resonance in Chemistry*, B. Pesce, Ed., Academic Press, N. Y. (1965), pp. 125–132.
215. D. Grant and J. R. van Wazer, *J. Amer. Chem. Soc. 86*, 3012 (1964).
216. D. N. Hall, A. A. Oswald, and K. Griesbaum, *J. Org. Chem. 30*, 3829 (1965).
217. D. N. Hall, *J. Org. Chem. 32*, 2082 (1967).
218. J. F. Harris, *J. Org. Chem. 32*, 2063 (1967).
219. R. T. Hobgood, G. S. Reddy, and J. H. Goldstein, *J. Phys. Chem. 67*, 110 (1963).
220. K. Isenberg and H. F. Herbrandson, *Tetrahedron 21*, 1067 (1965).
221. S. Kabuss, A. Lutringhaus, H. Friebolin, and R. Mecke, *Z. Naturforsch. 21b*, 320 (1966).
222. P. K. Korver, P. J. van der Haak, H. Steinberg, and Th. J. de Boer, *Rec. Trav. Chim. 84*, 129 (1965).
223. D. J. Martin and R. H. Pearce, *Anal. Chem. 38*, 1604 (1966).
224. R. V. Moen, Determination of *Cis-Trans* Ratio in Propenyl Sulfides by NMR. Talk presented at Second Middle Atlantic Regional meeting, Society for Applied Spectroscopy, New York City, Feb. 6–7, 1967.
225. W. H. Mueller, *J. Org. Chem. 31*, 3075 (1966).
226. W. H. Mueller and P. E. Butler, *J. Amer. Chem. Soc. 90*, 2075 (1968).
227. W. H. Mueller and P. E. Butler, *J. Org. Chem. 33*, 2642 (1968).
228. A. A. Oswald, K. Griesbaum, and B. E. Hudson, Jr., *J. Org. Chem. 28*, 2355 (1963).
229. A. A. Oswald, K. Griesbaum, D. N. Hall, and W. Naegele, *Can. J. Chem. 45*, 1173 (1967).
230. A. A. Oswald, B. E. Hudson, Jr., G. Rodgers, and F. Noel, *J. Org. Chem. 27*, 2439 (1962).
231. A. A. Oswald and W. Naegele, *J. Org. Chem. 31*, 830 (1966).

232. M-P. Simonnin, *Compt. Rend. 257*, 1075 (1963).
233. W. A. Thaler, A. A. Oswald, and B. E. Hudson, Jr., *J. Amer. Chem. Soc. 87*, 311 (1965).
234. M. R. Willcott, University of Houston, Private Communication.

Sulfones and Sulfonates

235. I. B. Douglass, F. J. Ward, and R. V. Norton, *J. Org. Chem. 32*, 324 (1967).
236. D. C. F. Garbutt, K. G. R. Pachler, and J. R. Parrish, *J. Chem. Soc. 1965*, 2324.
237. K. J. Ivin & J. B. Rose, Polysulfones, in: *Advances in Macromolecular Chemistry*, W. M. Pasika, Ed., Vol. I, Academic Press, N. Y. (1968) pp. 344–346.
238. R. J. Mulder, A. M. van Leusen, and J. Strating, *Tetrahedron Letters 1967*, 3061.
239. T. Nagai, K. Nishitomi, and N. Tokura, *Tetrahedron Letters 1966*, 2419.
240. R. W. Ohline, A. L. Allred, and F. G. Bordwell, *J. Amer. Chem. Soc. 86*, 4641 (1964).
241. W. E. Truce and R. W. Campbell, *J. Amer. Chem. Soc. 88*, 3599 (1966).

Sulfites

242. F. Kaplan and J. D. Roberts, *J. Amer. Chem. Soc. 83*, 4666 (1961).
243. P. C. Lauterbur, J. G. Pritchard, and R. L. Vollmer, *J. Chem. Soc. 1963*, 5307.
244. J. G. Pritchard and P. C. Lauterbur, *J. Amer. Chem. Soc. 83*, 2105 (1961).
245. Q. E. Thompson, M. M. Crutchfield, and M. W. Dietrich, *J. Org. Chem. 30*, 2696 (1966).
246. G. Wood and M. Miskow, *Tetrahedron Letters 1966*, 4433.

Sulfonium Salts and Ylides

247. C. R. Johnson and M. P. Jones, *J. Org. Chem. 32*, 2014 (1967).
248. K. Konds and K. Mislow, *Tetrahedron Letters 1967*, 1325.
249. N. J. Leonard and C. R. Johnson, *J. Amer. Chem. Soc. 84*, 3701 (1962).
250. H. Nozaki, M. Takaku, and K. Kondo, *Tetrahedron 22*, 2145 (1966).
251. H. Nozaki, D. Tunemoto, Z. Morita, K. Nakamura, K. Watanabe, M. Takaku, and K. Kondo, *Tetrahedron 23*, 4279 (1967).
252. K. W. Ratts, *Tetrahedron Letters 1966*, 4707.
253. K. W. Ratts and A. N. Yao, *J. Org. Chem. 31*, 1185 (1966).
254. W. E. Truce and G. D. Madding, *Tetrahedron Letters 1966*, 3681.

Sulfur Pentafluorides

255. N. Boden, J. W. Emsley, J. Feeney, and L. H. Sutcliffe, *Trans. Faraday Soc. 59*, 620 (1963).
256. F. W. Hoover and D. D. Coffman, *J. Org. Chem. 29*, 3567 (1964).

Thiophenes

257. P. Cagniant, P. Faller, and D. Cagniant, *Bull. Soc. Chim. France 1966*, 3055.
258. P. Fournari, R. Guilard, and M. Person, *Bull. Soc. Chim. France 1967*, 4115.
259. R. Guilard, P. Fournari, and M. Person, *Bull. Soc. Chim. France 1967*, 4121.

260. R. A. Hoffman and S. Gronowitz, *Arkiv Kemi 16*, 515 (1960).

261. A. Kergomard and S. Vincent, *Bull. Soc. Chim. France 1967*, 2197.

262. K. Takahashi, T. Kanda, and Y. Matsuki, *Bull. Chem. Soc. Japan 37*, 768 (1964).

263. K. Takahashi, T. Kanda, and Y. Matsuki, *Bull. Chem. Soc. Japan 38*, 1799 (1965).

264. K. Takahashi, T. Kanda, F. Shoji, and Y. Matsuki, *Bull. Chem. Soc. Japan 38*, 508 (1965).

265. K. Takahashi, I. Ito, and Y. Matsuki, *Bull. Chem. Soc. Japan 39*, 2316 (1966).

266. K. Takahashi, I. Ito, and Y. Matsuki, *Bull. Chem. Soc. Japan 40*, 605 (1967).

267. K. Takahashi, T. Sone, Y. Matsuki, and G. Hazato, *Bull. Chem. Res. Inst. Non-Aqu. Soln. Tohoku Univ. 15*, 1 (1966).

Other

268. C. E. Holloway and M. H. Gitlitz, *Can. J. Chem. 45*, 2659 (1967).

269. J. Jonas, W. Derbyshire, and H. S. Gutowsky, *J. Phys. Chem. 69*, 1 (1965).

270. A. Mathias, *Tetrahedron 21*, 1073 (1965).

271. R. L. Middaugh and R. S. Drago, *J. Amer. Chem. Soc. 85*, 2575 (1963).

272. R. M. Moriarty, *Tetrahedron Letters 1964*, 509.

273. W. H. Mueller and P. E. Butler, *J. Org. Chem. 33*, 2111 (1968).

F. PHOSPHORUS COMPOUNDS

General

274. J. J. Brophy and M. J. Gallagher, *Aust. J. Chem. 20*, 503 (1967).

275. A. J. Carty and R. K. Harris, *Chem. Commun. 1967*, 234.

276. R. Freymann, in: *Nuclear Magnetic Resonance in Chemistry*, B. Pesce, Ed., Academic Press, N. Y. (1965), pp. 13–34.

277. C. E. Griffin, *Tetrahedron 20*, 2399 (1964).

278. J. B. Hendrickson, M. L. Maddox, J. J. Sims, and H. D. Kaesz, *Tetrahedron 20*, 449 (1964).

279. G. Mavel, in: *Progress in Nuclear Magnetic Resonance Spectroscopy*, J. W. Emsley, J. Feeny and L. H. Sutcliffe, Eds., Vol. I, Pergamon Press, N. Y. (1966) pp. 251–373.

280. R. Schmutzler and G. S. Reddy, *Z. Naturforsch 20b*, 832 (1965).

281. K. Takahashi, T. Yamasaki, and G. Miyazima, *Bull. Chem. Soc. Japan 39*, 2787 (1966).

282. J. F. Nixon and R. Schmutzler, *Spectrochim. Acta 22*, 565 (1966).

Phosphines and Their Oxides, Sulfides, and Halides

283. A. M. Aguiar and D. Daigle, *J. Org. Chem. 30*, 3527 (1965).

284. J. K. Becconsall, R. M. Canadine, and R. Murray, *Chem. Commun. 1966*, 425.

285. T. L. Charlton and R. G. Cavell, *Chem. Commun. 1966*, 763.

286. C. Charrier, W. Chodkiewicz, and P. Cadiot, *Bull. Soc. Chim. France 1966*, 1002.

287. W. Drenth and D. Rosenberg, *Rec. Trav. Chim. 86*, 26 (1967).

288. B. Fontal, H. Goldwhite, and D. G. Roswell, *J. Org. Chem. 31*, 2424 (1966).

289. R. K. Harris and R. G. Hayter, *Can. J. Chem. 42*, 2282 (1964).

290. L. Maier, *Helv. Chim. Acta. 49*, 842 (1966).

291. S. L. Manatt, G. L. Juvinall, R. I. Wagner, and D. D. Elleman, *J. Amer. Chem. Soc. 88*, 2689 (1966).

292. R. W. Rudolph and R. W. Perry, *Inorg. Chem. 4*, 1339 (1965).

293. F. Seel, K. Rudolph, and R. Budenz, *Z. Anorg. Allg. Chem. 341*, 196 (1965).

294. M.-P. Simonnin and B. Borecka, *Bull. Soc. Chim. France 1966*, 3842.

295. F. J. Welch and H. J. Paxton, *J. Polymer Sci. A3*, 3439 (1965).

Phosphine Oxides and Phosphonates

296. C. E. Griffin, R. B. Davison, and M. Gordon, *Tetrahedron 22*, 561 (1966).

297. T. H. Siddall and C. A. Prohaska, *Appl. Spectry. 21*, 9 (1967).

Combinations of Phosphorus Esters, Thioesters, and Amidoesters

298. K. D. Bartle, R. S. Edmundson, and D. W. Jones, *Tetrahedron 23*, 1701 (1967).

299. K. D. Berlin, D. M. Hellwege, and M. Nagabhushanam, *J. Org. Chem. 30*, 1265 (1965).

300. E. Duval and E. A. C. Lucken, *Mol. Phys. 10*, 499 (1966).

301. J. R. Ferraro and D. F. Peppard, *J. Phys. Chem. 67*, 2639 (1963).

302. D. Gagnaire and J-B. Robert, *Bull. Soc. Chim. France 1967*, 2240.

303. D. Gagnaire, J-B. Robert, and J. Verrier, *Bull. Soc. Chim. France 1966*, 3719.

304. R. K. Harris, A. R. Katritzky, S. Musierowicz, and B. Ternai, *J. Chem. Soc. (A) 1967*, 37.

305. J. Herweh, *J. Org. Chem. 31*, 4308 (1966).

306. F. Kaplan and C. O. Shulz, *Chem. Commun. 1967*, 376.

307. G. Mavel and R. Favelier, *J. Chim. Phys. 64*, 627 (1967).

308. K. Pilgrim, *Tetrahedron 22*, 1241 (1966).

Phosphates and Phosphorothioates (Thiophosphates)

309. R. C. Axtmann, W. E. Shuler, and J. H. Eberly, *J. Chem. Phys. 31*, 850 (1959).

310. P. E. Butler, W. H. Mueller, and J. J. R. Reed, *Environ. Sci. Technol. 1*, 315 (1967).

311. R. C. de Selms and T-W. Lin, *J. Org. Chem. 32*, 2023 (1967).

312. W. H. Mueller and A. A. Oswald, (a) *J. Org. Chem. 31*, 1894 (1966); (b) *Preprints, Div. Petrol. Chem., Am. Chem. Soc. National Meeting, Miami Beach, April 9–14, 1967*, p. 69.

313. W. H. Mueller, R. M. Rubin, and P. E. Butler, *J. Org. Chem. 31*, 3537 (1966).

314. A. A. Oswald, K. Griesbaum, D. N. Hall, and W. Naegele, *Can. J. Chem. 45*, 1173 (1967).

315. A. A. Oswald, K. Griesbaum, and B. E. Hudson, *J. Org. Chem. 28*, 1262 (1963).

316. F. Ramirez, *Pure Appl. Chem. 9*, 337 (1964).

Phosphonates

317. F. A. Cotton and R. A. Schunn, *J. Amer. Chem. Soc. 85,* 2394 (1963).
318. C. Benezra, S. Nseic, and G. Ourisson, *Bull. Soc. Chim. France 1967,* 1140.
319. C. Benezra and G. Ourisson, *Bull. Soc. Chim. France 1966,* 2270.
320. K. D. Berlin and H. A. Taylor, *J. Amer. Chem. Soc. 86,* 3862 (1964).
321. M. Y. DeWolf, *J. Mol. Spectry. 18,* 59 (1965).
322. M. Gordon and C. E. Griffin, *J. Org. Chem. 31,* 333 (1966).
323. R. G. Harvey, *Tetrahedron 22,* 2561 (1966).
324. G. L. Kenyon and F. H. Westheimer, *J. Amer. Chem. Soc. 88,* 3557 (1966).
325. M. Lenzi, G. Sturtz, and G. Lavielle, (a) *Compt. Rend. C264,* 1329 (1967); (b) *Compt. Rend. C264,* 1425 (1967).

Phosphinates

326. K. D. Berlin and R. V. Pagilagan, *J. Org. Chem. 32,* 129 (1967).
327. H. Goldwhite and D. D. Rowsell, *J. Amer. Chem. Soc. 88,* 3572 (1966).

Amides of Phosphorus

328. C. Charrier and M-P. Simonnin, *Compt. Rend. C264,* 995 (1967).
329. A. H. Cowley and R. P. Pinnell, *J. Amer. Chem. Soc. 87,* 4454 (1965).
330. R. Keat and R. A. Shaw, *J. Chem. Soc. 1965,* 4802.
331. J. Parello, *Bull. Soc. Chim, France 1964,* 2033.
332. R. Schmutzler, *J. Chem. Soc. 1965,* 5630.
333. P. E. Sonnet and A. B. Borkovec, *J. Org. Chem. 31,* 2962 (1966).

Phosphonium Salts

334. D. W. Allen and I. T. Miller, *Tetrahedron Letters 1968,* 745.
335. M. Gordon and C. E. Griffin, *J. Chem. Phys. 41,* 2570 (1964).
336. C. E. Griffin and M. Gordon, *J. Organometal Chem. 1965,* 414.
337. K. Khaleeluddin and J. M. W. Scott, *Chem. Commun. 1966,* 511.
338. G. Singh and H. Zimmer, *J. Org. Chem. 30,* 417 (1965).

Other Phosphorus Compounds

339. G. Bergerhoff and F. Knoll, (a) *Angew. Chem. 77,* 1016 (1965); (b) *Monatsh. Chem. 97,* 808 (1966). Potassium Phosphide.
340. F. Ramirez, A. V. Patwardhan, N. Ramanathan, N. B. Desai, C. V. Greco, and S. R. Heller, *J. Amer. Chem. Soc. 87,* 543 and 549 (1965). Oxyphosphoranes.
341. H. Weitkamp and F. Korte, *Z. Anal. Chem. 204,* 245 (1964). Phospholines.

G. COUPLING CONSTANTS ONLY

Hydrogen–Hydrogen

342. F. A. L. Anet, *Can. J. Chem. 39,* 789 (1961).
343. A. A. Bothner-By and R. K. Harris, *J. Amer. Chem. Soc. 87,* 3451 (1965).

344. J. C. Davis and T. V. Van Auken, *J. Amer. Chem. Soc. 87,* 3900 (1965).
345. B. Gestblom, S. Gronowitz, R. A. Hoffman, B. Mathiasson, and S. Rodmar, *Arkiv Kemi 23,* 483 and 501 (1965).
346. J. B. Lambert, R. G. Keske, and D. K. Weary, *J. Amer. Chem. Soc. 89,* 5921 (1967).
347. M. L. Martin, G. J. Martin, and R. Couffignal, *J. Mol. Spectry. 34,* 53 (1970).
348. I. D. Rae, *Aust. J. Chem. 19,* 1983 (1966).
349. B. L. Shapiro, S. J. Ebersole, G. J. Karabatsos, F. M. Vane, and S. L. Manatt, *J. Amer. Chem. Soc. 85,* 4041 (1963).
350. S. Sternhell, *Rev. Pure and Appl. Chem. 14,* 15 (1964).
351. H. Weitkamp and F. Korte, *Tetrahedron, Supp. 7, 1966,* 75.

Hydrogen–Carbon-13

352. A. W. Douglas, *J. Chem. Phys. 45,* 3465 (1966).
353. K. Frei and H. J. Bernstein, *J. Chem. Phys. 38,* 1216 (1963).
354. J. H. Goldstein and G. S. Reddy, *J. Chem. Phys. 36,* 2644 (1962).
355. G. A. Gray, P. D. Ellis, D. D. Traficante, and G. E. Maciel, *J. Magn. Resonance 1,* 41 (1969).
356. P. Haake, W. B. Miller, and D. A. Tyssee, *J. Amer. Chem. Soc. 86,* 3577 (1964).
357. G. J. Karabatsos and C. E. Orzech, *J. Amer. Chem. Soc. 87,* 560 (1965).
358. P. C. Lauterbur, *J. Amer. Chem. Soc. 83,* 1838 and 1846 (1961).
359. P. C. Lauterbur, *J. Chem. Phys. 38,* 1406, 1415, and 1432 (1963).
360. P. C. Lauterbur, *J. Chem. Phys. 43,* 360 (1965).
361. C. T. Mathis and J. H. Goldstein, *Spectrochim. Acta 20,* 871 (1964).
362. W. McFarlane, *Mol. Phys. 10,* 603 (1966).
363. R. L. Middaugh and R. S. Drago, *J. Amer. Chem. Soc. 85,* 2575 (1963).
364. N. Muller and D. E. Pritchard, *J. Chem. Phys. 31,* 768 and 1471 (1959).

Hydrogen–Nitrogen

365. J. D. Baldeschwieler, *J. Chem. Phys. 36,* 152 (1962).
366. J-F. Biellman and H. Callot, *Bull. Soc. Chim. France 1967,* 397.
367. G. Binsch, J. B. Lambert, B. W. Roberts, and J. D. Roberts, *J. Amer. Chem. Soc. 86,* 5564 (1964).
368. A. K. Bose and I. Kugajevsky, *Tetrahedron 23,* 1489 (1967).
369. A. J. R. Bourn, D. G. Gillies, and E. W. Randall, in: *Nuclear Magnetic Resonance in Chemistry,* B. Pesce, Ed., Academic Press, New York (1965), pp. 277–281.
370. D. Crepaux and J. M. Lehn, *Mol. Phys. 14,* 547 (1968).
371. E. F. Mooney and P. H. Winson in *Annual Review of NMR Spectroscopy,* Vol. 2, E. F. Mooney, Ed., Academic Press, London (1969) pp. 125–152.
372. R. A. Ogg and J. D. Ray, *J. Chem. Phys. 26,* 1515 (1957).

Hydrogen–Phosphorus

373. S. L. Manatt, G. L. Juvinall, and D. D. Elleman, *J. Amer. Chem. Soc. 85,* 2664 1963).

374. J. F. Nixon and A. Pidcock in *Annual Review of NMR Spectroscopy*, Vol. 2 E. F. Mooney, Ed., Academic Press, London (1969).

H. STRONG INTRAMOLECULAR HYDROGEN BONDS

375. G. Allen and R. A. Dwek, *J. Chem. Soc. (B) 1966*, 161.
376. J. D. Bartels-Keith and R. F. W. Cieciuch, *Can. J. Chem. 46*, 2593 (1968).

377. J. L. Burdett and M. T. Rogers, *J. Amer. Chem. Soc. 86*, 2105 (1964).
378. G. Klose and E. Uhlmann, in: *Nuclear Magnetic Resonance in Chemistry*, B. Pesce, Ed., Academic Press, New York (1965) pp. 237–242.
379. E. Marcus, J. K. Chan and C. B. Strow, *J. Org. Chem. 31*, 1369 (1966).
380. M. Saquet and A. Thuillier, *Bull. Soc. Chim. France 1967*, 2841.

Appendix
Filing System for NMR Spectra

I. INTRODUCTION

The purpose of this filing system is to provide rapid access to NMR spectra from all sources by individual compound or by compound class. It permits a search of each class rapidly and thoroughly without using the time-consuming procedure of consulting an index and then searching several volumes for the desired spectra. This system puts all spectra of similar compounds together in the same section of a single volume, and permits rapid location of the proper volume from classes to speed the search still more. With this system the spectrum of any compound can be located in about one minute or less, or it can be determined with assurance that the desired spectrum is not in the collection. In the latter event, one has at hand all the spectra of similar compounds by the use of which the desired spectrum can be approximated.

This arrangement makes it easy to observe errors and inconsistencies in interpretation and to correct them for a whole set of spectra at the same time. It is of great help in preparing correlations of chemical shift with structure.

The organization scheme is given in Section II. This is followed, in Section III, by a list of the major filing divisions to show how the scheme is actually used. Subdivisions of these major divisions are made to suit the needs and interests of the user and to fit the actual collection of spectra available at any given time. They are expanded or changed as needed.

The overall size of the catalog is reduced by several arbitrary decisions:

1. Esters and amides are filed by their acid parts only, except when the alcohol or amine part is substituted. This limits access to the alcohol or amine part, but cuts the size of each section in half.

2. An order of precedence is used for functional groups so that compounds containing two functional groups can be filed in only one place according to this order.

3. Acyclic compounds containing more than two functional groups are filed in a separate category, subdivided by heteroatom or atoms.

4. The heterocyclic ring is considered more important than the substituents; therefore, these rings are filed together regardless of the degree or kind of substitution. Subdivisions permit the substituent types to be filed together. This decision seems justified by the great variety of heterocyclic rings and the resulting scarcity of examples of any given type.

When examples of a particular compound type (such as aldehydes) are scarce, however, all available samples of it are filed together. Cross-filing (multiple filing) is used when necessary to preserve the integrity of the general filing scheme.

II. ORGANIZATION OF FILING SYSTEM FOR NMR SPECTRA

The purpose of the present organization is to permit rapid access to an individual compound or to a compound type with a minimum of cross-filing. Compounds are filed in a single location in general. Cross indexes in the front of each volume or in individual sections permit ready reference to other volumes or sections which include the same functional groups.

Filing priorities are based on the greatest convenience for the NMR Group at this time. The functional group priorities (order of precedence) used by *Chemical Abstracts* has been modified to suit the spectra actually on hand and the frequency of need currently experienced.

ALIPHATICS AND HOMOCYCLICS

Unsubstituted hydrocarbons and hydrocarbons substituted with only *one* type of nonhydrocarbon group are filed by compound type. Sections for alicyclic olefins, aromatic naphthenes, aromatic olefins, etc., make cross filing unnecessary.

Aliphatic and homocyclic compounds containing only *two* types of nonhydrocarbon groups are filed as "substituted" compounds under one of the two groups. The order of precedence is as follows:

Organometallics and metal complexes, but not salts.
Phosphorus compounds.
Sulfur compounds:
> Thiocarboxylic acids are filed under substituted acids, and their esters under substituted esters.
> Thiophosphates are filed in a separate section under phosphorus.
> Thiocyanates are filed under nitrogen, in the section on cyanates.
> Thioamides are filed under substituted amides.

Carboxylic acids and anhydrides: cross-filed if they contain metals, phosphorus, sulfur, or scarce groups.
Esters and salts of carboxylic acids.
Amides of carboxylic acids.
> *Note:* Acids, esters, and amides of inorganic acids (noncarboxylic acids) are filed under their primary heteroatom (phosphorus, nitrogen, sulfur, metal, etc.).

Phenols.
Nitro compounds.
Nitriles.
Ketones.
Ethers.
Alcohols.
Amines.
Halogens:
> hydrocarbons substituted with more than one kind of halogen are filed under mixed halogen compounds.

The following compound types are scarce, and are therefore cross-filed as necessary to include all useful members:

> Acetylenes
> Aldehydes
> Amidines
> Azines
> Azo compounds
> Cyanates
> Hydrazones
> Imines (acyclic)
> Isocyanates
> Oximes
> Other scarce types

Aliphatic and homocyclic compounds containing *three* or more nonhydrocarbon functional groups are filed under multifunctional compounds. The exceptions are those containing acid, ester, amide, or scarce groups. These are filed under substituted acids, substituted esters, substituted amides, or other appropriate sections.

Halogens are considered one type of functional group. (See halogens, above).

HETEROCYCLICS

In general, compounds containing a heterocyclic ring are filed according to the nature of that ring system. All compounds containing a single type of heterocyclic system are filed together regardless of the other functional groups which may be present.

Compounds containing two or more types of heterocyclic systems which are not condensed together are cross-filed under each heterocyclic type. When such systems are condensed together they are filed under mixed heterocyclics.

Rings containing only one kind of heteroatom are filed by ring size and by position and number of heteroatoms. The volume spine labels should clearly indicate the proper location of any ring type. Rings containing more than one kind of heteroatom are filed under mixed heterocyclics.

Cyclic ketones are not heterocyclics. They are filed separately only when they occur alone. Cyclic anhydrides, esters, amides, sulfones, etc., are also filed separately. Oxo derivatives of other heterocyclic systems are filed under the parent heterocyclic system.

SPECIAL CATEGORIES

Some compounds are separated from the main catalog because of their end use, their special interest to us, or their lack of interest to us. The categories now so separated are:

Polymers (hydrocarbon and nonhydrocarbon)
Organic ions
Commercial products such as additives, petroleum fractions and crudes, finished petroleum products, surface active agents.
Biochemicals
Multifunctional compounds

Tables of the chemical shifts or coupling constants of a variety of compounds are filed in a separate volume of Chemical Shift Tables. Tables of data on a single compound type are filed under that compound type.

A collection of useful techniques that are not used frequently is filed under special techniques.

Information on double irradiation is filed separately.

Data and spectra for resonances other than hydrogen-1 are filed separately. Current volumes are *Carbon-13 Resonance, Fluorine-19 Resonance, Phosphorus-31 Resonance,* and *other Non-H[1] Resonances.*

Nomenclature of the heterocyclic systems is outlined by structure diagrams and alternate names at the front of some of the volumes on heterocyclics. Reference to these diagrams will simplify searching of the section tabs.

Nomenclature of phosphorus, sulfur, and other compounds not commonly encountered in our work is also shown in the fronts of the appropriate volumes.

III. MAJOR DIVISIONS OF FILING SYSTEM FOR NMR SPECTRA

HYDROCARBONS (Spine Color White)

PARAFFINS by carbon number
PARAFFINS by structure
CYCLOPARAFFINS
ALIPHATIC OLEFINS
ALICYCLIC AND CYCLIC OLEFINS
ACETYLENES (all compounds containing C≡C, cross filed if necessary)
AROMATIC NAPHTHENES
AROMATIC OLEFINS
ALKYL BENZENES
MULTIRING AROMATICS

OXYGEN COMPOUNDS (Spine Color Pink)

CARBOXYLIC ACIDS AND ANHYDRIDES, aliphatic
CARBOXYLIC ACIDS AND ANHYDRIDES, aromatic and alicyclic
CARBOXYLIC ACIDS AND ANHYDRIDES, oxygen substituted
CARBOXYLIC ACIDS AND ANHYDRIDES, nitrogen substituted
CARBOXYLIC ACIDS AND ANHYDRIDES, halogen substituted
CARBOXYLIC ACIDS AND ANHYDRIDES, other substituted (sulfur, heterocyclic, mixed)
ALCOHOLS, aliphatic
ALCOHOLS, alicyclic and aralkyl
ALCOHOLS, substituted (amino and halo substituents only)
ALDEHYDES (all examples, regardless of other substituents, except formates and aldehyde-acids)
ESTERS, saturated aliphatic (both parts saturated)
ESTERS, unsaturated aliphatic (either or both parts unsaturated)
ESTERS, alicyclic or mixed aliphatic and alicyclic
ESTERS, aromatic (either or both parts aryl or aralkyl)
ESTERS of substituted aliphatic acids
ESTERS of substituted alicyclic acids
ESTERS of substituted aryl aliphatic acids
ESTERS of substituted aromatic acids
> *Note:* The foregoing esters are filed according to their acid parts, in the following order of precedence:
> Increasing number of carboxyl groups on the acid part
> Increasing carbon number of the acid part
> Increasing branchiness of the acid part
> Increasing number of OH groups on the alcohol part
> Increasing carbon number of the alcohol part
> Increasing branchiness of the alcohol part

ESTERS of substituted alcohols or phenols, filed by alcohol part
ESTERS with both parts substituted, filed by acid part
ESTERS, special substituted (carbamates, thiocarbamates, carbazates)
ETHERS, aliphatic (both parts aliphatic)
ETHERS, alicyclic (either or both parts alicyclic)
ETHERS aromatic (either or both parts aryl or aralkyl)
ETHERS, substituted (hydroxy, halo, amino)
KETONES, subdivided into the same categories as the esters and ethers
PHENOLS, subdivided by number of OH groups and by alkyl, cycloalkyl, aralkyl, aryl substituents, and by number of condensed aromatic rings
PHENOLS, substituted (hydroxy, ether, keto, halo, amino, nitro, cyano)
PEROXIDES, all types (this is a scarce category)
CYCLIC ESTHERS
CYCLIC ETHERS
CYCLIC KETONES
> *Note:* The cyclic compounds are filed according to ring size and then by type and number of substituents.

QUINONES
COMPOUNDS WITH STRONG INTERNAL HYDROGEN BONDS (cross-filed from other categories)
KETO-ENOL TAUTOMERS (cross-filed from other categories)

NITROGEN COMPOUNDS (Spine Color Light Blue)

AMIDES, subdivided like esters insofar as possible with the smaller number of spectra

AMIDES, substituted (all substituents except metals and hetero-
cycles)
HYDRAZIDES
UREAS
SEMICARBAZIDES
AMINES AND THEIR SALTS, aliphatic, subdivided by primary,
secondary, tertiary, quaternary, and mixed order
AMINES AND THEIR SALTS, alicyclic, subdivided as above
AMINES AND THEIR SALTS, aromatic, subdivided as above
HYDRAZINES
AMINES, substituted (haloamines only)
HYDROXYLAMINES
NITRILES, subdivided by alkyl, alkenyl, alicyclic, aromatic, and
substituted
NITRO COMPOUNDS, subdivided by nitroalkanes, nitroalkenes,
nitrocycloalkanes, nitrocycloalkenes, nitroaromatics, and
substituted
ACYCLIC IMINES
HYDRAZONES
OXIMES
AMIDINES
AZINES
AXO AND AZOXY COMPOUNDS
CYANATES AND THIOCYANATES
ISOCYANATES AND ISOTHIOCYANATES
OTHER NONHETEROCYCLIC NITROGEN COMPOUNDS
CYCLIC AMIDES, subdivided by ring size and substituentts
IMIDES, subdivided by ring size and substituents
NITROGEN HETEROCYCLICS, subdivided by ring size and
number of nitrogens; the larger and more important classes
are filed by name, as shown below
PYRAZOLES
INDAZOLES
IMIDAZOLES
TRIAZOLES AND TETRAZOLES
HYDROPYRIDINES
PYRIDINES
PYRIDINE OXIDES
PYRIDINIUM SALTS
HOMOGENEOUSLY SUBSTITUTED PYRIDINES
MIXED SUBSTITUTED PYRIDINES
PYRIDAZINES and their hydrogenated forms
URACILS
BARBITURATES
PYRIMIDINES
PIPERAZINES
PYRAZINES
TRIAZINES and their hydrogenated forms
QUINOLINES and their hydrogenated forms
ISOQUINOLINES and their hydrogenated forms
QUINAZOLINES and their hydrogenated forms
QUINOXALINES and their hydrogenated forms
ACRIDINES

HALOGENATED HYDROCARBONS (Spine Color Green)

FLUORINATED HYDROCARBONS
CHLORINATED HYDROCARBONS
BROMINATED HYDROCARBONS
IODINATED HYDROCARBONS

PHOSPHORUS COMPOUNDS (Spine Color Red)

PHOSPHINES
PHOSPHONATES
PHOSPHONITES
PHOSPHINATES
PHOSPHORAMIDATES
PHOSPHATES
THIOPHOSPHATES
PHOSPHITES
PHOSPHORANES
PHOSPHONIUM SALTS
OTHER PHOSPHORUS COMPOUNDS

SULFUR COMPOUNDS (Spine Color Yellow)

THIOLS
SULFIDES, dialkyl, aralkyl, and aryl
SULFIDES, cyclic, subdivided by ring size and number of sulfur
atoms
SULFIDES, substituted
SULFOXIDES
SULFONES
SULFONATES
SULFONAMIDES
SULFITES
SULFATES
THIOPHENES
OTHER SULFUR COMPOUNDS

METAL-ORGANICS (NOT SALTS) (Spine Color Dark Blue)

ORGANOSILICON COMPOUNDS
ORGANOTIN COMPOUNDS
OTHER METAL-ORGANIC COMPOUNDS, subdivided by metal

MIXED COMPOUND TYPES (Multicolored Spine)

MIXED HETEROCYCLICS, subdivided by heteroatoms, ring
size, etc; illustrative subclasses are oxazolidines, oxazolines,
oxazoles, benzoxazoles, oxadiazoles, furazans, morpholines,
oxazines, isothiazoles, thiazoles, thiadiazoles, thiadiazines,
etc.
MIXED CONDENSED HETEROCYCLICS, subdivided by heteroatom or atoms and then by ring sizes
MULTIFUNCTIONAL COMPOUNDS (these are not heterocyclic), subdivided by heteroatom; i.e., O only, O and N, O
and halogens, etc.

OTHER CATEGORIES (Spine Color Buff)

POLYMERS, hydrocarbon, subdivided by monomer
POLYMERS, nonhydrocarbon, subdivided by monomer
ORGANIC IONS
BIOCHEMICALS, subdivided by alkaloids, antibiotics, nucleo-
sides, purines, steroids, sugars and their derivatives (except
nucleosides), other
INSECTICIDES
ADDITIVES such as antioxidants, UV absorbers, plasticizers,
rust preventives, etc.
LUBRICANTS
SOLVENTS
SURFACE ACTIVE AGENTS
DATA TABLES
C^{13} RESONANCES
F^{19} RESONANCES
p^{31} RESONANCES
OTHER NON-H^1 RESONANCES

Author Index

Abrahams, A., 406
Abramovitch, R. A., 408
Aguiar, A. M., 410
Allen, D. W., 411
Allen, G., 412
Allerhand, A., 408
Allinger, N. L., 406
Allred, A. L., 409
Altgelt, K. H., 65, 404
Anet, F. A. L., 103, 404, 406, 411
Angerman, N. S. 39, 404
Angiolini, J., 406
Ankel, T., 380, 404
Archer, E. D., 406
Arnal, E., 407
Arnaud, P., 406
Arndt, C., 409
Arnett, E. M., 406
Axtman, R. C., 410

Babiec, J. S., 35, 404
Baldeschwieler, J. D., 407, 411
Bansal, R. C., 406
Barrante, J. R., 35, 404
Bartels-Keith, J. D., 407, 412
Barth, D. E., 407
Bartle, K. D., 410
Bartz, K. W., 68-71, 73, 107, 191-195, 197, 198, 337-343, 352-355, 404, 405
Becconsall, J. K., 410
Becker, E. D., 405
Belanovic, K., 409
Bellavita, N., 34, 404
Benezra, C., 411
Bergerhoff, G., 411
Berlin, K. D., 410, 411
Bernstein, H. J., 411
Bhacca, N. S., 75, 405
Bible, R. H., 405
Biellmann, J. F., 408, 411
Binsch, G., 411

Boden, N., 407, 409
Boekelheide, V., 406
Bohlmann, F., 409
Böll, W. A., 406
Bollinger, J. M., 406
Booth, H., 405
Bordwell, F. G., 409
Borecka, B., 410
Borkovec, A. B., 411
Bose, A. K., 411
Bothner-By, A. A., 93, 94, 404, 405, 407, 411
Bourn, A. J. R., 407, 411
Bovey, F. A., 68, 75, 98, 404, 405, 407
Bray, N. F., 406
Bregman, J. M., 408
Bresler, W. E., 40, 404
Brink, M., 408
Brophy, J. J., 410
Brown, E. D., 408
Brown, R. F. C., 407
Brügel, W., 75, 365, 380, 404, 405, 408
Bucci, P., 407
Budenz, R., 410
Buehler, E., 408
Burdett, J. L., 412
Burdon, M. G., 409
Burke, J. J., 406
Burton, G. W., 408
Bushweller, C. H., 409
Butler, P. E., 36, 404, 408, 409, 410

Cadiot, P., 410
Cagniant, D., 406, 409
Cagniant, P., 409
Calder, J. C., 406
Callot, H., 408, 411
Campbell, J. R., 36, 404
Campbell, R. W., 409
Canadine, R. M., 410
Carel, A. B., 406

Carrington, A., 405
Carruthers, W., 406
Carter, R. E., 407
Carty, A. J., 410
Caserio, M. C., 407, 408
Casini, G., 408
Castellano S., 407, 408
Caubere, P., 407
Cavell, R. G., 410
Chamberlain, N. F., 57, 58, 62, 65, 67, 68, 75, 76, 107, 191-196, 200, 212-220, 337-343, 350, 404, 405, 408
Chan, J. K. 407, 412
Chapman, O. L., 40, 404
Charlton, T. L., 410
Charrier, C., 410, 411
Chodkiewicz, W., 410
Chow, H. S., 406
Chuche, J., 407
Chung, H., 406
Cieciuch, R. F. W., 407, 412
Clutter, D. R., 65, 404
Cobb, T. B., 406
Coffman, D. D., 409
Collins, P. J., 407
Convert, O., 407
Corey, E. J., 103, 404
Corio, P. L., 406
Cornu, A., 406
Cotton, F. A., 411
Couffignal, R., 411
Cowley, A. H., 411
Cram, D. J., 406
Crawford, T. H., 407
Crepaux, D., 411
Crutchfield, M. M., 409
Cywinski, N. F., 406

Daigle, D., 410
Danechpejouh, H., 407
Danyluk, S. S., 39, 404
Da Rooge, M. A., 406

Davis, H. E., 407, 408
Davis, J. B., 408
Davis, J. C., 411
Davis, R. E., 39, 404
Davison, R. B., 410
Day, R. J., 36, 404
de Boer, Th. J., 406, 407, 409
Defay, N., 406
Delavarenne, S., 406
Deljac, A., 409
Demuynck, M., 408, 409
Derbyshire, W., 410
de Roos, A. M., 407
Desai, N. B., 411
de Selms, R. C., 410
De Wolf, M. Y., 407, 411
Dietrich, M. W., 40, 76, 404, 405, 409
Dignan, J. C., 406
Doskocilova, D., 407
Douglas, A. W., 98, 103, 404, 407, 411
Douglass, I. B., 407, 408, 409
Drago, R. S., 408, 410, 411
Drenth, W., 410
Duval, E., 410
Dwek, R. A., 412

Easter, W. M., Jr., 406
Eberly, J. H., 410
Ebersole, S. J., 411
Edmundson, R. S., 410
Edwards, B. E., 409
Edwards, W. R., 67, 404
Eisenbraun, E. J., 406, 407
Elam, E. V., 407, 408
Elleman, D. D., 410, 412
Ellis, P. D., 411
Elvidge, J. A., 407
Empotz, G., 406
Emsley, J. W., 97, 404, 405, 407, 410
Engberts, J. B. F. N., 408
Ernst, R. R., 93, 404

Fales, H. M., 36, 404
Faller, P., 410
Farrar, T. C., 405
Favelier, R., 410
Feeney, J., 97, 404, 405, 407, 410
Ferraro, J. R., 410
Fields, E. K., 406
Fischer, U., 406
Flanagan, P. W., 406, 407
Fontal, B., 410
Forbes, J. W., 409
Fournari, P., 410
Francis, S. A., 406
Frazer, R. R., 407
Freeman, R., 93, 380, 404
Frei, K., 411
Freymann, R., 410
Friebolin, H., 406, 409

Gagnaire, D., 410
Gallagher, M. J., 410
Gaoni, Y., 406
Garbutt, D. C. F., 406, 409
Garner, R. J., 407
Gattuso, M. J., 406
Geerts-Evrard, F., 406
Gehring, D. G., 408
Gennaro, A. R., 407
Gestblom, B., 93, 404, 411
Gil, V.M.S., 408
Gillies, D. G., 407, 411
Gitlitz, M. H., 407, 410
Goldstein, J. H., 407, 409, 411
Goldwhite, H., 410, 411
Good, M. L., 408
Goodlett, V. W., 35, 404
Gordon, M., 410, 411
Grant, D., 408, 409
Gray, G. A., 411
Greco, C. V., 411
Griesbaum, K., 408, 409, 411
Griffin, C. E., 410, 411
Griffin, G. W., 406
Griffin, R. W., 406
Gronowitz, S., 410, 411
Guilard, R., 410
Gunning, H. E., 409
Gunther, H., 407
Gurudata (no initials), 406
Gutowsky, H. S., 76, 405, 410

Haake, P., 411
Haley, G. A., 65, 404
Hall, D. N., 409, 411
Hall, G. E., 407
Hamming, M. C., 407
Hansell, P. G., 406
Harris, J. F., 409
Harris, R. K., 410, 411
Hart, F. A., 38, 404
Harvey, R. G., 411

Hayamizu, K., 76, 405, 408
Hayter, R. G., 410
Hazato, G., 409
Heathcock, C., 233, 404
Helgeson, R. C., 406
Heller, S. R., 411
Hellwege, D. M., 410
Hendrickson, J. B., 410
Hepfinger, N. F., 406
Herbrandson, H. F., 409
Hermann, R. B., 406
Herweh, J., 410
Hikida, T., 409
Hill, E. A., 406
Hill, R. A., 408
Hinckley, C. C., 36, 404
Hinman, C. W., 407
Hirsch, E., 65, 404
Hobgood, R. T., 409
Hoffman, R. A., 410, 411
Hoffman, W., 408
Hogeveen, H., 408
Holland, R. J., 407, 408
Hollis, D. P., 75, 405
Holloway, C. E., 407, 410
Hoover, F. W., 410
Hranisavljevic-Jakovljevic, M., 408
Hudson, B. E., 408, 409, 411
Huisman, H. O., 408
Huitric, A. C., 407

Iqbal, S. M., 408
Isenberg, K., 409
Ishigure, K., 408
Ito, I., 410
Ivin, K. J., 409
Iwamura, H., 408

Jackman, L. M., 12, 404, 405, 407
Jensen, R. K., 65, 404
Johns, S. R., 407
Johnson, C. R., 409
Johnson, L. F., 75, 405
Jonas, J., 410
Jones, D. W., 410
Jones, K., 405
Jones, M. P., 409
Jung, J. A., 408
Jungnickel, J. L., 24, 404, 407, 409
Juvinall, G. L., 410, 412

Kabuss, S., 406, 409
Kaesz, H. D., 410
Kanda, T., 410
Kaplan, F., 409, 410
Karabatsos, G. J., 407, 408, 411
Karchmer, J. H., 405, 408
Karmer, F. A., 408
Karplus, M., 13, 404

Katritsky, A. R., 410
Keat, R., 411
Keller, R. E., 40, 76, 404, 405
Kelley, K. M., 406
Kenyon, G. L., 411
Kergomard, A., 410
Keske, R. G., 411
Kessler, H., 407
Khaleeluddin, K., 411
Kintzinger, J. P., 408
Kleinspehn, G. G., 408
Klink, R. E., 407
Klose, G., 412
Knight, S. A., 65, 404
Knoll, F., 411
Kondo, K., 409
Konds, K., 409
Korte, F., 411
Korver, O., 406
Korver, P. K., 406, 407, 409
Kossanyi, J., 380, 404, 407
Kostelnik, R., 408
Kruk, C., 408
Krükeberg, F., 380, 404
Krumel, K. L., 408
Kugajevsky, I., 411
Kuntz, I. D., 408
Kurland, R. J., 406

Lambert, J. B., 411
Lamberton, J. A., 407
Lande, S. S., 407
LaPlanche, L. A., 407
Lauer, D., 408
Lauterbur, P. C., 406, 409, 411
Lavielle, G., 411
Ledwith, A., 407
Lehn, J. M., 408, 411
Lelandais, D., 407
Lenzi, M., 411
Leonard, N. J., 409
Lin, T-W., 410
Lindberg, J. G., 76, 405
Longone, D. T., 406
Looker, J. J., 408
Lucken, E.A.C., 410
Lumbrose, C., 407
Lusebrink, T. R., 93, 404
Lutringhaus, A., 409
Luukkainen, T., 36, 404

Maccagnani, G., 408
Maciel, G. E., 411
Madding, G. D., 409
Maddox, M. L., 410
Maier, L., 410
Mair, B. J., 406
Maitte, P., 407
Manatt, S. L., 98, 404, 410, 411, 412
Marcus, E., 407, 412
Marcus, S. H., 409

Martin, D. J., 409
Martin, G., 407
Martin, G. J., 411
Martin M., 407, 409
Martin, M. L., 411
Martin, R. H., 406
Mathias, A., 407, 408, 410
Mathiasson, B., 411
Mathieson, D. W., 405
Mathis, C. T., 411
Matsuki, Y., 410
Matsuzaki, K., 408
Matter, U. E., 406
Mattox, J. R., 406
Mavel, G., 405, 410
Mayer, T. J., 406
Mayo, R. E., 407
McCarthy, W. C., 407
McFarlane, W., 411
McGreer, D. E., 407
McLachlan, A. D., 405
Mecke, R., 409
Melera, A., 406
Memory, J. D., 406
Meyer, L. H., 76, 405
Michel, R. H., 406
Middaugh, R. L., 408, 410, 411
Mikol, G. J., 408
Miller, A. H., 406
Miller, I. T., 411
Miller, J. B., 406
Miller, S. I., 409
Miller, W. B., 411
Miskow, M., 409
Mislow, K., 409
Miyazima, G., 410
Moen, R. V., 409
Moffatt, J. G., 409
Mohacsi, E., 76, 405
Mooney, E. F., 405, 411
Moriarty, R. M., 410
Morita, Z., 409
Morris, S. G., 407
Mortimer, F. S., 407
Moss, G. P., 38, 404
Mueller, W. H., 36, 404, 408, 409, 410
Mulder, R. J., 409
Muller, N., 411
Murray, R., 410
Musher, J. I., 103, 107, 404
Musierowicz, S., 410

Nachod, F. C., 406
Naegele, W., 409, 411
Nagabhushanam, M., 410
Nagai, T., 409
Nakamura, K., 409
Nash, J. S., 40, 404
Nesic, S., 408
Newsoroff, G. P., 406
Niedzielski, R. J., 408

Nishitomi, K., 409
Nist, B. J., 6, 111, 114, 404, 405
Nixon, J. F., 410, 412
Norton, R. V., 407, 409
Nozaki, H., 409
Nseic, S., 411
Nukada, K., 76, 405, 408

Ogg, R. A., 412
Ohline, R. W., 409
Ohnishi, M., 76, 405
Oki, M., 408
Olijnsma, T., 408
Orzech, C. E., 411
Oswald, A. A., 408, 409, 410, 411
Ourisson, G., 411
Owen, L. N., 408

Pachler, K. G. R., 406, 409
Page, T. F., 40, 404
Pagilagan, R. V., 411
Parello, J., 411
Parrish, J. R., 406, 409
Pascual, C., 406, 408
Pasika, W. M., 409
Patwardhan, A. V., 411
Pavia, A. A., 407
Paxton, H. J., 410
Pearce, R. H., 409
Pejkovic-Tadic, I., 408
Peppard, D. F., 410
Perry, R. W., 410
Persaud, K., 406
Person, M., 410
Pesce, B., 406, 407, 408, 409, 410, 411, 412
Peterson, L. I., 406
Petrakis, L., 65, 404
Pettit, G. R., 408
Pfau, M., 407
Pidcock, A., 412
Pier, E. A., 75, 405
Pierre, J-L., 406
Piette, L. H., 408
Pilgrim, K., 410
Pinnell, R. P., 411
Pople, J. A., 8, 404, 406
Pratt, L., 36, 404
Pratt, R. E., 407, 408
Pretsch, E., 406
Prinzbach, H., 406
Pritchard, D. E., 411
Pritchard, J. G., 409
Prohaska, C. A., 407, 410
Pross, A., 406
Prugh, J. D., 407
Purcell, J. M., 407, 408
Puskas, I., 406

Radom, L., 407
Rae, I.D, 407, 411
Ralph, P. D., 407

Ramanathan, N., 411
Ramirez, F., 411
Randall, E. W., 407, 411
Raper, A. H., 408
Ratts, K. W., 409
Ray, J. D., 412
Reddy, G. S., 408, 409, 410, 411
Reed, J. J. R., 75, 76, 244, 247, 405, 408, 410
Reilley, C. N., 36, 404, 407
Reilly, C. A., 407
Rieker, A., 407
Robert, J-B., 410
Roberts, B. W., 411
Roberts, J. D., 406, 409, 411
Rodmar, S., 411
Rogers, M. T., 407, 412
Rose, J. B., 409
Rosenberg, D., 410
Rossi, R., 407
Roswell, D. G., 410
Rothstein, E., 408
Rottendorf, H., 406
Rousselot, M-M., 34, 404, 409
Rowsell, D. D., 411
Rubin, R. M., 410
Rudolph, K., 410
Rudolph, R. W., 410
Russell, G. A., 408

Saika, A., 76, 405
Saito, H., 408
Salvi, M. L., 408
Saquet, M., 412
Savastianoff, D., 407
Schleyer, P. von R., 408
Schmidbaur, H., 409
Schmutzler, R., 410, 411
Schneider, B., 407
Schneider, W. G., 408
Schröck, W., 406
Schroder, G., 406
Schulz, C. O., 410
Schunn, R. A., 411
Scott, J. M. W., 411
Seel, F., 410
Shapiro, B. L., 406, 411
Shaw, R. A., 411
Shoji, F., 410
Shone, R. L., 406
Shoolery, J. N., 75, 405
Shuler, W. E., 410
Siddall, T. H., III, 407, 408, 410
Siebert, W., 409
Simon, W., 406, 408
Simonnin, M-P., 409, 410, 411
Simons, W. W., 406, 407
Sims, J. J., 410
Singh, G., 411
Sioumis, A. A., 407
Slichter, C. P., 405
Slomp, G., 76, 405, 407

Smutny, E. J., 406
Sondheimer, F., 406
Sone, T., 410
Sonnet, P. E., 411
Spotswood, T. McL., 408
Springer, J. M., 407
Staab, H. A., 408
Staniforth, M. L., 38, 404
Starnick, J., 409
Stefanac, Z., 409
Stefaniak, L., 408
Stehling, F. C., 69, 70, 71, 73, 76, 197, 198, 241, 352-355, 404, 405
Steinberg, H., 407, 409
Stenger, R. L., 65, 404
Sternhell, S., 12, 93, 404, 405, 406, 407, 411
Stewart, H.N.M., 406
St. Jacques, M., 406
Stothers, J. B., 406, 407
Strating, J., 408, 409
Strausz, O. P., 409
Strow, C. B., 407, 412
Studniarz, S. A., 408
Sturtz, G., 411
Sudmeier, J. L., 36, 404, 407
Sullivan, G. R., 406
Sun, C., 408
Sunners, B., 408
Susi, H., 407, 408
Sutcliffe, L. H., 97, 404, 405, 407, 410
Suzuki, T., 76, 405
Szymanski, H. A., 405

Taddei, F., 36, 404, 408
Takahashi, K., 409, 410
Takaku, M., 409
Takeuchi, M., 76, 405
Taller, R. A., 408
Talman, J. D. 406
Tanzer, C. I., 408
Taylor, H. A., 411
Ternai, B., 410
Thaler, W. A., 408, 409
Thompson, Q. E., 409
Thuillier, A., 412
Tidd, B. K., 408
Tiers, G.V.D., 75, 405
Tokura, N., 409
Traficante, D.D., 411
Trompen, W. P., 408
Truce, W. E., 409
Tunemoto, D., 409
Tyssee, D. A., 411

Uhlmann, E., 412
Ulrich, H., 406
Untch, K. G., 406
Urbanski, T., 408
Uryu, T., 408

Van Auken, T. V., 411

van der Haak, P. J., 406, 407, 408, 409
van der Linde, R., 406
van der Veen, J., 408
Vane, F. M., 411
van Gorkam, M., 407
van Leusen, A. M., 409
Van Wazer, J. R., 408, 409
Veenland, J. U., 406
Verrier, J., 410
Vialle, J., 408, 409
Vickers, G. D., 35, 404
Victor, T. A., 39, 404
Vidall, M., 406
Vincent, S., 410
Vollmer, R. L., 409
Vo-Quang, L., 406
Vo-Quang, Y., 406

Wagner, R. I., 410
Wagner, W. J., 408
Ward, F. J., 407, 409
Warnhoff, E. W., 408
Warren, C. L., 406
Watanabe, K., 409
Waugh, J. S., 405
Weary, D. K., 411
Weitkamp, H., 411
Welch, F. J., 410
West, R., 408
Westheimer, F. H., 411
White, H. F., 407
White, W. L., 406
Wiberg, K. B., 6, 111, 114, 404, 405, 407
Wiberley, S. E., 406
Wiemann, J., 407
Wiley, R. H., 406, 407
Wilhoit, R. C., 120
Willcott, M. R., 39, 404, 406, 409
Williams, R. B., 65, 404
Wilmshurst, J. R., 406
Wilshire, J.F.K., 408
Wilson, D. J., 406
Wilt, J. W., 408
Winson, P. H., 405, 411
Witanowski, M., 408
Wolovsky, R., 406
Wood, G., 409
Wood, T. F., 406
Wylde, J., 407
Wylde, R., 407

Yamamoto, O., 76, 405, 408
Yamasaki, T., 410
Yanagisawa, M., 76, 405
Yao, A. N., 409
Yelin, R., 405
Yew, F. F., 406

Zanger, M., 406, 407
Zauli, C., 408
Zimmer, H., 411
Zwolinski, B. J., 107, 404, 405

Subject Index

The Index Charts (pp. 118 and 294) and the Summary Charts (pp. 121–189 and 296–333) serve as their own indexes and are not generally suited to additional indexing. With a few exceptions, therefore, these charts are not referred to in this subject index. References to the Chemical Shift Detailed Charts and the Plates *are* included, however.

Page number formats indicate the type of material referred to as follows:

Roman—Text reference
Italic—Chemical shift data
Parenthesis—Coupling constant data
Brackets—Spectrum
Braces—Table

Acetylenes
  heterosubstituted *87*, (97), *221, 222, 279, 281, 286,* (316), (317), (330)
  hydrocarbons *87, 190,* (304), (316), (317)
Acids, carboxylic
  alkenyl *212, 214,* [380]
  alkyl [20], *210*
  alkynyl *221*
  aralkyl *228,* [383], [385], [386]
  aryl *230,* [383], [386]
  COOH group 34, *181*
Acids, thiocarboxylic *263*
Acids of phosphorus (*See* Phosphorus acids)
Acids of sulfur (*See* individual sulfur acids and sulfur esters)
Alcohols
  alkenyl *218*
  alkyl [4], [10], [20], [32], [34], [38], [55], 55, 56, *79,* 110, *210,* [368–377], [379]
  alkynyl *222*
  aralkyl *229*
  cycloalkyl *223*
  OH group [4], [20], [32], 33, [34], [35], 35, 37, [38], [39], 40, [55], 55, [72], 74, *87,* 91, (97), *181, 187, 210, 218, 222, 223, 229,* (310), [368-377]
Aldazines *262*

Aldehydes
  alkenyl *216*
  alkyl *210,* (301), (304)
  aryl [23], *231,* (309)
  CHO group *157,* (297), (301), (304), (309), (311), (312), (316), (317), (320), (326)
Amides, acid part
  alkenyl *248*
  alkyl *246, 250,* (311), (312), (327)
  aralkyl *248,* [379]
  aryl *251*
Amides, amine part
  alkyl 35, *246–250,* (311), (312), (325), (326)
  aralkyl *248,* (326)
  aryl 89, *251*
  NH group *185, 189, 246–254,* (311), (325), (326)
Amides, cyclic [55], *252,* (296), (300), (311)
Amides of phosphorus acids, amine part
  N-alkyl *289,* (332)
Amides of sulfur acids (*see* Sulfenamides, Sulfinamides, and Sulfonamides)
Amidines *262,* (326), (327)
Amines 37
  alkyl *242, 243,* (311), (326), (327)
  aralkyl *243*
  NH group 33, *87, 183, 189, 242-245,* (311), (325)

Amines (*cont'd*)
  phenyl *243,* (311), [382], [383], [385], [388], [389]
Ammonium salts 36
  alkenyl (326), (327)
  alkyl *242, 244,* (311), (326), (327)
  NH$_x$+ group 36, *183, 242, 244,* (325)
  phenyl *244*
Analytical procedures
  characterization 56
  molecular structure 45
  quantitative analysis 53
Anhydrides, cyclic (296)
Annulenes *202*
Aromatic hydrocarbons
  aromatic naphthenes *204, 206*
  aromatic olefins *207, 208*
  condensed rings *206,* [391]
  mixtures [57-59], 63, [64], [66]
  uncondensed rings *202,* [341], [367], [381–383], [385–389]
Aromatics, heterosubstituted *83, 84, 85, 87*
  chemical shift and spin coupling data are indexed at the entry for the specific substituent type
  disubstituted anthracene [42], 50
  disubstituted benzenes [23], [383-386]
  monosubstituted benzenes [382], [398]
  pentasubstituted benzenes [390]
  tetrasubstituted benzenes [389], [390]

Aromatics, heterosubstituted (*cont'd*)
trisubstituted benzenes [387-389]
Azo compounds
phenyl *262*

Band area (*See* also intensity) {5}, 10,
{10}, {46}, 54, {55}, [55], [60],
[62], 63, [64], {64}, [66], {66},
68, [72]
Band width and shape 13, 15, 28, 31, 41,
46, 56-58, 63, 65, 105-115
Benzenes, substituted (*See* Aromatic hydrocarbons and Aromatics, heterosubstituted)
Benzofurans *226*, (308), (309)
Benzopyrroles *259*
Benzothiophenes *278*
Bromine compounds
bromoalkanes *235*
bromoalkenes *236, 238*
bromoalkylaromatics *240, 241*
bromoaromatics *240, 241*
bromocycloalkanes *235*

Calibration 22, [23], 45, 53, 54
Characterization
of aromatic hydrocarbons 63, [64],
[66]
definition 56
by inspection 56
detergent alkylates [582]
lubricating oil [59]
petroleum fractions [57]
of olefinic hydrocarbons 62, [62]
of polymers 67, [67-73]
of saturated hydrocarbons 59, [60],
{61}
Chemical Shift
causes 8
correlations 75
for halogen compounds *235-241*
for hydrocarbons *57, 62, 66, 190-208*
for nitrogen compounds *242-262*
for oxygen compounds *208-234*
for phosphorus compounds *279-291*
for sulfur compounds *263-278*
definition [4], 7
effect of
chain length *79*
electronigativity and size 81, *80-84*
hybridization and substitution 84, *85*
ring size 90, [90]
solvent, concentration and temperature 75, 80, 107, *198* (*See* notes
on Charts 1-89)
steric factors 79, 87
strong intramolecular hydrogen bonds
91
list of correlation charts xv-xviii
measurement [4], 7, 41, [42], [53],
[69-71], [73]

Chemical Shift (*cont'd*)
ranges for functional groups *87*
relation to typical spectra 105-115
uses 9, 45, 46, 51, 52, 105
Chlorine Compounds
chloroalkanes *235*
chloroalkenes *236, 237*
chloroalkylaromatics *239, 241*, [386]
chloroaromatics *239, 241*, [382-390]
chlorocycloalkanes *235*
Coupling Constant
causes of spin coupling 12
correlations 93
hydrogen-carbon (314-319)
hydrogen-fluorine (320-324)
hydrogen-hydrogen (296-313)
hydrogen-nitrogen (325-327)
hydrogen-phosphorus (328-333)
definition [4], 12
effect of
electronegativity 98
heteroatoms 102
hybridization or bond order 97, (97)
isotope (12), (95), (96), 96
number of intervening bonds 100,
(100), (101)
ring size 102, (102)
steric factors 99
virtual coupling 103
list of correlation charts xviii
measurement [4], 12, [42], [51], [53],
[353-355], [368-377], [385],
[388]
uses 13, 45, (46), 51 (52), 105, 107

Dithiocarbamates *253*

Esters
alkenyl *218, 219*, (310), [381]
alkyl *210, 227, 233*, (310), [379]
alkynyl *221*
aralkyl *229*
cycloalkyl *223*
cyclic *225*, (296-300), (303-305),
(308), (309)
phenyl *232, 233*, (310), [382], [383],
[385-390]
Esters, acid part [16], [379], [380],
[382], [385], [398] (*See* also acids,
carboxylic, phosphorus acids, sulfur
acids)
Esters, cyclic
carboxylic *224*, (296), (300), (301),
(303), (304), (319)
phosphorus *289*, (331)
sulfur *277*, (296)
thiocarboxylic (301)
Esters of carboxylic acids, alcohol part 35
alkenyl *219*, (310)
alkyl *210, 221*, [379], [396]
aralkyl *227*

Esters of carboxylic acids (*cont'd*)
cycloalkyl *223*
phenyl *232*, [382]
Esters of carboxylic thioacids, alcohol or
thiol part
S-alkenyl *271*
O-alkyl *263*
S-alkyl *263*
S-benzyl *273*
S-phenyl *276*
S-styryl *274*
Esters of phosphorus acids, alcohol part
115
O-alkenyl *284*, (333)
O-alkyl *282*, 115, (331), [396], [397]
O-alkynyl *286*, (333)
O-aralkyl *227*
O-phenyl *232*, (333)
Esters of phosphorus thioacids, thiol part
S-alkenyl *284*, (333)
S-alkyl *282*, (332), (333)
S-alkynyl *286*
Esters of sulfur acids, alcohol part
alkenyl *271*
alkyl *265, 266*
aralkyl *273*
phenyl *232, 276, 277*
Esters of sulfur thioacids, thiol part
alkyl *266*
Esters of thiocarboxylic acids (*See* Esters
of carboxylic thioacids)

Filing system for spectra 413
Fluorine compounds
fluoroalkanes *235*, (320), (321)
fluoroalkenes *236*, (320), (322)
fluoroaromatics *241*, (323), [390]
fluorocycloalkanes (320), (321)
substituted (324)
Furans (*See* also Benzofurans)
alkyl *226*, (308), (309)
heterosubstituted *226*, (308), (309)

Halogen compounds (*See* fluorine, chlorine, bromine, and iodine compounds)
Hydrazides *253*
Hydrazones, carbonyl part
alkyl *255*
benzyl *255*
phenyl *255*, (327)
Hydrazones, hydrazine part
alkyl *255*
phenyl *255*

Imides (cyclic) *252*, (296), (300), (311)
Imines
acyclic *254*, (312), (325-327)
cyclic *245*, (296-300)
Integral 10, [10], [11], 22, [25], 26, 27,
[28], 30, [34], 40, [41], [42], 43,
[51], [53], [55], [60], [62], 63, [64],
[66], [69], [72]

Intensity (*See* also band area)
  definition 10
  measurement [4], 10, [10], [11], 30,
    37, 40, 41, 43, 45, {52}, 52, 63,
    65, {69}, {70}, 71, {73}
  of multiplet peaks 5, 6
  relation to
    resolution 25
    saturation 29
  of satellites 32, 33
  of sidebands 31
  uses 35, 37, 45, 51, 52
Iodine compounds
  iodoalkanes *235*
  iodoalkenes *236*
  iodoaromatics *241*
Isocyanates
  alkyl *262*
  phenyl [388]
Isothiocyanates
  alkyl *262*

Ketones
  alkenyl *216*, [380]
  alkyl 110, *210, 227*
  aralkyl *228*
  aryl *231*
  cyclic *224*
  cycloalkyl *223*

Molecular structure determination
  confirming proposed structure 51
  determining unknown structure 45
Multiplicity
  definition and cause 4
  measurement [4], [13], [29], [32],
    [33], [42], [51], [53], [55],
    [71], [337–399]
  multiplet rules 5
  number of lines 6
  other discussion 74, 105–115

Nitriles
  alkenyl [380]
  alkyl *257*
  aralkyl *257*
  phenyl *257*
Nitro-compounds
  alkyl *258*
  phenyl *258*, [388], [390]
Nitrosamines
  alkyl *262*
  aralkyl *262*
  phenyl *262*

Olefinic hydrocarbons
  acyclic *190, 197, 198*, [352–366]
  alicyclic *199*
  cyclic *201*

Olefins, substituted
  vinyl group 111, [380], [381]
Oximes
  alkyl *256*
  benzyl *256*
  phenyl *256*
Oxosulfonium salts *265*
Oxosulfonium ylides *265*
Oxyphosphoranes *282*

Paraffins
  acyclic *190–196*, [337–350]
  cyclic *199, 200*, [351]
Paraffins, substituted 110
  aliphatic alcohols 110, [368–377]
  aliphatic ketones 110
  n-butyl group 110, [378]
  X-CH$_2$-CH$_2$-Y group 110, [379]
  phosphorus compounds [392–397]
Phasing
  examples [27], [28], 40, [41]
  need for and method of achieving 26
Phenols *234*, [382–390]
Phosphates
  chemical shifts (*See* Phosphorus acids
    and Esters of phosphorus acids)
  spectra [396], [397]
Phosphinates (*See* Phosphorus acids and
    Esters of phosphorus acids)
Phosphine oxides 115, *281*
Phosphine sulfides *281* [395]
Phosphines 115, *279*, [392], [393]
Phosphites (*See* Phosphorus acids and Es-
    ters of phosphorus acids)
Phospholines *289*
Phosphonates
  chemical shifts (*See* Phosphorus acids
    and Esters of phosphorus acids)
  spectrum [382], [398]
Phosphonites (*See* Phosphorus acids and
    Esters of phosphorus acids)
Phosphonium salts *115, 291*, (329), (330),
    (332) [394]
Phosphorus acids
  alkenyl *284*, (330)
  alkyl *282*, (328–330)
  alkynyl *286*, (330)
  aralkyl *287*, (333)
  acyl *288*, (330)
  acylmethyl *288*, (329)
  haloalkyl *290*, (329)
  phenyl *287*, (333)
Phosphorus halides
  alkyl *279, 290*, (329), (330)
Polysulfides
  alkyl *264*
  aralkyl *273*
  cyclic *277*
  phenyl *276*
Pyridine oxides 261, (308)
Pyridines (308)
  alkyl *260*

Pyridines (*cont'd*)
  aralkyl *260*
  phenyl *260*
  pyridyl (bipyridyls) *261*
Pyridinium salts *261*, (308)
Pyrroles (*See* also Benzopyrroles)
  alkyl *259*
  heterosubstituted *259*

Quantitative analysis 53
Quinones *224*

Radiation damping 26
Resolution 25

Sample phases 15, 21
Saturation
  checking for 40, 113
  control 29, 113
  definition and cause 28
  effect on intensity 28, 29, 54
Semicarbazones
  alkyl *254*
Signal-to-noise ratio, S/N 24, [24], 29, 31,
    40, 74
Spectral characteristics 4 (*See* also Multi-
    plicity, Chemical shift, Band intensity,
    Coupling constant, and Band width and
    shape)
Sulfates
  alkyl *266*
  cyclic *277*
  phenyl *232*
Sulfenamides, amine part
  alkyl *253*
Sulfides
  alkenyl *268–270*, (313)
  alkyl *264*, (313), [379]
  aralkyl *272*
  cyclic *277*
  phenyl *274, 276*
  styryl *274*
Sulfinamides, amine part
  alkyl *253*
Sulfinates
  alkyl *265*
  aralkyl *273*
  phenyl *276*, [383]
Sulfites
  alkyl *265*
  aralkyl *273*
  cyclic *277*
Sulfonamides, acid part
  alkyl *266*
  aralkyl *273*
  phenyl *277*
Sulfonamides, amine part
  alkyl *253*
  phenyl *253*